Electron Energy-Loss Spectroscopy

in the Electron Microscope

Second Edition

Electron Energy-Loss Spectroscopy
in the Electron Microscope

Second Edition

R. F. Egerton

University of Alberta
Edmonton, Alberta, Canada

PLENUM PRESS • NEW YORK AND LONDON

Library of Congress Cataloging-in-Publication Data

Egerton, R. F.
   Electron energy-loss spectroscopy in the electron microscope / R.
F. Egerton. -- 2nd ed.
      p.   cm.
   Includes bibliographical references and index.
   ISBN 0-306-45223-5
   1. Electron energy loss spectroscopy.   2. Electron microscopy.
I. Title.
QC454.E4E34   1996
543'.0858--dc20   96-11645
                                                                 CIP

The following publishers gave permission to reproduce illustrations: Academic Press Inc., The American Institute of Physics, American Physical Society, The Australian Academy for Science, Claitor's Publishing Division, Les Editions de Physique, Elsevier Science Publishers, The Institute of Physics, Alan R. Liss, Inc., Journal of Microscopy, The Mineralogical Society of America, The Minerals, Metals and Materials Society, The Royal Society, The Royal Society of Chemistry, San Francisco Press, Scanning Microscopy International, Springer-Verlag, Taylor and Francis Ltd., and Wissenschaftliche Verlagsgesellschaft MbH.

ISBN 0-306-45223-5

© 1996, 1986 Plenum Press, New York
A Division of Plenum Publishing Corporation
233 Spring Street, New York, N. Y. 10013

Printed in the United States of America

Preface
to the Second Edition

Since the first (1986) edition of this book, the numbers of installations, researchers, and research publications devoted to electron energy-loss spectroscopy (EELS) in the electron microscope have continued to expand. There has been a trend towards intermediate accelerating voltages and field-emission sources, both favorable to energy-loss spectroscopy, and several types of energy-filtering microscope are now available commercially. Data-acquisition hardware and software, based on personal computers, have become more convenient and user-friendly. Among university researchers, much thought has been given to the interpretation and utilization of near-edge fine structure. Most importantly, there have been many practical applications of EELS. This may reflect an increased awareness of the potentialities of the technique, but in many cases it is the result of skill and persistence on the part of the experimenters, often graduate students.

To take account of these developments, the book has been extensively revised (over a period of two years) and more than a third of it rewritten. I have made various minor changes to the figures and added about 80 new ones. Except for a few small changes, the notation is the same as in the first edition, with all equations in SI units.

Chapter 1 retains the same format but has been updated. In Chapter 2, some historically important instruments such as the Möllenstedt analyzer have been omitted and the treatment of serial recording shortened to make room for a discussion of recent parallel-recording, energy-filtering, and spectrum-imaging systems. The section on prespectrometer lenses has been rewritten to provide a clearer account of the influence of chromatic aberration on quantitative spectroscopy.

Chapter 3 now includes discussions of the angular distribution of plural scattering, the effect of a collection aperture on Poisson's law, and the validity of the dipole approximation in core-loss spectroscopy. The concept of the plasmon wake is introduced and related to the begrenzungs effect.

The treatment of near-edge fine structure has been expanded. In Chapter 4, material has been added on curve-fitting procedures, difference spectra, and elemental analysis of thicker specimens.

Chapter 5 has been completely rewritten to accommodate a selection of new applications. My aim was to include practical details where these are important to the success of the technique, with references to previous chapters as necessary. Sections have been added on ELNES fingerprinting and the use of white-line ratios and chemical shifts. In the final section, I have tried to show how EELS, in combination with electron microscopy and diffraction, has contributed new information in several branches of applied science.

The Appendixes have also been updated. Relativistic Bethe theory is extended to provide a formula for the parameterization of small-angle inner-shell cross sections. The computer program for Fourier-log deconvolution has been improved to take account of instrumental resolution, and is now accompanied by programs for matrix deconvolution, Fourier-ratio deconvolution, and a separate program for Kramers–Kronig analysis. To test these algorithms (or for other purposes) programs are provided to generate plural-scattering spectra and Drude-model data (including surface scattering). The convergence-correction routine has been simplified, taking advantage of an analytical expression for the collection efficiency. The SIGMAK and SIGMAL cross-section programs have received a minor facelift designed to improve accuracy, and are supplemented by a program which calculates core-loss cross sections from tabulated oscillator strengths, based on experimental data and Hartree–Slater calculations. New programs have been added for the calculation of Lenz cross sections and for conversion between oscillator strength and cross section, and between mean energy loss and mean free path. As specified in Appendix B, the source code for these programs is also available via the Internet. In Appendix D, all energies are now threshold values, even in the case of the low-energy N_{67} and O-edges.

I am grateful to many colleagues for providing comments on the proposed changes and for reading selected passages, including Drs. P. Batson, J. Bruley, R. Brydson, G. Carpenter, P. Crozier, M. Disko, V. Dravid, P. Echenique, J. Fink, P. Goodhew, F. Hofer, D. Joy, O. Krivanek, M. Kundmann, R. Leapman, C. E. Lyman, D. Muller, S. Pennycook, P. Rez, R. Ritchie, P. Schattschneider, J. Spence, D. Williams, and K. Wong. I would welcome any further feedback from readers of the second edition.

Ray Egerton
University of Alberta
Edmonton, Canada
egerton@phys.ualberta.ca

Contents

1

An Introduction to Electron Energy-Loss Spectroscopy

Electron energy-loss spectroscopy (EELS) involves analyzing the energy distribution of initially monoenergetic electrons, after they have interacted with a specimen. This interaction may take place within a few atomic layers, as when a beam of low-energy (100–1000 eV) electrons is "reflected" from a solid surface. Because high voltages are not involved, the apparatus is relatively compact, but the low penetration depth implies the use of ultra-high vacuum; otherwise information is obtained mainly from the carbonaceous or oxide layers on the specimen's surface. At these low primary energies, a monochromator can be used to reduce the energy spread of the primary beam to a few millielectron volts (Ibach, 1991), and provided the spectrometer has a comparable resolution, the spectrum contains features characteristic of energy exchange with vibrational modes of surface atoms, as well as valence-electron excitation in these atoms. The technique is therefore referred to as *high-resolution* electron energy-loss spectroscopy (HREELS) and is used for studying the physics and chemistry of surfaces and of adsorbed atoms or molecules. Although it is an important tool of surface science, HREELS uses concepts which are substantially different to those involved in electron-microscope studies, so it will not be discussed further in the present volume. The physics and instrumentation involved are dealt with by Ibach and Mills (1992).

Surface sensitivity can also be obtained at higher electron energies, provided the electrons arrive at a glancing angle to the surface, so that they penetrate only a shallow depth before being scattered out. Reflection energy-loss spectroscopy (REELS) has been carried out with 30-keV electrons in a molecular beam epitaxy (MBE) chamber, where the relatively long working distances (30 cm) between the specimen, electron source, and spectrometer allowed elemental and structural analysis of the surface during crystal growth (Atwater *et al.*, 1993). REELS has also been carried out using electron microscopes and energies of 100 keV or more, as discussed in Section 3.3.5.

Provided the incident energy is high enough and the specimen sufficiently thin, practically all of the incident electrons are transmitted without reflection or absorption. Interaction takes place *inside* the specimen, and information about internal structure can be obtained by passing the transmitted beam into a spectrometer. For 100-keV incident energy, the specimen should be less than 1 μm thick and preferably no more than 100 nm.

Such specimens are self-supporting only over limited areas, so the incident electrons must be focused into a small diameter; in doing so we gain the advantage of analyzing small volumes. Electron lenses which can focus the electrons and guide them into an electron spectrometer are already present in a transmission electron microscope (TEM). Consequently, *transmission* EELS is usually carried out using a TEM, which also offers imaging and diffraction capabilities that can be used to identify the structure of the material being analyzed.

In this introductory chapter, we present a simplified account of the physical processes which occur while "fast" electrons are passing through a specimen, followed by an overview of energy-loss spectra and the instruments which have been developed to record such spectra. To identify the strengths and limitations of EELS, we conclude by considering alternative techniques for analyzing the chemical and physical properties of a solid specimen.

1.1. Interaction of Fast Electrons with a Solid

When electrons enter a material, they interact with the constituent atoms via electrostatic (Coulomb) forces. As a result of these forces, some of the electrons are scattered; the direction of their momentum is changed and in many cases they transfer an appreciable amount of energy to the specimen. It is convenient to divide the scattering into two broad categories: elastic and inelastic.

Elastic scattering involves Coulomb interaction with an atomic *nucleus*. Each nucleus represents a high concentration of charge; the electric field in its immediate vicinity is intense and an incident electron which approaches closely can be deflected through a large angle. Such high-angle scattering is referred to as Rutherford scattering; its angular distribution is the same as that calculated by Rutherford for the scattering of alpha particles. If the scattering angle exceeds 90°, the electron is said to be backscattered and may emerge from the specimen at the same surface which it entered (Fig. 1.1a).

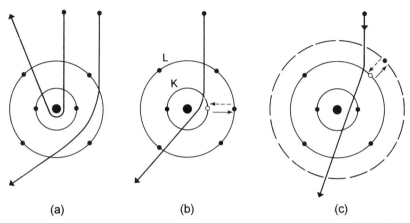

Figure 1.1. A classical (particle) view of electron scattering by a single atom (carbon). (a) Elastic scattering is caused by Coulomb attraction by the nucleus. Inelastic scattering results from Coulomb repulsion by (b) inner- or (c) outer-shell atomic electrons, which are excited to a higher energy state. The reverse transitions (deexcitation) are shown by broken arrows.

The majority of electrons travel further from the center of an atom, where the nuclear field is weaker as a result of the inverse-square law and the fact that the nucleus is partially shielded (screened) by the atomic electrons. Most incident electrons are therefore scattered through smaller angles, typically a few degrees (10–100 mrad) in the case of 100-keV incident energy. In a gas or (to a first approximation) an amorphous solid, the atoms or molecules can be regarded as scattering electrons independently of one another; but in a crystalline solid, the wave nature of the incident electrons cannot be ignored and interference between scattered electron waves changes the continuous distribution of scattered intensity into one which is sharply peaked at angles which are characteristic of the atomic spacing. The elastic scattering is then referred to as diffraction.

Although the term *elastic* usually denotes negligible exchange of energy, this condition applies only when the scattering angle is small. For a head-on collision (scattering angle = 180°), the energy transfer is given (in eV) by $E_{max} = 2148 (E_0 + 1.002)E_0/A$, where E_0 is the incident energy (in MeV) and A is the atomic weight of the target nucleus. For $E_0 = 100$ keV, E_{max} is greater than 1 eV and, in the case of a light element, may exceed the energy needed to displace the atom from its lattice site, resulting in *displacement damage* within a crystalline sample. But such high-angle collisions are comparatively rare; for the majority of elastic interactions, the energy transfer is limited to a small fraction of an eV and (in crystalline materials) is best described in terms of phonon excitation (vibration of the whole array of atoms).

Inelastic scattering occurs as a result of Coulomb interaction between a fast incident electron and the *atomic electrons* surrounding each nucleus. Some inelastic processes can be understood in terms of the excitation of a *single* atomic electron into a Bohr orbit (orbital) of higher quantum number (Fig. 1.1b) or, in terms of energy-band theory, to a higher energy level (Fig. 1.2).

Consider first the interaction of a fast electron with an *inner-shell* electron whose ground-state energy lies typically some hundreds or thousands of eV below the Fermi level of the solid. Unoccupied electron states exist only above the Fermi level, so the inner-shell electron can make an upward transition only if it absorbs an amount of energy comparable to or greater than its original binding energy. Because the total energy is conserved at each collision, the fast electron loses an equal amount of energy and is scattered through an angle which is typically of the order of 10 mrad for 100-keV incident energy. As a result of the inner-shell scattering, the target atom is left in a highly excited (or ionized) state and will quickly lose its excess energy. In the *deexcitation* process, an outer-shell electron (or an inner-shell electron of lower binding energy) undergoes a downward transition to the vacant "core hole" and the excess energy is liberated as electromagnetic radiation (x-rays) or as kinetic energy of another atomic electron (Auger emission).

Outer-shell electrons can also undergo single-electron excitation. In an insulator or semiconductor, a valence electron makes an interband transition across the energy gap; in the case of a metal, a conduction electron

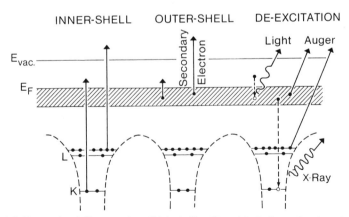

Figure 1.2. Energy-level diagram of a solid, including K- and L-shell core levels and a valence band of delocalized states (shaded); E_F is the Fermi level and E_{vac} the vacuum level. The primary processes of inner- and outer-shell excitation are shown on the left, secondary processes of photon and electron emission on the right.

makes a transition to a higher state, possibly within the same energy band. If the final state of these transitions lies above the vacuum level of the solid and if the excited atomic electron has enough energy to reach the surface, it may be emitted as a *secondary electron*. As before, the fast electron supplies the necessary energy (generally a few eV or tens of eV) and is scattered through an angle of typically 1 or 2 mrad (for $E_0 \simeq 100$ keV). In the deexcitation process, electromagnetic radiation may be emitted in the visible region (cathodoluminescence), although in many materials the reverse transitions are radiationless and the energy originally deposited by the fast electron appears as heat. Particularly in the case of organic compounds, not all of the valence electrons return to their original configuration; the permanent disruption of chemical bonds is described as radiation (ionization) damage.

As an alternative to the single-electron mode of excitation, outer-shell inelastic scattering may involve many atoms of the solid. This collective effect is known as a plasma resonance (an oscillation of the valence-electron density) and takes the form of a longitudinal traveling wave.* According to quantum theory, the excitation can also be described in terms of the creation of a pseudoparticle, the *plasmon*, whose energy is given by $E_p = \hbar\omega_p$, $\hbar$ being Planck's constant and ω_p being the plasmon frequency (in radians per second) which is proportional to the square root of the valence-electron density. For the majority of solids, E_p lies in the range 5–30 eV.

Plasmon excitation being a collective phenomenon, the excess energy is shared among many atoms when viewed over an extended period of time. At a given instant, however, the energy is likely to be carried by only one electron (Ferrell, 1957), which makes plausible the fact that plasmons can be excited in most insulators, E_p being generally higher than the excitation energy of the valence electrons (i.e., the band gap).† The lifetime of a plasmon is very short; in its decay (deexcitation) process, energy is deposited, via interband transitions, in the form of heat.

In addition to the volume or "bulk" plasmons which are excited *within* the specimen, a fast electron can create surface plasmons at each exterior surface. However, these surface excitations are numerically important only in very thin (<20 nm) samples.

Plasmon excitation and single-electron excitation represent alternative modes of inelastic scattering. In materials in which the valence electrons behave somewhat like free particles (e.g., the alkali metals), the collective

---

* Similar to a sound wave except that the restoring forces are electrostatic and collisions with the lattice cause damping.

† The essential requirement for plasmon excitation is that the participating electrons can communicate with each other and pass on their excess energy. This condition is fulfilled for an energy band of delocalized states but not for the atomic-like core levels.

form of response is predominant. In other cases (e.g., rare-gas solids), plasmon effects are weak or nonexistent. Most materials fall between these two extremes.

1.2. The Electron Energy-Loss Spectrum

The secondary processes of electron and photon emission from a specimen can be studied in detail by appropriate spectroscopies, as discussed in Section 1.4. In electron energy-loss spectroscopy we study directly the *primary* processes of electron excitation, each of which results in a fast electron losing a characteristic amount of energy. The beam of transmitted electrons is directed into a high-resolution electron spectrometer which separates the electrons according to their kinetic energy and produces an *electron energy-loss spectrum* showing the scattered intensity as a function of the decrease in kinetic energy of the fast electron.

A typical loss spectrum, recorded from a thin specimen over a range of about 1000 eV, is shown in Fig. 1.3. The first *zero-loss* or "elastic" peak represents electrons which are transmitted without suffering any measurable energy loss, including those which are scattered elastically in the forward direction and those which have excited phonon modes, for which the energy loss is less than the experimental energy resolution. In addition, the zero-loss peak includes electrons which can be regarded as unscattered, since they lose no energy and emerge undeflected after passing through the specimen.*

Inelastic scattering from outer-shell electrons is visible as a peak (or, in thicker specimens, a series of peaks) in the 5–50 eV region of the spectrum. At higher energy loss, the electron intensity decreases according to some fairly high power of energy loss, making it convenient to use a logarithmic scale for the recorded intensity, as in Fig. 1.3. Superimposed on this smoothly decreasing intensity are features which represent inner-shell excitation; they take the form of edges rather than peaks, the inner-shell intensity rising rapidly and then falling more slowly with increasing energy loss. The sharp rise occurs at the *ionization threshold*, whose energy-loss coordinate is approximately the binding energy of the corresponding atomic shell. Since inner-shell binding energies depend on the atomic number of the scattering atom, the ionization edges present in an energy-loss spectrum indicate which elements occur within the specimen. Quantitative

---

* The corresponding electron waves suffer a phase change, but the latter is of importance only in high-resolution imaging.

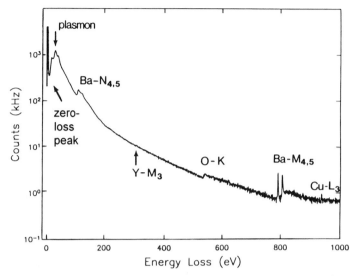

Figure 1.3. Electron energy-loss spectrum of a high-temperature superconductor ($YBa_2Cu_3O_7$) with the electron intensity on a logarithmic scale, showing zero-loss and plasmon peaks and ionization edges arising from each element (courtesy of D.H. Shin, Cornell University).

elemental analysis is possible by measuring an area under the appropriate ionization edge, making allowance for the background intensity.

When viewed in greater detail, both the valence-electron (low-loss) peaks and the ionization edges possess a *fine structure* which reflects the crystallographic or energy-band structure of the specimen. If an element or compound occurs in different forms, quite distinct fine structures are produced, as illustrated in Fig. 1.4.

If the energy-loss spectrum is recorded from a sufficiently thin region of the specimen, each spectral feature corresponds to a different excitation process. In thicker samples, there is a reasonable probability that a transmitted electron will be inelastically scattered more than once, giving a total energy loss which is the sum of the individual losses. In the case of plasmon scattering, the result is a series of peaks at multiples of the plasmon energy (Fig. 1.5). The plural (or multiple) scattering peaks have appreciable intensity if the specimen thickness approaches or exceeds the *mean free path* of the inelastic scattering process, which is typically 50–150 nm for outer-shell scattering at 100-keV incident energy. Electron-microscope specimens are typically of this thickness, so plural scattering is a significant but generally unwanted effect (since it distorts the shape of the energy-loss spectrum), which can fortunately be removed by appropriate deconvolution procedures.

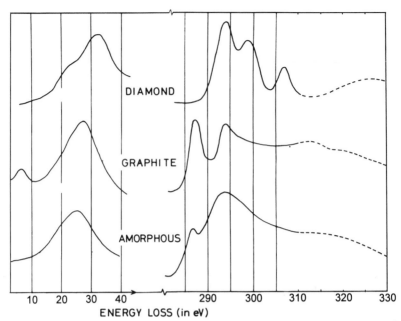

Figure 1.4. Low-loss and *K*-ionization regions of the energy-loss spectra of three allotropes of carbon (Egerton and Whelan, 1974a). The plasmon peaks occur at different energies (33 eV in daimond, 27 eV in graphite, 25 eV in amorphous carbon) because of the different valence-electron densities. The *K*-edge threshold is shifted upward by about 5 eV in diamond owing to the formation of an energy gap. The broad peaks shown by dashed lines are caused by electrons which undergo both inner-shell and plasmon scattering.

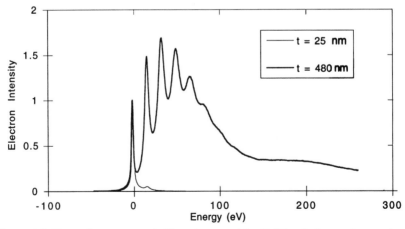

Figure 1.5. Energy-loss spectra of silicon (measured by K. Wong), for specimens of two different thicknesses. The thin sample gives a strong zero-loss peak and a weak first-plasmon peak; the thicker sample provides plural-scattering peaks at multiples of the plasmon energy.

On a classical (particle) model of scattering, the mean free path (MFP) is the average distance between scattering events. More generally, the MFP is inversely proportional to a scattering *cross section* which is a direct (rather than inverse) measure of the intensity of scattering from each atom (or molecule) and which can be calculated by the use of either classical physics or wave mechanics.

Inner-shell excitation gives rise to relatively low scattered intensity (low cross section) and therefore has a mean free path which is long compared to the specimen thickness. The probability that a fast electron produces more than one inner-shell excitation is therefore negligible. However, an electron which has undergone inner-shell scattering may (with fair probability) also cause outer-shell excitation. This "mixed" inelastic scattering again involves an energy loss which is the sum of the two separate losses and results in a broad peak above the ionization threshold, displaced from the threshold by approximately the plasmon-peak energy; see Fig. 1.4. If necessary, the mixed-scattering intensity can be removed from the spectrum by deconvolution.

1.3. The Development of Experimental Techniques

We describe now the evolution of techniques for recording and analyzing the energy-loss spectrum of fast electrons, particularly in combination with electron microscopy. More recent developments are dealt with in detail in later chapters.

In his doctoral thesis, published in 1929, Rudberg reported measurements of the kinetic energy of electrons which had been reflected from the surface of a metal such as copper or silver. The kinetic energy was determined using a magnetic-field spectrometer which bent the electron trajectories through 180° (in a 25-mm radius) and gave a resolution of about one part in 200, adequate for the low primary energies which were used (40–900 eV). By measuring currents with a quadrant electrometer, the electron intensity could be plotted as a function of energy loss, as in Fig. 1.6. Rudberg showed that the loss spectrum was *characteristic* of the chemical composition of the sample and independent of the primary energy and the angle of incidence. In these experiments, oxidation of the surface was minimized by preparing the sample *in situ* by evaporation onto a Mo or Ag substrate which could be electrically heated. Similar measurements were later carried out on a large number of elements by Powell, Robins, and Best at the University of Western Australia. The reflection technique has since been refined to give energy resolution of a few meV at incident energies of a few hundred eV (Ibach and Mills, 1982) and has also been implemented

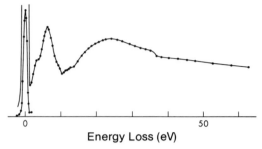

Energy Loss (eV)

Figure 1.6. Energy-loss spectrum of 204-eV electrons reflected from the surface of an evapo-rated specimen of copper (Rudberg, 1930). The zero-loss peak is shown on a reduced intensity scale. From E. Rudberg, Characteristic energy losses of electrons scattered from incandescent solids, *Proc. R. Soc. London* **A127**, 111–140 (1930) by permission of The Royal Society.

in a field-emission scanning electron microscope where ultrahigh vacuum can be achieved (Cowley, 1982).

The first measurement of the energy spectrum of *transmitted* electrons was reported by Ruthemann (1941), who used higher incident energy (2–10 keV), an improved magnetic spectrometer (bend radius = 175 mm, re-solving power 1 in 2000), and photographic recording. Figure 1.7a shows

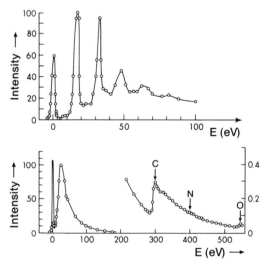

Figure 1.7. (a) Energy-loss spectrum of 5.3-keV electrons transmitted through a thin foil of aluminum (Ruthemann, 1941), exhibiting plasmon peaks at multiples of 16 eV loss. (b) Energy-loss spectrum of 7.5-keV electrons transmitted through a thin film of collodion, show-ing *K*-ionization edges arising from carbon, nitrogen, and oxygen. Reprinted from Ruthemann (1941) with permission.

Ruthemann's energy-loss spectrum of a thin self-supporting film of Al, displaying a series of peaks at multiples of 16 eV which were later interpreted in terms of plasma oscillation (Bohm and Pines, 1951). In 1942, Ruthemann reported the observation of *inner-shell* losses in a thin film of collodion, a form of nitrocellulose (Fig. 1.7b).

The first attempt to use the inner-shell losses for elemental microanalysis was made by Hillier and Baker (1944), who constructed an instrument which could focus 25–75 keV electrons into a 20-nm probe and which could function as either a microprobe or a shadow microscope. Two condenser lenses were used to focus the electrons onto the specimen, a third lens served to "couple" the transmitted electrons into a 180° magnetic spectrometer (Fig. 1.8). Because of poor vacuum and the resulting hydrocarbon contamination, the 20-nm probe caused specimens to become opaque in a few seconds. Therefore an incident beam diameter of 200 nm was used, corresponding to analysis of 10^{-16}–10^{-14} g of material. K-ionization edges were recorded from several elements (including Si), as well as L- and M-edges of iron. Spectra of collodion showed K-edges of carbon and oxygen

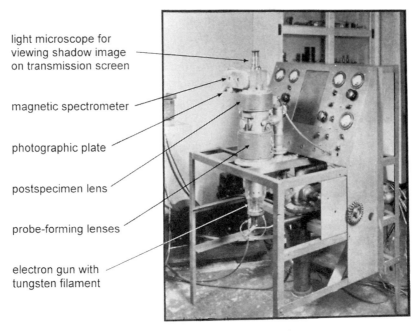

light microscope for viewing shadow image on transmission screen

magnetic spectrometer

photographic plate

postspecimen lens

probe-forming lenses

electron gun with tungsten filament

Figure 1.8. Photograph of the first electron microanalyzer (Hillier and Baker, 1944). Two magnetic lenses focused electrons onto the specimen, which was located within the bore of the second lens. A third lens focused the transmitted beam into a 180° spectrometer or produced a shadow image of the specimen when the spectrometer field was turned off.

but the nitrogen edge was usually absent, an observation which can be explained in terms of preferential removal of nitrogen by the electron beam (Egerton, 1980f).

In 1949, Möllenstedt published the design of an electrostatic energy analyzer in which electrons are slowed down by passing them between two cylindrical electrodes connected to the electron-source voltage. The deceleration results in high off-axis chromatic aberration (i.e., high dispersion) and an energy resolution of about 1 part in 50,000, allowing high-resolution spectra to be recorded on photographic plates. The Möllenstedt analyzer was subsequently added to conventional electron microscopes (CTEMs) in several laboratories, for example by Marton (at NBS, Washington), Boersch (Berlin), and Watanabe (Tokyo), and at the Cavendish Laboratory (Cambridge). It was usually attached to the bottom of the TEM column, allowing the microscope to retain its full range of magnification and diffraction facilities. Because the analyzer is nonfocusing in a direction parallel to the axis of the cylinders, a long (but narrow) entrance slit can be used; the spectrum is then recorded as a function of position in the specimen or, if a diffraction pattern is present on the TEM screen, as a function of electron scattering angle.

An alternative use of the deceleration principle was employed by Blackstock *et al.* (1955), Klemperer and Shepherd (1963), Kincaid *et al.* (1978) and Ritsko (1981). The electron source and analyzer were at ground potential, the electrons being accelerated toward the specimen and decelerated afterwards. This design allows good energy resolution but would be difficult to apply to an electron microscope, where the specimen cannot easily be raised to a high potential.* A retarding-field spectrometer was also used by Raether and colleagues in Hamburg, who conducted in-depth studies of the angular distribution and dispersion of bulk and surface plasmons in a wide variety of materials (Raether, 1980).

A combination of electric and magnetic fields (Wien filter) was first used for transmission energy-loss measurements by Boersch *et al.* (1962) in Berlin. For a given energy resolution, the entrance slit can be much wider than for the Möllenstedt analyzer (Curtis and Silcox, 1971), allowing the angular distribution of both strong and weak energy-loss peaks to be studied in detail (Silcox, 1977, 1979). By placing a second Wien filter before the specimen (to act as a monochromator), an energy resolution of 2 meV has been achieved at 25 keV incident energy (Geiger *et al.*, 1970).

---

* Fink (1989) has used a spectrometer system in which the sample is at ground potential, and the electron source, monochromator, analyzer, and detector are at −170 kV.

1.3.1. Energy-Selecting (Energy-Filtering) Electron Microscopes

Instead of displaying the loss spectrum from a particular region of specimen, it is sometimes preferable to display a magnified image of the specimen (or its diffraction pattern) at a *selected* energy loss. This can be done by utilizing the imaging properties of a magnetic field produced between prism-shaped polepieces, as first demonstrated by Castaing and Henry (1962) at the University of Paris. The plasmon-loss image was found to contain diffraction contrast (as present in unfiltered images) but, in suitable specimens, it can also convey "chemical contrast" which has been used for identifying different phases (Castaing, 1975). The Castaing–Henry filter was installed in other laboratories and used to obtain spectra and images from inner-shell energy losses (Colliex and Jouffrey, 1972; Henkelman and Ottensmeyer, 1974a; Egerton *et al.*, 1974). At the University of Toronto, Ottensmeyer reduced the aberrations of his filter by curving the prism edges. In 1984, the Zeiss instrument company marketed an 80-kV TEM (model 902) incorporating the straight-edged version.

In order to maintain a straight electron-optical column, the Castaing–Henry filter uses an electrostatic "mirror" electrode at the electron-gun potential, an arrangement not well suited to a high-voltage microscope. For their 1-MeV microscope at Toulouse, Jouffrey and colleagues adopted the purely magnetic "omega filter," based on a design by Rose and Plies (1974). In Berlin, Zeitler's group improved this system by correcting various aberrations, resulting in a commercial product, the Zeiss EM-912 energy-filtering microscope (Bihr *et al.*, 1991).

An alternative method of energy filtering is based on the *scanning* transmission electron microscope (STEM). In 1968, Crewe and co-workers in Chicago built the first high-resolution STEM and later used an energy analyzer to improve the images of single heavy atoms on a thin substrate. At the same laboratory, Isaacson, Johnson, and Lin measured fine structure present in the energy-loss spectra of amino acids and nucleic acid bases, and used this structure as a means of assessing electron irradiation damage to these biologically important compounds.

In 1974, the Vacuum Generators Company marketed a field-emission STEM (the HB5), on the prototype of which Burge and colleagues installed an electrostatic energy analyzer (Browne, 1979). Ferrier's group at Glasgow University investigated the practical advantages and disadvantages of adding post specimen lenses to the STEM, including their effect on spectrometer performance. Isaacson designed an improved magnetic spectrometer for the HB5, while Batson demonstrated that superior energy resolution can be obtained by using a retarding-field Wien filter. Various groups (for example, Colliex and co-workers in Paris, Brown and Howie in Cambridge,

Spence and colleagues in Arizona) have used the HB5 to explore the high-resolution possibilities of EELS. Leapman and co-workers at NIH (Washington) have employed STEM energy-selected imaging to obtain elemental maps of biological specimens at 10-nm resolution, and have shown that energy-loss spectra can reveal concentrations (of lanthanides and transition metals) down to 10 ppm within 50-nm-diameter areas of an inorganic test specimen. Using a STEM with a high-excitation objective lens, Pennycook and colleagues demonstrated that atomic columns in a suitably oriented crystalline specimen can be imaged using high-angle elastic scattering, while the low-angle energy-loss signal is simultaneously used to extract chemical information at atomic resolution.

1.3.2. Spectrometers as Attachments to Electron Microscopes

From the mid-1970s onward, EELS began to attract the attention of electron microscopists as a method of light-element microanalysis. For this purpose a single-prism magnetic spectrometer mounted beneath a conventional TEM is usually sufficient; Marton (1946) appears to have been the first to assemble such a system. At the Cavendish laboratory, Wittry (1969) devised an electron-optical arrangement which is now widely used: the prism spectrometer operates with the projector-lens crossover as its object point, allowing a spectrometer entrance aperture (just below the TEM screen) to select the region of the specimen (or its diffraction pattern) being analyzed (Fig. 1.9). *Serial* recording of the spectrum was achieved by deflecting it across an energy-selecting slit in front of a single-channel electron detector (scintillator and photomultiplier tube). At the University of California, Krivanek employed similar principles and made the magnetic spectrometer small enough to easily fit below any conventional TEM. This basic design was marketed by the Gatan company as their Model 607 serial-recording EELS system. Joy and Maher (at Bell Laboratories) and Egerton (in Alberta) built data-analysis systems for electronic storage of energy-loss spectra and background subtraction at core edges, as required for quantitative elemental analysis.

The energy-loss spectrum can be recorded simultaneously (rather than sequentially) by using a position-sensitive detector, such as a photodiode or charge-coupled diode (CCD) array. Following development work in several university laboratories, Gatan introduced in 1986 their Model 666 spectrometer, using quadrupole lenses to project the spectrum onto a YAG transmission screen and photodiode array. Parallel-recording spectrometers greatly reduce the time needed to record inner-shell losses, resulting in less drift and electron irradiation of the specimen.

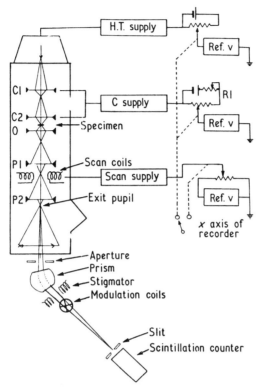

Figure 1.9. TEM energy-analysis system with a magnetic-prism spectrometer (45° deflection angle) which used the projector crossover as its object point (Wittry, 1969). The spectrum was scanned across the detection slit by applying a ramp voltage to the high-tension supply, a similar voltage being applied to the condenser-lens supplies in order to keep the diameter of illumination constant. Alternatively, scan coils could be used to deflect the TEM image or diffraction pattern across the spectrometer entrance aperture to obtain a line scan at fixed energy loss.

With further quadrupole and sextupole lenses added to correct for spectrometer aberrations and a two-dimensional CCD array as the detector, the Gatan 666 spectrometer became the Gatan imaging filter (GIF). Installed beneath a conventional TEM, it performs similar functions to an in-column (Castaing–Henry or omega) imaging filter, even in higher-voltage microscopes. The ability to produce an elemental map (by selecting an appropriate ionization edge and making corrections for the pre-edge background) increases the analytical power of EELS by enabling elemental segregation to be imaged in a semiquantitative manner.

More complete information about the specimen is contained in its *spectrum-image*, obtained using a STEM and a parallel-recording detector to record an entire energy-loss spectrum at each pixel, or its *image-spectrum* (a sequence of energy-selected images recorded using a two-dimensional detector in an energy-filtering TEM). Several groups (led by Colliex in Orsay, L'Espérance in Montreal, Hunt at Lehigh University) developed software for spectrum-image acquisition and processing, and this technique will likely become more widespread in the future.

1.4. Alternative Analytical Methods

Electron energy-loss spectroscopy is only one of many techniques available for determining the structure and/or chemical composition of a solid. Some of the techniques which are capable of high spatial resolution are listed in Table 1.1. As pointed out by Wittry (1980), each method is based on a well-known physical phenomenon but attains usefulness as a microanalytical tool only when suitable instrumentation becomes available.

Some of these analytical techniques, such as Auger spectroscopy, are surface-sensitive: they characterize the first monolayer (or few monolayers)

Table 1.1. Microanalysis Techniques Employing Electron, Ion, and Photon Beams, with Estimates of the Currently Available Spatial Resolution for a Near-Ideal Specimen (see text)

Incident beam	Detected signal	Examples	Resolution (nm)
Electron	Electron	Electron microscopy (TEM, STEM)	0.2
		Electron diffraction (SAED, CBED)	10
		Electron energy-loss spectroscopy (EELS)	~1
		Auger-electron spectroscopy (AES)	~2
	Photon	X-ray emission spectroscopy (XES)	~10
		Cathodoluminescence (CL)	
Ion	Ion	Rutherford backscattering spectroscopy (RBS)	
		Secondary-ion mass spectrometry (SIMS)	50
	Photon	Proton-induced x-ray emission (PIXE)	500
Photon	Photon	X-ray diffraction (XRD)	50
		X-ray absorption spectroscopy (XAS)	50
		X-ray fluorescence spectroscopy (XRF)	
	Electron	X-ray photoelectron spectroscopy (XPS)	
		Ultraviolet photoelectron spectroscopy (UPS)	1000
	Ion	Laser microprobe mass analysis (LAMMA)	1000

of atoms. Others, such as EELS and high-energy electron diffraction, probe deeper into the bulk or (in the case of a thin specimen) give information integrated over specimen thickness. Which category of technique is preferable depends on the kind of information required.

Analysis techniques can also be classified as destructive or nondestructive. Secondary-ion mass spectrometry (SIMS) and laser-microprobe mass analysis (LAMMA) are examples of techniques which are necessarily destructive, a property which may sometimes be a disadvantage but which, in the case of SIMS, can be utilized to give three-dimensional information by "depth profiling." Electron-beam methods may also be destructive, since inelastic scattering of the incident electrons can result in radiation damage. The extent of this damage depends on the electron dose needed to give a useful signal. Elemental analysis by EELS, Auger, or x-ray emission spectroscopy relies on *inner-shell* scattering, which is comparatively weak, so radiation damage can be a serious problem. Transmission electron microscopy and electron diffraction utilize elastic scattering, which is relatively strong. However, these latter techniques are often used to determine structure at an atomic level, which is more sensitive to irradiation than is the chemical composition, so damage is still a problem in some cases.

The different techniques can also be grouped according to the type of incident and detected particle, as in Table 1.1. We now outline some of the important characteristics of each technique, leaving electron energy-loss spectroscopy until the end of the discussion.

1.4.1. Ion-Beam Methods

In *secondary-ion mass spectrometry* (SIMS), a specimen is bombarded with 1–20 keV ions, causing surface atoms to be sputtered away, some as secondary ions whose atomic number is determined by passing through a mass spectrometer. Since the surface is being steadily eroded, elemental concentrations are obtained as a function of depth. A spatial resolution of 1 μm is routine, and 50–100 nm is possible with a liquid–metal source (Levi-Setti, 1983). All elements are detectable, including hydrogen, which has been measured in silicon at concentrations below 1% (Magee, 1984). Imaging of the secondary ions is possible, at sub-μm spatial resolution (Grivet and Septier, 1978). Quantification is complicated by the fact that sputtering gives rise mainly to *neutral* atoms: the yield of ions is low and highly dependent on the chemical composition of the matrix. To avoid these difficulties, the sputtered atoms can be ionized (by an electron gun, laser or r.f. cavity), the technique then being known as SNMS.

If the incident beam consists of high-energy light ions (e.g., 1-MeV protons or He ions), the sputtering rate is low but some of the ions are elastically scattered in the backward direction. The energy-loss spectrum of the escaping primary ions contains peaks or edges whose energy is characteristic of each backscattering element and whose width is an indication of the depth distribution, since ions which are backscattered deep in the sample lose energy on their way out. *Rutherford backscattering spectroscopy* (RBS) therefore provides a nondestructive method of performing three-dimensional elemental analysis; however, the lateral spatial resolution is poor, being limited by the current density available in the incident beam.

Very low elemental concentrations (0.1–10 ppm) can be determined from *proton-induced x-ray emission* (PIXE). Most of the incident protons (typically of energy 1–5 MeV) are only slightly deflected by the nuclear field; therefore the bremsstrahlung background to the characteristic x-ray peaks is lower than when using incident electrons. Spatial resolution down to 1 μm has been demonstrated (Johansson, 1984).

Field-ion microscopy (FIM) relies on the field-induced impingement of helium ions to image individual atoms at the tip of a specimen, usually a sharpened wire. Raising the field allows atoms to be desorbed and chemically identified in a time-of-flight mass spectrometer, the instrument then being known as an *atom probe* (Miller and Smith, 1989). By accumulating data during the process of field evaporation, one can obtain a three-dimensional image of each atomic species. Although unsurpassed in terms of spatial resolution, the atom probe does have the disadvantage that it examines only an extremely small area (2- to 50-nm in diameter) of the tip, which may provide inadequate sampling in the case of inhomogeneous specimens.

1.4.2. Incident Photons

X-ray diffraction is a convenient laboratory technique and has been used for many years to determine the symmetry of crystals and to measure lattice parameters to high accuracy. With conventional x-ray sources, a relatively large volume of specimen (many cubic micrometers) is necessary to record diffraction spots which stand out above the instrumental background, but synchrotron sources offer improved spatial resolution.

In *x-ray absorption spectroscopy* (XAS), the intensity of a transmitted beam of x-rays is measured as a function of incident wavelength. To obtain sufficient intensity, a high-brightness source is generally used, such as a synchrotron. X-ray absorption edges occur at an incident energy close to binding energy of each atomic shell and are the closest analog to the ionization edges seen in electron energy-loss spectra. Fine structure is present up to several hundred eV beyond the edge (EXAFS) and in the near-edge region (XANES or NEXAFS) and can be used to determine interatomic distances relative to a known element, even in the case of amorphous specimens.

Soft x-rays from a synchrotron can be focused by means of zone plates to yield an *x-ray microscope* with a resolution of the order of 50 nm. Contrast due to differences in absorption coefficient can be used for mapping individual elements or even (via XANES fine structure) different chemical environments of the same element (Ade *et al.*, 1992). Although spatial resolution is inferior to that of the electron microscope, radiation damage can be less and the specimen need not be in vacuum. In the case of thicker specimens, three-dimensional information is obtainable via tomographic techniques.

Photoelectron spectroscopy can be carried out with incident x-rays (XPS) or ultraviolet radiation (UPS). In the former case, electrons are released from inner shells and enter an electron spectrometer which produces a spectrum containing peaks at energy losses (incident photon energy minus the kinetic energy of the photoelectron) equal to the inner-shell binding energy of the elements present in the sample. Besides providing elemental analysis, the XPS peaks may display chemical shifts which reveal the oxidation state of each element. In UPS, valence electrons are excited by the incident radiation and the electron spectrum is characteristic of the valence-band states. Photoelectron microscopy is possible by immersing the sample in a strong magnetic field and imaging the photoelectrons, with energy discrimination; 2-μm resolution has been demonstrated and 0.1 μm might be possible (Beamson *et al.*, 1981).

In *laser-microprobe mass analysis* (LAMMA), light is focused into a small-diameter probe (≥ 500 nm diameter) in order to volatilize a small region of a sample, releasing ions which are analyzed in a mass spectrometer. All elements can be detected and measured with a sensitivity of the order of 1 ppm. The analyzed volume is typically 0.1 $(\mu m)^3$, so a sensitivity of 10^{-19} g is possible (Schmidt *et al.*, 1980). As in the case of SIMS, quantification is complicated by the fact that the ionization probability is matrix dependent. If necessary, different isotopes can be distinguished.

1.4.3. Electron-Beam Techniques

Transmission electron microscopy (TEM) is capable of atomic resolution, using either a conventional (CTEM) or scanning transmission (STEM) instrument. In the case of crystalline specimens, CTEM "chemical lattice images" (phase-contrast images obtained under particular conditions of specimen thickness and defocus) allow columns of different elements to be distinguished (Ourmadz *et al.*, 1990).

Transmission electron diffraction also offers high spatial resolution. In the case of the selected-area technique (SAED), resolution is limited by spherical aberration of the objective lens, but in convergent-beam diffraction (CBED) the analyzed region is defined by the diameter of the incident beam, typically 20 nm. Besides giving information about crystal symmetry, CBED has been used to measure small (0.1%) changes in lattice parameter arising from compositional gradients. Some metal alloys contain precipitates whose CBED pattern is sufficiently characteristic to enable their chemical composition to be identified through a fingerprinting procedure (Steeds, 1984).

In general, electron microscopy and diffraction provide structural information which is complementary to the structural and chemical information provided by EELS. All three techniques can be combined in a single instrument without any sacrifice of performance. In addition, use of an electron spectrometer to produce *energy-filtered* images and diffraction patterns can greatly increase their information content (Auchterlonie *et al.*, 1989; Spence and Zuo, 1992; Midgley *et al.*, 1995).

Auger electron spectroscopy (AES) may be carried out with incident x-rays or charged particles, but the highest spatial resolution is obtained by using an electron beam, which also permits scanning Auger microscopy (SAM) on suitable specimens. Auger-peak energies are characteristic of each element present at the sample, but AES is particularly sensitive to low-Z elements which have high Auger yields. Quantitation is more complicated than for EELS and may have to rely on the use of standards (Seah, 1983). The detected Auger electrons have energies in the range 20–500 eV, where the escape depth is of the order 1 nm. The technique is therefore highly surface sensitive and bulk specimens can be used, although (because of contributions from backscattering) the spatial resolution is then limited to about 100 nm. Use of a thin specimen and improved electron optics allows a spatial resolution of around 2 nm (Hembree and Venables, 1992), and identification of single atoms may be a future possibility, given a sufficiently radiation-resistant specimen (Cazaux, 1983). In practice, a resolution of 50 nm is more typical for a 35-keV, 1-nA incident beam (Rivière, 1982); the usable resolution is generally limited by statistical noise present in the signal, a situation which can be improved by use of a parallel-recording electron detector. Even then, the measured signal will remain less than for EELS because only a fraction of the excited Auger electrons (those generated within an escape depth of the surface) are detected.

X-ray emission spectroscopy (XES) can be performed on bulk specimens, for example using an electron-probe microanalyzer (EPMA) fitted with a wavelength-dispersive (WDX) spectrometer. As a method of elemental analysis, the EPMA technique has been refined to give good accuracy, about 1% with appropriate absorption and fluorescence corrections. A detection limit of 100 ppm (0.01%) is routine, and 1 ppm is possible with 50-keV accelerating voltage, high probe current (0.5 μA), and computer control (Robinson and Graham, 1992), corresponding to a detection limit $\approx 10^{-17}$ g in an analyzed volume of a few $(\mu m)^3$. The WDX spectrometer detects all elements except H, He, and Li and has an energy resolution ≈ 10 eV. Alternatively, bulk specimens can be analyzed in a scanning electron microscope fitted with an energy-dispersive (EDX) spectrometer, offering shorter recording times but worse resolution (≈ 100 eV), which sometimes limits the accuracy of analyzing overlapping peaks.

Higher spatial resolution is available by using a thin specimen and a TEM fitted with an EDX detector, particularly if it has an immersion objective lens which can focus electrons into probes below 100 nm in diameter. Characteristic x-rays are emitted isotropically from the specimen, resulting in a geometrical collection efficiency of typically 1% (for 0.13-sr collection solid angle). For a tungsten-filament electron source, the detection limit for medium-Z element in a 100-nm specimen has been estimated to be about 10^{-19} g (Shuman *et al.*, 1976; Joy and Maher, 1977). Metal-catalyst particles of mass below 10^{-20} g have been analyzed using a field-emission source (Lyman *et al.*, 1995), although radiation damage is a potential problem (Dexpert *et al.*, 1982). Estimates of the minimum detectable concentration vary between 10 mmol/kg (0.04% by weight) for potassium in biological tissue (Shuman *et al.*, 1976) to values in the range 0.05%–3% (dependent on probe size) for medium-Z elements in a 100-nm Si matrix (Joy and Maher, 1977; Williams, 1987). With a windowless or ultrathin window (UTW) detector, elements down to boron can be detected, although the limited energy resolution of the EDX detector may give rise to peak-overlap problems at low photon energies; see Fig. 1.10a. For quantitative analysis of light elements, extensive absorption corrections may be necessary (Chan *et al.*, 1984).

The ultimate spatial resolution of thin-film x-ray analysis is limited by elastic scattering, which causes a broadening of the transmitted beam; see Fig. 1.11. For a 100-nm-thick specimen and 100-keV incident electrons, this broadening is about 4 nm in carbon and increases (with atomic number) to 60 nm for a gold film of the same thickness (Goldstein *et al.*, 1977). Inelastic scattering also degrades the spatial resolution, since it results in the production of fast secondary electrons which generate characteristic x rays, particularly from light elements. For energy losses between 1 and 10 keV, the fast secondaries are emitted almost perpendicular to the incident-beam direction and have a range of the order of 10–100 nm (Joy *et al.*,

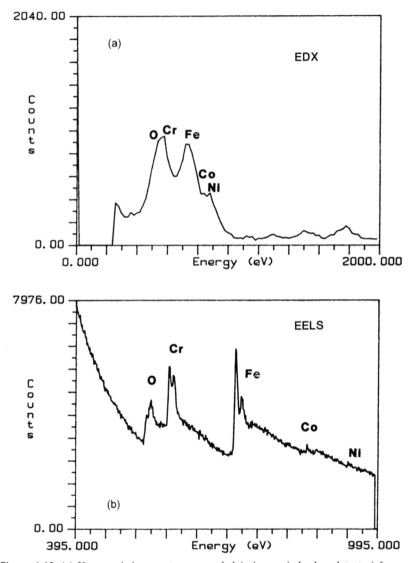

Figure 1.10. (a) X-ray emission spectrum recorded (using a windowless detector) from an oxidized region of stainless steel, showing overlap of the oxygen *K*-peak with the *L*-peaks of chromium and iron. (b) Ionization edges are more clearly resolved in the energy-loss spectrum as a result of the better energy resolution of the electron spectrometer (Zaluzec *et al.*, 1984). From *Analytical Electron Microscopy—1984*, p. 351, © San Francisco Press, Inc., by permission.

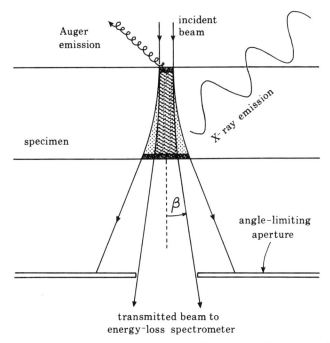

Figure 1.11. Spreading of an electron beam within a thin specimen. X-rays are emitted from the dotted region, whereas the energy-loss signal is recorded mainly from the hatched region. Auger electrons are emitted within a small depth adjacent to each surface.

1982). An x-ray signal is also generated (at some distance from the incident beam) by backscattered electrons and by secondary fluorescence (Bentley *et al.*, 1984).

1.5. Comparisons of EELS and EDX Spectroscopy

Prior to the 1980s, most EDX detectors were protected (from water vapor and hydrocarbons in the microscope vacuum) by a 10-μm-thick beryllium window, which strongly absorbs photons of energy less than 1000 eV and precludes analysis of elements of atomic number less than 11. With the development of ultrathin (UTW) or atmospheric-pressure (ATW) windows, elements down to boron can be routinely detected, making EDX competitive with EELS for the microanalysis of light elements in a TEM specimen. Table 1.2 indicates several factors relevant to a comparison of the two techniques.

Table 1.2. Comparison of EELS with Windowless EDX
Spectroscopy of a TEM Specimen

Advantages of EELS	Disadvantages of EELS
Higher core-loss signal	Higher background
Higher ultimate spatial resolution	Very thin specimen needed
Absolute, standardless quantitation	Quantitation errors possible in crystals
Structural information available	More operator intensive

1.5.1. Detection Limits and Spatial Resolution

For the same incident-beam current, EELS yields a count rate of core-loss electrons which exceeds that of characteristic x-rays for two reasons. First, the generation rate of low-energy x-rays is reduced because of the low x-ray fluorescence yield, which falls below 2% for $Z < 11$ (see Fig. 1.12). Second, transmitted electrons are concentrated into a limited angular range: the collection efficiency of the electron spectrometer is typically 20% to 50%, whereas characteristic x-rays are emitted isotropically and the fraction recorded by an EDX detector is about 1%.

Unfortunately, the background in an energy-loss spectrum, which arises from inelastic scattering from all atomic electrons whose binding energy is less than the edge energy, is generally higher than the background in an EDX spectrum, which arises largely from bremsstrahlung production. In addition, the characteristic features in an energy-loss spectrum are not

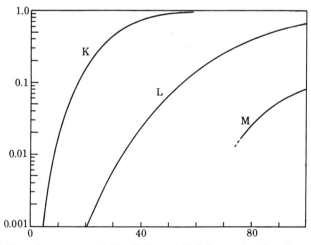

Figure 1.12. X-ray fluorescence yield for K-, L-, and M-shells, as a function of atomic number.

peaks but edges; the core-loss intensity is spread over an extended energy range beyond the edge, which tends to make the edges less visible than corresponding peaks in an EDX spectrum. It is possible to define a signal/noise ratio (SNR) which takes account of the edge shape, and the *minimum detectable concentration* of an element can be shown to depend on SNR, not signal/background ratio (Section 5.6.3). On this basis, Leapman and Hunt (1991) have compared the sensitivity of EELS and EDX spectroscopy and shown that EELS is capable of detecting smaller concentrations of elements of low atomic number; see Fig. 1.13. Using a field-emission STEM with a parallel-recording spectrometer, Leapman and Newbury (1993) were able to detect concentrations of transition metals and lanthanides down to 10 ppm in powdered-glass samples. Shuman *et al.* (1984) reported a sensitivity of 20 ppm for Ca in organic test specimens. In many specimens, the ultimate sensitivity will be limited by mass loss of light elements, but in this respect EELS is preferable to EDX spectroscopy since a larger proportion of the inner-shell excitations can be recorded by the spectrometer.

EELS can also offer slightly better *spatial resolution* than x-ray emission spectroscopy because the volume of specimen giving rise to the energy-loss signal can be defined to some extent by means of an angle-limiting

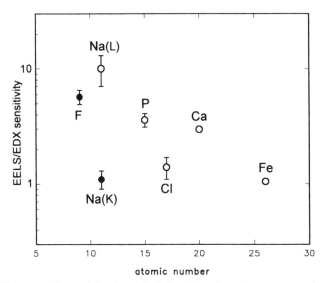

Figure 1.13. Elemental sensitivity of EELS relative to EDX spectroscopy, assuming a parallel-recording electron spectrometer, 0.18-sr UTW x-ray detector, and second-difference multiple-least-squares processing of both spectra, and based on signal-background ratios measured with 100-keV incident electrons (Leapman and Hunt, 1992). The increase in sensitivity between Cl and for Ca is due to the emergence of white-line peaks at the L_{23} edge.

aperture, as shown in Fig. 1.11. The effects of beam broadening (due to elastic scattering) and beam tails (due to spherical aberration of the probe-forming lens) should therefore be less, as confirmed experimentally (Collett *et al.*, 1984; Titchmarsh, 1989). In addition, the energy-loss signal is unaffected by backscattering, secondary fluorescence, and the generation of fast secondary electrons within the specimen.

Spatial resolution is a major factor determining the *minimum detectable mass* of an element. With a field-emission source to form a 1-nm probe with about 1 nA current, clusters of thorium atoms (on a thin carbon film) down to one or two atoms in size have been identified (Krivanek *et al.*, 1991). The only alternative technique capable of single-atom identification is the field-ion atom probe, whose applications are limited by problems associated with specimen preparation (p. 18).

1.5.2. Specimen Requirements

If the specimen is too thick, plural scattering can greatly increase the background to ionization edges below 1000 eV, making these edges invisible for specimens thicker than 100 nm or even 50 nm. This requirement places stringent demands on specimen preparation, which can sometimes be met by ion milling (of inorganic materials) or ultramicrotome preparation of ultrathin sections; see page 29. The situation is eased somewhat by the use of higher accelerating voltage, although in some materials (particularly above 300 kV) this introduces radiation damage by displacement processes. EDX spectroscopy can usually tolerate specimens which are thin enough for regular TEM imaging (up to a few hundred nm), although absorption corrections for light-element quantification become severe in specimens thicker than 100 nm.

1.5.3. Accuracy of Quantification

In EELS, the signal intensity depends only on the physics of the primary excitation and is independent of the spectrometer. Quantification need not involve the use of standards; measured core-loss intensities can be converted to elemental ratios using cross sections, calculated for the collection angle, range of energy loss, and incident-electron energy involved in the analysis. These cross sections are known to within 5% for most K-edges and 15% for most L-edges, the accuracy for other edges being highly variable (Egerton, 1994). EELS analysis of 45-nm NiO films distributed to four laboratories yielded elemental ratios within 10% of stoichiometry (Bennett and Egerton, 1995); analysis of small areas of less-ideal specimens might give more variable results.

In contrast, the intensities of EDX peaks depend on the properties of the detector. For thin specimens, this problem is addressed by using a sensitivity factor (k-factor) for each characteristic peak. Because detector parameters are not precisely known, these k-factors cannot be calculated with high accuracy. For the same reasons, k-factors measured in other laboratories can only be used as a rough guide. To achieve an accuracy of better than 15%, the appropriate k-factors must be measured for each analyzed element, using test specimens of known composition and the same x-ray detector and microscope (operating at the same accelerating voltage).

In the case of *low-energy* x-rays (e.g., K-peaks from light elements) the k-factors are highly dependent on absorption in the front end and protective window of the detector and on absorption of x-rays within the specimen itself. Although it is possible to correct for such absorption, the accuracy of such correction is highly dependent on the specimen geometry. Therefore, the accuracy of light-element EDX analysis is not expected to be as good as for heavier elements.

In certain cases, *overlap* of peaks can prevent meaningful EDX analysis, as in the case of light-element quantification using K-peaks when there are heavier elements whose L-peaks occur within 100 eV; see Fig. 1.10a. In the energy-loss spectrum, the corresponding edges overlap but may be more easily distinguished because of the higher energy resolution, of the order of 1 eV rather than 100 eV. The problem of background subtraction (e.g., at the Cr edge in Fig. 1.10b) can be overcome by MLS fitting to reference standards.

1.5.4. Ease of Use and Information Content

Changing from TEM imaging to the recording of an EDX spectrum typically involves positioning the incident beam, ensuring that probe current is not excessive (to avoid detector saturation), withdrawing any objective aperture, and operating switches to insert the EDX detector and start acquisition. Once set up, the EDX detector and electronics require a minimum of maintenance, beyond ensuring that the detector remains cooled by liquid nitrogen. EDX software has been developed to the point where elemental ratios (correct or incorrect) are produced in a routine fashion. Only in some cases are problems of peak overlap, absorption, and fluorescence important. For these reasons, EDX spectroscopy remains the technique of choice for *the majority* of TEM microanalysis problems.

Obtaining an energy-loss spectrum involves setting the spectrometer excitation (positioning the zero-loss peak), choosing an energy dispersion and collection solid angle, and verifying that the specimen is suitably thin. For some measurements, low drift of the spectrum is important, and this condition may involve waiting for the microscope high voltage and spectrometer power supplies to stabilize. Recent improvements in software have made spectral analysis more convenient; however, the success of basic operations such as the subtraction of instrumental and pre-edge backgrounds still depends on the skill of the operator and some understanding of the physics involved.

To summarize, EELS is more demanding than EDX spectroscopy in terms of the expertise and knowledge required. In return for this investment, energy-loss spectroscopy provides greater sensitivity in certain specimens and yields more information: in addition to elemental analysis, it can provide a rapid estimate of the local thickness of a TEM specimen, as well as information about its crystallographic and electronic structure. In principle, EELS yields similar data to x-ray, ultraviolet, visible, and (in some cases) infrared spectroscopy, all carried out with the same instrumentation and with the possibility of very high spatial resolution. Obtaining this information is the subject of the remainder of this book; examples of practical applications are given in Chapter 5.

1.6. Further Reading

The following chapters assume some familiarity with the operation of a transmission electron microscope, a topic which is covered in several books. Reimer (1993) gives a thorough account of the physics and electron optics involved, while Hirsch *et al.* (1977) remains a seminal guide to diffraction-contrast imaging. Phase-contrast imaging is dealt with by Spence (1988a) and by Buseck *et al.* (1988); reflection imaging and diffraction are reviewed by Wang (1993, 1996).

Analytical electron microscopy is described in a very practical way by Williams (1987) and in several multiauthor volumes: Hren *et al.* (1979), Joy *et al.* (1986), and Lyman *et al.* (1990). A detailed review of alternative analytical techniques (AES, EELS, EDX, PIXE, RIMS, SIMS, SNMS, and XPS), based on lectures from the 40th Scottish Universities Summer School in Physics, is available in book form (Fitzgerald *et al.*, 1992).

Several review papers provide an introduction to EELS: for example, Silcox (1979), Joy (1979), Joy and Maher (1980c), Isaacson (1981), Leapman (1984), Zaluzec (1988), and Egerton (1992b). Colliex (1984a) offers an extensive review, while Marton *et al.* (1955) present an account of early progress in the field. The physics and spectroscopy of outer-shell excitation is dealt with by Raether (1965), Daniels *et al.* (1970), and Raether (1980); basic theory of inelastic scattering is treated by Schattschneider (1986). The benefits of energy filtering in TEM, as well as an account of the instrumentation and physics involved, are dealt with comprehensively in a recent volume edited by Reimer (1995). The effect of inelastic scattering on TEM images and diffraction patterns is treated in depth by Spence and Zuo (1992) and by Wang (1995).

EELS has been the subject of several workshops, which are represented by collected papers in *Ultramicroscopy* (vol. 28, April 1989: Aussois; vol. 59, July 1995: Leukerbad) and in *Microscopy, Microanalysis, Microstructures* (vol. 6, no. 1, Feb. 1995: applied papers from Leukerbad; vol. 2, nos. 2/3, April/June 1991: Lake Tahoe). Similar workshops on electron spectroscopic imaging (ESI) are documented in *Ultramicroscopy* (vol. 32, no. 1, 1990: Tübingen) and *Journal of Microscopy* (April 1991: Dortmund; vol. 6, Pt. 3, 1992: Munich). Materials-science applications of EELS are described in some detail in Disko *et al.* (1992).

The present volume does not deal with reflection-mode high-resolution spectroscopy (HREELS) at low incident-electron energy, which is covered by Ibach and Mills (1982). For the related field of transmission energy-loss spectroscopy of gases, see Bonham and Fink (1974) and Hitchcock (1989, 1994).

Specimen preparation is always important when using a transmission electron microscope. Chemical and electrochemical thinning of bulk materials is covered by Hirsch *et al.* (1977). Ion milling is applicable to a wide range of materials and is especially useful for cross-sectional specimens (Bravman and Sinclair, 1984) but can result in changes in chemical composition within 10 nm of the surface (Ostyn and Carter, 1982; Howitt, 1984). To avoid this, mechanical methods are attractive. They include ultramicrotomy (Cook, 1971; Ball *et al.*, 1984; Timsit *et al.*, 1984; Tucker *et al.*, 1985; see also papers in *Microscopy Research and Technique*, vol. 31, 1995, 265–310), cleavage techniques using tape or thermoplastic glue (Hines, 1975), small-angle cleavage (McCaffrey, 1993), and abrasion into small particles which are then dispersed on a support film (Moharir and Prakash, 1975; Reichelt *et al.*, 1977; Baumeister and Hahn, 1976). Extraction replicas can be employed to remove precipitates lying close to a surface; plastic film is normally used but other materials are more suitable for the extraction of carbides (Garratt-Reed, 1981; Chen *et al.*, 1984; Duckworth *et al.*, 1984; Tatlock *et al.*, 1984).

2

Instrumentation for Energy-Loss Spectroscopy

2.1. Energy-Analyzing and Energy-Selecting Systems

Complete characterization of a specimen in terms of its inelastic scattering would involve recording the scattered intensity $J(x, y, z, \theta_x, \theta_y, E)$ as a function of position (coordinates x, y, z) within the specimen and as a function of scattering angle (components θ_x and θ_y) and energy loss E. Even if technically feasible, such a procedure would involve storing a vast amount of information, so in practice the acquisition of energy-loss data is restricted to the following categories (see Fig. 2.1).

(a) An *energy-loss spectrum $J(E)$* recorded at a particular point on the specimen or (more precisely) integrated over a circular region defined by the incident electron beam or an area-selecting aperture. Such spectroscopy (also known as energy analysis) can be carried out using a conventional transmission electron microscope (CTEM) or a scanning transmission electron microscope (STEM) fitted with a double-focusing spectrometer such as the magnetic prism (Sections 2.1.1 and 2.2).

(b) A *line spectrum $J(y, E)$ or $J(\theta_y, E)$*, where distance perpendicular to the E-axis represents a single coordinate in the image or diffraction pattern. This mode can be obtained by using a spectrometer which focuses only in the direction of dispersion, such as the Wien filter (Section 2.1.3).

(c) An *energy-selected image $J(x, y)$ or filtered diffraction pattern $J(\theta_x, \theta_y)$* recorded for a given energy loss E (actually, a restricted range of energy loss) using CTEM or STEM techniques, as discussed in Section 2.6.

(d) A *spectrum-image $J(x, y, E)$* obtained by acquiring an energy-loss spectrum at each pixel as a STEM probe is rastered over the specimen. To make the recording times realistic, a parallel-recording spectromer is needed, as discussed in Section 2.6.4. Using a conventional TEM fitted with

(a) selected-area SPECTRUM   (c) ENERGY-SELECTED IMAGE (or DP)

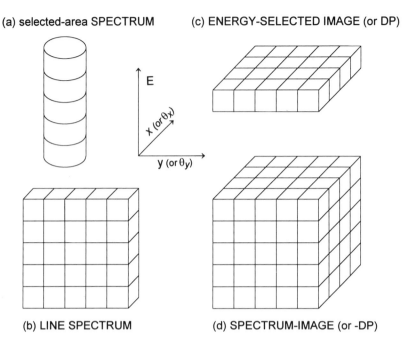

(b) LINE SPECTRUM   (d) SPECTRUM-IMAGE (or -DP)

Figure 2.1. Energy-loss data obtained from (a) TEM fitted with a double-focusing spectrometer; (b) TEM with a line-focus spectrometer; (c) CTEM fitted with an imaging filter or STEM with serial-recording spectrometer; (d) STEM fitted with a parallel-recording spectrometer.

an imaging filter, the same information can be obtained by recording a series of energy-filtered images at successive values of energy loss, sometimes known as an *image-spectrum* (Lavergne *et al.*, 1992). In fig. 2.1d, this corresponds to acquiring the information from successive layers rather than column by column as in the STEM method. Through filtering a diffraction pattern or the use of a rocking beam (in STEM) it is possible to record $J(\theta_x, \theta_y, E)$, corresponding to a *diffraction-spectrum* or *spectrum-diffraction*, respectively.

Details of the operation of these energy-loss *systems* are discussed in Sections 2.3–2.6. In this section, we review the types of spectrometer which have been used for energy-loss measurements in combination with transmission electron microscopy, where an incident energy of the order of 10^5 eV is required to avoid excessive scattering in the specimen. Since an energy resolution of 1 eV is desirable, the choice of spectrometer is limited to those types which can offer high resolving power, thereby ruling out techniques such as time-of-flight analysis which are used successfully in other branches of spectroscopy.

2.1.1. The Magnetic-Prism Spectrometer

In the magnetic-prism spectrometer, electrons traveling at a speed v in the z-direction are directed between the poles of an electromagnet whose magnetic field B is in the y-direction, perpendicular to the incident beam. Within the field, the electrons travel in a circular orbit whose radius of curvature R is given by

$$R = (\gamma m_0/eB)v \qquad (2.1)$$

where $\gamma = 1/(1 - v^2/c^2)^{1/2}$ is a relativistic factor and m_0 is the rest mass of an electron. The electron beam emerges from the magnet having been deflected through an angle of approximately ϕ; for convenience, ϕ is often chosen to be 90°.

As Eq. (2.1) indicates, the precise angular deflection of an electron depends on its velocity within the field. Electrons which have lost energy in the specimen will have a lower value of v and smaller R, so they leave the magnet with a slightly larger deflection angle (Fig. 2.2a).

Besides introducing bending and dispersion, the magnetic field focuses the electron beam. Electrons which originate from a point object O (a distance u from the entrance of the magnet) and which deviate from the central trajectory (the optic axis) by some angle γ_x (measured in the radial direction) will be focused into an image I_x a distance v_x from the exit of the magnet; see Fig. 2.2a. This focusing action occurs because electrons with positive γ_x travel a greater distance within the magnetic field and therefore undergo a slightly larger angular deflection so that they return towards the optic axis. Conversely, electrons with negative γ_x travel a shorter distance in the field, are deflected less, and converge towards the same point I_x. To a first approximation, the difference in path length is proportional to γ_x, giving first-order focusing in the x–z plane. If the edges of the magnet are perpendicular to the entrance and exit beam ($\varepsilon_1 = \varepsilon_2 = 0$), points O, I_x, and C (the center of curvature) lie in a straight line (Barber's rule); the prism is then properly referred to as a *sector* magnet and focuses *only* in the x-direction. If the entrance and exit faces are tilted through positive angles ε_1 and ε_2 (in the direction shown in Fig. 2.2a), the differences in path length will be less and the focusing power in the x–z plane is reduced.

Focusing can also take place in the y–z plane (i.e., in the axial direction, parallel to the magnetic-field axis), but this requires a component of magnetic field in the radial (x) direction. Unless a gradient-field design is adopted (Crewe and Scaduto, 1982), such a component is absent within the interior of the magnet, but in the fringing field at the polepiece edges there is a component of field B_n (for $y \neq 0$) which is normal to each polepiece edge (see Fig. 2.2a). Provided the edges are not perpendicular

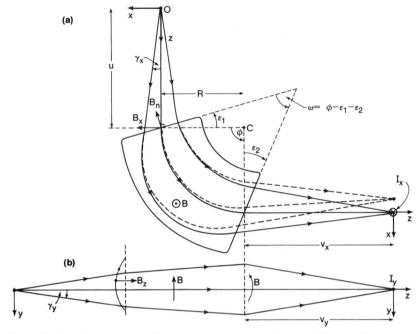

Figure 2.2. Focusing and dispersive properties of a magnetic prism. The coordinate system rotates with the electrons, so the x-axis always represents the "radial" direction and the z-axis is the direction of motion of the central, zero-loss trajectory (the optic axis). Radial focusing in the x–z plane (sometimes called the first principal section) is represented in (a); the trajectories of electrons which have lost energy are indicated by dashed lines and the normal component B_n of the fringing field is shown for the case $y > 0$. Axial focusing in the y–z plane (a flattened version of the second principal section) is illustrated in (b).

to the optic axis ($\varepsilon_1 \neq 0 \neq \varepsilon_2$), B_n itself has a radial component B_x in the x-direction, in addition to its component B_z along the optic axis. If ε_1 and ε_2 are positive (so that the wedge angle ω is less than the bend angle ϕ), $B_x > 0$ for $y > 0$, and the magnetic forces at both the entrance and exit edges are in the negative y-direction, returning the electron towards a point I_y on the optic axis. Each boundary of the magnet therefore behaves like a convex lens for electrons traveling in the y–z plane (Fig. 2.2b).

In general, the focusing powers in the x- and y-directions are unequal, so that *line foci* I_x and I_y are formed at different distances v_x and v_y from the exit face; in other words, the device exhibits astigmatism. For a particular combination of ε_1 and ε_2, however, the focusing powers can be made equal and the spectrometer is said to be *double-focusing*. In the absence of aberrations, electrons originating from O would all pass through a single point I, a distance $v_x = v_y = v$ from the exit. A double-focusing spectrometer therefore behaves like a normal lens; if an extended object were placed in the x–y plane at point O, its image would be produced in the x–y plane passing through I. But unlike the case of an axially symmetric lens, this two-dimensional imaging occurs only for a single value of the object distance u. If u is changed, a different combination of ε_1 and ε_2 is required to give double focusing.

Moreover, like any optical element, the spectrometer suffers from aberrations. For example, *aperture* aberrations cause a point image to be broadened into an *aberration figure* (Castaing *et al.*, 1967). For the straight-edged prism shown in Fig. 2.2a, these aberrations are predominantly second-order; in other words, the dimensions of the aberration figure depend on γ_x^2 and γ_y^2. Fortunately, it is possible to correct second-order aberrations by curving the edges of the magnet to appropriate radii, as discussed in Section 2.2.

For energy analysis in the electron microscope, the single magnetic prism is the most frequently used type of spectrometer. This popularity arises largely from the fact that it can be manufactured as a compact, add-on attachment to either a conventional or a scanning transmission microscope, the basic performance and operation of the microscope remaining unaffected. Operation of the spectrometer does not involve connection to the microscope high-voltage supply, so the magnetic prism can be used even when the accelerating voltage exceeds 500 keV (Darlington and Sparrow, 1975; Perez *et al.*, 1975). However, good energy resolution demands a magnet-current supply of very high stability and depends upon the high-voltage supply of the microscope being equally stable. The dispersive power is rather low, values of around 2 μm eV^{-1} being typical for 100-keV electrons, so good energy resolution requires finely machined detector slits (for serial acquisition) or postspectrometer magnifying optics (in the case of a parallel-recording detector). On the other hand, the dispersion is fairly linear over a range of several thousand eV, making the magnetic prism well suited to parallel recording of inner-shell losses.

2.1.2. Energy-Selecting Magnetic-Prism Devices

As discussed in Section 2.1.1, the edge angles of a magnetic prism can be chosen such that electrons coming from a point object will be imaged to a point on the exit side of the prism, for a given electron energy. In other words, there are points R_1 and R_2 which are fully stigmatic and lie within a real object and image, respectively (see Fig. 2.3a). Owing to the dispersive properties of the prism, the plane through R_2 will contain the object intensity convolved with the electron energy-loss spectrum of the specimen. Electron-optical theory (Castaing *et al.*, 1967; Metherell,

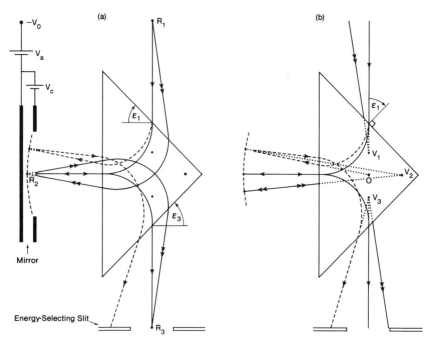

Figure 2.3. Castaing-Henry filter, showing the real image points (R_1, R_2, and R_3), the virtual image points (V_1, V_2, and V_3), and the achromatic point O at the center of the prism. Bias voltages V_a and V_c adjust the apex and curvature of the electrostatic mirror.

1971) indicates that, for the same prism geometry, there exists a second pair of stigmatic points V_1 and V_2 (Fig. 2.3b) which generally lie within the prism and so correspond to *virtual* image points. Electrons which are focused so as to converge on V_1 would appear (after deflection by the prism) to emanate from V_2. If an electron lens were used to produce an image of the specimen at the plane passing through V_1, a second lens focused on V_2 could project a real image of the specimen from the electrons which have passed through the prism. An aperture or slit placed at R_2 would transmit only those electrons whose energy loss lies within a certain range, so the final image would be an *energy-selected* (or energy-filtered) image.

Ideally, the image at V_2 should be achromatic (see Fig. 2.3), a condition which can be arranged by suitable choice of the object distance (location of point R_1) and prism geometry. In that case, the prism does not introduce additional chromatic aberration, regardless of the width of the energy-selecting slit.

In order to limit the angular divergence of the rays at R_1 (so that spectrometer aberrations do not degrade the energy resolution) while at the same time ensuring a reasonable field of view at the specimen, the prism is best located in the *middle* of a CTEM column, between the objective and projector lenses. A single magnetic prism is then at a disadvantage, since it bends the electron beam through a large angle; the mechanical stability of a vertical lens column would be lost. Instead, a multiple-deflection system is used, such that the net angle of deflection is zero.

Prism–Mirror System

A filtering device first developed at the University of Paris (Castaing and Henry, 1962) consists of a uniform-field magnetic prism and an electrostatic mirror. Electrons are deflected through 90° by the prism, emerge in a horizontal direction and are reflected through 180° by the mirror so that they pass a second time into the magnetic field. Because their velocity is now reversed, the electrons are deflected downwards and emerge from the device traveling in their original direction along the vertical axis of the microscope.

R_2 and V_2 are required to act as real and virtual *objects* for the second magnetic deflection in the lower half of the prism, producing real and virtual images R_3 and V_3, respectively. To achieve this, the mirror must be located such that its apex is at R_2 and its center of curvature is at V_2, electrons being reflected from the mirror back towards V_2 (Fig. 2.3b). Provided the prism itself is symmetrical ($\varepsilon_1 = \varepsilon_3$ in Fig. 2.3), the virtual image V_3 will be achromatic and at the same distance from the midplane as V_1 (Castaing *et al.*, 1967; Metherell, 1971). In practice, the electrostatic mirror consists of a planar and an annular electrode, both biased some hundreds of volts negative with respect to the gun potential of the microscope. The apex of the mirror depends on the bias applied to both electrodes; the curvature can be adjusted by varying the voltage difference between the two (Henkelman and Ottensmeyer, 1974b).

Although not essential (Castaing, 1975), the position of R_1 can be chosen (for a symmetric prism, $\varepsilon_1 = \varepsilon_3$) such that the point focus at R_3 is located at the same distance as R_1 from the midplane of the system. Since the real image at R_2 is chromatic (as discussed earlier) and since the dispersion is additive during the second passage through the prism, the image at R_3 is also chromatic; if R_1 is a point object, R_3 contains an energy-loss spectrum of the sample and an energy-selecting aperture or slit placed at R_3 will define the range of energy loss contributing to the image V_3. The latter is converted into a real image by means of magnetic lenses, in practice the intermediate and projector lenses of the CTEM column. Because V_3 is achromatic, the resolution in the final image is (to first order) independent of the width of the energy-selecting slit, which ensures that the range of energy loss can be made sufficiently large to give good image intensity and that (if desired) the energy-selecting aperture can be withdrawn so as to produce an unfiltered image.

In the usual mode of operation of an energy-selecting CTEM, a low-excitation "postobjective" lens forms a magnified image of the specimen at V_1 and a demagnified image of the back-focal plane of the objective lens at R_1. In other words, the object at R_1 is a portion (selected by the objective aperture) of the diffraction pattern of the specimen: the central part, for bright-field imaging. With suitable operation of the lens column, the location of the specimen image and diffraction pattern can be reversed, so that energy-filtered diffraction patterns are obtained (Henry et al., 1969; Egerton et al., 1975; Egle et al., 1984).

In addition, the intermediate-lens excitation can be changed so that the intensity distribution at R_3 is projected onto the CTEM screen. The energy-loss spectrum can then be recorded in a parallel mode (using TEM film or a CCD camera) or serially (by scanning the spectrum past an aperture and electron-detection system). If the system is slightly misaligned, a line spectrum is produced (Henkelman and Ottensmeyer, 1974a; Egerton et al., 1975) rather than a series of points, and is more convenient for photographic or CCD recording.

The original Castaing-Henry system has been improved by curving the prism edges to reduce second-order aberrations (Andrew et al., 1978; Jiang and Ottensmeyer, 1993), allowing a greater angular divergence at R_1 and therefore a larger field of view in the energy-filtered image, for a given energy resolution.

Although the mirror potential is tied to the microscope high-voltage, the dispersion of the system arises entirely from the magnetic field. Therefore, good energy resolution is dependent upon stability of the high-voltage supply, unlike electrostatic or retarding-field analyzers.

The Omega Filter

A second approach to energy filtering in CTEM takes the form of a purely magnetic device known as the omega filter. After passing through the specimen, the objective lens, and a low-excitation postobjective lens, the electrons pass through a magnetic prism and are deflected through an angle ϕ, typically 90–120°. Instead of being reflected by an electrostatic mirror (as in the Castaing-Henry system) they then enter a second prism whose magnetic field is in the reverse direction, so that the beam is deflected downwards. A further two prisms are located symmetrically with respect to the first pair and the complete trajectory takes the form of the greek letter Ω (Fig. 2.4). The beam emerges from the device along its original

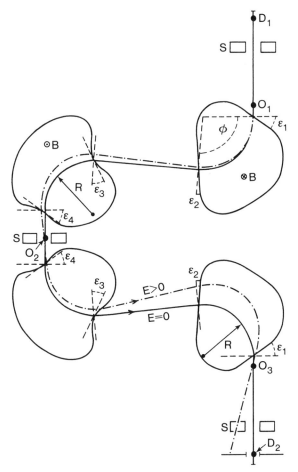

Figure 2.4. Optics of an aberration-corrected omega filter (Pejas and Rose, 1978). Achromatic images of the specimen are formed at O_1, O_2, and O_3; the plane through D_2 contains an energy-dispersed diffraction pattern. Sextupole lenses (S) are placed close to D_1, O_2, and D_2 in order to correct for image-plane tilt.

axis, allowing vertical alignment of the lens column to be preserved. The dispersion within each magnetic prism is additive and an energy-loss spectrum is formed at a position D_2 which is conjugate with the object point D_1; see Fig. 2.4. For energy-filtered imaging, D_1 and D_2 contain diffraction patterns while planes O_1 and O_2 (located just outside or inside the first and last prisms) contain real or virtual images of the specimen. An energy-selected image of the specimen is produced by using an energy-selecting slit in plane D_2 and a second intermediate lens to image O_3 onto the CTEM screen (via the projector lens). Alternatively, the intermediate lens can be focused on D_2 in order to record the energy-loss spectrum. If the postobjective lens provides an image and diffraction pattern at D_1 and O_1, an energy-filtered diffraction pattern is provided at O_3.

As a result of the symmetry of the omega filter about its midplane, second-order aperture aberration and second-order distortion vanish if the system is properly aligned (Rose and Plies, 1974; Krahl et al., 1978; Zanchi et al., 1977b). The remaining second-order aberrations can be compensated by curving the polefaces of the second and third prisms, using sextupole coils symmetrically placed about the midplane (Fig. 2.4) and operating the system with line (instead of point) foci between the prisms (Pejas and Rose, 1978; Krahl et al., 1978; Lanio, 1986).

Unlike the prism–mirror system, the omega filter does not require connection to the microscope accelerating voltage. It is therefore well suited for use with a high-voltage CTEM (Zanchi et al., 1975, 1977a, 1982). Since a magnetic field of the same strength and polarity is used in the second and third prisms, these two can be combined into one (Zanchi et al., 1975), although this design does not allow a sextupole at the midplane.

Another type of all-magnetic energy filter consists of two magnets whose field is in the same direction but of different strength; the electrons execute a trajectory in the form of the greek letter α. An analysis of the first-order imaging properties of the alpha filter is given by Perez et al. (1984).

2.1.3. The Wien Filter

A dispersive device employing both magnetic and electrostatic fields was reported in 1897 by W. Wien and first used with high-energy electrons by Boersch et al. (1962). The magnetic field (induction B in the y-direction) and electric field (strength E, parallel to the x-axis) are each perpendicular to the entrance beam (the z-axis). The polarities of these fields are such that the magnetic and electrostatic forces on an electron are in opposite directions; their relative strengths obey the relationship $E = v_1 B$ such that an electron moving parallel to the optic axis with a velocity v_1 and kinetic energy E_1 continues in a straight line, the net force on it being zero. Electrons traveling at some angle to the optic axis or with velocities other than v_1 execute a cycloidal motion (Fig. 2.5) whose rotational component is at the cyclotron frequency: $\omega = eB/\gamma m_0$, where e and m_0 are the electronic charge and rest mass and $\gamma = 1/(1 - v_1^2/c^2)^{1/2}$ is a relativistic factor. Electrons starting from a point ($z = 0$) on the optic axis and initially traveling in the

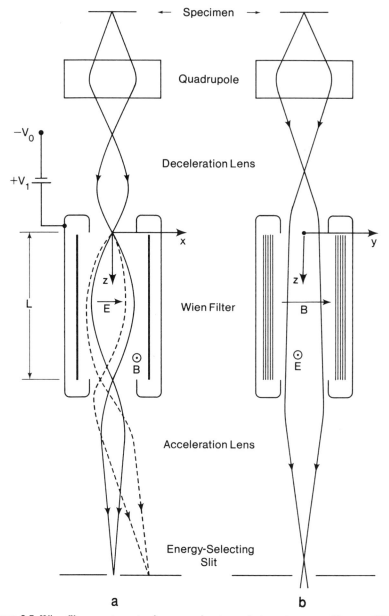

Figure 2.5. Wien-filter spectrometer for a scanning-transmission microscope (Batson, 1985b). Electron trajectories are shown (a) in the dispersive $(x–z)$ plane and (b) in the nondispersive $(y–z)$ plane; the dashed lines represent electrons which have lost energy in the specimen. A quadrupole lens has been added to make the system approximately double-focusing.

x–z plane return to the z-axis after one complete revolution; in other words, an achromatic focus occurs at $z = 2L$, where $L = (\pi v_1/\omega) = \pi \gamma m_0 E/eB^2$. In addition, an inverted chromatic image of unit magnification ($M_x = -1$) is formed at $z = L$ (i.e., after half a revolution), its energy dispersion being $L/(\pi E_1)$ (Curtis and Silcox, 1971). Velocity components along the y-axis (magnetic-field direction) are unaffected by the magnetic and electrostatic fields, so both the chromatic and achromatic images are actually line foci.

The Wien filter is generally used with decelerated electrons. In other words, the filter is operated at a potential $-V_0 + V_1$ which is close to the negative potential $-V_0$ of the electron source. The positive bias V_1 is obtained from a power supply connected to the high-voltage line; its value, typically in the range 100–1000 eV, determines the energy (eV_1) of electrons which are moving in a straight line through the filter. The retarding and accelerating fields at the entrance and exit of filter act as electrostatic lenses (Fig. 2.5), whose effect must be taken into account in the design of the system.

Although retardation involves the inconvenience of handling high voltages, it offers several advantages. First of all, the dispersion at the chromatic focus is increased by a factor V_0/V_1 for a given length L of the filter; values of 100 μm/eV or more are typical. The electrostatic lens at the exit of the filter can be used to project the spectrum onto the detection plane, with either a decrease or a further increase in the dispersion, depending on the distance of the final image. Secondly, the required magnitudes and stabilities of B and E are reduced and the mechanical tolerances of the polepieces and electrodes are relaxed. Thirdly, because the electron velocity for straight-line transmission depends on V_1 rather than V_0, fluctuations and drift in V_0 do not affect the energy resolution. This factor is particularly important where high resolution is to be combined with long recording times, for example when recording inner-shell losses using a field-emission STEM (Batson, 1985b). A Wien filter used in conjunction with a monochromator (Section 2.1.4) achieved an energy resolution of 2 meV (Geiger et al., 1970).

Since the system just discussed does not focus in the y-direction, the energy-loss spectrum can be produced as a function of distance along a straight line in the entrance plane, this line being defined by an entrance slit. If a diffraction pattern (or a magnified image) of the specimen is projected onto the entrance plane, using the lenses of a CTEM, the final image will contain a map of electron intensity as a function of both energy loss and scattering angle (or specimen coordinate). A two-dimensional sensor placed at the final-image plane can therefore record a large amount of information about the specimen (Batson and Silcox, 1983).

The Wien filter can become double-focusing if either E or B is made nonuniform, for example by curving the electric-field electrodes, by tilting the magnetic polepieces to create a magnetic-field gradient, or by shaping both the electric and magnetic fields to provide a quadrupole action (Andersen, 1967). The device is then suitable for use as an imaging filter in CTEM mode (Andersen and Kramer, 1972). Second-order aberrations of the chromatic focus can be controlled by using two independently excited magnetic-field coils, equivalent to introduction of an internal sextupole (Andersen, 1967).

2.1.4. Electron Monochromators

Besides being dependent on the spectrometer, the energy resolution of an energy-analysis system is limited by energy spread in the electron beam incident on the specimen. If the electrons are produced by a thermal source operating at a temperature T_s, the energies of the electrons leaving the cathode will follow a Maxwellian distribution, whose full width at half maximum (FWHM) is $\Delta E_s = 2.45(kT_s)$ (Reimer, 1993). For a tungsten filament whose emission surface is at a temperature of 2800 K, $\Delta E_s \cong 0.6$ eV; for a lanthanum hexaboride source at 1700 K, $\Delta E_s = 0.3$ eV.

The Boersch Effect

The energy spread ΔE_0 measured in an electron microscope is always larger than the above values, 1–2 eV being typical in the case of an instrument which uses a tungsten-filament source. The discrepancy between ΔE_s and ΔE_0 is generally referred to as the Boersch effect, since Boersch (1954) first investigated the dependence of the measured spread on physical parameters of the electron microscope: cathode temperature, Wehnelt geometry, Wehnelt bias, accelerating voltage, vacuum conditions, and the deployment of magnetic and electrostatic lenses. He found that ΔE_0 increases with the emission current and is further increased when the beam is focused into a crossover. Subsequent experimental work (Martin and Geissler, 1972; Ditchfield and Whelan, 1977; Bell and Swanson, 1979) confirmed these findings.

When electrons are rapidly accelerated to an energy E_0, their *energy* spread δE remains unaltered, in accordance with the conservation of energy. However, the *axial* spread δv_z in velocity is reduced as the axial velocity v_z increases, since (nonrelativistically) $\delta E = \delta(m_0 v_z^2/2)$, giving $\delta v_z = \Delta E_s/(m_0 v_z)$. The equivalent axial beam temperature attained is $T_z = (kT_s/E_0)T_s$ (Knauer, 1979), and is very low (<0.1 K) for $E_0 > 10$ keV. If the electrons spend enough time in sufficiently close proximity to one another, so that they interact via Coulomb forces, the difference between the axial and transverse temperatures will be reduced, raising δv_z and increasing the measured energy spread. This is known as the "thermal" Boersch effect; the resulting value of ΔE_0 depends on the path length of the electrons and on the current density (Knauer, 1979).

In addition, electrons which are focused into a crossover can suffer "collision broadening" through interaction between their transverse velocity components. The energy broadening depends on the current density at the crossover and on the divergence angle (Crewe, 1978; Knauer, 1979; Rose and Spehr, 1980). The beam current is highest in the electron gun, so appreciable broadening can occur at a gun crossover. Although a low-temperature field-emission source provides the lowest energy spread (0.3 eV) at low emission currents (<10 nA), ΔE_0 increases to typically 1 eV at 100 nA emission due to Coulomb interaction of electrons just outside the tip (Bell and Swanson, 1979).

Types of Monochromator

The Wien filter has high dispersion and good energy resolution (a few meV) when operated with low-velocity electrons (Section 2.1.3); it can therefore be used to produce an incident beam of small energy width if an energy-selecting aperture is placed in an image of its chromatic focus (Boersch et al., 1962, 1964). A second Wien filter (after the specimen) acts as an energy analyzer, making possible energy-loss spectroscopy of vibrational modes (Boersch et al., 1962; Katterwe, 1972; Geiger, 1981). An energy resolution below 3 meV was achieved at 30-keV incident energy. For analysis of small areas of a TEM specimen, a higher accelerating voltage is needed and a field-emission source is also desirable in order to maximize the fraction of electrons allowed through the monochromator (approximately the required resolution divided by the energy width of the source). The instrument shown in Fig. 2.6a has been operated at 80 keV incident energy with an energy resolution of 80 meV (Terauchi et al., 1994) and has achieved a resolution down to 15 meV for energy losses below 5 eV.

A scheme using two identical nondecelerating filters for the monochromator and analyzer (Krivanek et al., 1991b) is shown in Fig. 2.6b. The blades of the energy-selecting slit (following the monochromator) are electrically isolated and connected to a differential amplifier, any out-of-balance signal being used to correct the high-voltage power supply. This negative feedback is designed to keep the electron beam centered between the slits and thereby avoid fluctuations in beam current at the specimen.

(Similar beam stabilization has been applied to a parallel-recording spectrometer: Kruit and Shuman, 1985a). The energy resolution would still be compromised by any drift in the monochromator or analyzer excitations. However, this drift is compensated by connecting the two omega filters in series: any increase in monochromator current (giving rise to an increase in beam energy because of feedback from the slit) is accompanied by an equal increase in analyzer current such that the dispersed output beam remains stationary on the detector. To achieve energy resolution below 10 meV, the central region of the instrument would have to be well shielded against stray magnetic fields, which would otherwise destroy the compensation between monochromator and analyzer.

Dispersion Compensation

One disadvantage of conventional monochromator systems is that the monochromator reduces the beam current by a large factor if the energy spread of the electron source (including the Boersch effect) greatly exceeds the required energy resolution. As a result, the beam current at the specimen is low and long recording times are required. The remedy is to eliminate the energy-selecting slit of the monochromator and use a completely symmetrical system of monochromator and analyzer.

For example, if the length of a Wien filter is extended to a value $2L = 2\pi\gamma m_0 E/eB^2$, an *achromatic* focus is formed at the exit plane, as discussed in Section 2.1.3. Because the final image is achromatic, its width is independent of the incident energy spread ΔE_0, a result of the fact that the chromatic aberration of the second half of the system exactly compensates that of the first half. If a specimen is now introduced at the *chromatic* focus ($z = L$), the resulting energy losses will give rise to an energy-loss spectrum at $z = 2L$, but the width of each spectral line remains independent of the value of ΔE_0. In addition, second-order aperture aberrations of the second half cancel those of the first, so the energy resolution of the system (if perfectly symmetrical) depends only on the higher-order aberrations and on the object size of the electron source. In practice, the two halves of the double Wien filter can be separated by a short distance (to allow room for inserting the specimen) but great care has to be taken to keep the system symmetrical. An experimental prototope based on this principle (Andersen and Le Poole, 1970) achieved an energy resolution of 50 meV (measured without a specimen) using 10-keV transmitted electrons. Scattering in the sample degraded the resolution to about 100 meV but a transmitted current of up to 0.1 μA was available. The system can be made double-focusing at the chromatic image by using two independently excited field coils in each Wien filter (Andersen, 1967), thereby reducing the y-spreading of the

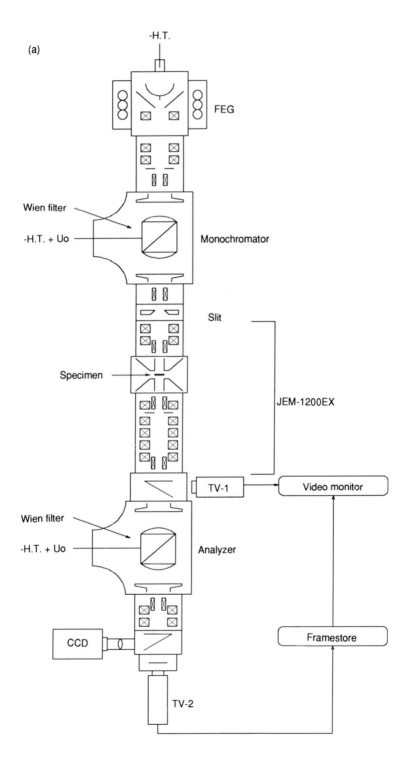

(a)

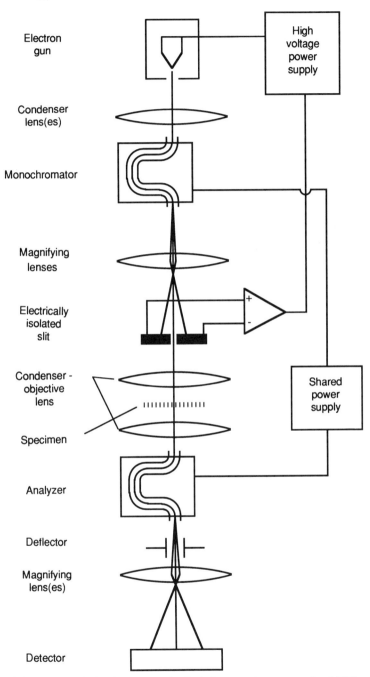

Figure 2.6. Energy-analyzing (or energy-selecting) electron microscopes using (a) Wien filters (Terauchi *et al.*, 1994) and (b) omega filters for the monochromator and analyzer (Krivanek *et al.*, 1991b). See page 44 for details.

beam at the specimen. Some spreading in the x-direction is unavoidable, as a result of the energy spread of the electron source and the dispersion produced by the first Wien filter, and may limit the spatial resolution of analysis.

The same principle (known as dispersion compensation or dispersion matching) has been employed in nuclear physics, the target being placed between a pair of magnetic-sector spectrometers which bend a high-energy beam of electrons in the same direction (Schaerf and Scrimaglio, 1964). It has been applied to reflection spectroscopy of low-energy (e.g., 5 eV) electrons, using two identical spherical-electrostatic sectors (Kevan and Dubois, 1984).

2.2. Optics of the Magnetic-Prism Spectrometer

As discussed in Section 2.1.1, a magnetic-prism spectrometer produces three effects on a beam of electrons: bending, dispersion, and focusing. Focusing warrants the most attention, since the attainable energy resolution depends on the width of the exit beam at the detection plane. Provided the spatial distribution of the magnetic field is known, the behavior of an electron within the spectrometer can be predicted by applying the equation of motion, Eq. (2.1), to each region of the trajectory. Details are given in several references, for example Penner (1961), Brown *et al.* (1964), Brown (1967), and Enge (1967). The aim of this section is to summarize the important results of such analysis and to provide an introduction to the use of computer programs for spectrometer design.

We will adopt the coordinate system and notation of Brown *et al.* (1964), which is widely used in nuclear physics. For a negative particle such as the electron, the y-axis is antiparallel to the direction of magnetic field. The z-axis always represents the direction of motion of an electron traveling along the central trajectory (the optic axis); in other words, the coordinate system can be imagined to rotate about the y-axis as the electron proceeds through the magnetic field. The x-axis is perpendicular to the y- and z-axes and points radially outwards, away from the center of curvature of the electron trajectories. Using this "curvilinear" coordinate system, the y-axis focusing can be represented on a flat y–z plane (Fig. 2.2b).

The behavior of electrons at the entrance and exit edges of the magnet is simpler to calculate if the magnetic field is assumed to remain constant up to the polepiece edges, dropping abruptly to zero outside this region. This assumption is known as the SCOFF (sharp cut-off fringing field) approximation and is more likely to be realistic if the gap between the polepieces (measured in the y-direction) is very small. In practice, the field strength just inside the magnet is less than in the interior and a fringing field extends some distance (of the order of the gap length) outside the geometrical boundaries. The z-dependence of the field strength adjacent to the magnet boundaries can be specified by one or more coefficients; this is known as the EFF (effective fringing field) approximation and leads to more accurate predictions of the spectrometer focusing.

An important distinction is between first- and second-order focusing. Formation of an image is primarily a first-order effect, so first-order theory can be used to predict object and image distances, image magnifications, and dispersive power, the latter being first-order in energy loss. Second- or higher-order analysis is needed to describe image aberration and distortion, together with certain other properties such as the dispersion-plane tilt.

2.2.1. First-Order Properties

We first consider the "radial" focusing of electrons which originate from a point object O located a distance u from the entrance face of a magnetic prism (Fig. 2.2). For a particular value of u, all electrons which arrive at the *center* of the magnet (after deflection through an angle $\phi/2$) will be traveling parallel to the optic axis before being focused by the second half of the prism into a crossover (or image) I_x located a distance v_x from the exit edge. We can regard these particular values of u and v_x as being the focal lengths f_x of the first and second halves of the prism, and their reciprocals as the corresponding focusing powers. In the SCOFF approximation, these focusing powers are given by (Wittry, 1969)

$$1/f_x = [\tan(\phi/2) - \tan \varepsilon]/R \qquad (2.2)$$

where $\varepsilon = \varepsilon_1$ for the first half of the prism and $\varepsilon = \varepsilon_2$ for the second half, ε_1 and ε_2 being the tilt angles of the prism edges. Note that a positive value of ε, which gives convergent focusing in the y-direction, *reduces* the radial focusing power, leading to longer object and image distances. In other words, the boundaries of the magnet have a divergent focusing action whereas the effect of the uniform field in the center of the magnet is to gradually return the electrons toward the optic axis, as illustrated in Fig. 2.7.

In contrast, focusing in the axial (y-) direction takes place *only* at the entrance and exit of the magnet. Each boundary can be characterized by a focusing power which is given, in the SCOFF approximation, by

$$1/f_y = \tan(\varepsilon)/R \qquad (2.3)$$

Because $1/f_x$ and $1/f_y$ change in opposite directions as ε is varied, the entrance- and exit-face tilts can be chosen such that the net focusing powers in the radial and axial directions are equal; the prism is then double-

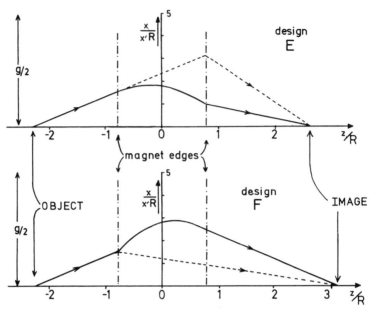

Figure 2.7. Trajectories of electrons through a magnetic-prism spectrometer. Solid lines represent the component of motion in the x–z plane (first principal section); dashed lines represent motion in the y–z plane (second principal section). The horizontal axis indicates distance along the optic axis, relative to the center of the prism. For design E, $\varepsilon_1 = 0$ and $\varepsilon_2 = 45°$; for design F, $\varepsilon_1 = 45°$ and $\varepsilon_2 = 10°$ (Egerton, 1980b).

focusing. Although not essential, approximate double-focusing is generally desirable because it minimizes the width (in the y-direction) of the image at the energy-selecting plane, making the energy resolution of the system less dependent on the precise orientation of the detector about the z-axis or (in the case of serial recording) on the mechanical perfection of the detector slit.

For a bend angle of 90°, the most symmetrical solution of Eq. (2.2) and Eq. (2.3) corresponds to the double-focusing condition: $u = v_x = v_y = 2R$ and $\tan \varepsilon_1 = \tan \varepsilon_2 = 0.5$ (i.e., $\varepsilon = 26.6°$). In practice, the object distance u may be dictated by external constraints, such as the location of the projector-lens crossover in an electron-microscope column. The spectrometer will still be double-focusing provided the prism angles ε_1 and ε_2 satisfy the relation (valid in the SCOFF approximation)

$$\tan \varepsilon_2 = \frac{1}{2}\left[\frac{1 - (\tan \varepsilon_1 + R/u)/\tan \phi}{\tan \varepsilon_1 + R/u + \cot \phi} - \frac{\tan \varepsilon_1 - R/u}{1 - \phi(\tan \varepsilon_1 - R/u)}\right] \quad (2.4)$$

As ε_1 increases, the required value of ε_2 increases, as illustrated in Fig. 2.8. The image distance $v_x = v_y = v$ is given by

$$\frac{v}{R} = \left[\frac{\tan \varepsilon_1 - R/u}{1 - \phi(\tan \varepsilon_1 - R/u)} + \tan \varepsilon_2\right]^{-1} \tag{2.5}$$

A large difference between ε_1 and ε_2 leads to stronger focusing, reflected in a shorter image distance (Fig. 2.8). The dispersive power D at the image plane is (Livingood, 1969)

$$\begin{aligned}
D &= \frac{dx}{dE_0} \\
&= \frac{R}{E_0}\left(\frac{E_0 + m_0 c^2}{E_0 + 2 m_0 c^2}\right) \\
&\quad \times \frac{\sin \phi + (1 - \cos \phi)(\tan \varepsilon_1 + R/u)}{\sin \phi[1 - \tan \varepsilon_2(\tan \varepsilon_1 + R/u)] - \cos \phi\,(\tan \varepsilon_1 + \tan \varepsilon_2 + R/u)}
\end{aligned} \tag{2.6}$$

where E_0 represents the kinetic energy of electrons entering the spectrometer and $m_0 c^2 = 511$ keV is the electron rest energy. If the spectrometer is to be reasonably compact and not too heavy, the value of R cannot exceed 10–20 cm and D is limited to a few μm/eV for $E_0 = 100$ keV.

The Effect of Fringing Fields

The SCOFF approximation is convenient for discussing the general properties of a magnetic prism and is useful in the initial stages of spectrometer design, but it cannot provide accurate predictions of the focusing. The effects of a spatially extended fringing field have been described by Enge (1964) as follows.

Firstly, the exit beam is displaced in the radial direction compared to the SCOFF trajectory. This effect can be taken into account by shifting the magnet slightly in the $+x$-direction or by increasing the magnetic field by a small amount.

Secondly, the focusing power in the axial (y-) direction is decreased, whereas the radial (x-) focusing remains practically unaltered. As a result, either ε_1 or ε_2 must be increased (compared to the SCOFF prediction) in order to maintain double focusing. The net result is a slight increase in image distance; see Fig. 2.8.

The third effect of the extended fringing fields is to add a convex component of curvature to the entrance and exit edges of the magnet, the magnitude of this component varying inversely with the polepiece width w. Such curvature affects the spectrometer aberrations, as discussed in Section 2.2.2.

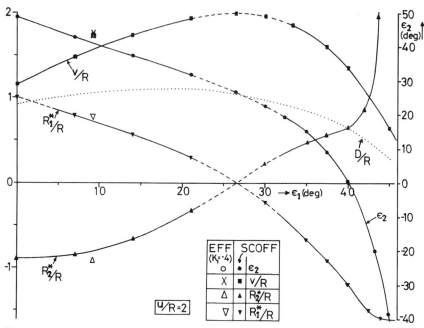

Figure 2.8. Double-focusing parameters of a magnetic prism, for a fixed object distance ($u = 2R$) and bend angle $\phi = 90°$. The curves were calculated using the SCOFF approximation; dashed lines indicate the region in which correction of second-order aperture aberration requires excessive edge curvatures, as determined by Eq. (2.18). One set of points is given for an extended fringing field with $K_1 = 0.4$ (Egerton, 1980b).

Fourthly, the extended fringing field introduces a discrepancy between the "effective" edge of the magnet (which serves as a reference point for measuring object and image distances) and the actual "mechanical" edge, the former generally lying outside the latter.

In order to define the spatial extent of the fringing field, so that it can be properly taken into account in EFF calculations and is less affected by the surroundings of the spectrometer, plates made of a soft magnetic material ("mirror planes") are placed parallel to the entrance and exit edges, to "clamp" the field to a low value at the required distance from the edge. If the plate–polepiece separation is chosen as $g/2$, where g is the length of the polepiece gap in the y-direction and, in addition, the polepiece edges are beveled at 45° to a depth $g/2$ (see Fig. 2.9), the magnetic field decays almost linearly over a distance g along the optic axis. More importantly, the position, angle, and curvature of the magnetic-field boundary more nearly coincide with those of the polepiece edge. However, the correspon-

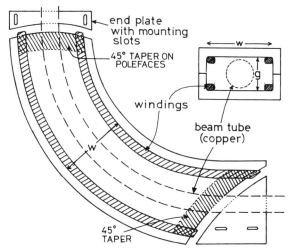

Figure 2.9. Cross sections (in the *x–z* and *x–y* planes) through an aberration-corrected spectrometer, showing the curved and tapered pole edges, soft-magnetic mirror plates, and window-frame coils. The magnet operates in air, the vacuum being confined within a nonmagnetic beam (or "drift") tube (Egerton, 1980b).

dence is not likely to be exact, partly because the fringing field penetrates to some extent into the holes which must be provided in the mirror plates to allow the electron beam to enter and leave the spectrometer (Fig. 2.9). The remaining discrepancy between the effective and mechanical edge depends on the polepiece gap, on the separation of the field clamps from the magnet, and on the radius of curvature of the edges (Heighway, 1975).

The effect of the fringing field on spectrometer focusing can be specified in terms of the gap length g and a shape parameter K_1 defined by

$$K_1 = \int_{-\infty}^{\infty} \frac{B_y(z')[B - B_y(z')]}{gB^2} \, dz' \tag{2.7}$$

where $B_y(z')$ is the y-component of induction at $y = 0$ and at a perpendicular distance z' from the polepiece edge; B is the induction between the polepieces within the interior of the spectdrometer. The SCOFF approximation corresponds to $K_1 = 0$; the use of tapered polepiece edges and mirror plates, as specified above, gives $K_1 \cong 0.4$. If the fringing field is not clamped by mirror plates, the value of K_1 is higher: approximately 0.5 for a square-edged magnet and 0.7 for tapered polepiece edges (Brown *et al.*, 1977).

If the polepiece gap is large, a second coefficient K_2 may be necessary to properly describe the effect of the fringing field; however, its effect is small for $g/R < 0.3$ (Heighway, 1975).

Matrix Calculations

Particularly when fringing fields are taken into account, the equations needed to describe the focusing properties of a magnetic prism become quite complicated. Their form can be simplified and the method of calculation made more systematic by using a matrix notation, as is done in the design of light-optical systems. The optical path between object and image is divided into sections and a "transfer matrix" written down for each section. The first stage of the electron trajectory corresponds to "drift" in a straight line through the field-free region between the object plane and the entrance edge of the magnet. The displacement coordinates (x, y, z) of an electron change but not its angular coordinates $(x' = dx/dz, y' = dy/dz)$. Upon arrival at the entrance edge of the magnet, these four coordinates are therefore given by the following matrix equation:

$$
\begin{pmatrix} x \\ x' \\ y \\ y' \end{pmatrix} = \begin{pmatrix} 1 & u & 0 & 0 \\ 0 & 1 & 0 & 0 \\ 0 & 0 & 1 & u \\ 0 & 0 & 0 & 1 \end{pmatrix} \begin{pmatrix} x_0 \\ x_0' \\ y_0 \\ y_0' \end{pmatrix}
\tag{2.8}
$$

Here, x_0 and y_0 are the components of electron displacement at the object plane, x_0' and y_0' are the corresponding angular components, and the 4×4 square matrix is the transfer matrix for drift over a distance u (measured along the optic axis).

The electron next encounters the focusing action of the tilted edge of the magnet. In the SCOFF approximation, the focusing powers can be written as $1/f_x = -(\tan \varepsilon)/R$ and $1/f_y = (\tan \varepsilon)/R$. Since the focusing is of equal magnitude but opposite sign in the x- and y-directions, the magnet edge is equivalent to a quadrupole lens. Allowing for extended fringing fields, the corresponding transfer matrix is (Brown, 1967)

$$
\begin{pmatrix} 1 & 0 & 0 & 0 \\ R^{-1}\tan \varepsilon_1 & 1 & 0 & 0 \\ 0 & 0 & 1 & 0 \\ 0 & 0 & -R^{-1}\tan(\varepsilon_1 - \psi_1) & 1 \end{pmatrix}
\tag{2.9}
$$

where ψ_1 represents a correction for the extended fringing field, given by

$$
\psi_1 \cong (g/R)K_1(1 + \sin^2 \varepsilon_1)/\cos \varepsilon_1
\tag{2.10}
$$

The third part of the trajectory involves bending of the beam within the interior of the magnet. As discussed in Section 2.1.1, the uniform magnetic field has a positive (convex) focusing action in the x-direction but no focusing action in the y-direction. The effect is equivalent to that of a dipole field, as produced by a "sector" magnet with $\varepsilon_1 = \varepsilon_2 = 0$. The corresponding transfer matrix can be written in the form (Penner, 1961):

$$
\begin{pmatrix}
\cos\phi & R\sin\phi & 0 & 0 \\
-R^{-1}\sin\phi & \cos\phi & 0 & 0 \\
0 & 0 & 1 & 0 \\
0 & 0 & 0 & 1
\end{pmatrix}
\tag{2.11}
$$

where ϕ is the bend angle.

Upon arrival at the exit edge of the prism, the electron again encounters an effective quadrupole, whose transfer matrix is specified by Eq. (2.9) and Eq. (2.10) but with ε_2 substituted for ε_1. Finally, after leaving the prism, the electron drifts to the image plane, its transfer matrix being identical to that in Eq. (2.8) but with the object distance u replaced by the image distance v.

Following the rules of matrix manipulation, the five transfer matrices are multiplied together to yield a 4×4 transfer matrix which relates the electron coordinates and angles at the image plane (x_i, y_i, x_i', and y_i') to those at the object plane. However, the first-order properties of a magnetic prism can be specified more completely by introducing two additional parameters. One of these is the total distance or path length l traversed by an electron, which is of interest in connection with time-of-flight measurements (e.g., in nuclear physics applications of bending magnets) but is not relevant to dispersive operation of a spectrometer. The other additional parameter is the fractional momentum deviation δ of the electron, relative to that required for travel along the optic axis (which can be taken to correspond to a kinetic energy E_0 and zero energy loss). This last parameter is related to the energy loss E by

$$
\delta = -E(v/c)^{-2}(E_0 + m_0 c^2)^{-1}
\tag{2.12}
$$

The first-order properties of the prism are then represented by the equation

$$
\begin{pmatrix}
x_i \\
x_i' \\
y_i \\
y_i' \\
l_i \\
\delta_i
\end{pmatrix}
=
\begin{pmatrix}
R_{11} & R_{12} & 0 & 0 & 0 & R_{16} \\
R_{21} & R_{22} & 0 & 0 & 0 & R_{26} \\
0 & 0 & R_{33} & R_{34} & 0 & 0 \\
0 & 0 & R_{43} & R_{44} & 0 & 0 \\
R_{51} & R_{52} & 0 & 0 & 1 & R_{56} \\
0 & 0 & 0 & 0 & 0 & 1
\end{pmatrix}
\begin{pmatrix}
x_0 \\
x_0' \\
y_0 \\
y_0' \\
l_0 \\
\delta_0
\end{pmatrix}
\tag{2.13}
$$

Many of the elements in the 6×6 matrix of Eq. (2.13) are zero as a result of the mirror symmetry of the spectrometer about the x–z plane. Of the other coefficients, R_{11} and R_{33} describe the lateral image magnifications (M_x

and M_y) in the x- and y-directions. In general, $R_{11} \neq R_{33}$, so the image produced by the prism suffers from rectangular distortion. For a real image, R_{11} and R_{33} are negative, denoting the fact that the image is inverted about the optic axis. R_{22} and R_{44} are the angular magnifications, approximately equal to the reciprocals of R_{11} and R_{33}, respectively.

Provided the spectrometer is double-focusing and the value of the final drift length used in calculating the R-matrix corresponds to the image distance, R_{12} and R_{34} are both zero. If the spectrometer is not double-focusing, $R_{12} = 0$ at the x-focus and the magnitude of R_{34} gives an indication of the length of the line focus in the y-direction. To obtain good energy resolution from the spectrometer, R_{12} should be zero at the energy-selection plane and R_{34} should preferably be small. The other matrix coefficient of interest in connection with energy-loss spectroscopy is $R_{16} = \partial x_i / \partial(\delta_0)$, which relates to the energy dispersion of the spectrometer. Using Eq. (2.12), the dispersive power $D = -\partial x_i / \partial E$ is given by

$$D = (v/c)^{-2}(E_0 + m_0 c^2)^{-1} R_{16} \tag{2.14}$$

The R-matrix of eq. (2.13) can be evaluated by multiplication of the individual transfer matrices, provided the values of u, ε_1, ϕ, ε_2, K_1, g, and v are specified. Such tedious arithmetic is best done by computer, for example by running the TRANSPORT program (Brown *et al.* 1977), which has been used extensively in nuclear physics applications and which is freely available.* This program consists of about 3000 lines of FORTRAN, requires approximately 300 kbytes of memory, and computes (in addition to the R-matrix) the second-order focusing (such as aberration coefficients) and a beam matrix which specifies the diameter and divergence of a particle beam at any point within the system, given suitable values of the input coordinates (x_0, x_0', etc.). If desired, the program will also calculate the effects of other elements (e.g., quadrupole or sextupole lenses), of a magnetic-field gradient or inhomogeneity, and of stray magnetic fields.

2.2.2. Higher-Order Focusing

The matrix notation is well suited to the discussion and calculation of second-order properties of a magnetic prism. Using the same six "coordinates" (x, x', y, y', l, and δ), second derivatives in the form (for example) $\partial^2 x_i / \partial x_0 \partial x_0'$ can be defined and arranged in the form of a $6 \times 6 \times 6$ "T-matrix," analogous to the first-order R-matrix. Many of the $6^3 = 216$ second-order T-coefficients are zero or are related to one another by the midplane

---

* For example, by anonymous FTP to www.amc.anl.gov (EMMPDL/Eels files in ANL Software Library).

symmetry of the magnet. For energy-loss *spectroscopy*, where the beam diameter at the object plane (i.e., the source size) is small and where image distortions and off-axis astigmatism are of little significance, the most important second-order matrix elements are $T_{122} = \partial^2 x_i/\partial(x_0')^2$ and $T_{144} = \partial^2 x_i/\partial(y_0')^2$. These coefficients represent second-order *aperture* aberrations which increase the image width in the x-direction and therefore degrade the energy resolution, particularly in the case of a large spread of incident angles (x_0' and y_0').

Whereas the first-order focusing of a magnet boundary depends on its effective quadrupole strength (equal to $-\tan \varepsilon = -\partial z/\partial x$ in the SCOFF approximation), the second-order aperture aberration depends on the effective sextupole strength: $-(2\rho \cos^3 \varepsilon)^{-1}$ in the SCOFF approximation (Tang, 1982a). The aberration coefficients can therefore be varied by adjusting the angle ε and the curvature $\rho = \partial^2 z/\partial x^2$ of the boundary. With convex boundaries it is possible to correct the aberration for electrons traveling in the x–y plane ($T_{122} = 0$); but if one boundary is made concave, the aberration for electrons traveling out of the radial plane can also be corrected ($T_{122} = T_{144} = 0$). Alternatively, the correction can be carried out by means of magnetic or electrostatic sextupole lenses placed before and after the spectrometer (Parker *et al.*, 1978).

A second-order property which is of importance if electron lenses follow the spectrometer is the angle ψ between the dispersion plane (the plane of best chromatic focus for a point object) and the x–y plane adjacent to the image; see Fig. 2.10. This tilt angle is related to the matrix element $T_{126} = \partial^2 x_i/\partial x_0'\partial(\delta)$ by*

$$\tan \psi = -T_{126}/(R_{22}R_{16}) \qquad (2.15)$$

Another second-order coefficient of some relevance is T_{166}, which does not affect the energy resolution but specifies the nonlinearity of the energy-loss axis.

The matrix method has been extended to third-order derivatives, including the effect of extended fringing fields (Matsuda and Wollnik, 1970; Matsuo and Matsuda, 1971; Tang, 1982a, b). Many of the 1296 third-order coefficients are zero as a result of the midplane symmetry, and only a limited number of the remaining ones are of interest for energy-loss spectroscopy. The coefficients $\partial^3 x_i/\partial(x_0')^3$ and $\partial^3 x_i/\partial x_0'\partial(y')^2$ represent aperture aberrations and (like T_{122} and T_{144}) may have either the same or opposite signs (Scheinfein and Isaacson, 1984). Tang (1982a) has suggested that

---

* If TRANSPORT is used to calculate the matrix elements, a multiplying factor of 1000 is required on the right-hand side of Eq. (2.15) as a result of the units (x in cm, x' in mrad, δ in %) used in that program.

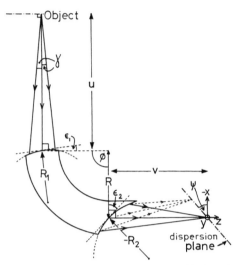

Figure 2.10. Electron optics of an aberration-corrected double-focusing spectrometer. The *y*-axis and the applied magnetic field are perpendicular to the plane of the diagram. The polepiece-tilt angles (ε_1 and ε_2) refer to the central trajectory (the optic axis). Exit trajectories of energy-loss electrons are shown by dashed lines (Egerton, 1980b).

correction of second-order aberrations by curving the entrance and exit edges of the magnet often increases these third-order coefficients, such that the latter may limit the energy resolution for entrance angles above about 10 mrad.

The chromatic term $\partial^3 x_i/\partial(\delta)^3$ causes additional nonlinearity of the energy-loss axis but is likely to be important only for energy losses of several keV. The coefficients $\partial^3 x_i/\partial(x_0')^2\partial(\delta)$, $\partial^3 x_i/\partial(y_0')^2\partial(\delta)$, and $\partial^3 x_i/\partial(x_0')\partial(\delta)^2$ introduce tilt and curvature of the dispersion plane, which may degrade the resolution if a parallel-detection system is used.

If the third-order aberrations can be successfully corrected, for example by the use of octupole lenses outside the spectrometer (Tang, 1982b), the energy resolution would be limited by fourth-order aberrations: $\partial^4 x_i/\partial(x_0')^4$, $\partial^4 x_i/\partial(y_0')^4$, and $\partial^4 x_i/\partial(x_0')^2\partial(y_0')^2$. However, fourth-order matrix theory has not been developed. Instead, ray-tracing programs can be used to predict the focusing of electrons with high accuracy (Carey, 1978), although a complete analysis using this method requires considerable computing time.

2.2.3. Design of an Aberration-Corrected Spectrometer

We now outline a procedure for designing a double-focusing uniform-field spectrometer with aperture aberrations corrected to second order by curving the entrance and exit edges. It is first of all necessary to choose the prism angles so as to obtain suitable first-order focusing. As discussed below, the value of ε_1 should either be fairly large (close to 45°) or quite small (<10°, or even negative). Knowing the location of the spectrometer object point and the required bend radius R (which determines the energy dispersion D and the size and weight of the magnet), approximate values of ε_2 and v can be calculated using Eq. (2.4) and Eq. (2.5). If either v or ε_2 turns out to be inconveniently large, a different value of ε_1 should be chosen.

These first-order parameters are then refined to take account of extended fringing fields. This requires a knowledge of the integral K_1 (which depends on the shape of the polepiece corners and on whether magnetic-field clamps are to be used) and the polepiece gap g (typically $0.1R$–$0.2R$). If the TRANSPORT program is used, a fitting procedure can be implemented to find the exact image distance v which corresponds to an x-focus ($R_{12} = 0$). The value of R_{34} will then be nonzero, indicating a line focus; if $|R_{34}|$ is excessive (>1 μm/mrad), either ε_1 or ε_2 should be changed slightly to obtain a closer approach to double focusing. The dispersive power of the spectrometer may be estimated from Eq. (2.6) or obtained more accurately using Eq. (2.14).

The next stage is to determine values of the edge curvatures R_1 and R_2 which make the second-order aperture aberrations zero. This is most easily done by recognizing that T_{122} and T_{144} both vary linearly with the edge curvatures. In other words,

$$-T_{122} = a_0 + a_1(R/R_1) + a_2(R/R_2) \qquad (2.16)$$

$$-T_{144} = b_0 + b_1(R/R_1) + b_2(R/R_2) \qquad (2.17)$$

where a_0, a_1, a_2, b_0, b_1, and b_2 are constants for a given first-order focusing. In general, a_0 is positive but a_1 and a_2 are negative; T_{122} can therefore be made zero with R_1 and R_2 both positive, implying convex entrance and exit edges. However, b_0, b_1, and b_2 are usually all positive so $T_{144} = 0$ requires that either R_1 or R_2 be negative, indicating a concave edge (Fig. 2.10). The required edge radii (R_1^* and R_2^*) are found empirically by using a matrix program to calculate T_{122} and T_{144} for three arbitrary pairs of R/R_1 and R/R_2, such as (0, 0), (0, 1), and (1, 0). This generates six simultaneous equations which can be solved for a_0, a_1, a_2, b_0, b_1, and b_2. R_1^* and R_2^* are then deduced by setting T_{122} and T_{144} to zero in Eq. (2.16) and Eq. (2.17).

Not all spectrometer geometries yield reasonable values of R_1^* and R_2^*. For example a completely symmetric case ($u = v = 2R$, $\varepsilon_1 = \varepsilon_2 = 26.6°$ for $\phi = 90°$, in the SCOFF approximation) gives $R_1^* = R_2^* = 0$, corresponding to infinite curvature. As $|\varepsilon_1 - \varepsilon_2|$ increases, the necessary

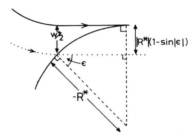

Figure 2.11. Geometry of a concave pole edge, of radius R^* and tilt angle ε. The dotted trajectory is the optic axis; the solid-line trajectory represents the maximum displacement $(x = w^*/2)$ at which an electron can pass through the edge (Egerton, 1980b).

edge curvatures should be kept reasonably low because the maximum effective width w^* of the polepieces at the entrance or exit edge is given (see Fig. 2.11) by

$$w^*/2 = |R^*|(1 - \sin|\varepsilon|) \tag{2.18}$$

for a concave edge and by $w^*/2 = R^* \cos \varepsilon$ for a convex edge. In practice, the concave edge corresponds to the higher value of ε, so (for small R^*)

Table 2.1. Design Parameters for Aberration-Corrected Spectrometers

ϕ (deg)	ε_1 (deg)	ε_2 (deg)	u/R	v/R	R_1^*/R	R_2^*/R	g/r	Reference
60	14.64	18.07	∞	1.46	1.351	−2.671	0.07	Fields (1977)
90	0	45.0	1.45	2.16	0.807	−1.357	0.2	Egerton (1980b)
70	11.75	28.79	3.60	2.38	0.707	−0.603	0.137	Shuman (1980)
90	17.5	45.0	5.5	0.98	1.0	−0.496	0.125	Krivanek and Swann (1981)
66.6	−15	45.8	2.25	2.06	2.34	−1.30	0.18	Tang (1982a)
90	15.9	46.5	4.52	1.08	0.867	−0.500	0.19	Reichelt and Engel (1984)
80	14.6	35.1	3.5	1.82	0.728	−0.576	0.25	Scheinfein and Isaacson (1984)
90	16	47	6.2	0.9	1.0	−0.42	0.125	Krivanek et al. (1995a)

Eq. (2.18) imposes an upper limit on the angular range (x') of electrons which can pass through the magnet.

In the above analysis, the object distance u and bend radius R were assumed to be fixed by the geometry of an electron-microscope column and the space available for the spectrometer. If the ratio u/R can be varied, however, there is freedom to adjust a further second-order matrix element, such as T_{126}. Parker et al. (1978) have shown that there may be two values of image (or object) distance for which $T_{126} = 0$, resulting in zero tilt of the dispersion plane. This condition is desirable if an electron lens is to be used after the spectrometer, to magnify the dispersion onto a parallel-recording detector for instance.

Table 2.1 gives several examples of designs in which second-order aperture aberrations have been corrected. One of them (Tang, 1982a) also minimizes two of the third-order coefficients: $\partial^3 x_i / \partial(x_0')^3$ and $\partial^3 x_i / \partial x' \partial(y_0')^2$. The recent designs are all similar in that they require a convex entrance face ($R_1^* > 0$) with a small tilt angle and a concave exit face with a relatively large tilt angle.

2.2.4. Practical Considerations

The main aim in designing an electron spectrometer is to achieve high energy resolution even in the presence of a large spread γ of entrance angles, which may be necessary if the spectrometer system is to have a high collection efficiency (see Section 2.3). In practice, a resolution of 1 eV is obtainable for an entrance divergence as large as 10 mrad and energy losses up to 1 keV (Krivanek and Swann, 1981; Colliex, 1982; Scheinfein and Isaacson, 1984). Even if R_1^* and R_2^* are large, so that Eq. (2.18) does not limit x_0', the value of γ is restricted by the internal diameter of the "drift" tube (Fig. 2.9), which is necessarily less than the magnet gap g. Recent designs have therefore been based on relatively large g/R ratios (see Table 2.1), even though this makes accurate calculation of the fringing-field properties more difficult and may lead to appreciable discrepancies between the effective and mechanical edges of the magnet (Heighway, 1975). A cylindrical drift tube also limits the maximum energy loss observable (without the zero-loss electrons colliding with the inner wall) to a few keV; if necessary, the energy range could be increased by using a drift tube of rectangular cross section.

The accuracy with which the second-order aberrations can be corrected depends on the mechanical accuracy of the entrance and exit bend radii. The tolerable error in R_1^* and R_2^* depends on the design but is generally in the range 1%–10%, for 1-eV resolution and $\gamma = 10$ mrad (Egerton, 1980b; Tang, 1982a). Inaccuracies within this range can be corrected for by incorporating weak sextupole lenses close to the approximate edge. Quadruopole coils may also be added in order to adjust the first-order focusing (Krivanek and Swann, 1981) and octupole lenses employed to minimize the effect of third-order aberrations (Tang, 1982a). The image-plane tilt cannot be independently controlled by means of a sextupole; 1° accuracy in ψ may require that the exit-edge curvature be within 0.1% of the design value (Scheinfein and Isaacson, 1984).

The designs listed in Table 2.1 all assume that the magnetic induction B within the magnet is completely uniform. More generally, the induction might vary with distance x from the optic axis according to

$$B(x) = B(0)[1 - n(x/R) - m(x/R)^2 + \cdots] \qquad (2.19)$$

where the coefficients n, m, ... introduce multipole components in the focusing. This property is exploited in gradient-field spectrometers where a quadrupole term (dependent on the value of n) is used to produce first-order focusing in the x- and y-directions, as an alternative to tilting the entrance and exit faces. Likewise, a sextupole component (dependent on m) could be added to control second-order aberrations (Crewe and Scaduto, 1982); but where these aberrations have been set to zero by appropriate choice of edge curvatures, any x^2 variation in magnetic field serves to increase the aberration and degrade the energy resolution. Matrix calculations suggest that m should typically be less than 2×10^{-6} to ensure 1-eV resolution at $\gamma = 10$ mrad, implying that $B(x)$ should be constant to 0.02% within $x = \pm R/10$ (Egerton, 1980b). A "C-core" magnet (where the magnetic field is generated by a "bobbin" of wire connected by side arms to the polepieces) cannot give the required degree of uniformity. However, a more symmetrical arrangement with "window-frame" coils placed on either side of the gap (Fig. 2.9) can give a sufficiently uniform field, particularly if the separation between the planes of the two windings is carefully adjusted (Tang, 1982a).

Further requirements for uniform field are that the magnetic material be sufficiently homogeneous and adequately thick. Homogeneity is achieved by annealing the magnet after machining and by choosing a material with high relative permeability μ and low coercivity at low field strength, such as mu-metal. The minimum thickness t can be estimated by requiring the magnetic reluctance ($\propto w/t\mu$) of each polepiece (in the x-direction) to be much less than the reluctance of the gap ($\propto g/w$), giving

$$t \gg w^2/(\mu g) \qquad (2.20)$$

Equation (2.20) precludes the use of thin magnetic sheeting, which would otherwise be attractive in terms of reduced weight of the spectrometer.

It might appear that B-uniformity of 0.02% would require the polepiece gap to be uniform to within 2 μm over an x-displacement of 1 cm, for $g = 1$ cm. However, the allowable variation in field strength is that averaged over the whole electron trajectory, variations in the z-direction having little effect on the focusing. Also, provided they are small, the x- and x^2-terms in Eq. (2.19) can be corrected by external quadrupole and sextupole coils.

A substantial loss of energy resolution can occur if stray magnetic fields penetrate into the spectrometer. Field penetration can be reduced by enclosing the magnet and (more importantly) the entrance and exit drift spaces in a soft magnetic material such as mu-metal. Such screening is usually not completely effective but the influence of a remaining alternating field can be canceled by injecting an ac signal into the spectrometer scan coils, as described in Section 2.2.5.

The magnetic induction within the spectrometer is quite low (<0.01 T for 100 keV operation) and can be provided by window-frame coils of about 100 turns carrying a current of the order of 1 A. To prevent drift of the spectrum due to changes in temperature and resistance of the windings, the power supply must be *current*-stabilized to within three parts in 10^6 for 1-eV stability at 100-keV incident energy. Stability is improved if the power supply can be left running continuously.

2.2.5. Spectrometer Alignment

Like all electron-optical elements, a magnetic prism performs to its design specifications only if it is correctly aligned relative to the incoming beam of electrons. Since the energy dispersion is small for high-energy electrons, the alignment must be done carefully if the optimum energy resolution is to be achieved.

Initial Alignment

When a spectrometer is installed for the first time, or if the alignment of the spectrometer or the microscope column has been disturbed, the electron beam may travel in a path which is far from the optic axis of the spectrometer (defined by the prism orientation and the value of the magnetic induction B). In this situation, a rough alignment of the system can be carried out in much the same way as alignment of an electron-microscope column. Beam-limiting apertures, such as the spectrometer entrance aperture, are withdrawn and the entrance beam broadened, for example by defocusing the illumination at the specimen plane. It may also be useful to sweep the magnetic field periodically by applying a fast ramp (e.g., 1 cycle per second) to the spectrometer scan coils, so as to deflect the exit beam over a range of several millimeters in the x-direction. The x-coordinate of the exit beam can be monitored using the usual electron-detection system, but it is desirable to know also the beam location in the y-direction. The easiest way of doing this is to remove any detector slit and observe the beam position directly on a transmission phosphor screen. By alternately focusing and defocusing the prespectrometer lenses, it is possible to discover if the electron beam is passing through the center of the drift tube or is cut off asymmetrically by the tube walls or fixed apertures. To ensure that the beam travels close to the *mechanical* axis of the spectrometer, it may be necessary to shift or tilt the magnet so that the positions (on the phosphor screen) where the exit beam is cut off are symmetric with respect to the intended location of the detector.

The Aberration Figure

For optimum performance from the spectrometer, the beam must travel close to the *magnetic* axis of the prism and the spectrometer focusing must be correct. The desirable conditions can be recognized from the *shape* of the beam at the detector plane. The first-order focusing is correct when the exit beam at the detector plane has minimum width in the direction of dispersion. This condition could be found by moving the detector or phosphor screen in the z-direction, but it is more convenient to use electrical controls, keeping the detector fixed and varying the current in a prespectrometer lens or a quadrupole coil placed in front of the spectrometer. The focusing can be set more accurately if the depth of focus is made small, by arranging for a large angular divergence γ of the electrons entering the spectrometer. If the spectrometer is *diffraction*-coupled to the specimen by means of electron lenses (see Section 2.3), large γ is achieved by defocusing the illumination into a circle of large diameter at the specimen plane, the presence of a specimen being optional. If there are no lenses between the specimen and spectrometer or if the lenses provide *image* coupling (Section 2.3), large divergence is obtained by focusing the electrons into a very small probe at the specimen plane or by inserting a specimen which is sufficiently thick to give appreciable scattering over large angles.

If there were no aberrations and if the spectrometer were precisely double-focusing, the exit beam would appear as a point or as a circle of very small diameter at the detector plane. Spectrometer aberrations spread the beam into an *aberration figure* which can be observed directly on a fluorescent screen if the entrance divergence and the aberration coefficients are large enough. Second-order aberrations produce a figure whose shape (Fig. 2.12) can be deduced from the equations

$$x_i = T_{122}(x_0')^2 + T_{144}(y_0')^2 \tag{2.21}$$

$$y_i = T_{324} x_0' y_0' \tag{2.22}$$

T_{122}, T_{144}, and T_{324} are matrix coefficients representing the second-order aperture aberrations; x_0' and y_0' represent the angular coordinates of an electron entering the spectrometer. For a fixed x_0' and a range of y_0' (or vice versa), the relationship between the image-plane coordinates x_i and y_i is a parabola. In practice, both x_0' and y_0' take a continuous range of values: $-\gamma$ to $+\gamma$, where γ is the maximum entrance angle (defined by a spectrometer entrance aperture, for example). The image-plane intensity is then represented by the shaded area in Fig. 2.12. When the magnet is correctly aligned, this figure is symmetric about the vertical (x-) axis; second-order aberrations are properly corrected when its width in the x-direction is a minimum. Because the aberration figure has very small dimensions, it is difficult to observe unless the spectrometer is followed by magnifying electron lenses or a TV camera.

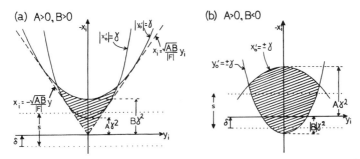

Figure 2.12. Aberration figures of a properly aligned magnetic prism whose resolution is determined by the second-order aberration coefficients $T_{122} = -A$ and $T_{144} = -B$ which are (a) of the same sign and (b) of opposite sign (Egerton, 1981b). The entrance angles x_0' and y_0' are assumed to be limited to values in the range $-\gamma$ to γ by a square entrance aperture; the result for a circular aperture is similar except that the top of figure (a) is convex. Dotted lines indicate the position of a detector slit when recording an alignment figure. The matrix element T_{324} is denoted by F. (From *Ultramicroscopy*, R. F. Egerton, Alignment and characterization of an electron spectrometer, © 1981, pp. 93–96 with permission from Elsevier Science B.V., Amsterdam, The Netherlands.)

The Alignment Figure

An alternative way of observing the aberration properties of a spectrometer is to place a narrow slit in the image plane and measure the electron flux through the slit by means of a fast-electron detector (e.g., scintillator and photomultiplier), as in the case of serial recording of energy-loss spectra. Instead of scanning the exit beam across the slit, however, the entrance angle is varied by rocking the entrance beam about the spectrometer object point (Fig. 2.10). In the case of a CTEM fitted with a scanning attachment, the incident probe can be scanned over the specimen plane in the form of a two-dimensional raster; if the object plane of the spectrometer contains a diffraction pattern of the specimen (at the projector-lens crossover, for example), the beam entering the spectrometer is swept in angle in both the x- and y-directions. Applying voltages proportional to x_0' and y_0' to the horizontal and vertical channels of an oscilloscope and using the signal from the electron detector to modulate the brightness of the oscilloscope beam (z-modulation), a display is obtained whose characteristics relate to those aberrations which directly affect the resolving power of the spectrometer. To avoid confusion with the aberration figure formed by the spectrometer exit beam, this oscilloscope display is referred to as an *alignment* figure.

An electron arriving at the image plane will pass through the detection slit provided

$$-\delta < x_0 < s - \delta \tag{2.23}$$

where s is the slit width in the x-direction and δ specifies the position of the aberration figure relative to the slit; see Fig. 2.12. For the case where second-order aberrations are dominant, the shape of the alignment figure is specified by Eq. (2.21) and Eq. (2.23). If $\delta = 0$ and if T_{122} and T_{144} are both negative (as in the case of a typical straight-edged magnet), a solid ellipse is formed, whose dimensions depend on the values of T_{122}, T_{144}, and s (Fig. 2.13a). As the current in the spectrometer field coils is increased, ($\delta > 0$), the pattern shrinks inwards and eventually disappears; if the spectrometer excitation is decreased, the pattern expands in outline but develops a hollow center. If T_{122} and T_{144} were both positive, this same sequence would be observed as the spectrometer current were *decreased.*

When T_{122} and T_{144} are of opposite sign, the alignment figure consists of a pair of hyperbolas (Fig. 2.13b). If $T_{144} > 0$, the hyperbolas come together as the spectrometer current is increased, and then separate in the y-direction, as in Fig. 2.14. The sequence would be reversed if T_{122} were the positive coefficient.

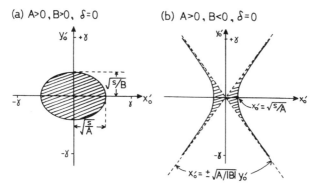

Figure 2.13. Alignment figures of a magnetic prism whose resolution is limited by second-order aberration coefficients $T_{122} = -A$ and $T_{144} = -B$ which are (a) of the same sign and (b) of opposite sign. In (a), the scan range (2γ) is assumed to be greater than the major axis of the ellipse (Egerton, 1981b). From *Ultramicroscopy*, R. F. Egerton, Alignment and characterization of an electron spectrometer, © 1981, pp. 93–96 with permission from Elsevier Science B.V., Amsterdam, The Netherlands.

When third-order aberrations are dominant, the aberration figure has three lobes (Fig. 2.15) but retains its mirror-plane symmetry about the x-axis, in accordance with the symmetry of the magnet about the x–z plane.

Following from the above discussion, one can list several uses of the alignment figure.

(1) In order to optimize the energy resolution, the spectrometer should be aligned such that the pattern is symmetrical about its x_0' axis. The most sensitive alignment is usually the tilt of the magnet about the exit-beam

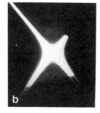
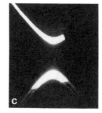

Figure 2.14. Change in shape of the alignment figure as the spectrometer excitation is increased, for the case $T_{122} < 0$ and $T_{144} > 0$ (Egerton, 1981b). In (a), $\delta < 0$; in (b), $\delta \approx 0$; in (c), $\delta > 0$. From *Ultramicroscopy*, R. F. Egerton, Alignment and characterization of an electron spectrometer, © 1981, pp. 93–96 with permission from Elsevier Science B.V., Amsterdam, The Netherlands.

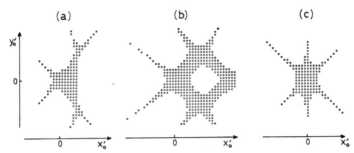

Figure 2.15. Calculated alignment figures for a magnetic-prism spectrometer having (a) third-order + residual second-order aberrations (T_{122} and T_{144} are of opposite sign), (b) fourth-order + residual third-order aberrations, and (c) pure third-order aberrations (Scheinfein and Isaacson, 1984).

direction; however, it is not easy to provide this rotation in the form of a single mechanical control.

(2) The alignment figure enables the current flow in sextupole or octupole coils to be set to compensate for residual second- or third-order aberrations. To maximize the collection efficiency of the spectrometer for a given energy resolution, the current should be adjusted so that the pattern is as large in area and as near-circular as possible. During this adjustment, it may be necessary to change the spectrometer excitation slightly to prevent the display disappearing from the screen or developing a hollow center.

(3) The symmetry of the alignment figure indicates the order of the uncorrected aberrations and the relative signs of the dominant aberration coefficients. The absolute signs can be deduced from the change in the pattern as the spectrometer excitation is varied. The relative magnitude of the coefficients can be estimated by measuring the aspect ratio (Fig. 2.13a) or angle between the asymptotes (Fig. 2.13b). Absolute magnitudes could be obtained if the display were calibrated in terms of entrance angle.

(4) If a spectrometer entrance aperture is inserted, limiting the angular range of electrons entering the prism, its image should appear in outline on the oscilloscope display. To achieve the best combination of energy resolution and collection efficiency, the aperture is centered such that as little as possible of the alignment figure is cut off from the display.

(5) The influence of stray ac magnetic fields can be detected as a blurring or waviness of the edges of the alignment figure, dependent on the scanning frequency. Imperfections in an energy-selecting slit (due to mechanical irregularity or contamination) show up as a streaking of the pattern.

Stray-Field Compensation

It is difficult to prevent stray magnetic fields affecting the performance of an electron spectrometer. In the case of a magnetic prism attached to a CTEM, for instance, external fields may penetrate into the viewing chamber and deflect the electron beam before it enters the spectrometer. Little can be done about slowly varying or dc fields, except to ensure that relays (for example) are not located too close to the spectrometer and that movable magnetic objects, such as steel chairs, are replaced by nonmagnetic ones. Some of the external interference comes from mains-frequency fields and can be compensated by a simple circuit which applies a mains-frequency current of adjustable amplitude and phase to the spectrometer excitation coils (Egerton, 1978b). The amplitude and phase controls are adjusted so that the width of the zero-loss peak is minimized. External fields are less likely to be troublesome if the TEM viewing chamber is made of a magnetically shielding material such as steel or soft iron rather than a nonmagnetic material such as brass.

2.3. The Use of Prespectrometer Lenses

A single-prism electron spectrometer fitted to a conventional (fixed-beam) TEM is located beneath the image-recording chamber. Because electrons which emerge from the specimen must pass through the microscope imaging lenses before reaching the spectrometer, the performance of the EELS system (energy resolution, collection efficiency, spatial resolution of analysis) depends considerably on the characteristics of these lenses and the way in which they are operated.

Several papers (Johnson, 1980a,b; Egerton, 1980a; Krivanek *et al.*, 1995a) have discussed the influence of postspecimen lenses on the performance of an electron spectrometer. Good energy resolution requires that an electron-beam crossover of small diameter be placed at the spectrometer object plane, and this crossover can be either a low-magnification image of the specimen or a portion (usually the central beam) of its diffraction pattern. These two possibilities were originally referred to as image (or spot) mode and diffraction mode of spectrometer operation, but the terms *image coupling* and *diffraction coupling* are preferable, in order to avoid confusion with the *operating mode of the microscope*, which denotes the presence of an image or diffraction pattern *at the TEM screen*.

In the case of a dedicated STEM, there may be no imaging lenses between the specimen and spectrometer, but the postfield of the objective lens can appreciably affect the performance of the system. This situation is dealt with in Section 2.3.5.

2.3.1. Deployment of CTEM Lenses

Some early spectrometer systems (Pearce-Percy, 1976; Joy and Maher, 1978; Egerton, 1978b) operated with the final imaging lens turned off; electrons were focused into a small crossover at the level of the TEM screen, which was also the spectrometer object plane. The region of specimen (diameter d) giving rise to the energy-loss spectrum was determined by the diameter of electron beam incident on the specimen or by inserting a selected-area diffraction (SAD) aperture. The energy resolution available in this mode has been analyzed by Johnson (1980a,b) and Egerton (1980a).

Wittry (1969) designed a spectrometer to operate with the projector lens on (as in normal TEM operation) and this arrangement is utilized by the Gatan serial- and parallel-recording spectrometers. The projector forms an optical crossover just below its lens bore, a distance h (typically 30–40 cm) above the TEM viewing screen, and this crossover acts as the object point O of the spectrometer; see Fig. 2.16. Because the projector lens is designed to produce a large-diameter image or diffraction pattern, the solid angle of divergence at O is large and (because electron-optical brightness is conserved) the crossover has a very small diameter.

If the microscope is operated in *image mode*, with an image of the specimen of magnification M formed at the viewing screen, the spectrometer is *diffraction-coupled* because the projector-lens crossover contains a small diffraction pattern of the specimen. The size of this diffraction pattern is represented by a camera length: $L_o = h/M$, and may be as small as 1 μm. The angular range of scattering (collection semiangle β) allowed into the spectrometer is controlled by varying the size of the objective-lens aperture. The region of specimen giving rise to the energy-loss spectrum is determined by the spectrometer entrance aperture (SEA) and corresponds to a portion of the image close to the center of the TEM viewing screen, before the screen is lifted to allow electrons through to the spectrometer. More precisely, the diameter of analysis is $d = 2R/M'$, where R is the SEA radius and $M' = M(h'/h)$ is the image magnification at the SEA plane, h' being distance of the *aperture* below O. Because of the large depth of field, an image which is in focus at the TEM screen is effectively in focus at the SEA plane, allowing the SEA to be used as an area-selecting aperture.

If the TEM is operated in *diffraction mode*, with a diffraction pattern of camera length L projected onto the viewing screen, the spectrometer is *image-coupled* because the projector crossover contains a small image of the illuminated area of the specimen. The image magnification at O is

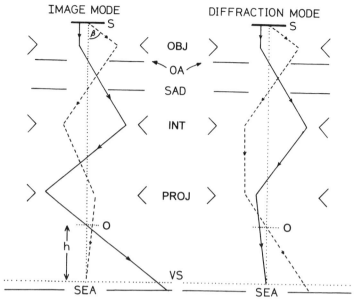

Figure 2.16. Simplifed optics for CTEM image and diffraction modes. S represents the specimen; OBJ, INT, and PROJ represent the objective lens, intermediate-lens system, and final imaging (projector) lens; O and VS are the spectrometer object point and viewing screen; OA, SAD, and SEA are the objective, selected-area-diffraction, and spectrometer entrance apertures.

$M_o = h/L$ and is typically of the order of unity. Unless a microscope objective aperture limits it to a smaller value, the collection semiangle is given by $\beta = R/L'$, where $L' = L(h'/h)$ is the camera length at the SEA plane. By using deflection coils to shift the diffraction pattern around on the TEM screen, different regions of the diffraction pattern can be analyzed by the spectrometer. The area of specimen being analyzed is determined by the electron-beam diameter at the specimen plane or by a SAD aperture if such an aperture is inserted to define a smaller area.

2.3.2. Effect of Lens Aberrations on Spatial Resolution

Because of chromatic aberration of the imaging lenses, the TEM image cannot be in focus for all energy losses. Most of this aberration occurs at the *objective* lens, where the image-plane angular divergence is higher than in subsequent lenses (Reimer, 1993). If an objective lens (chromatic aberration coefficient C_c, magnification M_o) is focused for zero-loss electrons, an electron of energy loss E and scattering angle θ arrives at the first image plane with a radial displacement $R = M_o\theta\Delta f$ relative to the optic axis, where $\Delta f = C_c(E/E_0)$ and E_0 is the incident energy. Because R is proportional to θ, the Lorentzian distribution of inelastic intensity $dJ/d\Omega$ per unit *solid* angle (Chapter 3) gives rise to a Lorentzian distribution of intensity dJ/dA per unit *area* in the image plane. The equivalent intensity at the specimen plane, the chromatic *point-spread function* (PSF) is given by

$$\text{PSF} \propto (r^2 + r_E^2)^{-1} \tag{2.24}$$

Here r is the radial coordiante at the specimen and $r_E = \theta_E\,\Delta f$, where $\theta_E \approx E/2E_0$ is the characteristic angle of inelastic scattering. For two such functions displaced by a distance $2r_E$, the intensity falls by about 20% between each maximum (in accordance with the Rayleigh criterion) so the point resolution r_i in an energy-selected image is

$$r_i \approx 2\theta_E\,\Delta f \approx C_c(E/E_0)^2 \tag{2.25}$$

In the case of a TEM image which is formed from a range Δ of energy loss, point-spread functions of different defocus are superimposed. If the image is in focus for electrons *at the center of this range*, the average deviation from the mean energy loss is $\Delta/4$, giving an effective defocus of $\Delta f \approx C_c(\Delta/4E_0)$ so that

$$r_i \approx 2\theta_E\,\Delta f \approx (C_c/4)(\Delta/E_0)^2 \tag{2.26}$$

Equation (2.26) gives the point resolution for elemental (atomic number) contrast in a core-loss image obtained using an imaging filter, where the microscope high voltage is raised by an amount equal to a characteristic loss of the selected element (Krivanek *et al.*, 1992). Equations (2.25) and (2.26) hold even when an angle-limiting objective aperture is present (provided $\beta > \theta_E$), but they assume that the specimen is sufficiently thin so that elastic scattering contributes only to the *tail* of the chromatic PSF and degrades the overall contrast rather than point resolution (Egerton and Wong, 1995).

For energy-loss *spectroscopy* carried out in TEM image mode, a spectrometer entrance aperture (in the final-image plane) defines the region of analysis but its precision is limited by chromatic aberration. Because typically only 10% of the inelastic scattering occcurs below the angle θ_E (see page 214), the aperture collects information from a region of specimen whose radius is considerably larger than r_i. Typically 97% of the scattering occurs below the Bethe-ridge angle: $\theta_r \approx (E/E_0)^{1/2}$, so the *total* chromatic broadening (expressed as a radius in the specimen plane) is approximately

$$r_c \approx \theta_r\,\Delta f \tag{2.27}$$

However, if an objective aperture removes scattering beyond an angle β ($<\theta_r$), the chromatic radius is reduced to

$$r_c(\beta) \approx \beta \, \Delta f \tag{2.28}$$

where $\Delta f = C_c(E/E_0)$ if the screen-level image is focused for zero-loss electrons and $\Delta f \approx C_c(\Delta/4E_0)$ if it is focused for electrons at the center of the energy range being analyzed. A typical-sized aperture provides considerable improvement in spatial resolution, as shown in Fig. 2.17b.

For $\Delta f = C_c(E/E_0)$ and $C_s \approx C_c$, Eq. (2.27) also gives the total broadening from *spherical* aberration of the *inelastically* scattered electrons; see Fig. 2.17b. If an objective aperture if present, this spherical aberration can be neglected in comparison with chromatic aberration if $\beta \ll (E/E_0)^{1/2}$, a condition which holds for inner-shell losses when the aperture selects only the dipole region of scattering (Yang and Egerton, 1992).

To obtain higher spatial resolution than the limit imposed by postspecimen-lens aberrations, area-selecting apertures must be avoided. The area of analysis may instead be defined by the incident beam, whose diameter can be made as small as 1 nm in a modern TEM through the use of a strong objective-lens prefield.

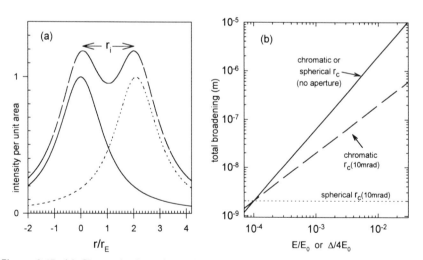

Figure 2.17. (a) Chromatic-aberration point-spread function (solid and dotted curves) for two object points separated by a distance $2r_E$. The total intensity (dashed curve) falls by 19% between the two maxima. (b) Radial broadening r_c (assuming $C_c = C_s = 2$ mm) for electrons of energy loss E with and without a 10-mrad objective aperture (Egerton and Wong, 1995).

2.3.3. Effect of Lens Aberrations on Collection Efficiency

In TEM image mode, postspecimen-lens aberrations produce a blurring of all image features, including the edge of the illumination disk. If the diameter of illumination on the TEM screen is much larger than the diameter of the spectrometer entrance aperture, this blurring occurs well outside the SEA perimeter (see Fig. 2.18a) and will not affect the inelastic signal collected by the aperture. Considering only chromatic aberration (broadening r_c at the specimen plane, as discussed in the last section), this condition requires that

$$M'r > R_a + M'r_c \qquad (2.29)$$

where M' is the final magnification at the SEA plane, r is the radius of illumination at the specimen, and R_a is the SEA radius. In other words, the magnified radius of illumination must exceed the SEA radius by an amount at least equal to the chromatic broadening in the image. Provided an objective aperture is used, r_c is normally below 1 μm (Fig. 2.17a) and Eq. (2.29) can be satisfied in CTEM mode by adjusting the condenser lenses so that the radius of illumination (observed on the TEM screen) is several times the SEA radius. Under these conditions, the loss of electrons (due to chromatic aberration) from points within the selected area is *com-*

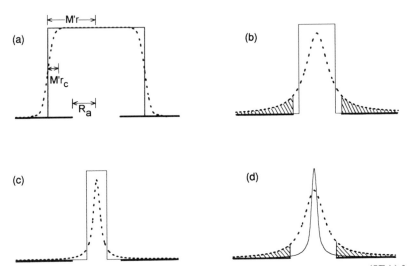

Figure 2.18. Electron intensity at the plane of the spectrometer entrance aperture (SEA) for microscope image mode (a–c) and diffraction mode (d). Chromatic aberration of postspecimen lenses changes each solid profile into the dashed one. Shaded areas represent electrons which are rejected by the SEA as a result of this aberration.

pensated by an equal gain from illuminated regions of specimen outside this area (Titchmarsh and Malis, 1989). However, this compensation will be exact only if the current density is uniform within the disk of illumination and if the specimen is uniform in thickness and composition within this region. These conditions are more easily approximated in the case of a homogeneous thin film than for a heterogeneous specimen thinned from bulk material.

If the illumination is focused so that its screen-level radius becomes comparable to that of the spectrometer entrance aperture, part of the aberration tails may be cut off by the aperture (Fig. 2.18b) and the collection efficiency of the spectrometer will be reduced. Since r_c is a function of energy loss, the decrease in collection efficiency due to chromatic aberration is *energy dependent* and will lead to an error in the elemental ratio deduced from measurements on two different ionization edges.

If r_c and r are small enough, however, the aberration tails may occur entirely *within* the SEA (Fig. 2.18c) and the collection efficiency will again be unaffected. The necessary condition is

$$M'r + M'r_c < R_a \qquad (2.30)$$

For $R_a = 3$ mm and $r = r_c = 100$ nm, Eq. (2.30) can be satisfied if $M' < 15,000$. In practice, the required condition is obtained only if the illumination can be accurately focused into the center of the spectrometer entrance aperture.

If the TEM is operated in *diffraction mode*, chromatic broadening of the angular distribution of inelastic scattering can cause a change in the inelastic signal collected by the spectrometer entrance aperture. Here the major chromatic effect is thought to arise from microscope *intermediate* lenses but would be significant for energy losses above 500 eV and analyzed areas, defined by the incident beam or SAD aperture, larger than 3 μm in diameter (Yang and Egerton, 1992). Errors in quantitative analysis should therefore be negligible in the case of sub-μm probes (Titchmarsh and Malis, 1989). However, this conclusion assumes that the imaging lenses are in good alignment, a condition which can be ensured by positioning the illumination (or SAD aperture) so that the voltage center of the *diffraction pattern* coincides with the center of the viewing screen.

If the energy-loss spectrum is acquired by *serial recording*, the effects of chromatic aberration are entirely avoided by keeping the spectrometer at a fixed excitation and scanning through increasing energy loss by applying a ramp signal to the microscope high-voltage generator. In the case of *parallel recording*, the microscope voltage can be raised by an amount equal to the energy loss being analyzed, but chromatic aberration will cause the collection efficiency to be lower for energy losses sufficiently different from this value. The lens system therefore acts as a bandpass filter and could introduce artifacts in the form of broad peaks in the spectrum (Kruit and Shuman, 1985b), particularly if the optic axis does not coincide with the center of the spectrometer entrance aperture (Yang and Egerton, 1992).

2.3.4. Effect of Lenses on Energy Resolution

The resolution in an energy-loss spectrum depends on several factors: the energy spread ΔE_0 of the electrons before they reach the specimen (reflecting the energy width of the electron source and the Boersch effect), a spectrometer resolution ΔE_{so} which depends on the electron optics, and the spatial resolution s of the electron detector (equal to the slit width in the case of serial recording). It is usual to treat these components as independent and add them in quadrature, so that the measured resolution ΔE is given by

$$(\Delta E)^2 \approx (\Delta E_0)^2 + (\Delta E_{so})^2 + (s/D)^2 \qquad (2.31)$$

where D is the spectrometer dispersion. In general, ΔE_{so} is a function of energy loss E, since both the spectrometer focusing and the angular width of inelastic scattering are E-dependent. As a result, the energy resolution at an ionization edge may be worse than the resolution which is measured as the full width at half maximum (FWHM) of the zero-loss peak.

The chromatic image produced by a double-focusing spectrometer is actually a *convolution* of the energy-loss spectrum with the image (or diffraction) intensity at the spectrometer object plane. If the object has a width d_o, an ideal spectrometer with magnification M_x (in the direction of dispersion) would produce an image of width $M_x d_o$. If the spectrometer has aberrations of order n, the image is broadened further by an amount $C_n \gamma^n$, where γ is the divergence semiangle of the electrons entering the spectrometer; the aberration coefficient C_n depends on the appropriate nth-order matrix elements (Section 2.2.2). The spectrometer resolution is then given by

$$(\Delta E)^2 \approx (M_x d_o/D)^2 + (C_n \gamma^n/D)^2 \qquad (2.32)$$

The values of d_o and γ depend on how the postspectrometer lenses are deployed.

In the absence of a spectrometer entrance aperture, the product $d_o \gamma$ would be constant, since electron-optical brightness is conserved (Reimer, 1993). If the lens conditions are changed so as to reduce d_o and thereby decrease the source-size contribution to ΔE_{so}, the value of γ and of the spectrometer-aberration term increases, and *vice versa*. As a result, there is a particular combination of d_o and γ which minimizes ΔE_{so}. This combination corresponds to optimum L_o or M_o at the spectrometer object plane and to optimum values of M or L at the TEM viewing screen; see Fig. 2.19.

The effect of a spectrometer entrance aperture (radius R_a, distance h' below O) is to limit γ to a value R_a/h', such that the second term in Eq. (2.32) cannot exceed the value corresponding to the circle of illumination just filling the SEA; in other words, $M'd/2 = R_a$ in image mode and $\beta L' = R_a$ in diffraction mode, M' and L' being the magnification and camera length at the SEA plane. If the microscope is operated in image mode and the SEA is used to select the area of analysis, the spectrometer resolution is typically 1 eV (excluding broadening from the detector and electron source) for a spectrometer with second-order aberrations corrected; see

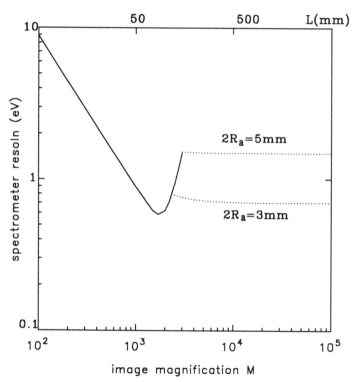

Figure 2.19. Spectrometer resolution ΔE_{so} as a function of microscope magnification (in image mode) or camera length (in diffraction mode), calculated assuming an illumination diameter $d = 1$ μm at the specimen, $\beta = 10$ mrad, and a Gatan spectrometer with $C_2 \approx 0$, $C_3 \approx 50$ m, $D = 1.8$ μmeV^{-1} (at $E_0 = 100$ keV). The dotted lines correspond to the situation in which the diameter of illumination (or central diffraction disk) at the SEA plane exceeds the diameter of a 3-mm or 5-mm spectrometer entrance aperture.

Fig. 2.19. At very low screen magnifications or camera lengths, the energy resolution becomes significantly worse, due to an increase in spectrometer object size. The only cure for this degradation is to reduce β (by using a small objective aperture) in image mode or to reduce the diameter d of the analyzed region in diffraction mode.

The above analysis neglects aberrations of the postspecimen lenses, which could affect the energy resolution if a specimen image is present at the spectrometer object plane (Johnson, 1980a; Egerton, 1980a). In serial acquisition, chromatic aberration is avoidable by scanning the high voltage; in parallel recording, each energy loss might be brought into focus by appropriately tilting the plane of the electron detector, but this has not been practically demonstrated.

2.3.5. STEM Optics

If energy-loss spectroscopy is carried out in a dedicated scanning trans-mission electron microscope (STEM), such as the VG-Scientific HB-501 or -601, there may be no imaging lenses between the spectrometer and specimen. However, the specimen is immersed in the field of the probe-forming objective lens, whose postfield reduces the divergence of the elec-tron beam entering the spectrometer by an angular compression factor $M = \beta/\gamma$; see Fig. 2.20a. This postfield creates a *virtual* image of the illuminated area of the specimen, which acts as the object point O for the spectrometer (Fig. 2.21a). The spectrometer is therefore *image-coupled* with an object-plane magnification equal to the angular compression factor.

Because there are no image-plane (area-selecting) apertures, the spa-tial resolution is defined by the incident-probe diameter d. Although this diameter is influenced by spherical and chromatic aberration of the *pre*field, it is unaffected by aberration coefficients of the postfield, so the spatial resolution for EELS is independent of energy loss.

The focusing power of a magnetic lens is roughly proportional to the reciprocal of the electron kinetic energy (Reimer, 1993). The angular compression factor therefore changes with energy loss, but only by about 1% per 1000 eV (for $E_0 = 100$ keV), so the corresponding variation in collection efficiency is unimportant. But in the case of an instrument which has additional lenses between the specimen and spectrometer, chromatic aberration in these lenses can cause the collection efficiency to substantially increase or decrease with energy loss, leading to errors in quantitative analysis (Buggy and Craven, 1981; Craven *et al.*, 1981).

The energy resolution of the spectrometer/postfield combination can be analyzed as for a TEM with image coupling. The contribution $MM_x d/D$, representing geometric source size, is negligible if the incident beam is

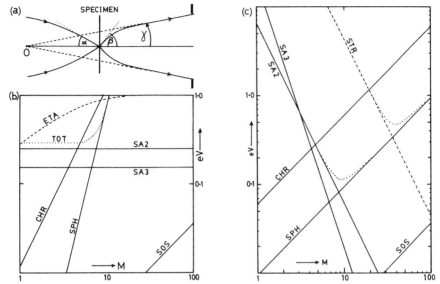

Figure 2.20. (a) Geometry of a STEM probe when the objective is fully focused and the beam diameter at the specimen plane is very small. Point O acts as a virtual object for the spectrometer. (b) Contributions to the energy resolution from spectrometer object size (SOS), from second- and third-order spectrometer aberrations (SA2 and SA3), and from spherical (SPH) and chromatic (CHR) aberration ($E = 400\,eV$) of the objective-lens postfield, calculated as a function of angular compression factor M ($= \beta/\gamma$) for the case where γ is fixed at 5 mrad. Also shown are the collection efficiency (ETA) at $E = 400$ eV and the total energy broadening (TOT), assuming that the effect of chromatic aberration has been corrected. Values assumed are relevant to a VG-HB5 high-excitation polepiece and an aberration-corrected spectrometer: $C_s' = 1.65$ mm, $C_c' = 2$ mm, $d = 1$ nm, $E_0 = 100$ keV, $D = 2\,\mu m/eV$, $M_x = 0.73$, $C_2 = 2$ cm, and $C_3 = 250$ cm (Scheinfein and Isaacson, 1984). (c) The same contributions to the energy resolution, calculated for the case where M and γ are varied to maintain $\beta = 25$ mrad (ETA = 0.7 for $E = 400$ eV). Ignoring chromatic aberration of the postfield, the total energy broadening is a minimum at $M \approx 10$. For comparison, the dashed curve (STR) shows the performance of the equvalent straight-edged magnetic spectrometer with second-order aberrations uncorrected ($C_2 = 140$ cm).

fully focused (the use of a field-emission source allows values of d below 1 nm), but it becomes appreciable if the probe is defocused to several hundred nm or scanned over a similar distance with no descanning applied. Spherical and chromatic aberrations of the objective postfield (aberration coefficients C_s' and C_c') broaden the spectrometer source size, resulting in contributions $MM_xC_s'\beta^3/D$ and $MM_xC_c'\beta(E/E_0)D$, which become significant if M exceeds 10; see Fig. 2.20b. Finally, spectrometer-aberration terms, of the form $(C_n/D)(\beta/M)^n$, can be made small provided second-order aberration is corrected by design of the spectrometer; see Fig. 2.20b,c. As seen

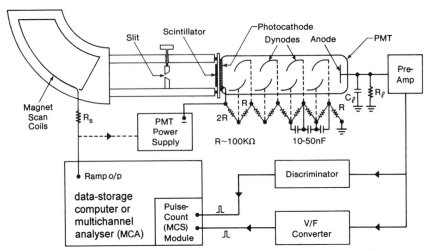

Figure 2.21. A typical serial-acquisition system for energy-loss spectroscopy. An energy-selecting slit is located at the spectrometer image plane. Electrons passing through the slit cause luminescence in a transmission scintillator; the light is turned into an electrical signal and amplified by the photomultiplier tube (PMT). The PMT output is fed into a multichannel scaling (MCS) circuit attached to a multichannel analyzer (MCA), whose ramp output scans the spectrum across the energy-selecting slit.

from Fig. 2.20c, good resolution is possible with high collection efficiency ($\beta = 25$ mrad) if M is of the order of 10; in practice ΔE is limited to about 0.5 eV by high-voltage fluctuations and energy spread of the field-emission source (Scheinfein and Isaacson, 1984).

2.4. Serial Recording of the Energy-Loss Spectrum

For quantitative measurements, the energy-loss spectrum is recorded as a sequence of *channels*, the electron intensity at each channel being represented by a number stored in computer memory. There are two general strategies for converting the intensity distribution into stored numbers.

In *parallel recording*, a position-sensitive electron detector simultaneously records all of the incident electrons, resulting in relatively short recording times and therefore less likelihood of drift or radiation damage to the specimen during spectrum acquisition. Although the method is somewhat complicated and costly to implement, it is the preferred option for recording inner-shell losses. The procedures involved are discussed in Section 2.5.

In *serial recording*, the spectrum at the image plane of the electron spectrometer is examined sequentially, the required spatial resolution (≈ 1 μm) being obtained by placing a narrow slit in front of the electron detector. Because electrons which do not enter the slit are wasted, this method is inefficient and requires longer recording times to avoid excessive statistical (shot) noise at high energy loss. However, it allows the use of a low-noise electron detector which can record a large range of electron intensity in a single readout. Because the same detector is used to record all energy-loss channels, serial recording avoids certain artifacts (due to variation in sensitivity between channels and coupling of signal between channels) which arise in parallel recording. The serial mode is entirely adequate for recording most low-loss spectra.

A serial recording system contains four essential components: (1) the detection slit, which selects electrons of a particular energy loss; (2) the electron detector; (3) a method of scanning the loss spectrum across the detection slit; and (4) a means of converting the output of the electron detector into binary numbers for electronic storage. These components are shown schematically in Fig. 2.21 and are discussed in Sections 2.4.1–2.4.5.

2.4.1. Design of the Detection Slit

The requirement of an energy-selecting slit are as follows. Since the energy dispersion of an electron spectrometer is only a few μm/eV at 100 keV primary energy, the edges of the slit must be smooth (on a μm scale) over a distance equal to the horizontal width of the electron beam at the spectrometer-image plane. For a double-focusing spectrometer, this width is in principle very small, but stray magnetic fields can result in appreciable y-deflection and cause a modulation of the detector output if the edges are not sufficiently straight and parallel (Section 2.2.5). The slit blades are usually constructed from materials such as brass or stainless steel, which are readily machinable, but gold-coated glass fiber has also been used (Metherell, 1971).

The slit width s (in the vertical direction) needs to be adjustable to suit different circumstances. For examining fine structure present in the loss spectrum, a small value of s ensures good energy resolution of the recorded data, limited only by the spectrometer performance and the energy width of the electron source. For the measurement of elemental concentrations from ionization edges, high energy resolution is not essential and a larger value of s may be necessary to obtain a satisfactory signal ($\propto s$) and signal/noise ratio. For large s, the energy resolution becomes approximately s/D, where D is the dispersive power of the spectrometer.

The slit blades should be designed so that electrons which strike them produce negligible response from the detector. Fulfilment of this condition is not trivial. It is relatively easy to prevent the incident electrons being transmitted through the slit blades, since the stopping distance (electron range) of 100-keV electrons is less than 100 μm for solids with atomic number greater than 14. On the other hand, the x-rays which are generated when an electron is brought to rest are more penetrating; in iron or copper, the attenuation length of 60-keV photons (the distance required to reduce the x-ray intensity by a factor of $e = 2.72$) is about 1 mm. Transmitted x-rays which reach the detector give rise to a spurious signal (Kihn *et al.*, 1980) which, unlike the true signal, is independent of the slit opening s. In addition, half of the x-ray photons are generated in the backwards direction and a small fraction of these will be reflected so that they pass through the slit and strike the detector (Craven and Buggy, 1984). Equally important, an appreciable number of fast electrons which strike the slit blades will be backscattered and after one or more subsequent back-reflections they too may pass through the slit and arrive at the detector, having retained much of their original kinetic energy. Coefficients of electron backscattering are typically in the range 0.1–0.6, so the stray-electron signal, which depends on both the vertical width and the horizontal length of the slit opening, can be appreciable.

The stray electrons and x-rays are observed as an instrumental or *spectrometer* background to the energy-loss spectrum (analogous to the "hole count" in EDX spectroscopy), resulting in reduced signal/background and signal/noise ratio especially at high energy loss. Most of the energy-loss intensity occurs within the low-loss region of the spectrum (particularly the zero-loss peak), so most of the stray electrons and x-rays are generated close to the point where the zero-loss beam strikes the "lower" slit blade when recording higher energy losses.

Requirements for a low spectrometer background are as follows:

(a) The slit material should be conducting (to avoid electrostatic charging in the beam) and thick enough to prevent x-ray penetration. For 100-keV operation, 5 mm thickness of brass or stainless steel appears to be sufficient. At higher accelerating voltages, it might be necessary to back the lower slit with a material of higher atomic number such as lead.

(b) The angle of the slit edges should be close to 80° so that the zero-loss beam is absorbed by the full thickness of the slit material when recording energy losses above a few hundred eV.

(c) The defining edges should be in the same plane so that, when the slit is almost closed to the spectrometer exit beam, there is no oblique path available for scattered electrons (traveling at some large angle to the optic axis) to reach the detector.

(d) The length of the slit in the horizontal (nondispersive) direction should be restricted in order to reduce the probability of scattered electrons and x-rays reaching the detector. A length of a few hundred micrometers may be necessary to facilitate alignment and allow for deflection of the beam by stray dc and ac magnetic fields.

(e) Since the coefficient η of electron backscattering is a direct function of atomic number, the "lower" slit blade should be coated with a material such as carbon ($\eta \simeq 0.05$). The easiest procedure is to "paint" the slit blades with an aqueous suspension of colloidal graphite. The porous structure of the coating helps further in absorbing scattered electrons.

(f) The lower slit blade should be flat within the region over which the zero-loss beam is scanned. Sharp steps or protuberances can give rise to sudden changes in the scattered-electron background, which could be mistaken for real spectral features (Joy and Maher, 1980a). For a similar reason, the use of "spray" apertures in front of the energy-selecting slit should be avoided.

(g) Moving the detector further away from the slits and minimizing its exposed area (so that the latter is just sufficient to accommodate the angular divergence of the spectrometer exit beam) decreases the fraction of stray electrons and x-rays which reach the detector (Kihn *et al.*, 1980). Employing a scintillator whose thickness is just sufficient to stop fast electrons (but not hard x-rays) would further reduce the x-ray contribution.

(h) Use of an electron-counting technique for the higher energy losses reduces the intensity of the instrumental background because many of the backscattered electrons and x-rays produce output pulses which fall below the threshold level of the discriminator circuit (Kihn *et al.*, 1980).

Measurement of the Spectrometer Background

Because most of the stray electrons and x-rays are generated by the zero- and low-loss electrons, an instrumental background similar to that present in a "real" spectrum can be measured by carrying out serial acquisition with no specimen in the beam. The resulting "background spectrum" enables the spectrometer contribution to be assessed; if found to be significant, it can be subtracted from real spectral data (Craven and Buggy, 1984). For this background correction to be accurate, the real and background spectra should be recorded using the same slit width and their total areas scaled to be equal before subtraction.

The magnitude of the spectrometer background can also be judged from the *jump ratio* of an ionization edge, defined as the ratio of maximum and minimum intensities just after and just before the edge. A very thin carbon foil (<10 nm for $E_0 = 100$ keV) provides a convenient test sample (Joy and Newbury, 1981); if the spectrum is recorded with a collection semiangle β less than 10 mrad, a jump ratio of 15 or more at the K-edge indicates a low instrumental background (Egerton and Sevely, 1983).

A further test for spectrometer background is to record the K-ionization edge of a thin (<50 nm) aluminum or silicon specimen in the usual "bright-field" condition (with the collection aperture centered about the optic axis) and in dark field, where the collection aperture is shifted or the incident illumination tilted so that the undiffracted beam is intercepted by the aperture. In the latter case, stray electrons and x-rays are generated mainly at the collection aperture and have little chance of reaching the detector, particularly in a CTEM system in which an objective-lens aperture functions as the collection aperture. As a result, the jump ratio (or signal/ background ratio) of the edge may be higher in dark field, the amount of improvement reflecting the magnitude of the bright-field instrumental background (Oikawa *et al.*, 1984; Cheng and Egerton, 1985). If the spectrometer background is high, the measured jump ratio may actually increase with increasing specimen thickness (Hosoi *et al.*, 1984; Cheng and Egerton, 1985), up to a thickness at which plural scattering in the specimen imposes an opposite trend (Section 3.5.4).

Although not usually a severe problem, electron scattering within the microscope column may contribute to the instrumental background (Joy and Maher, 1980a). For example, insertion of an area-selecting aperture may degrade the jump ratio of an edge. Presumably some electrons are scattered through small angles by the edge of the aperture, so that they enter the spectrometer and contribute a diffuse background to the energy-loss spectrum.

2.4.2. Electron Detectors for Serial Recording

Serial recording has sometimes been carried out with solid-state (silicon-diode) detectors but counting rates are limited to about 10^4 cps (Kihn *et al.*, 1980) and radiation damage is a potential problem at high doses (Joy, 1984a). Windowless electron multipliers may be usable up to 100 Mcps but require excellent vacuum to prevent contamination (Joy 1984a); channeltrons give a relatively noisy output for electrons whose energy exceeds 10 keV (Herrmann, 1984). The preferred detector for serial recording consists of a scintillator (which emits visible photons in response to each incident electron) and a photomultiplier tube (PMT) which converts some of these photons into electrical pulses or an output current.

Scintillator Materials

Properties of some useful scintillators are listed in Table 2.2. Polycrystalline scintillators are usually prepared by sedimentation of a phosphor powder from aqueous solution or an organic liquid such as chloroform, sometimes with an organic or silicate binder to improve the adhesive properties of the layer.

To maximize the signal/noise ratio, as much light as possible should enter the photomultiplier tube, which requires an efficient phosphor of suitable thickness. If the thickness is less than the electron range, some kinetic energy of the incident electron is wasted; if the scintillator is too thick, light may be lost by absorption or scattering within the scintillator, particularly in the case of a transmission screen (Fig. 2.21). The optimum mass-thickness for P-47 powder appears to be about 10 mg/cm^2 for 100-keV electrons (Baumann *et al.*, 1981).

To increase the fraction of light entering the photomultiplier and to prevent electrostatic charging in the beam, the entrance surface of the scintillator is given a reflective metallic coating. Aluminum can be evaporated directly onto the surface of glass, plastic, and single-crystal scintillators. In the case of a powder-layer phosphor, the metal would penetrate between the crystallites and actually reduce the light output, so an aluminum film is prepared separately and floated onto the phosphor or else the phosphor layer is coated with a smooth layer of plastic (e.g., collodion) before being aluminized.

Table 2.2. Properties of Several Scintillator Materials[a]

Material	Type	Peak wavelength (nm)	Principal decay constant (ns)	Energy conversion efficiency (%)	Dose for damage (Mrad)
NE 102	Plastic	420	2.4, 7	3	1
NE 901	Li-glass	395	75	1	10^3
ZnS(Ag)	Polycrystal	450	200	12	
P-47	Polycrystal	400	60	7	10^2
P-46	Polycrystal	550	70	3	>10^4
CaF$_2$(Eu)	Crystal	435	1000	2	10^4
YAG	Crystal	560	80		>10^4
YAP	Crystal	380	30	7	>10^4

[a]P-46 and P-47 are yttrium aluminum garnet (YAG) and yttrium silicate, respectively, each doped with about 1% of cerium. The data were taken from several sources, including Blasse and Bril (1967), Pawley (1974), Autrata *et al.* (1983), and Engel *et al.* (1981); the efficiencies and radiation resistance should be regarded as approximate, since measurements by different workers often disagree by a factor of 2 or more.

The efficiency of many scintillators decreases with time as a result of irradiation damage (Table 2.2). This process is particularly rapid in plastics (Oldham *et al.*, 1971), but since the electron penetrates only about 100 μm the damaged layer can be removed by grinding and polishing, allowing a scintillator which is a few millimeters thick to be refurbished several times. In the case of inorganic crystals, the loss of efficiency is due mainly to darkening of the material (creation of color centers), resulting in absorption of the emitted radiation, and can sometimes be reversed by annealing the crystal (Wiggins, 1978).

The decay time of a scintillator is of particular importance to electron counting. Plastics generally have time constants below 10 ns, allowing pulse counting up to at least 20 MHz with appropriate electronics. However, most scintillators have several time constants, extending sometimes up to several seconds. By shifting the effective base line at the discriminator circuit, the longer time constants increase the "dead time" between counted pulses (Craven and Buggy, 1984).

P-46 (cerium-doped $Y_3Al_5O_{12}$) can be grown as a single crystal (Blasse and Bril, 1967; Autrata *et al.*, 1983) and combines high quantum efficiency with excellent radiation resistance. The light output is in the yellow region of the spectrum but can be efficiently detected by most photomultiplier tubes. Electron counting up to a few megahertz has been reported with this material (Craven and Buggy, 1984).

Photomultiplier Tubes

A photomultiplier tube contains a photocathode (which emits electrons in response to incident photons), several "dynode" electrodes (which accelerate the photoelectrons and increase their number by a process of secondary emission), and an anode which collects the amplified electron current so that it can be fed into a preamplifier; see Fig. 2.21.

To produce photoelectrons from visible photons, the photocathode must have a low work function. Cesium antimonide is a popular choice, particularly for scintillation counting and for photon wavelengths below 500 nm. More recently, single-crystal semiconductors such as gallium arsenide have been used for the photocathode. The semiconductor is *p*-type but its surface is treated with Cs to produce an *n*-type inversion layer and a "negative electron affinity" (the bottom of the conduction band lies above the vacuum level), thereby increasing the photoelectron escape depth and ensuring high responsive quantum efficiency (RQE), i.e., a large number of photoelectrons emitted for each incident photon.

The spectral response of a PMT depends not only on the material of the photocathode but also on its treatment during manufacture and on the type of glass used in constructing the tube. Both the sensitivity and spectral response of the tube may change with time as gas is liberated from internal surfaces and becomes adsorbed on the cathode. Photocathodes whose spectral response extends to longer wavelengths tend to have more "dark emission," leading to a higher dark current at the anode and increased output noise. The dark current decreases by typically a factor of 10 when the PMT is cooled from room temperature to $-30°C$, but is considerably increased if the cathode is exposed to room light (even with no voltages applied to the dynodes) or to strong light from the scintillator, and can take several hours to return to its original value.

The dynodes consist of a staggered sequence of electrodes and are traditionally constructed from BeCu or CsSb with a secondary-electron yield of about 4, giving an overall gain of 10^6 or more if there are 10 electrodes. Gallium phosphide has also been used for the first dynode; the higher secondary-electron yield (typically 40) offers improved signal/noise ratio and easier discrimination against noise pulses in the electron-counting mode (Engel *et al.*, 1981).

The anode of a photomultiplier tube is usually operated at ground potential, the photocathode being at a negative voltage (typically -700 to -1500 V). The dynode potentials are supplied by a chain of low-noise resistors (Fig. 2.21). For *analog operation*, where the anode signal is treated as a continuous current, the dynode resistors are usually equal in value (typically 100 kΩ) except for the resistance between the cathode and first anode, which is about twice as large. The PMT acts as an almost-ideal current generator, the negative voltage developed at the anode being proportional to the load resistor R_l and (within the linear region of operation) to the light input. Linearity is believed to be within 3% provided the anode current does not exceed one tenth of that flowing through the dynode-resistance chain (Land, 1971). The electron gain can be controlled over a wide range by varying the voltage applied to the tube. Since the gain depends sensitively on applied potential, the voltage stability of the power supply needs to be an order of magnitude better than the required stability of the output current.

An electron whose energy is 10 keV or more produces some hundreds of photons within a typical scintillator. Even allowing for some light loss before reaching the photocathode, the resulting negative pulse at the anode is well above the noise level of the PMT, so energy-loss electrons can be individually counted. The maximum counting frequency is determined by the decay time of the scintillator, the characteristics of the PMT, and the output circuitry. To ensure that the dynode potentials (and therefore the secondary-electron gain) remain constant during the pulse interval, capacitors are placed across the final-dynode resistors (Fig. 2.21). To maximize the pulse amplitude and avoid overlap of output pulses, the time constant

$R_l C_l$ of the anode circuit must be less than the average time between output pulses. The capacitance to ground C_l is kept low by locating the preamplifier close to the PMT (Craven and Buggy, 1984).

2.4.3. Noise Performance of a Serial Detector

In addition to the energy-loss signal (S), the output of the electron detector contains a certain amount of noise (N). The quality of the *signal* can be expressed in terms of a signal-to-noise ratio: SNR = S/N. However, some of this noise is already present within the electron beam in the form of shot noise; if the mean number of electrons recorded in a given time is n, the actual number recorded under the same conditions follows a Poisson distribution whose variance is $m = n$ and whose standard deviation is $\sqrt{m}$, giving an inherent signal/noise ratio: $(SNR)_i = n/\sqrt{m} = \sqrt{n}$. The noise performance of a *detector* is therefore represented as a ratio known as the *detective quantum efficiency* (DQE), defined by

$$DQE \equiv [SNR/(SNR)_i]^2 = (SNR)^2/n \qquad (2.33)$$

For an "ideal" detector which adds no noise to the signal: $SNR = (SNR)_i$, giving DQE = 1.0. In general, DQE is not constant for a particular detector but depends on the incident-electron intensity (Herrmann, 1984).

The measured DQE of a scintillator/PMT detector is typically in the range 0.5–0.9 for incident-electron energies between 20 and 100 keV (Pawley, 1974; Baumann *et al.*, 1981; Comins and Thirlwall, 1981). One reason why DQE is below unity has to do with the statistics of photon production within the scintillator and photoelectron generation of the PMT photocathode. Assuming Poisson statistics, it can be shown that DQE is limited to a value given by

$$DQE \leq p/(1 + p) \qquad (2.34)$$

where p is the average number of photoelectrons produced for each incident fast electron (Browne and Ward, 1982). For optimum noise performance, p should be kept reasonably high by using an efficient scintillator, metalliz-ing its front surface to reduce light losses and providing an efficient light path to the PMT. However, Eq. (2.34) shows that the DQE is only seriously affected if p falls below about 10.

DQE is also reduced as a result of the statistics of electron multiplica-tion within the PMT and dark emission from the photocathode. These effects can be minimized by using a material with a high secondary-electron yield (e.g., GaP) for the first dynode and by using pulse-counting of the output signal to discriminate against the dark current. In practice, the pulse-height distributions of the noise and signal pulses overlap (Engel *et al.*,

1981), so that even at its correct setting a discriminator rejects a fraction f of the signal pulses, reducing the DQE by the factor $(1 - f)$. The overlap occurs as a result of a high-energy tail in the noise distribution and because some signal pulses are weaker than the rest. For example, an appreciable fraction of incident electrons are backscattered within the scintillator and their light emission is less by an amount which depends on the angle of backscattering and the depth at which it occurred. Also, incident electrons lose differing amounts of energy in the metallizing layer before entering the scintillator and so generate different number of photons, thereby increasing the spread of pulse heights at the PMT anode. In electron-counting mode, the discriminator setting therefore represents a compromise between loss of signal and increase in detector noise, both of which reduce the DQE.

If the PMT is used in analogy mode together with a V/F converter (see Section 2.4.4), the DQE will be slightly lower than for pulse counting with the discriminator operating *at its optimum setting*. In the low-loss, high-intensity region of the spectrum, a lower DQE is unimportant since the signal/noise ratio is more than adequate. In the case of an A/D converter used in conjunction with a filter circuit whose time constant is comparable to the dwell time per channel, there is an additional noise component. Besides variation in the number of fast electrons which arrive within a given dwell period, the contribution of a given electron to the sampled signal depends on its time of arrival (Tull, 1968). As a result, the DQE is halved compared to the value obtained using a V/F converter, which integrates the charge pulses without the need of an input filter.

The preceding discussion relates to the DQE of the electron detector alone. When this detector is part of a serial-acquisition system, one can define a detective quantum efficiency $(DQE)_{syst}$ for the recording system as a whole, taking n in Eq. (2.33) to be the number of electrons analyzed by the spectrometer during the acquisition time rather than the number which pass through the detection slit. At any instant, the detector samples only those electrons which are passing through the slit (width s) so that, evaluated over the entire acquisition, the fraction of analyzed electrons which reach the detector is $s/\Delta x$, where Δx is the distance (in the image plane) over which the spectrometer exit beam is scanned. The overall DQE in serial mode can therefore be written as

$$(DQE)_{syst} = (s/\Delta x)(DQE)_{detector} \qquad (2.35)$$

The energy resolution ΔE in the recorded data cannot be better than s/D, allowing Eq. (2.35) to be rewritten in the form

$$(DQE)_{syst} \le (\Delta E/E_{scan})(DQE)_{detector} \qquad (2.36)$$

where E_{scan} is the energy width of the recorded data. Typically ΔE is in the range 1–10 eV while E_{scan} may be in the range 100–5000 eV, so the overall DQE is usually below 1%. In a serial-detection system, $(DQE)_{syst}$ can always be improved by widening the detection slit, but at the expense of degraded energy resolution.

2.4.4. Signal Processing and Storage

In this section we discuss methods for converting the output of a serial-mode detector into numbers stored in computer memory. The detector will be assumed to consist of a scintillator and PMT, although similar principles apply in the case of a solid-state detector.

Electron Counting

Photomultiplier tubes have low noise and high sensitivity; some types can even be used to count incident *photons*. In a suitable scintillator, a high-energy electron produces a rapid burst of light containing *many* photons, so it is relatively easy to detect and count individual electrons. If well implemented, this procedure results in a roughly one-to-one relationship between the stored counts per channel and the number of energy-loss electrons reaching the detector. The intensity scale is then linear down to low count rates and the sensitivity of the detector is unaffected by changes in PMT gain arising from tube aging or power-supply fluctuations. Ideally, the sensitivity is also independent of the incident-electron energy.

In order to extend electron counting down to low arrival rates, a lower-level discriminator is used to eliminate anode signals generated by stray light, low-energy x-rays, or noise sources within the PMT (mainly dark emission from the photocathode). If the PMT has single-photon sensitivity, dark emission produces discrete pulses at the anode, each containing G electrons, where G is the electron gain of the dynode chain. To accurately set the discriminator threshold, it is useful to measure the distribution of pulse amplitudes at the output of the PMT preamplifier, using either an instrument with pulse-height analysis (PHA) facilitates or a fast oscilloscope (Engel *et al.*, 1981). The pulse-height distribution should contain a maximum (at zero or low pulse amplitude) arising from noise and a second maximum which represents signal pulses; the discriminator threshold is placed between the peaks (Engel *et al.*, 1981). If these peaks are well separated, the discriminator can also be set by increasing its threshold until the output-pulse rate reaches a plateau.

A major limitation of pulse counting is that (owing to the distribution of decay times of the scintillator) the maximum count rate is only a few megahertz for P-46 (Ce-YAG) and of the order of 20 MHz for a plastic scintillator (or less if the scintillator has suffered radiation damage), rates which correspond to an electron current below 4 pA. Since the incident-beam current is typically in the range 1 nA–1 μA, alternative arrangements are made to record the low-loss region of the spectrum.

Analog/Digital Conversion

At high incident rates, the charge pulses produced at the anode of a PMT merge and the output of the preamplifier can be considered as a continuous current or voltage, whose level is related to the electron flux falling on the scintillator. There are two ways of converting this voltage into binary form for digital storage. The most popular procedure is to feed the preamplifier output into a voltage-to-frequency (V/F) converter (Maher *et al.*, 1978; Zubin and Wiggins, 1980), which is essentially a voltage-controlled oscillator; its output consists of a continuous train of pulses which can be counted using the same scaling circuitry as employed for electron counting. The output frequency is proportional to the input voltage between (typically) 10 μV and 10 V, providing excellent linearity over a large dynamic range. V/F converters are compact and inexpensive and are available with output rates as high as 20 MHz. Unfortunately the output frequency is slightly temperature dependent, but this drift can be accommodated by providing a "zero level" frequency-offset control which is adjusted from time to time to keep the output rate down to a few counts per channel with the electron beam turned off. The minimum output rate should not fall to zero, since this condition would imply a lack of response at low electron intensity, resulting in recorded spectra whose channel contents vanish at some value of the energy loss (Joy and Maher, 1980a). Any remaining background within each spectrum (measured, for example, to the left of the zero-loss peak) can be subtracted digitally in computer memory.

An alternative method of digitizing the analog output of a PMT is via an analog-to-digital (A/D) converter, the main disadvantage being limited dynamic range (typically 4096:1 for a high-speed device) and the fact that, whereas the V/F converter effectively integrates the detector output over the dwell time per data channel, an A/D converter usually samples the voltage level only once per channel. To eliminate contributions from high-frequency noise, the PMT output must therefore be smoothed with a time constant approximately equal to the dwell period per channel (Egerton and Kenway, 1979; Zubin and Wiggins, 1980), which involves resetting a switch in the filter circuit each time the dwell period is changed. Even if this is done, the smoothing introduces some smearing of the data between adjacent channels. The situation could be improved by sampling the data many times per channel and taking an average.

2.4.5. Scanning the Energy-Loss Spectrum

Several methods are available for scanning the energy-loss spectrum across the detection slit (as required during serial recording) or for shifting it relative to the detector (as is sometimes necessary in parallel recording).

(a) *Ramping the Magnet.* The magnetic field in a single-prism spectrometer is easily changed by varying the main excitation current or by applying a current ramp to separate window-frame coils. A digital-to-analog converter can be used to turn the channel-advance pulses of the multichannel-recording circuitry into a voltage ramp; a series resistor R_s (Fig. 2.21) then determines the scan range (and eV/channel) for a given primary energy. Because the detection slit does not move, the recorded electrons always have the same radius of curvature within the spectrometer. From Eq. (2.1), the magnetic induction required to record electrons of energy E_0 is

$$B = (eR)^{-1}[(E_0 + E_0^2/(2m_0c^2)] \tag{2.37}$$

where $m_0c^2 = 511$ keV is the electron rest energy. Even assuming ideal properties of the magnet ($B \propto$ ramp current), a linear ramp will give an energy axis which is slightly nonlinear (see Fig. 2.22) because of the quadratic term in Eq. (2.37). However, the nonlinearity is only about 0.4% over a 1000eV scan at $E_0 = 100$ keV.

(b) *Pre- or Postspectrometer Deflection.* Inductance of the magnet windings causes a lag between the change in B and the applied voltage, limiting the scan rate to typically 1/s. Higher rates, which are convenient for adjusting the position of the zero-loss peak and setting the width of the detection slit, are achieved by injecting a ramp signal into small dipole coils

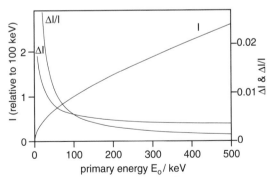

Figure 2.22. Spectrometer current I (relative to its value for 100-keV electrons) and the change $\Delta I = I' - I$ and fractional change $\Delta I/I$ in current for a 1000-eV change in energy of the detected electrons, as a function of the primary energy E_0. These quantities were calculated from Eq. (2.37) assuming $B \propto I$ (Meyer *et al.*, 1995).

located just before or after the spectrometer. Even faster control is possible by using electrostatic deflection plates (Fiori *et al.*, 1980). Particularly if the tilt angle ψ of the disperion plane is nonzero, the spectrum becomes slighly defocused as the scan proceeds but the effect is unimportant over a small range of energy loss.

(c) *Ramping the High Voltage.* The spectrum can also be shifted by applying a ramp voltage to the microscope high-voltage generator, thereby varying the potential of the electron source. The spectrometer and detection slit are then used as a filter which responds to electrons of a fixed kinetic energy, thereby avoiding the unwanted effects of chromatic aberration in the postspecimen lenses (Section 2.3). The variation in energy of the electrons passing through the condenser lenses results in a change in illumination focus, but this effect can be compensated by applying a suitable fraction of the scanning signal to the condenser-lens power supply (Wittry, 1969; Krivanek *et al.*, 1992).

(d) *Drift-Tube Scanning.* If the drift tube of a magnetic spectrometer is electrically isolated from ground, applying a voltage to it changes the kinetic energy of electrons traveling through the magnet and shifts the energy-loss scale. The applied potential also produces a weak electrostatic lens at the entrance and exit of the drift tube, tending to defocus and possibly deflect the spectrum, but these effects can be shown to be negligible provided the internal diameter of the drift tube and its immediate surroundings are not too small (Batson *et al.*, 1981). In the absence of an electrostatic-lens effect, a given voltage applied to the drift tube will displace the energy-loss spectrum by the same number of eV, allowing the energy-loss axis to be accurately calibrated. Meyer *et al.* (1995) have shown that if a known voltage (e.g., 1000 eV) is applied to the drift tube and the zero-loss peak is returned to its original position by changing the magnet current from I to I', a measurement of I'/I can be used to determine the accelerating voltage to an accuracy of about 50 V; see Fig. 2.22.

The following general remarks apply to all methods of scanning. Particularly where long recording times are used, it is convenient to scan through the required range of energy loss several (or many) times and add the individual spectra together in computer memory, a technique known as multiscanning or signal averaging. The accumulated data can be continuously displayed, allowing broad features to be discerned after only a few scans, so that the acquisition can be aborted if necessary; otherwise, the scans are repeated until the SNR in the spectrum becomes acceptable. For a given total time T of acquisition, the SNR is the same as for a spectrum acquired in a single scan of duration T. However, the effect of instrumental instability, such as a fluctuation in the accelerating voltage V_0, shows up differently. With a single scan, slow drift in V_0 causes a distortion (nonlinearity) of the energy-loss scale and a more rapid fluctuation produces an artifact (e.g., a peak or edge) in the spectrum. In the case of very rapid scans which are directly added to computer memory, both types of instability cause a blurring of the spectral features, equivalent to a loss of energy resolution. This loss of resolution can be largely eliminated within the computer; before being added to the accumulated spectrum, the data from each scan can be shifted so that a particular spectral feature (e.g., the zero-loss peak) always occurs at the same MCA channel (Batson *et al.*, 1971; Egerton and Kenway, 1979).

If the scan is computer controlled, "segmented" scans are possible. For example, the dwell time per channel might be increased at a chosen energy loss, which is equivalent to increasing the detector gain except that slower scanning results in better statistics of the recorded data (improved signal/noise ratio). Alternatively, the scan output may "jump over" regions of energy loss which contain no useful information, thereby saving time and minimizing the radiation dose to the specimen. Another way of introducing a gain change, and thereby compressing the dynamic range of the data which has to be stored, is to alter the voltage applied to the photomultiplier tube (or change from analog detection to pulse counting) at a preset energy loss. This change can be done automatically by sensing the ramp output or by programming the ramp generator to send a flag signal to the PMT power supply and/or input-processing circuitry (Egerton, 1980e; Krivanek and Swann, 1981).

As discussed in Section 2.4.2, some phosphors have a prolonged "afterglow" and photomultipler tubes display "memory" effects (generally an increase in dark current) after being exposed to strong irradiation. These effects can lead to distortion of the loss spectrum after scanning through an intense zero-loss peak (Joy and Maher, 1980a). One solution is to scan from high energy loss downwards and to wait a sufficient period of time between scans.

2.4.6. Coincidence Counting

If an energy-dispersive x-ray (EDX) detector is operated in conjunction with the energy-loss spectrometer, it is possible to improve the signal/background and (in principle) the signal/noise ratio of an ionization edge. By applying both detector outputs to a gating circuit which gives an output pulse only when an energy-loss electron and a characteristic x-ray of the same energy are received simultaneously (within a certain resolution time), the background to an ionization edge can be largely eliminated (Wittry, 1976). Some "false coincidences" will occur (due to x-rays and energy-loss electrons which come from separate scattering events) at a rate proportional to the product of the resolution time, the x-ray signal and the energy-loss signal. To keep their contribution small, the incident-beam current must be kept low, resulting in a low overall count rate. A small contribution from false coincidences can be recognized and subtracted from the coincidence spectrum, since it has the same energy dependence as the energy-loss signal before coincidence gating (Kruit *et al.*, 1984). A peak due to bremsstrahlung loss also appears in the coincidence energy-loss spectrum, but below the ionization-edge threshold (Fig. 2.23).

Measured coincidence rates amounted to only a few counts per second, so the procedure is not very useful at present (Kruit *et al.*, 1984, Nicholls *et al.*, 1984). To obtain higher count rates, the collection efficiency of the EDX detector must be improved, perhaps by using a large-area room-temperature detector (good energy resolution in the EDX channel is not important). It would also help if the energy-loss spectrum were recorded using a parallel-detection system operated in an electron-counting mode.

Particularly in the case of light-element samples, the x-ray detector could be replaced by an Auger-electron detector as the source of the gating

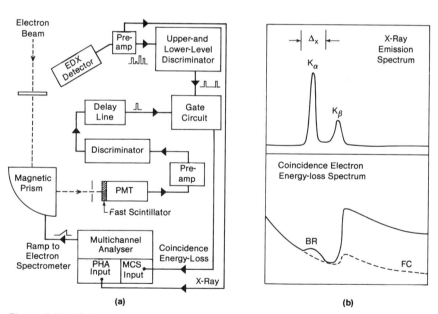

Figure 2.23. (a) Scheme for simultaneous measurement of x-ray emission and coincident energy-loss electrons (recorded serially). The energy window Δ_x for x-ray gating is selected by a dual-level discriminator. (b) The coincidence energy-loss spectrum contains a background FC due to false coincidences and a small peak BR arising from bremsstrahlung energy losses at the x-ray gating window Δ_x, here chosen to coincide with the lowest-energy x-ray line.

signal (Wittry, 1980). Alternatively, by using the energy-loss signal for gating, the coincidence technique could be used to improve the energy resolution of the Auger spectrum (Cazaux, 1984; Haak *et al.*, 1984). Coincidence between energy loss and secondary-electron generation can also be monitored and has yielded absolute probabilities of secondary-electron emission resulting from bulk-plasmon and surface-plasmon excitation (Müllejans *et al.*, 1993).

2.5. Parallel Recording of the Energy-Loss Spectrum

A parallel-recording system utilizes a *position-sensitive* detector which is exposed to a broad range of energy loss. Because there is no energy-selecting slit, the detective quantum efficiency (DQE) of the recording *system* is the same as that of the *detector*, rather than being limited by Eq. (2.36). As a result, spectra can be recorded in shorter times (and with less radiation dose) than with serial acquisition for the same noise content. These advantages are of particular importance for the spectroscopy of ionization edges at high energy loss, where the electron intensity is low.

Photographic film constitutes the simplest parallel-recording device and was used extensively in early EELS work. With suitable emulsion thickness, the DQE exceeds 0.5 over a limited exposure range (Herrmann, 1984; Zeitler, 1992). Its main disadvantages are a limited dynamic range, the need for chemical processing, and the fact that densitometry is required to obtain quantitative data.

Modern systems utilize a self-scanning array of silicon diodes: either a photodiode array (PDA) or charge-coupled diode (CCD) array. These two types differ in their internal mode of operation, but both provide a pulse-train output which can be fed into an electronic data-storage system, just as in serial acquisition. PDAs have been used mainly in their one-dimensional (linear) form, as in the Gatan PEELS spectrometer. CCD devices are normally employed as two-dimensional (area) arrays, which are also used for the recording of TEM images and diffraction patterns.

2.5.1. Operation of Self-Scanning Diode Arrays

A linear photodiode array contains a large number (e.g., 1024) of silicon diodes, each connected to a common video output line through its own field-effect transistor (FET), as illustrated in Fig. 2.24. When the device is exposed to radiation (photons or charged particles), holes and electrons generated within about one diffusion length of the surface augment the thermal leakage current of the nearest $p-n$ junction, discharging the corres-

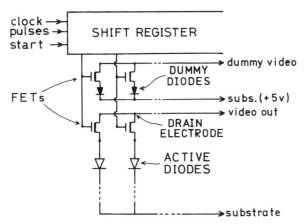

Figure 2.24. Equivalent circuit of a photodiode array, showing components for the first two diode elements.

ponding diode at a rate which depends on its internal capacitance and on the local irradiation level. At the end of its *integration period* τ, each diode is recharged by applying a bias voltage to the gate of the FET. The amount of applied charge can be measured as an input current or change in voltage at the FET drain electrode, which is connected directly to a common output line. The bias signal is applied sequentially to each photodiode via a shift register fabricated on the same piece of silicon. Recharge pulses therefore appear in sequence on the output line as a video signal, which is subsequently amplified and processed. In practice, alternate diodes are driven by different shift registers and their outputs appear on separate output lines, but these "odd" and "even" outputs are combined in the external circuitry. "Dummy" diodes are incorporated along the array and produce their own odd and even outputs which are used to cancel switching transients (in the video output) arising from parasitic coupling within the array.

Charge-coupled diode arrays are manufactured using the same silicon planar technology as photodiode arrays. Charge on each diode is depleted by incident radiation but the mechanism of charge readout is different; charge packets are moved from the diodes into an adjacent shift register and then shifted to an output electrode. A related device is the CID array, where charge is injected into the silicon substrate and the readout process is nondestructive.

The external electronics needed to operate an array consists of timing circuitry, which determines the integration and readout times, and a video section which processes the output pulses. The video circuit contains either a charge-sensitive amplifier or a sample-and-hold circuit whose output voltage represents the amount of discharge of each diode. The timing circuit is driven by an oscillator, usually running at between 10 kHz and 1 MHz. In the case of the photodiode array, the oscillator supplies clock pulses to operate the on-chip shift registers and may also drive an integration counter whose output restarts the shift registers and initiates a new readout after a predetermined number of clock pulses. The oscillator then determines the *diode-sampling frequency* (the pulse rate within each video train) while the integration counter controls the integration time between successive readouts. The counting circuit can be programmed so that the video output is either continuous (one readout being followed almost immediately by the next) or separated by periods in which the diodes are responding to radiation but the shift registers are inactive.

In Gatan PEELS systems, the diode sampling frequency is typically 50 kHz, so readout of a 1024-channel photodiode array takes about 20 ms which, in continuous readout, is therefore the minimum integration period. For electronic storage in computer memory, each video pulse is converted into a binary number by use of a digital-to-analog converter (DAC) preceded by a sample-and-hold circuit. Direct memory access (DMA) allows the digital data to be transferred rapidly into computer memory without passing through the microprocessor. The computer also controls the voltage applied to the drift tube of the spectrometer, allowing the spectrum to be displaced (with the zero-loss peak off the array) by up to 1000 eV.

Diode arrays are also available as two-dimensional (area) arrays, made for the electronic recording of light-optical images. By using only a thin vertical strip of the array to record the spectrum and horizontally shifting the data into a storage area, the diode integration time can be reduced to about 10 μs (Zaluzec and Strauss, 1989). It is also possible to bin the data into a single row of pixels, so that as many as 290 spectra can be stored in the array before readout is necessary (Berger and McMullan, 1989). The Gatan imaging filter (Section 2.5) contains an area CCD array fiber-optically coupled to a thin YAG scintillator, and can be used to record energy-loss spectra if the preceding quadrupole lenses are appropriately adjusted.

2.5.2. Indirect Exposure Systems

Diode arrays are designed as light-optical sensors and are used extensively in video applications and in astronomy. They can therefore be used in combination with a conversion screen (imaging scintillator) in an *indirect exposure* system. Figure 2.25 shows a commercial spectrometer which employs a YAG scintillator, coupled by fiber-optic plate to a thermoelectrically cooled photodiode array (Krivanek *et al.*, 1987). To provide sufficient energy resolution to examine fine structure in a loss spectrum, the spectrom-

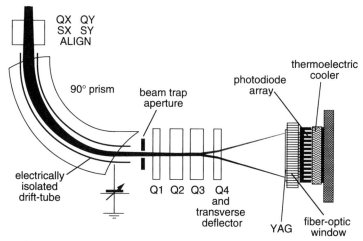

Figure 2.25. Schematic diagram of the Gatan model 666 parallel-detection system, showing the prespectrometer focusing and alignment coils, magnetic-prism spectrometer, quadrupole array (Q1–Q4), and YAG scintillator, which is fiber-optically coupled to a linear photodiode array (O. Krivanek, personal communication).

eter dispersion is increased by quadrupole lenses Q1–Q4. The main components of an indirect exposure system will now be discussed in sequence.

Magnification of the Dispersion

The spatial resolution (interdiode spacing) of a typical PDA is 25 μm, while the energy dispersion of a compact magnetic spectrometer is of the order of 2 μm/eV for 100-keV electrons. To achieve a resolution better than 12 eV, it is therefore necessary to magnify the spectrum onto the detector plane. A round lens can be used for this purpose (Johnson *et al.*, 1981c) but it introduces a magnification-dependent rotation of the spectrum unless compensated by a second lens (Shuman and Kruit, 1985).

A magnetic quadrupole lens provides efficient and rotation-free focusing in the vertical (dispersion) plane but no focusing in the horizontal direction, so a line spectrum is produced. In fact, a line spectrum is preferable because it involves lower current density and therefore less risk of radiation damage to the scintillator, besides averaging out local variations in light output due to imperfections in the scintillator and optical coupling. The simplest system therefore consists of a single quadrupole (Egerton and Crozier, 1987), but the vertical magnification is limited to about 15 if the horizontal width is not to exceed the width (2.5 mm) of a large-aperture photodiode array.

The Gatan parallel-recording spectrometer system uses four quadru-poles, allowing the final dispersion to be varied while keeping the horizontal width at the array constant (for a given spectrometer entrance aperture). Quadrupole Q2 focuses in the horizontal plane, transferring the horizontal crossover produced by the spectrometer at Q1 onto the midplane of Q3; see Fig. 2.26a. Therefore Q1 and Q3 act as *field lenses* in the nondispersion plane; they have practically no effect on the horizontal width of the final spectrum. Q3 is varied to change the final dispersion (using the vertical crossover as a virtual object; Fig. 2.26b) and Q1 is readjusted by a small amount to ensure that the spectrum remains focused. Q4 adjusts the hori-zontal width of the spectrum, helps to increase dispersion, and gives greater flexibility under low-dispersion conditions. To provide convenient pushbut-ton operation, the quadrupole currents are controlled by a microprocessor.

Other quadrupole designs (Scott and Craven, 1989; Stobbs and Boothroyd, 1991; McMullan *et al.*, 1992) allow the spectrometer to form a real image, at which an energy-selecting slit can be introduced in order to perform serial recording or energy-filtered imaging if necessary.

A simple way of ensuring adequate energy resolution, without the use of magnifying optics, is to tilt the scintillator and diode array through an angle ψ_d, thereby increasing the dispersion by a factor $\sec(\psi_d)$. However, this would lead to loss of focus across the array unless the spectrometer is designed to have its dispersion plane at the same angle.

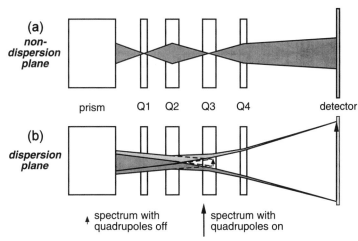

Figure 2.26. Electron trajectories in (a) the horizontal plane and (b) the vertical (dispersion) plane of a four-quadrupole dispersion-magnification system (Krivanek *et al.*, 1987).

Conversion Screen

The fluorescent screen used in a parallel-recording system performs the same function as the scintillator in a serial-recording system and has similar requirements in terms of sensitivity and radiation resistance (see page 85), but, since it is an *imaging* device, spatial resolution and uniformity are also important. Good resolution is achieved by making the scintillator thin, thereby reducing lateral spreading of an incident electron beam. Resolution is specified in terms of a line-spread or point-spread function (PSF), which is the response of the detector to an electron beam whose width is less than the interdiode spacing. The modulation transfer function (MTF) is the Fourier transform of the PSF and represents the response of the scintillator to sinusoidally varying illumination of varying spatial frequency.

Uniformity is most easily achieved by use of an amoprhous material, such as NE 102 plastic, or a single-crystal such as CaF_2, NaI, or Ce-doped yttrium aluminum garnet (YAG). Since organic materials and halides suffer radiation damage under a focused electron beam, YAG has become a preferred conversion-screen material in parallel-recording spectrometers (Krivanek *et al.*, 1987; Egerton and Crozier, 1987; Strauss *et al.*, 1987; Batson, 1988; Yoshida *et al.*, 1991; McMullan *et al.*, 1992), and CCD-camera electron-imaging systems (Daberkow *et al.*, 1991; Ishizuka, 1993; Krivanek and Mooney, 1993).

Single-crystal YAG is uniform in its light-emitting properties, emits yellow light to which a silicon-diode array is highly sensitive, and is relatively resistant to radiation damage (see Table 2.2). It can be thinned to 50 μm or less and polished by standard petrographic techniques. The YAG can be bonded directly to a fiber-optic plate, using a material of high refractive index to ensure good transmission of light in the forward direction. Even so, some light is multiply reflected between the two surfaces of the YAG and may travel some distance before entering the array, giving rise to extended tails on the point-spread function; see Fig. 2.27a. These tails can be reduced by incorporating an antireflection coating between the front face of the YAG and its aluminum coating. Long-range tails on the PSF are indicated by the low-frequency behavior of the MTF; see Fig. 2.27b. Similar measurements have been made on CCD-imaging systems (Kujawa and Krahl, 1992; Krivanek and Mooney, 1993).

Powder phosphors can be more efficient than YAG, and light scattering at grain boundaries reduces the multiple internal reflection which gives rise to the tails on the point-spread function. However, variations in light output between individual grains give rise to fixed-pattern noise and a consequent reduction in effective DQE (Daberkow *et al.*, 1991).

Backscattering of electrons from the conversion screen reduces the detective quantum efficiency of the system (see page 89) and subsequent

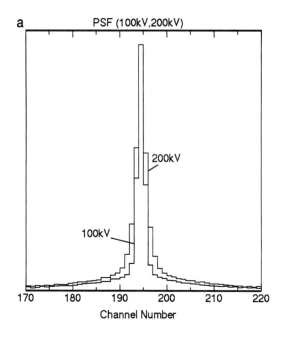

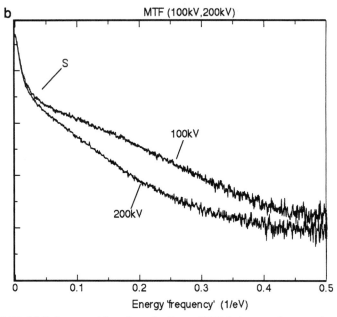

Figure 2.27. (a) Point-spread function of a Gatan PEELS detector (measured as the zero-loss peak with no specimen and 2-eV/channel dispersion) for 100-kV and 200-kV accelerating voltage, showing the narrow central peak and extended tails. (b) Modulation transfer function, evaluated as the square root of the PSF power spectrum; the rapid fall to the shoulder S arises from the PSF tails (Egerton *et al.*, 1993).

scattering of these electrons back to the scintillator gives rise to a *spectrometer* background, analogous to that encountered in serial recording (Section 2.4.2). This background augments the *detector* background which arises from diode leakage current. Whereas the latter is measured (for subtraction from each recorded spectrum) by excluding electrons from the spectrometer, the spectrometer background can be measured with an electron beam but no specimen, and a scaled version subtracted if necessary (Egerton *et al.*, 1993). In the Gatan PEELS system, the spectrometer background is reduced by absorbing the intense zero-loss beam at a beam-trap aperture (see Fig. 2.25) when the zero-loss peak is positioned off the array.

Light Coupling to the Array

A convenient means of transferring the conversion-screen image (representing variation in energy-loss intensity) to the diode array is by imaging (coherent) fiber optics. The resulting optical system requires no focusing, has a large f-number ($\approx f/1$), has no field aberrations (e.g., distortion) and is compact and rigid, thereby minimizing its sensitivity to mechanical vibration. The fiber-optic plate can be bonded with transparent adhesive to the scintillator and with silicon oil to the diode array; minimizing the differences in refractive index reduces light loss by internal reflection at each interface.

Fiber-optic coupling is less satisfactory for electrons of higher energy (>200 keV), some of which penetrate the scintillator and cause radiation damage (darkening) of the fibers or generate x-rays which could damage a nearby diode array. Some electrons are backscattered from the fiber plate, causing light emission into adjacent diodes and thereby augmenting the tails on the response function (Gubbens *et al.*, 1991). These problems are alleviated by using a self-supporting scintillator and glass lenses to transfer the image from the scintillator to the array. Lens optics allows the sensitivity of the detector to be varied (by means of an aperture stop) and makes it easier to introduce magnification or demagnification, so that the resolution of the conversion screen and the detector can be matched in order to optimize the energy resolution and DQE (Batson, 1988). However, the light coupling is less efficient, resulting in decreased noise performance of the system at low incident intensities.

2.5.3. Direct Exposure Systems

Although diode arrays are designed to detect visible photons, they also respond to charged particles such as electrons. In fact, a single 100-keV electron generates about 27,000 electron–hole pairs within the silicon, well above the readout noise, allowing a directly exposed photodiode or

CCD array (with glass window removed) to achieve high DQE at low electron intensities (Egerton, 1984a). At very low intensity (less than one electron/diode within the integration period) there is even the possibility of operation in an electron-counting mode.

This high sensitivity could be a disadvantage, since the saturation charge of even a large-aperture PDA is equivalent to only a few hundred directly incident electrons, giving a dynamic range $\approx 10^2$ for a single readout. However, the sensitivity can be reduced by shortening the integration time; by accumulating 10^4 readouts, the dynamic range could be increased to 10^6 (Egerton, 1984a). It should be possible to continuously read out a PDA with a clocking frequency ≈ 1 MHz, allowing energy losses above 100 eV to be recorded with an incident-beam current of 1 nA. But to record the *entire* spectrum with a reasonable incident-beam current (>1 pA), a dual system might be necessary, serial recording being employed to record the low-loss region and zero-loss peak (Bourdillon and Stobbs, 1986).

Direct exposure systems have not been exploited commercially, largely because of the possibility of radiation damage to the array. To prevent rapid damage to field-effect transistors located along the edge of a photodiode array, this area (and preferably the edges of the diodes) must be masked from the beam (Jones *et al.*, 1982). Even with appropriate masking, radiation damage causes a gradual increase in dark current, resulting in increased diode shot noise and reduced dynamic range; eventually, diode discharge by the dark current would reduce the dynamic range to zero (Shuman, 1981). The damage mechanism is believed to involve creation of electron–hole pairs within the SiO_2 passivating layer which covers the diodes (Snow *et al.*, 1967). The electrons have higher mobility than the holes; with voltages applied to the array, they drift out of the oxide, leaving it with a positive charge which inverts the underlying *p*-type silicon, creating a transition region of very small width and high leakage current. The damage rate is found to be higher at 20-keV incident energy compared to 100 keV, consistent with effects taking place in the oxide layer (Roberts *et al.*, 1982). When bias voltages are removed, the electrons move back into the oxide and the device starts to recover, especially if the electron beam is left on (Egerton and Cheng, 1982).

Since the dark current diminishes with decreasing temperature, cooling the array reduces the symptoms of electron-beam damage. Measurements on a photodiode array cooled to $-30°C$ suggested an operating lifetime of at least 1000 hours (Egerton and Cheng, 1982); Jones *et al.* (1982) reported no observable degradation for an array kept at $-40°C$, provided the current density was below 10 $\mu A/m^2$. Photodiode arrays have been operated at temperatures as low as $-150°C$ for low-level photon detection (Vogt *et al.*, 1978).

An array which has suffered radiation damage can be at least partially revived by various forms of annealing. Thermal annealing for a few minutes at 300°C removes most of the oxide charge (Snow *et al.*, 1967; Roberts *et al.*, 1982). Irradiation by ultraviolet light or electrons also has an annealing effect (Gordon, 1967; Egerton and Cheng, 1982) and might make possible *in situ* restoration. Irradiation of a chemically thinned device from its back surface has been proposed as a way of avoiding oxide-charge buildup (Imura *et al.*, 1971; Hier *et al.*, 1979) and future changes in manufacturing technique could possibly lead to devices which are immune to electron-irradiation damage.

2.5.4. Noise Performance of a Parallel Detection System

Consider a one- or two-dimensional array which is uniformly irradiated by fast electrons, N being the average number recorded by each element during the integration period. Random fluctuations (electron-beam shot noise) will contribute a root-mean-square (rms) variation between channels of magnitude $\sqrt{N}$, according to Poisson statistics. However, the point-spread function of the detector may be wider than the interdiode spacing, so electrons arriving at a given location are spread over several output channels, reducing the recorded electron-beam shot noise to

$$N_b = N^{1/2}/s \qquad (2.38)$$

where s is a smoothing (or mixing) factor which can be determined experimentally (Yoshida, 1991; Ishizuka, 1993; Egerton *et al.*, 1993).

In the case of indirect recording, N fast electrons generate (on average) Np photons in the scintillator. If all electrons followed the same path, there would be a statistical variation $(Np)^{1/2}$ in the number of photons produced. In fact, each electron behaves differently; for example, some penetrate only a short distance before being backscattered and therefore produce significantly fewer photons. Allowing for channel mixing and dividing by p so that the photon noise component N_p is expressed in units of fast electrons, one obtains

$$N_p = s^{-1}N^{1/2}(\sigma_p/p) \qquad (2.39)$$

where σ_p/p is the fractional rms variation in light output. Monte Carlo simulations and measurements of the height distribution of photon pulses have given $\sigma_p/p \approx 0.31$ for a YAG scintillator which is thick enough to absorb the incident electrons, and $\sigma_p/p \approx 0.59$ for a 50-μm YAG scintillator exposed to 200-keV electrons (Daberkow *et al.*, 1991).

Each photon produces an electron–hole pair in the diode array, but random variation (shot noise) in the diode leakage current and electronic noise (whose components may include switching noise, noise on supply and ground lines, videoamplifier noise, and adc digitization error of an A/D converter) add a total readout noise N_r, expressed here in terms of fast electrons. It is also possible to include a term N_v, representing the rms fractional variation v in gain between individual diode channels, which arises from differences in sensitivity between individual diodes, nonuniformities of the optical coupling, and variations in sensitivity of the scintillator (large in the case of powder phosphor). Adding all noise components in quadrature, the total noise N_t is given by

$$N_t^2 = N_b^2 + N_p^2 + N_r^2 + N_v^2 \qquad (2.40)$$

From Eqs. (2.38)–(2.40), the signal/noise ratio (SNR) of the diode-array output is

$$\text{SNR} = N/N_t = N(N/s^2 + N\sigma_p^2 p^{-2} s^{-2} + N_r^2 + v^2 N^2)^{1/2} \qquad (2.41)$$

The signal/noise ratio increases with signal level N, tending to asymptotically to $1/v$. Measurements on a first-generation Gatan PEELS detector, based on a Hamamatsu S2304 PDA, gave $N_r = 60$ and a limiting SNR of 440, implying $v = 0.23\%$ (Egerton *et al.*, 1993). Since the measurements used broad illumination, this value of v could be a factor of 2 or 3 lower than for the case of spectra (see Section 2.5.5). The shape of the SNR versus N_i curve could be fitted quite well with $s = 5$, as shown by the solid curve in Fig. 2.28a.

As in the case of a serial-detector system, a measure of the noise performance of the detector is its DQE, defined as

$$\text{DQE} = [(\text{SNR})/(\text{SNR})_i]^2 \qquad (2.42)$$

$(\text{SNR})_i$ now represents the signal/noise ratio of an ideal detector having the *same energy resolution* (same PSF) but with $N_p = N_r = N_v = 0$, so that

$$(\text{SNR})_i = N/N_b = sN^{1/2} \qquad (2.43)$$

Making use of Eqs. (2.41)–(2.43), the detective quantum efficiency is

$$\text{DQE} = s^{-2}(\text{SNR})^2/N = 1/[1 + \sigma_p^2/p^2 + s^2 N_r^2/N + v^2 s^2 N] \qquad (2.44)$$

Electron-beam shot noise is represented by the unity term in brackets in Eq. (2.44); the other terms cause DQE to be less than 1. For low N, the third term becomes large and DQE is reduced by readout noise. At large N, the fourth term predominates and DQE is reduced as a result of gain variations. Between these extremes, DQE reaches a maximum, as illustrated in Fig. 2.28b. Note that if the smoothing effect of the PSF is ignored, the *apparent* DQE = $(\text{SNR})^2/N$ may exceed unity.

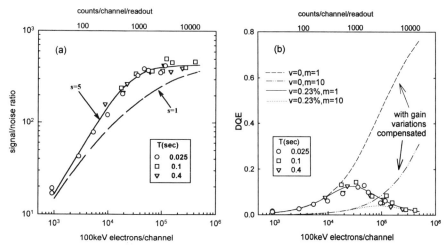

Figure 2.28. (a) Signal/noise ratio and (b) DQE for a Gatan PEELS linear-array detector, as a fuction of signal level (up to saturation). The experimental data are for three different integration times and are matched to a calculated curve with $s = 5$. Dashed curves show the reduced DQE expected when combining 10 readouts.

Provided the gain variations are reproducible, they constitute a fixed pattern which can be removed by suitable signal processing (Section 2.5.5), making the last term in Eq. (2.44) zero. The detective quantum efficiency then increases to a limiting value at large N, equal to $[1 + \sigma_p^2/p^2]^{-1} \approx 0.9$ for a 50-μm YAG scintillator (thick enough to absorb 100-keV electrons), and this high DQE allows the detection of small fraction changes in electron intensity, corresponding to elemental concentrations below 0.1 at.% (Leapman and Newbury, 1993).

Multiple Readouts

If a given recording time T is divided into m periods by accumulating m readouts in computer memory, the electron-beam and diode-array shot noise components are unaltered, since they depend only on the total integration time. The noise due to gain variations is also unaltered, since it depends only on the total recorded signal. But readout noise (exclusive of diode shot noise) is increased by a factor of $\sqrt{m}$, assuming that noise adds independently at each readout, increasing the N_r^2 term in Eq. (2.44) by a factor of m. As a result, the DQE is expected to be lower, as shown in Fig. 2.28b.

The advantage of multiple readouts is an increased dynamic range. The *minimum* signal which can be detected is $S_{min} = FN_t$, where N_t is the total noise and F is the minimum acceptable signal/noise ratio, often taken as 5 (Rose, 1970). Since some of the noise components are not increased by multiple readout, the total noise increases by a factor *less* than $\sqrt{m}$. The *maximum* signal which can be recorded in m readouts is

$$S_{max} = m(Q_{sat} - I_d T/m) = mQ_{sat} - I_d T \qquad (2.45)$$

where Q_{sat} is the diode-saturation charge and I_d is the diode thermal-leakage current. S_{max} is increased by more than a factor of m compared to the largest signal $(Q_{sat} - I_d T)$ which could be recorded in the same time with a single readout. Therefore the dynamic range S_{max}/S_{min} of the detector is increased by *more than* a factor of $\sqrt{m}$ by use of multiple readouts.

To obtain sufficient dynamic range with the least penalty in terms of readout noise, the integration time per readout should be adjusted so that the array output almost saturates at each readout. This procedure minimizes the number of readouts in a given acquisition period and optimizes the signal/noise ratio.

2.5.5. Dealing with Diode-Array Artifacts

The extended tails on the point-spread function affect all spectral features recorded with a diode array. Their effect can be largely removed by a deconvolution procedure (sharpening) involving division of the Fourier components of the spectrum and of the PSF, the latter being taken as the zero-loss peak. To avoid noise amplification, the central portion of the zero-loss peak can be used as a reconvolution function; see Section 4.1.1. The same procedure can be applied to images or spectra recorded using two-dimensional CCD arrays (Mooney *et al.*, 1993).

Some types of photodiode array suffer from incomplete discharge: each readout contains a partial memory of previous ones. Under these conditions, it is advisable to discard several readouts if acquisition conditions, such as the beam current or the region of specimen analyzed, are suddenly changed. For the Hamamatsu array used in the original Gatan PEELS system, the time constant associated with this latent-memory effect is about 2.5 readouts with the thermoelectric cooler on but less than one readout if the photodiode array is operated at room temperature. For an integration time T less than 0.5 s, room-temperature operation causes little increase (due to photodiode shot noise) in readout noise, as evidenced by the lack of T-dependence in Fig. 2.28.

Longer-term memory effects arise from electrostatic charging within the YAG, resulting in a local increase in sensitivity in regions of high electron intensity. This condition is alleviated by prolonged exposure to a broad, undispersed electron beam.

The thermal leakage current varies slightly between individual diodes, so the dark-current background is not quite constant across the spectrum. The result is a form of fixed-pattern noise, which is easily removed by subtracting a readout (scaled to the same integration time) recorded at the same array temperature but with the electron beam excluded from the detector. In photodiode arrays, another type of fixed pattern (repeating every two or four channels) arises from the method of internal scanning; it is independent of integration time and can be stored in computer memory (e.g., by subtracting the dark spectrum acquired in multiple readouts from a single readout acquired in the same total time) and subtracted from each recorded spectrum.

Gain Normalization

Variations in sensitivity (gain) across the array are more difficult to deal with. The simplest procedure is to record the response (gain spectrum) of the array when illuminated by a broad undispersed beam, then divide each real spectrum by the gain spectrum. This *gain-normalization* or *flat-fielding* procedure can achieve a precision of better than 1% for two-dimensional (CCD) detectors (Krivanek *et al.*, 1995). However, it makes only a small improvement in the case of one-dimensional photodiode arrays (compare curves a and b in Fig. 2.29) because sensitivity variations in the *horizontal* direction (due to nonuniformities in the YAG, in the optical

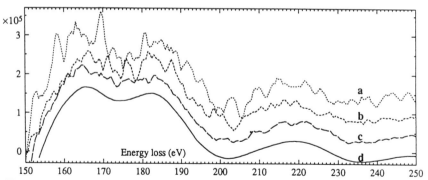

Figure 2.29. (a) Background-subtracted spectrum of Ni_3Al acquired from Gatan PEELS. (b) Same spectrum after dividing by a gain spectrum recorded with flat-field illumination. (c) Average spectrum $\bar{S}_1(E)$ obtained by shifting and adding eight of the original spectra. (d) Spectrum obtained from $\bar{S}_1(E)$ after two further iterations of gain averaging (Boothroyd *et al.*, 1990).

coupling or in the diodes themselves) are averaged out over the entire width of each diode during exposure to a broad beam but averaged over a smaller width (dependent on the diameter of the spectrometer entrance aperture) when recording a spectrum.

To remove the effect of gain variations from an ionization edge, so that its fine structure can be analyzed, it is sometimes possible to shift the spectrum so as to record a structureless region of the pre-edge background, then divide the core-loss by the pre-edge spectrum (Bourdillon and Stobbs, 1986). This success of this procedure probably depends on the fact that the horizontal widths of the two spectra are approximately the same. The following more general methods have also been used to correct for gain nonuniformities introduced by linear photodiode arrays.

Difference Spectra

A widely implemented method (Shuman and Kruit, 1985) for reducing the effect of gain variations is to record up to three spectra, $J_1(E - \varepsilon)$, $J_2(E)$, and $J_3(E + \varepsilon)$, displaced in energy by applying a small voltage ε to the spectrometer drift tube. The data-acquisition computer calculates a first-difference $FD(E)$ or second-difference $SD(E)$ spectrum according to

$$FD(E) = J_1(E - \varepsilon) - J_2(E) \tag{2.46}$$

$$SD(E) = J_1(E - \varepsilon) - 2J_2(E) + J_3(E + \varepsilon) \tag{2.47}$$

Writing the original spectrum as $J(E) = G(E)[A + BE + C(E)]$, where $G(E)$ represents gain modulation by the detector, and assuming that ε is small gives

$$\begin{aligned} FD(E) &= G(E)[-B\varepsilon + C(E - \varepsilon) - C(E)] \\ &\approx G(E)[-B\varepsilon + \varepsilon^{-1}(dC/dE)] \end{aligned} \tag{2.48}$$

$$\begin{aligned} SD(E) &= G(E)[C(E - \varepsilon) - 2C(E) + C(E + \varepsilon)] \\ &\approx G(E)[\varepsilon^{-2}(d^2C/dE^2)] \end{aligned} \tag{2.49}$$

Because component A is absent from $FD(E)$, gain modulation of any constant background is removed when forming a first-difference spectrum. Likewise, gain modulation of any linearly varying "background" component is removed when forming $SD(E)$. Consequently, the signal/background ratio of genuine spectral features is enhanced. However, the resulting spectra, which resemble first and second derivatives of the original data, are highly sensitive to spectral fine structure, making quantitative treatment of the data more difficult. FD(E) can be integrated digitally, the integration constant A being estimated by matching to the original spectra, but the *statistical* noise content (excluding gain variations) in the jth channel of the integrated spectrum $J(j)$ is increased relative to a directly acquired spectrum by a factor of $[2(j - 1)]^{1/2}$ for $\varepsilon = 1$ channel.

Dynamic Calibration Method

Shuman and Kruit (1985) proposed an alternative method in which the gain $G(j)$ of the jth channel of the detector is calculated from two difference spectra, $J_1(i)$ and $J_2(i)$, shifted by one channel:

$$\frac{G(j)}{G(1)} = \prod_{i=2}^{i=j} J_1(i) \bigg/ \prod_{i=1}^{i=j-1} J_2(i) \tag{2.50}$$

where Π represents a product of the contents $J(i)$ of all channels between the stated limits of i. By multiplying one of the original spectra by $G(j)$, the effect of gain variations is removed without generating a first or second derivative. Unfortunately, the noise content of $J(j)$ is increased by a factor of $[2(j - 1)]^{1/2}$, as for integrated first-difference spectra.

Gain Averaging

This method of eliminating the effect of sensitivity variations was first applied to low-energy EELS (Hicks *et al.*, 1980) and involves scanning a spectrum along the array in one-channel increments, with a readout at each scan voltage. The resulting spectra are electronically shifted into register before being added together in data memory. Within the range of energy loss which has been sampled by *all* diodes within the array, the effect of gain variations should exactly cancel. If the array contains N elements and gain-corrected data is required in M channels, $N + M$ spectra must be accumulated. Because M may be greater than N, the recorded spectral range can be greater than that covered by the array. However, some electrons fall outside the array and are not recorded, so the system DQE is reduced by a factor of $N/(N + M)$ compared to parallel recording of $N + M$ spectra which are stationary on the array. The method has been implemented for a 512-channel photodiode array with M up to 300 (Batson, 1988); extensive computing was required to correct for the fact that the scan step was not equal to the interdiode spacing. The same procedure has been used with a smaller number of readouts ($M < N$), in which case the gain variations are reduced but not eliminated; see Fig. 2.29c.

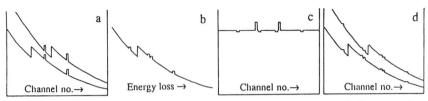

Figure 2.30. Principle of iterative gain averaging (Boothroyd *et al.*, 1990), illustrated for $M = 2$. (a) Recorded spectra, showing two channels with higher gain; (b) initial result $\bar{S}_1(E)$ of shifting and averaging; (c) first estimate $G_1(E)$ of the gain profile; (d) first iteration $S_m^1(E)$ obtained by dividing (b) by (c).

Iterative Gain Averaging

A variation on the above method is to record M spectra, each of the form $J_m(E)G(E)$, with successive energy shifts Δ between each spectrum. Two such spectra are shown in Fig. 2.30a. The spectra are then electronically shifted back into register and added to give a single spectrum:

$$\bar{S}_1(E) = \frac{1}{M} \sum_{m=0}^{m=M} J_m(E)G(E - m\Delta) \tag{2.51}$$

where the gain variations are spread over the spectrum (Fig. 2.30b and Fig. 2.29c) and reduced in amplitude by a factor $\approx \sqrt{M}$. Each original spectrum is then divided by $\bar{S}_1(E)$ and the result averaged over all M spectra:

$$G_1(E) = \frac{1}{M} \sum_m \frac{S_m(E)}{\bar{S}_1(E)} = \sum_m \frac{J_m(E)G(E)}{\bar{S}_1(E)} \tag{2.52}$$

to give a first estimate $G_1(E)$ of the gain profile of the detector (Fig. 2.30c) so that each original spectrum can be corrected for gain variations:

$$S_m^1(E) = \frac{S_m(E)}{G_1(E)} = \frac{J_m(E)G(E)}{G_1(E)} \tag{2.53}$$

The process can then be repeated, with the M spectra $S_m(E)$ replaced by $S_m^1(E)$, to obtain revised data $\bar{S}_2(E)$, $G_2(E)$, and $S_m^2(E)$, and the process repeated until the effect of gain variations becomes negligible; see Fig. 2.29d. Schattschneider and Jonas (1993) have analyzed the procedure in detail and shown that the variance due to gain fluctuations decreases proportional to the square root of the number of iterations.

2.6. Energy-Selected Imaging (ESI)

As discussed on page 31, the information carried by inelastic scattering can be acquired and displayed in several ways, one of these being the energy-loss spectrum. In a transmission electron microscope, the volume of material giving rise to the spectrum can be reduced to less than 10^{-20} by concentrating the incident electrons into a small-diameter probe (or by use of an area-selecting aperture), and the TEM image or diffraction pattern can be used to define this analyzed volume relative to its surroundings. Alternatively, energy analysis can be carried out in a way which uses directly the imaging or diffraction capabilities of the microscope. An image-forming spectrometer can be used as a filter which accepts energy losses within a specified range, giving an energy-selected image or diffraction pattern. In this section, we outline several experimental arrangements which achieve this.

2.6.1. Postcolumn Energy Filter

As discussed in Section 2.1.1, a magnetic-prism spectrometer acts some-what like an electron lens, producing a (chromatic) image of the spectrometer object O at the plane of an energy-selecting slit or diode-array detector. A conventional TEM provides a suitable small-diameter object at its projector-lens crossover, but it also produces a magnified image or diffraction pattern at the level of the viewing screen, closer to the spectrometer entrance. The spectrometer therefore forms a second image, further from its exit, which will correspond to an *energy-selected* image or diffraction pattern *if* an energy-selecting slit is inserted at the first-image (spectrum) plane.

In general, this energy-filtered image suffers from several defects. Its magnification is different in the x- and y-directions, although such rectangular distortion can easily be corrected electronically if the image is recorded into a computer. It exhibits axial astigmatism (the spectrometer is double-focusing only at the spectrum plane), which may be correctable by using the objective-lens stigmators (Shuman and Somlyo, 1982). More seriously, it is (like the energy-loss spectrum) a *chromatic* image, blurred in the x-direction by an amount dependent on the width of the energy-selecting slit. By using a 20-μm slit (giving an energy resolution of 5 eV) and a single-prism spectrometer of conventional design (see page 60), Shuman and Somlyo (1982) obtained a spatial resolution of 1.5 nm over a 2-μm field of view at the specimen plane.

The Gatan imaging filter (Fig. 2.31) uses postspectrometer quadrupoles and sextupoles to correct for these image defects. The first two quadrupoles

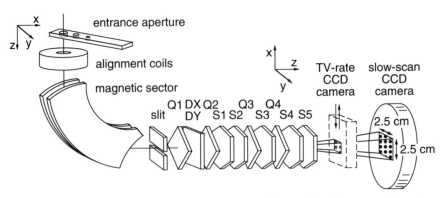

Figure 2.31. Gatan GIF 200 imaging filter, based on a 10-cm-radius electron spectrometer followed by quadrupole and sextupole optics (Krivanek *et al.*, 1991).

provide a modest increase ($\times 5$) in energy dispersion and ensure that the spectrum is double-focused at the energy-selecting slit. The remaining quadrupoles project onto the area-array CCD detector an energy-filtered and magnified ($\times 15$) version of the screen-level image or diffraction pattern, with energy dispersion corrected to first order. Five sextupole lenses (S1 to S5) correct for second-order aberrations and for geometric distortion. To compensate for the additional image magnification, the TEM projector lens is run at reduced excitation; as a result, the image resolution is more easily degraded by stray alternating magnetic fields below the projector lens (a possible problem if the viewing chamber is made of nonmagnetic material).

The excitation of quadrupoles Q3–Q6 can be adjusted to project the slit plane onto the CCD array, in order to focus or record the loss spectrum (with the slit wide open) or to inspect the region of energy loss defined by the slit. To minimize the effects of chromatic aberration of postspectrometer lenses, the selected energy loss is changed by varying the microscope high voltage. A retractable TV-rate CCD camera is used for rapid observations or adjustment of the optics; a slow-scan CCD camera allows permanent recording of images or spectra, with integration times up to several minutes.

The postcolumn filter has been simplified for use in biological applications, where the TEM is normally used at magnifications below 10^5 and with accelerating voltages not exceeding 120 kV, allowing only four quadrupoles to be used (Krivanek *et al.*, 1995a). The first-order electron optics are shown in Fig. 2.32, with the nondispersive $y–z$ plane flattened as in Fig. 2.2. In imaging mode, quadrupoles Q1 and Q2 focus (in both x- and y-directions) the spectrometer's achromatic plane onto the CCD detector.

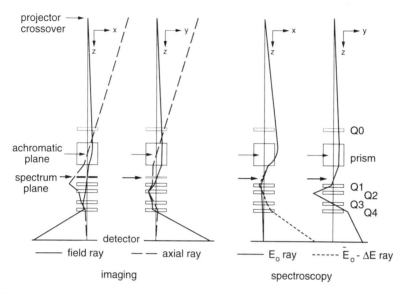

Figure 2.32. Electron trajectories through a Gatan GIF 100 energy filter in imaging and spectroscopy models.

By correcting second-order aberrations, the transmissivity of the filter (if defined as the product of maximum specimen area and collection solid angle for 1-eV resolution) is increased to 10^{-3} $(\mu m)^2$ sr. In spectroscopy mode, Q1 is strengthened so that its object becomes the energy-loss spectrum at the slit plane and Q2 is set to give an $x–y$ crossover in the center of the third quadrupole. Q3 can be adjusted to achieve a final dispersion between 0.05 and 1 eV/channel), Q4 being varied simultaneously to keep the horizontal (y-direction) width of the spectrum constant. A 1024×1024-element CCD camera provides 16-bit resolution in slow-scan mode and 8-bit resolution for rapid image readout (2 frame/s). To avoid the need for beam blanking, the frame-transfer direction of the CCD array is chosen to be perpendicular to the direction of energy dispersion (Strauss *et al.*, 1987). Because the pixel size is smaller than the phosphor resolution, each pixel is characterized uniquely by a "gain" value which can be measured by exposure to uniform illumination. The resulting gain-normalized spectra have high dynamic range (up 3×10^5 if 256-element columns are combined into a single readout channel) and DQE values should be higher than those obtainable from a linear photodiode array.

The Gatan design has also been adapted for use at primary energies up to 1.25 MeV by increasing the bending radius to 25 cm, replacing the

YAG scintillator with a thin layer of P20 or P43 phosphor deposited on the beam-exit side of a thin (<20 μm) aluminum foil, and incorporating lens and mirror coupling between the phosphor and CCD camera (Gubbens *et al.*, 1995). The imaging filter has also been modified for attachment to a 100-keV dedicated STEM in order to give energy-selected diffraction patterns (Krivanek *et al.*, 1994).

2.6.2. Prism–Mirror and Omega Filters

The Castaing–Henry and omega filters (Section 2.1.2) were developed specifically for producing energy-filtered images and diffraction patterns. Since they are located in the middle of a TEM column, they must be designed as part of a complete system rather than as add-on attachments to an existing microscope. Following the construction of such systems in several laboratories (Castaing *et al.*, 1967; Henkelman and Ottensmeyer, 1974; Egerton *et al.*, 1974; Zanchi *et al.*, 1977a; Krahl *et al.*, 1978), energy-selecting microscopes are now produced commercially (Egle *et al.*, 1984; Bihr *et al.*, 1991). Electron optics of the Zeiss omega-filter system is shown in Fig. 2.33; the various possible operating modes of this instrument have been described by Reimer (1991).

Resulting from the symmetry of these multiple deflection systems, the image or diffraction pattern produced is achromatic and has no distortion or axial astigmatism. Midplane symmetry also precludes second-order aperture aberrations and image-plane distortion (Rose, 1989). Reduction of axial aberration in the *energy-selecting plane* and of field astigmatism and tilt of the *final image* requires sextupoles or curved pole faces (Rose and Pejas, 1979; Jiang and Ottensmeyer, 1993) or optimization of the optical parameters (Lanio, 1986; Krahl *et al.*, 1990). Without such measures, the selected energy loss changes over the field of view: with a narrow energy-selecting slit and no specimen in the microscope, the zero-loss electrons form a pattern on the final screen which is equivalent to the alignment pattern of a single magnetic prism (Section 2.2.5); with an aluminum specimen, a low-magnification energy-selected image exhibits concentric rings corresponding to different multiples of the plasmon energy (Zanchi *et al.*, 1977a).

2.6.3. Energy Filtering in STEM Mode

Energy filtering is relatively straightforward in the case of a scanning transmission electron microscope, where the incident electrons are focused into a very small probe which is scanned over the specimen in the form of

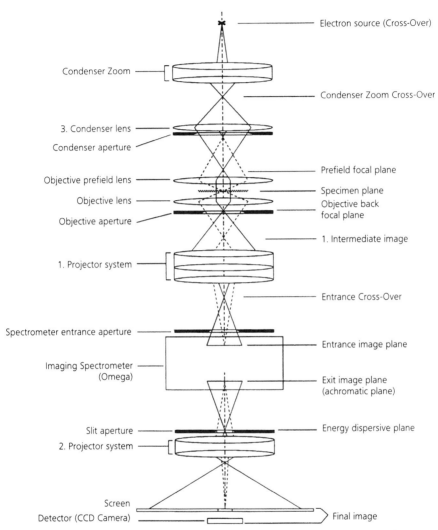

Figure 2.33. Zeiss EM912 energy-filtering microscope: solid lines show field-defining rays, dotted lines represent image-defining trajectories. The omega filter produces a unit-magnification achromatic image of the specimen image (or diffraction pattern) created by the first group of postspecimen lenses and generates (at the plane of the energy-selecting slit) a unit-magnification energy-dispersed image of its entrance crossover. From *Carl Zeiss Topics,* Issue No. 4, p. 4.

a raster pattern. A filtered image can be viewed simply by directing the output of a single-channel electron detector (as used in serial recording of the energy-loss spectrum) to the image-display monitor. If the image intensity needs processing, it is preferable to scan the probe digitally and store the detector signal as an array in computer memory.

In STEM mode, the spatial resolution of a filtered image is independent of the spectrometer; it is determined by the incident-probe diameter (spot size) and by scattering in the specimen. For an instrument fitted with a field-emission source, the probe diameter can be below 1 nm; in the case of a conventional TEM (fitted with a scanning attachment) using a tungsten filament or LaB_6 source, comparable beam current is available only for spot sizes above 10 nm.

If electron lenses are present between the specimen and spectrometer, the mode of operation of these lenses affects the available energy resolution (Section 2.3.4). In addition, the scanning action of the electron beam causes a corresponding motion of the spectrum, due to movement of the spectrometer object point (image coupling) or as a result of spectrometer aberrations (for diffraction coupling). Different regions of the image then correspond to different energy loss, which limits the field of view for a given range of energy loss. The field can be increased by widening the energy-selecting slit, but this limits the energy resolution to $MM_x d/D$ (see Section 2.3.5), where d represents the field of view. A preferable solution is to *descan* the electron beam by applying the raster signal to a second set of scan coils placed directly after the specimen (Fig. 2.34) or to focusing (dipole) coils of the spectrometer. Ideally, the field of view would then be unlimited and the energy resolution the same as for a stationary probe. Since there is no equivalent of descanning in the conventional TEM, the use of STEM mode makes it easier to obtain good energy resolution in low-magnification energy-filtered images.

The STEM spectrometer can also be used to obtain an energy-selected *diffraction pattern* by using x- and y-deflection coils to scan the latter across the spectrometer entrance aperture. However, this method is relatively inefficient, since a large proportion of the electrons are rejected by the angle-defining aperture. If the required energy resolution is $\Delta\beta$ and the scan range is $\pm\beta$, the collection efficiency (and system DQE) is $(\Delta\beta/\beta)^2/4$. Shorter recording times and lower radiation dose are possible by using an imaging filter to process the whole diffraction pattern simultaneously. Krivanek *et al.* (1994) describe a filter which is compatible with the vacuum requirements of a dedicated STEM and which can display, in addition to an energy-selected diffraction pattern, a line spectrum corresponding to an energy-dispersed section through the pattern.

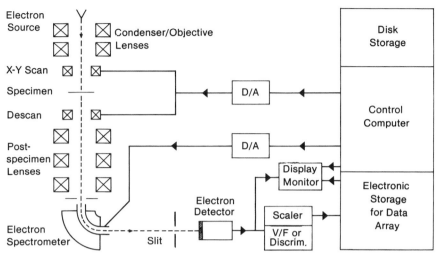

Figure 2.34. Energy-filtered imaging in scanning transmission mode. The energy loss is determined by an analog signal applied to the spectrometer power supply, to a postspectrometer deflection system or (as shown) to the spectrometer drift tube. Spectrum and specimen scanning can be computer-controlled, via digital-to-analog converters. In the case of a dedicated STEM, postspecimen lenses may be absent.

2.6.4. Spectrum Imaging

If the single-channel electron detector in Fig. 2.34 is replaced by a parallel-recording system and the energy-loss spectrum is read out at each picture point, a four-dimensional data array is created, which corresponds to electron intensity within a three-dimensional volume of (x, y, E) space (Fig. 2.1) and is referred to as a spectrum-image (Jeanguillaume and Colliex, 1989). The number of intensity values involved is large: 4M for a 64×64-pixel image or 250M for a 512×512 image, assuming a 1024-channel spectrum. If each spectral intensity is recorded to a depth of 16 bits (64K gray levels), the total information content for a 512×512 spectrum-image is 0.5 Gbyte if the data is stored as integers (without data compression). Recent advances in storage media have made the acquisition and storage of such data feasible: a 4-mm digital audio tape (DAT) allows storage of 1.3 Gbyte, while optical disks represent a faster (although currently more expensive) alternative.

A similar collection of data can be obtained from an imaging filter, by reading out a series of energy-selected images at different energy loss, to give what has been called an image-spectrum (Lavergne *et al.*, 1992). This process is less efficient than spectrum-imaging (electrons which are intercepted by the energy-selecting slit do not contribute) and therefore involves a higher electron dose to the specimen, for the same information content. However it may involve a shorter recording time, if the incident-beam current divided by the number of energy-selected images exceeds the probe current used in spectrum-imaging; see page 125.

The main attraction of the spectrum-image concept is that more of the information provided by electron/specimen interaction is recorded (only the angular dependence of scattering is ignored) and can be recalled into a data-handling computer for subsequent processing to extract information which might otherwise have been lost. Examples of such processing include the calculation of local thickness, pre-edge background subtraction, MLS fitting to reference spectra, Fourier-log and Fourier-ratio deconvolution, and Kramers–Kronig analysis. The resulting information can be displayed as line scans (Tencé *et al.*, 1995) or two-dimensional images (Botton and L'Esperance, 1994) of specimen thickness, elemental concentration, complex permittivity, and other quantities (Hunt and Williams, 1991). In addition, instrumental artifacts such as nonuniformities of the diode-array detector, drift of the microscope high voltage, and beam current can be corrected by postacquisition processing; specimen drift is best corrected "on the fly" during acquisition (Hunt and Williams, 1991).

The acquisition time of a spectrum-image is typically quite long. Direct memory access (DMA) can be used to speed up the transmission of data from the detector (see Fig. 2.35), but the minimum pixel time is limited by the array readout time: 25-ms readout gives a minimum frame time of 102 s for a 64×64 image (almost 2 hr for 512×512). Because spectrum channels are read out sequentially, continuous readout could result in a 1-pixel blurring of the image (Hunt and Williams, 1991). To eliminate this and to avoid array-memory effects (Section 2.5.5), the array can be cleared of old data by means of a dummy readout between each acquisition, although this further increases the frame time (and specimen dose) by a factor of 2. In order to accommodate the large dynamic range of a 1024-channel spectrum, a *prespecimen* beam blanker may be used to provide array-exposure times much shorter than the readout time, a procedure which is more efficient (in terms of specimen dose) than use of a postspectrometer "electron attenuator" (Hunt and Williams, 1991).

2.6.5. Elemental Mapping

An important form of energy-filtered image is obtained by selecting a range of energy loss (10 eV or more in width) corresponding to an inner-shell ionization edge. Since each edge is characteristic of a particular element in the specimen, the core-loss image provides information about

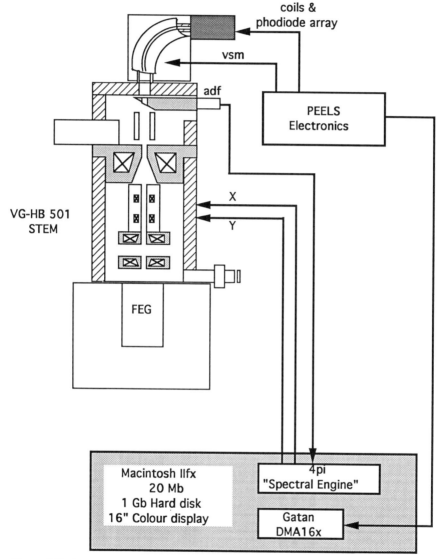

Figure 2.35. Spectrum-imaging system based on a field-emission STEM fitted with a parallel-recording spectrometer, interfaced via data-acquisition and beam-control boards to a personal computer (Tencé *et al.*, 1995). From *Ultramicroscopy*, M. Tencé, M. Quartuccio, and C. Colliex, PEELS compositional profiling and mapping at nanometer spatial resolution, © 1995, pp. 42–54, with permission from Elsevier Science B.V., Amsterdam, The Netherlands.

the spatial distribution of a known element. However, every ionization edge is superimposed on a spectral background due to other energy-loss processes. To obtain an image which represents the characteristic-loss intensity alone, the background contribution I_b within the core-loss region of the spectrum must be subtracted, as in the case of spectroscopy (Section 4.4). The background intensity often decreases smoothly with energy loss E, approximating to a power-law form $J(E) = AE^{-r}$, where A and r are parameters which can be determined by examining $J(E)$ at energy losses just below the ionization threshold (Section 4.4.2). Unfortunately, both A and r may vary across the specimen as a result of changes in thickness and composition (Leapman and Swyt, 1983; Leapman *et al.*, 1984c), so a separate estimation of I_b is required at every picture element (pixel).

In the case of STEM imaging, where each pixel is measured sequentially, local values of A and r can be obtained through a least-squares fit to the pre-edge intensity recorded over several channels preceding the edge. With electrostatic deflection of the spectrometer exit beam and fast electronics, the necessary data processing can be done within each pixel dwell period ("on the fly") and the system can provide a live display of the appropriate part of the spectrum (Gorlen *et al.*, 1984). Alternatively, A and r are estimated by measuring two or more energy-loss channels preceding the edge (Jeanguillaume *et al.*, 1978), for example using the procedure shown in Fig. 4.10. By recording a complete line of the picture before changing the spectrometer excitation, the need for fast beam deflection is avoided; background subtraction and fitting can be done off-line after storing the images (Tencé *et al.*, 1984).

In the case of energy filtering in a conventional TEM, the simplest method of background subtraction is to record one image at an energy loss just below the ionization edge of interest and subtract some constant fraction of its intensity from a second image recorded just above the ionization threshold. This was originally done by recording images on photographic film and using optical subtraction (Ottensmeyer and Andrew, 1980) or digital subtraction in computer memory, following densitometry (Bazett-Jones and Ottensmeyer, 1981). With an image-recording CCD camera feeding its output directly into a computer, image subtraction is more accurate and convenient. However, this simple procedure assumes that the exponent r which describes the *energy dependence* of the background intensity is constant across the image or that the background is always proportional to the core-loss intensity. In practice, r may vary as the local composition or thickness of the specimen changes (Fig. 3.35) and the background subtraction is no longer accurate (Leapman *et al.*, 1984c). This variation can be taken into account by electronically recording *two* background-loss images at slightly different energy loss and determining A and r at each pixel, as in STEM mode. However, the reduction in systematic error of background fitting occurs at the expense of an increased statistical error (Section 4.4.3), so for an acceptable signal/noise ratio each image requires a longer recording time (Leapman and Swyt, 1983; Pun and Ellis, 1983).

Influence of Diffraction Contrast

Even if the background intensity is correctly subtracted, the core-loss image may be modulated by diffraction (or aperture) contrast, which arises from variations in the amount of *elastic* scattering intercepted by the angle-limiting aperture. A simple test is to examine an *unfiltered* bright-field image, recorded using the same collection aperture; the intensity modulation in this image is a measure of the amount of diffraction contrast. Several methods have been proposed for removing aperture contrast, in order to obtain a true elemental map.

(a) Dividing the core-loss intensity I_k by a pre-edge background level, as suggested by Johnson (1979b), to form a jump-ratio image.

(b) Dividing by the intensity of a low-loss (e.g., first-plasmon) peak, which is also modulated by diffraction contrast.

(c) Dividing I_k by the intensity I_l measured over an equal energy window in the low-loss region of the spectrum (Egerton, 1978a; Butler *et al.*, 1982). According to Eq. (4.65), the ratio I_k/I_l is proportional to areal density (number of atoms of an analyzed element per unit area of the specimen).

(d) Taking a ratio of the core-loss intensities of two elements, giving an image which represents their elemental ratio; see Eq. (4.66).

(e) Recording the filtered images with a large collection semiangle, so that most of the inelastic and elastic scattering enters the spectrometer.

Methods (a), (b), and (d) have the advantage that they correct also for variations in specimen thickness (at least in very thin specimens), giving an image intensity which reflects the *concentration* of the analyzed element (atoms/volume) rather than its areal density.

2.6.6. Comparison of Energy-Filtered TEM and STEM

To compare the advantages of the fixed-beam TEM and STEM procedures of energy filtering, we will presume an equal collection efficiency (same β^*; see page 286) in each case and similar spectrometer performance in terms of energy resolution over the field of view. We assume electron detectors with similar noise properties, a reasonable assumption now that CCD detectors with DQE approaching unity are being used for recording both energy-filtered images and energy-loss spectra. We also take the spatial resolution to be the same in both methods, made plausible by the fact that a resolution of 1 nm is possible (for some specimens) using either procedure. Under these assumptions, the accumulated inelastic signal and the information content recorded from each resolution element will be identical, for the same incident-electron dose.

We begin by considering *elemental mapping* and compare the TEM imaging filter with a STEM system which has a *serial-recording* spectrometer. At any instant, a single energy loss is recorded, energy-selecting slits being present in both systems; see Fig. 2.36. For two-parameter background fitting, three complete images are acquired for each element in either mode, as discussed in the previous section. For the same amount of information, the *electron dose* is therefore the *same* in both methods. The only difference is that the electron dose is delivered continuously in the fixed-beam TEM but for only a small fraction of the frame time in STEM mode.

If a *field-emission* STEM is used, the dose *rate* (current density in the probe) can be considerably higher than in a TEM which uses a tungsten filament or LaB_6 electron source. Whether an equal dose implies the same amount of radiation damage to the specimen therefore depends on whether the radiation damage is dose-rate-dependent. For polymer specimens, temperature rise in the beam could increase the amount of structural damage

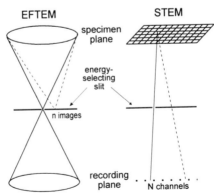

Figure 2.36. Comparison of TEM and STEM modes of energy-selected imaging. The STEM system can be envisaged either *with* an energy-selecting slit and single-channel detector (serial-recording spectrometer) or *without* the slit, allowing use of a diode-array detector (spectrum-imaging mode).

for a given dose (Payne and Beamson, 1993). In some inorganic oxides, mass loss (hole drilling) occurs only at the high current densities such as are possible in a field-emission STEM (see page 396). For such specimens, the STEM procedure could be more damaging for the same recorded information.

The *recording time* for a single-element map is generally longer in STEM, since the probe current (even with a field-emission source) is typically below 1 nA, whereas the beam current in conventional TEM can exceed 1 μA. STEM recording times of the order of 1 hour may be necessary to achieve adequate statistics in a 512×512-pixel image and may be inconvenient or expensive in terms of instrument time. The usual solution is to reduce the amount of information recorded by decreasing the number of pixels. In the conventional TEM, the recording time is independent of the number of pixels, so the advantage of EFTEM in terms of *information rate* is maximized by using a CCD camera with a large number of pixels or by recording the image on photographic film.

When a diode-array detector is used after the STEM spectrometer, the three images required for each element are recorded simultaneously, reducing both the time and the dose by a factor of 3. This factor becomes $3n$ in a case where n ionization edges are recorded simultaneously. The STEM with a *parallel-recording* spectrometer should therefore produce less radiation damage than an energy-filtering TEM, unless dose-rate effects outweigh this advantage.

The advantage of STEM is increased further when an extended range of energy loss is recorded by the diode array, as in spectrum-imaging. If use is made of the information recorded by all N detector channels, a given electron dose to the specimen yields N times as much information as in energy-filtering TEM, where N energy-selected images would have to be acquired sequentially to form an image-spectrum of equal information content. STEM spectrum-imaging also makes it easier to perform on- or off-line correction for specimen drift, so that long recording times (although inconvenient) do not necessarily compromise the spatial resolution of analysis. The acquisition time in STEM would actually be shorter if the product NI_p, where I_p is the probe current, exceeds the beam current used for EFTEM imaging.

Both scanning and fixed-beam modes of operation are possible in a CTEM fitted with a scanning attachment and (preferably) a field-emission source and parallel-recording spectrometer. With such an instrument, single-element imaging could be performed faster in EFTEM mode, but for multielement imaging or use of an extended range of energy-loss data, the STEM mode is more efficient in terms of specimens dose and (possibly) acquisition time.

2.6.7. Z-Ratio Imaging

A *Z*-ratio image is obtained by taking the ratio of a total-inelastic image and a dark-field image formed from electrons which have been scattered through larger angles. Although the procedure has been implemented in a conventional TEM (Ottensmeyer and Arsenault, 1983), it can be carried out more efficiently in a STEM. The inelastic intensity I_i is measured after the electron spectrometer, using a wide energy-selecting slit which excludes the zero-loss beam; the dark-field intensity I_d is recorded (without energy analysis) using an annular detector placed before the spectrometer; see Fig. 2.37. Because both signals are available simultaneously, signal division can be performed on-line by analog circuitry (Crewe *et al.*, 1975), but the two images can also be stored in computer memory and then digitally processed (Reichelt *et al.*, 1984).

Because of its relatively broad angular distribution (Section 3.2.1), elastic scattering contributes substantially to I_d, whereas the main contribution to I_i comes from electrons which have suffered only inelastic scatter-

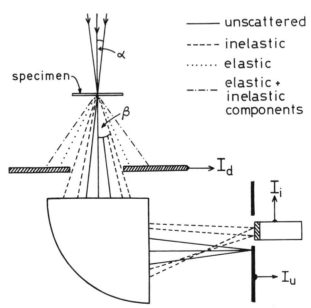

Figure 2.37. *Z*-contrast imaging in STEM mode. The inner radius of the annular detector subtends an angle β at the specimen, the outer radius being considerably larger. The electron spectrometer separates inelastic and unscattered electrons.

ing. If the specimen is amorphous and very thin, the amounts of elastic and inelastic scattering are proportional to specimen thickness and to the appropriate atomic cross sections σ_e and σ_i. On the assumption that all elastically scattered electrons are collected by the dark-field detector and all electrons which escape elastic scattering are analyzed by the spectrometer, the ratio signal is given by

$$R = I_d/I_i = \sigma_e/\sigma_i = \lambda_i/\lambda_e \qquad (2.54)$$

where λ_e and λ_i are the mean free paths for elastic and inelastic scattering. The intensity in the ratio image is therefore independent of specimen thickness (and incident-beam current) but is proportional to the local value of the elastic/inelastic scattering ratio which, according to the simple Lenz model of atomic scattering (Section 3.2.1), is proportional to atomic number Z. The ratio technique was first used to display images of mercury atoms ($Z = 80$) while suppressing unwanted contrast due to variations in mass-thickness of a very thin (<10 nm) carbon support film (Crewe _et al._, 1975). It has subsequently been used to increase the contrast and apparent resolution of thin (about 30-nm) sections of stained and unstained tissue (Carlemalm and Kellenberger, 1982; Reichelt _et al._, 1984).

In the case of specimens which are not thin in comparison with the elastic or inelastic mean free path, the dark-field and inelastic signals are given (Lamvik and Langmore, 1977; Egerton, 1982d) by

$$I_d = I[1 - \exp(-t/\lambda_e)] \qquad (2.55)$$

$$I_i = I \exp(-t/\lambda_e)[1 - \exp(-t/\lambda_i)] \qquad (2.56)$$

where I is the incident-beam current and t is the local thickness of the specimen. The exponential functions in Eqs. (2.55) and (2.56) occur because some electrons are scattered more than once. As a result of this plural scattering, neither I_i nor I_d is proportional to specimen thickness. In fact, I_i eventually decreases with increasing thickness (Fig. 2.38) because an increasing fraction of the inelastically scattered electrons undergoes elastic scattering and therefore contribute to I_d rather than to I_i. As a result, the ratio image become increasingly thickness dependent as t increases; see Fig. 2.38.

Equations (2.55) and (2.56) are approximations since, depending on the entrance semiangle β of the spectrometer and the width of the energy-selecting slit, not all of the inelastic scattering contributes to I_i and not all the elastic scattering to I_d. A Monte Carlo method can be used as a basis of more realistic calculations (Reichelt and Engel, 1984). For 100-keV electrons, $\beta \approx 10$ mrad offers reasonable discrimination between the elastic and inelastic scattering; for these conditions Monte Carlo calculations and

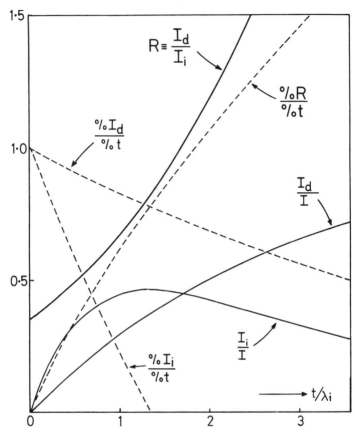

Figure 2.38. Thickness, dependence of the dark-field, inelastic, and ratio signals, together with the percentage change in these quantities for a 1% change in specimen thickness t (dashed curves). The calculations (Egerton, 1982d) assume that all elastically scattered electrons contribute to I_d, all electrons which are inelastically (but not elastically) scattered contribute to I_i, and that $\lambda_e/\lambda_i = 2.8$, typical of an organic specimen. For amorphous carbon $\lambda_i \approx 70$ nm for 100-keV incident electrons.

experimental measurements on polystyrene (Reichelt and Engel, 1984) give results similar to those of Fig. 2.38.

Even in thicker samples, it is possible to obtain an image whose intensity is proportional to the scattering ratio by combining I_i and I_d in a more complicated way, using logarithmic amplifiers (Egerton, 1982d) or off-line signal processing (Jeanguillaume *et al.*, 1992). The noise level may increase as a result of the larger number of mathematical operations required and the need to compute a difference between similar quantities (Reichelt *et al.*, 1984). In principle, a dark-field detector is not required for Z-ratio imaging, since

$$I_d = I - I_i - I_u \tag{2.57}$$

where I_u is the unscattered intensity, which could be measured as the zero-loss component of the spectrum by using a two-channel detector or a parallel-recording diode array. The probe current I could be measured as the value of I_d for no specimen in the beam. However, $I_u \approx I$ for very thin specimens, so the value of I_d derived from Eq. (2.57) would be highly sensitive to fluctuations in probe current. The advantage of a dark-field detector is that changes in probe current cancel when computing the ratio image.

Z-ratio imaging was originally investigated in connection with amorphous specimens. If the sample is crystalline, both the elastic and inelastic images are strongly influenced by diffraction contrast, which may increase rather than cancel if the ratio is taken (Donald and Craven, 1979). But by using a high-resolution STEM fitted with an annular detector which records only *high-angle* (>50 mrad) scattering of 100-keV electrons, individual columns of atoms can be resolved for a crystalline specimen set at a zone-axis orientation (Pennycook and Jesson, 1991). In addition, channeling of the incident electrons limits beam spreading, allowing core-loss spectroscopy to be used to identify the atomic number of each column (Browning *et al.*, 1993; Pennycook *et al.*, 1995a).

3

Electron Scattering Theory

It is convenient to divide the scattering of fast electrons into elastic and inelastic components. Experimentally, we can distinguish between the two on an empirical basis, the term *elastic* being taken to mean that the energy loss to the sample is less than some experimental resolution limit. Such a criterion results in electron scattering by phonon excitation being classified as elastic (or quasielastic) if measurements have been made using an electron microscope, where the energy resolution is rarely better than 0.5 eV.

3.1. Elastic Scattering

Elastic scattering represents interaction of incident electrons with the electrostatic field of an atomic nucleus. The atomic electrons are involved only to the extent that they terminate the nuclear field and therefore determine its range and magnitude. Because a nucleus is some thousands of times more massive than an electron, the energy transfer involved in elastic collisions is usually negligible. However, for the small fraction of electrons which are scattered through large angles (including head-on collisions for which the scattering angle $\theta = \pi$ radians), the transfer can amount to some tens of eV, as evidenced by the occurrence of displacement damage at high incident energies (Chadderton, 1965).

Although not a subject for direct study by electron energy-loss spectroscopy, elastic scattering is of relevance for the following reasons:

1. Because some electrons undergo both elastic and inelastic interactions within the sample, elastic scattering modifies the angular distribution of the inelastically scattered electrons.

2. In a crystalline material, elastic scattering can redistribute the electron flux (current density) within each unit cell and thereby alter the probability of certain types of inelastic scattering (see Section 3.1.4).
3. The intensity of elastic scattering (as a ratio of the inelastically scattered intensity) can provide an estimate of the local atomic number or chemical composition of a specimen (Section 2.6.7).

3.1.1. General Formulas

A quantity of basic importance in scattering theory is the differential cross section $d\sigma/d\Omega$, which represents the probability of an incident electron being scattered (per unit solid angle Ω) by a given atom. For elastic scattering, one can write

$$d\sigma/d\Omega = |f|^2 \qquad (3.1)$$

where f is the (complex) scattering amplitude or scattering factor, which is a function of the scattering angle θ or the scattering vector $\mathbf{q}$. The phase component of f is important in high-resolution phase-contrast microscopy (Spence, 1988a) but for the calculation of scattered intensity only the amplitude is required, as implied in Eq. (3.1). Within the first Born approximation (equivalent to assuming only single scattering within each atom), f is proportional to the three-dimensional Fourier transform of the atomic *potential* $V(r)$.

Alternatively, the differential cross section can be expressed in terms of an elastic *form factor* $F(q)$:

$$\frac{d\sigma}{d\Omega} = \frac{4}{a_0^2 q^4} |F(q)|^2 = \frac{4\gamma^2}{a_0^2 q^4} |Z - f_x(q)|^2 \qquad (3.2)$$

Here $a_0 = 4\pi\varepsilon_0\hbar^2/m_0 e^2 = 0.529 \times 10^{-10}$ m is the first Bohr radius and $\gamma = (1 - v^2/c^2)^{-1/2}$ is a relativistic factor (see Appendix E); $f_x(q)$ is the atomic scattering factor (or form factor) for an incident x-ray photon and is equal to the Fourier transform of the *electron density* within the atom. The atomic number Z which occurs in Eq. (3.2) represents the nuclear charge and denotes the fact that incident electrons are scattered by the entire electrostatic field of an atom whereas x-rays interact mainly with the atomic electrons.

3.1.2. Atomic Models

The earliest and simplest model for elastic scattering of charged particles is based on the unscreened electrostatic field of a nucleus and was first used by Rutherford to account for the scattering of α-particles seen by

Geiger and Marsden. Whereas an α-particle is repelled by the nucleus, an incident electron is attracted; by applying classical mechanics, the trajectories can be shown to be hyperbolic (Fig. 3.1). Both classical and wave-mechanical theory leads to the same expression for the differential cross section, which can be obtained by setting the electronic term $f_x(q)$ to zero in Eq. (3.2), giving

$$d\sigma/d\Omega = 4\gamma^2 Z^2/a_0^2 q^4 \tag{3.3}$$

Here q is the magnitude of the "scattering vector" and is given by $q = 2k_0 \sin(\theta/2)$, as illustrated in Fig. 3.2; $\hbar k_0 = \gamma m_0 v$ is the momentum of the incident electron and $\hbar q$ is the momentum transferred to the nucleus. For light elements, Eq. (3.3) is a reasonable approximation at large scattering angles (see Fig. 4.22) and can be useful for estimating rates of backscattering ($\theta > \pi/2$) in solids (Reimer, 1989). But because no allowance has been made for screening of the nuclear field by the atomic electrons, the Rutherford model greatly overestimates the elastic scattering at small θ (corresponding to large impact parameter b) and gives an infinite cross section if integrated over all angles.

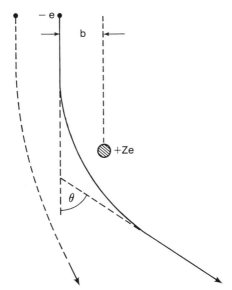

Figure 3.1. Rutherford scattering of an electron by the electrostatic field of an atomic nucleus, viewed from a classical (particle) perspective. Each scattering angle θ corresponds to a particular impact parameter b; as b increases, θ decreases. For small θ, $d\sigma/d\Omega$ is proportional to θ^{-4}.

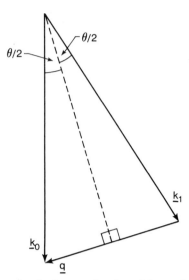

Figure 3.2. Vector diagram for elastic scattering; $\mathbf{k}_0$ and $\mathbf{k}_1$ are the wavevectors of the fast electron before and after scattering. From geometry of the right-angled triangles, the magnitude of the scattering vector is given by $q = 2k_0 \sin(\theta/2)$. Its direction has been taken as that of momentum transfer *to* the specimen (*opposite* to the wavevector change of the fast electron) so as to be identical with the quantity $\mathbf{q}$ involved in models of elastic and inelastic scattering.

The simplest way of incorporating screening is to assume a Wentzel (or Yukawa) expression in which the nuclear potential is attenuated exponentially as a function of distance r from the nucleus:

$$V(r) = (Ze/4\pi\varepsilon_0 r) \exp(-r/r_0) \tag{3.4}$$

where r_0 is a screening radius. Equation (3.4) leads to an angular distribution given by

$$\frac{d\sigma}{d\Omega} = \frac{4\gamma^2}{a_0^2}\left(\frac{Z}{q^2 + r_0^{-2}}\right)^2 \approx \frac{4\gamma^2 Z^2}{a_0^2 k_0^4}\frac{1}{(\theta^2 + \theta_0^2)^2} \tag{3.5}$$

where $\theta_0 = (k_0 r_0)^{-1}$ is the characteristic angle of *elastic* scattering. The fraction of elastic scattering which lies within the angular range $0 < \theta < \beta$ is (for $\beta \ll 1$ rad)

$$\frac{\sigma_e(\beta)}{\sigma_e} = \frac{1}{1 + [2k_0 r_0 \sin(\beta/2)]^{-2}} \approx \left(1 + \frac{\theta_0^2}{\beta^2}\right)^{-1} \tag{3.6}$$

Following Lenz (1954), an estimate of r_0 is obtained by taking the atomic radius given by the Thomas–Fermi statistical model:

$$r_0 = a_0 Z^{-1/3} \tag{3.7}$$

Integrating Eq. (3.5) over all scattering angles gives

$$\sigma_e = \int_0^{\pi} \frac{d\sigma}{d\Omega} 2\pi \sin\theta\, d\theta = \frac{4\pi\gamma^2}{k_0^2} Z^{4/3} = (1.87 \times 10^{-24}\, \text{m}^2) Z^{4/3}(v/c)^{-2} \qquad (3.8)$$

where v is the velocity of the incident electron and c is the speed of light in vacuum. For an element of low atomic number such as carbon, Eq. (3.8) gives cross sections which are accurate to about 10%. For a heavy element such as mercury, the Lenz model underestimates small-angle scattering by an order of magnitude (see Fig. 3.3), owing largely to the neglect of electron exchange; in this case, Eq. (3.8) gives (for 100-keV incident electrons) only about 60% of the value obtained from more sophisticated calculations (Langmore et al., 1973). Some authors use a coefficient of 0.885 in Eq. (3.7) or take $r_0 = 0.9 a_0 Z^{-1/4}$; however, the main virtue of the Lenz model is that it provides a rapid estimate of the *angular dependence* of scattering, as in the computer program listed in Appendix B.

More accurate cross sections are achieved by calculating the atomic potential from an iterative solution of the Schrödinger equation, as in the Hartree–Fock and Hartree–Slater methods (Ibers and Vainstein, 1962; Hanson et al., 1964). Alternatively, electron spin and relativistic effects within the atom can be included by using the Dirac equation (Cromer

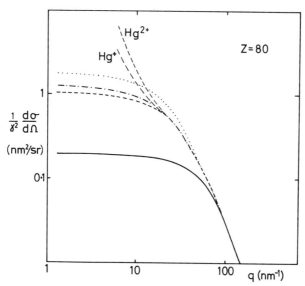

Figure 3.3. Angular dependence of the differential cross section for elastic scattering from a mercury atom, calculated using the Lenz model with a Wentzel potential (solid curve), and on the basis of Hartree-Fock (dotted curve), Hartree-Slater (chained curve), and Dirac-Slater (dashed curve) wavefunctions. Dirac-Slater results are also shown for the singly and doubly ionized atoms. (After Langmore et al., 1973, courtesy of Optik).

and Waber, 1965), leading to so-called Mott cross sections. Partial-wave methods may be used to avoid the Born approximation (Rez, 1984), which fails if Z approaches or exceeds $137(v/c)$, in other words for heavy elements or low incident energies.

Langmore *et al.* (1973) proposed the following equation for estimating the total elastic cross section of an atom of atomic number Z:

$$\sigma_e = \frac{(1.5 \times 10^{-24} \text{ m}^2)Z^{3/2}}{(v/c)^2} \left[1 - \frac{Z}{596(v/c)}\right] \qquad (3.9)$$

The coefficient and Z-exponent are based on Hartree–Slater calculations; the term in brackets represents a correction to the Born approximation. The accuracy of Eq. (3.9) is limited to about 30% because the graph of σ_e against Z is in reality not a smooth curve but displays irregularities which reflect the outer-shell structure of each atom; see Fig. 3.4. A compilation

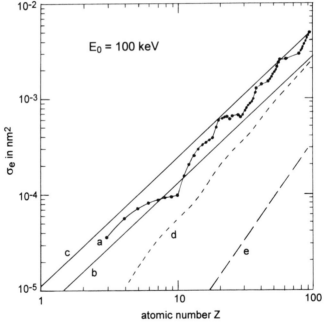

Figure 3.4. Cross section σ_e for elastic scattering of 100-keV electrons, calculated by Humphreys, Hart–Davis, and Spencer (1974). Curve (a) shows individual data points derived from Doyle–Turner scattering factors, based on relativistic Hartree–Fock wavefunctions. Lines (b) and (c) represent the Lenz model, with and without a multiplying factor of 1.8 added to Eq. (3.8). Curves (d) and (e) give cross sections $\sigma_e - \sigma_e(\beta)$ for elastic scattering *above* an angle β of 24 mrad and 150 mrad, respectively; note that the atomic-shell periodicity in the Z-dependence disappears as β becomes large and the scattering approximates to Rutherford collisions with small impact parameter, for which screening by valence electrons is unimportant.

of elastic cross sections ($d\sigma/d\Omega$ and σ_e) is given by Riley *et al.* (1975), based on relativistic Hartree–Fock wave functions.

For an *ionized* atom, the atomic potential remains partially unscreened at large r, so $d\sigma/d\Omega$ continues to increase with increasing impact parameter (i.e., decreasing θ); see Fig. 3.3. As a result, the amount of elastic scattering can appreciably exceed that from a neutral atom, particularly in the range of low scattering angles (Anstis *et al.*, 1973; Fujiyoshi *et al.*, 1982).

The scattering calculations which have been mentioned are all based on properties of a single isolated atom. In a molecule, the cross section per atom is reduced at low scattering angles, typically by 10%–20%, as a result of chemical bonding (Fink and Kessler, 1967). In a crystalline solid, the angular dependence of elastic scattering is altered dramatically by Bragg reflection from atomic planes. In *amorphous* solids, Bragg reflection is weak and an atomic model can be used as a guide to the magnitude and angular distribution of elastic scattering. As an alternative to describing the amount of scattering in terms of a cross section (σ_e per atom), one can use an inverse measure: $\lambda_e = (\sigma_e n_a)^{-1}$, where n_a is the number of atoms per unit volume of the specimen. The *mean free path* λ_e represents the mean distance between elastic collisions in the amorphous material.

3.1.3. Diffraction Effects

In crystalline materials, the regularity of the atomic arrangement requires that the phase difference between waves scattered from each atom be taken into account by introducing a *structure factor* $F(\theta)$ defined by

$$F(\theta) = \sum_j f_j(\theta) \exp(- i\mathbf{q} \cdot \mathbf{r}_j) \tag{3.10}$$

Here, r_j and f_j are the coordinate and scattering amplitude of atom j; $\mathbf{q} \cdot \mathbf{r}_j$ is the associated phase factor and the integration is carried out over all atoms ($j = 1, 2$, etc.) in the unit cell. Equation (3.10) can also be expressed in the form

$$F(\theta) \propto \int V(r) \exp(-\mathbf{q} \cdot \mathbf{r}) \, d\tau \tag{3.11}$$

where $V(r)$ is the scattering potential and the integration is carried out over all volume elements within a unit cell. Equation (3.11) indicates that the structure factor is related to the Fourier transform of the lattice potential.

The intensity scattered in a direction θ relative to the incident beam is $|F(\theta)|^2$ and peaks at values of θ for which the scattered waves are in phase with one another. Each diffraction maximum (Bragg beam) can also be regarded as arising by "reflection" from atomic planes, whose spacing d depends on the unit-cell dimensions and on the Miller indices. Bragg reflection occurs when the angle between the incident beam and the diffracting planes coincides with a Bragg angle θ_B defined by

$$\lambda = 2d \sin \theta_B \qquad (3.12)$$

where $\lambda = 2\pi/k_0$ is the incident-electron wavelength. The scattering angle θ is twice θ_B, so Eq. (3.12) is equivalent (for small θ_B, i.e., large v) to $\lambda = \theta d$. For 100-keV incident electrons, $\lambda = 3.7$ pm and the scattering angles corresponding to Bragg reflection usually exceed 10 mrad. Larger values of θ_B correspond to reflection from planes of smaller separation or to higher-order reflections whose phase difference is a *multiple* of 2π.

The Bragg-reflected beams can be recorded by a two-dimensional detector such as photographic film or a CCD array. For a single-crystal specimen, the diffraction pattern consists of an array of sharp spots whose symmetry and spacing are closely related to the crystal symmetry and lattice constants. In the case of a polycrystalline sample whose crystallite size is much less than the incident-beam diameter, random rotational averaging produces a diffraction pattern which consists of a series of concentric rings rather than a spot pattern.

In a crystalline solid, it is difficult to apply the concept of a mean free path for elastic scattering; the intensity of each Bragg spot depends on the crystal orientation relative to the incident beam and is not proportional to crystal thickness. Instead, each reflection is characterized by an *extinction distance* ξ_g, which is typically in the range 25–100 nm for 100-keV electrons and low-order (small θ_B) reflections. In the "ideal two-beam" case, where Eq. (3.12) is approximately satisfied for only one set of reflecting planes and where effects of *inelastic* scattering are negligible, 50% of the incident intensity is diffracted at a crystal thickness of $t = \xi_g/4$ and 100% at $t = \xi_g/2$. For $t > \xi_g/2$, the diffracted intensity *decreases* with increasing thickness and would go to zero at $t = \xi_g$ (and multiples of ξ_g) if inelastic scattering could be neglected. This oscillation of intensity gives rise to the "thickness" or Pendellösung fringes which are seen in transmission-microscope images.

3.1.4. Electron Channeling

Solution of the Schrödinger equation for an electron moving in a periodic potential results in wavefunctions known as Bloch waves, which are plane waves whose amplitude is modulated by the periodic lattice potential. Inside the crystal, a transmitted electron is represented by the sum of a number of Bloch waves, each having the same total energy. Because each Bloch wave propagates independently without change of form, this representation is sometimes preferable to describing electron propagation in terms of direct and diffracted beams whose relative amplitudes vary with the depth of penetration. In the ideal two-beam situation referred to in Section 3.1.3, there are only two Bloch waves, whose flux-density distributions (as a function of atomic coordinate) are illustrated in Fig. 3.5. The type-2 wave has its intensity maximum located halfway between the planes of Bragg-reflecting atoms and propagates parallel to these planes. The type-1 wave propagates in the same direction but has its current density peaked exactly *on* the atomic planes. Because of the attractive force of the atomic nuclei, the type-1 wave has a more negative potential energy and therefore a higher kinetic energy and larger wavevector than the type-2 wave. Resulting from this difference in wavevector between the Bloch waves, their combined intensity exhibits a "beating" effect, which provides an alternative but equivalent explanation for the occurrence of thickness fringes in the TEM image.

The relative amplitudes of the Bloch waves depend on the crystal orientation. For the two-beam case, both amplitudes are equal at the Bragg condition (Fig. 3.5), but if the crystal is tilted towards the "zone-axis" orientation (such that the angle between the incident beam and the atomic planes is less than the Bragg angle) more intensity occurs in Bloch wave 1. Conversely, if the crystal is tilted in the opposite direction, Bloch wave 2 becomes more prominent. Away from the Bragg orientation the current-density distributions within the Bloch waves become more uniform, such that they more nearly resemble plane waves.

Besides having a larger wave vector, the type-1 Bloch wave has a greater probability of being scattered by *inelastic* events occurring close to the center of an atom, such as inner-shell and thermal-diffuse (phonon)

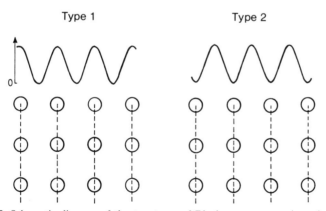

Figure 3.5. Schematic diagram of the two types of Bloch wave present in a simple cubic crystal at the exact Bragg-reflecting orientation, assuming two-beam conditions. The Bloch wave intensities are shown at the top of the diagram, in relation to the planes of reflecting atoms. (After Hirsch *et al.*, 1977, but following the more recent numbering scheme.)

excitation. Electron microscopists refer to this inelastic scattering as *absorption*, meaning that the scattered electron is absorbed by an angle-limiting "objective" aperture introduced to enhance image contrast or limit lens aberrations. The effect is incorporated into diffraction-contrast theory by adding to the lattice potential an imaginary component $V_0^i = \hbar v/2\lambda_i$, where λ_i is an appropriate inelastic mean free path. The variation of this "absorption" with crystal orientation is called anomalous absorption and is characterized by an imaginary potential V_g^i. In certain directions the crystal appears more "transparent" in a bright-field transmission microscope image; in other directions it is more opaque owing to increased inelastic scattering outside the objective aperture. This behavior is analogous to the Borrmann effect in x-ray penetration and is similar in many respects to the channeling of nuclear particles through solids. Anomalous absorption is also responsible for the Kikuchi bands which appear in the background to an electron-diffraction pattern (Kainuma, 1955; Hirsch *et al.*, 1977).

The orientation dependence of the Bloch-wave amplitudes also affects the intensity of inner-shell edges visible in the energy-loss spectrum. As the crystal is tilted through a Bragg orientation, an ionization edge will become either more or less prominent, depending upon the location (within the unit cell) of the atoms being ionized, relative to those which lie on the Bragg-reflecting planes (Taftø and Krivanek, 1982a). Inner-shell ionization is followed by deexcitation of the atom, involving the emission of Auger electrons or characteristic x-ray photons. So as a further result of the orientation dependence of absorption, the amount of x-ray emission varies with crystal orientation, provided the incident beam is sufficiently parallel (Hall, 1966; Cherns *et al.*, 1973). This variation in x-ray signal is utilized in the ALCHEMI method of determining the site of an emitting atom (Spence and Taftø, 1983).

In the more typical situation in which a number of Bragg beams are excited simultaneously, there are an equally large number of Bloch waves whose current-density distribution may be more complicated than in the two-beam case. The current density associated with each Bloch wave has a two-dimensional distribution when viewed in the direction of propagation; see Fig. 3.6.

Decomposition of the total intensity of the beam-electron wavefunction into a sum of the *intensities* of separate Bloch waves is a useful approximation for channeling effects in thicker crystals. In thin specimens, however, this *independent Bloch wave model* fails because of interference between the Bloch waves, as observed by Cherns *et al.* (1973). For even thinner crystals ($t \ll \xi_g$) the current density is uniform over the unit cell and the incident electron can be approximated by a plane wave, allowing atomic theory to be used to describe inelastic processes, as in Section 3.6.

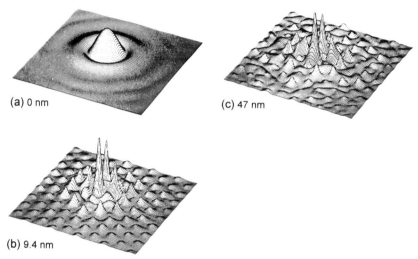

(a) 0 nm

(c) 47 nm

(b) 9.4 nm

Figure 3.6. Amplitude of the fast-electron wavefunction (square root of current density) for (a) 100-keV STEM probe randomly placed at the {111} surface of a silicon crystal, (b) the same electrons after penetration to a depth of 9.4 nm, and (c) after penetration by 47 nm. Channeling has largely concentrated the electron flux along $\langle 111 \rangle$ rows of Si atoms (Loane et al., 1988).

3.1.5. Phonon Scattering

Because of thermal (and zero-point) energy, atoms in a crystal vibrate about their lattice sites and this vibration acts as a source of electron scattering. An equivalent statement is that the transmitted electrons generate (and absorb) phonons while passing through the crystal. Since phonon energies are of the order of kT (k = Boltzmann's constant, T = absolute temperature) and do not exceed kT_D (T_D = Debye temperature), the corresponding energy losses (and gains) are below 0.1 eV and are not resolved by the usual electron-microscope spectrometer system. There is, however, a wealth of structure in the vibrational-loss spectrum, which can be observed using reflected low-energy electrons (Willis, 1980; Ibach and Mills, 1982).

Except near the melting point, the amplitude of atomic vibration is small compared to the interatomic spacing and (as a consequence of the uncertainty principle) the resulting scattering has a wide angular distribution. Particularly in the case of heavier elements, phonon scattering provides a major contribution to the diffuse background of an electron-diffraction pattern. This extra scattering occurs at the expense of the purely elastic scattering; the intensity of each elastic (Bragg-scattered) beam is reduced by the Debye–Waller factor: $\exp(-2M)$, where $M = 2(2\pi \sin \theta_B/$

Table 3.1. Values of Phonon Mean Free Path ($1/\mu_0$, in nm) Calculated by Hall and Hirsch (1965) for 100-keV Electrons

T (K)	Al	Cu	Ag	Au	Pb
300	340	79	42	20	19
10	670	175	115	57	62

$\lambda)^2\langle u^2\rangle$, λ is the electron wavelength, and $\langle u^2\rangle$ is the component of mean-square atomic displacement in a direction perpendicular to the corresponding Bragg-reflecting planes.

The *total* phonon-scattered intensity (integrated over the entire diffraction plane) is specified by an absorption coefficient $\mu_0 = 2V_0^i/(\hbar v)$, where V_0^i is the phonon contribution to the imaginary potential and v is the electron velocity. An inverse measure is the parameter $\xi_0^i/2\pi = 1/\mu_0$, which is roughly equivalent to a mean free path for phonon scattering. Typical values of $1/\mu_0$ are shown in Table 3.1) Unlike other scattering processes, phonon scattering is appreciably temperature dependent, increasing by a factor of 2–4 between 10 K and room temperature. Like elastic scattering, it increases with the atomic number Z of the scattering atom, roughly as $Z^{3/2}$; see Table 3.1. Seale and Scheinin (1993) have parameterized the angular dependence of phonon scattering in terms of Z-dependent atomic scattering factors calculated using the Einstein model.

Because the energy transfer is a fraction of an eV, phonon scattering is contained within the zero-loss peak and cannot be removed from TEM images or diffraction patterns by energy filtering. Inelastic scattering involving plasmon and single-electron excitation is concentrated at small angles (around each Bragg beam), but typically 20% or more occurs at scattering angles above 7 mrad (Egerton and Wong, 1995). Electronic scattering therefore contributes to the diffuse background of a diffraction pattern, and this component *is* removed by filtering since the average energy loss is some tens of eV. It varies only weakly with atomic number and for light elements ($Z < 13$) makes a larger contribution to the diffraction pattern than does phonon scattering (Eaglesham and Berger, 1994).

3.2. Inelastic Scattering

As discussed in Chapter 1, fast electrons are inelastically scattered through interaction with either outer- or inner-shell atomic electrons and these two processes predominate in different regions of the energy-loss spectrum. Before considering the separate contributions in detail, we deal briefly with theories which predict the *total* cross section for inelastic scattering by atomic electrons. In light elements, outer-shell scattering makes the dominant contribution to this cross section.

3.2.1. Atomic Models

For comparison with elastic scattering, we consider the angular dependence of the total inelastic scattering (integrated over all energy loss) as expressed by the differential cross section $d\sigma_i/d\Omega$. By modifying Morse's theory of elastic scattering, Lenz (1954) obtained a differential cross section which can be written in the form (Reimer, 1993)

$$\frac{d\sigma_i}{d\Omega} = \frac{4\gamma^2 Z}{a_0^2 q^4}\left\{1 - \frac{1}{[1 + (qr_0)^2]^2}\right\} \tag{3.13}$$

where $\gamma^2 = (1 - v^2/c^2)^{-1}$ and $a_0 = 0.529\times10^{-10}$ m is the Bohr radius; r_0 is a screening radius, defined by Eq. (3.4) for a Wentzel potential and equal to $a_0 Z^{-1/3}$ on a Thomas–Fermi model. The magnitude q of the scattering vector is given approximately by the expression

$$q^2 = k_0^2(\theta^2 + \bar{\theta}_E^2) \tag{3.14}$$

in which $k_0 = 2\pi/\lambda = \gamma m_0 v/\hbar$ is the magnitude of the incident-electron wavevector, θ is the scattering angle, and $\bar{\theta}_E = \bar{E}/(\gamma m_0 v^2)$ is a characteristic angle corresponding to the mean energy loss $\bar{E}$. Comparison with Eq. (3.3) shows that the first term $(4\gamma^2 Z/a_0^2 q^4)$ in Eq. (3.13) is simply a Rutherford cross section for scattering by the atomic electrons, taking the latter to be stationary free particles. The remaining term in square brackets is described as an inelastic form factor (Schnatterly, 1979).

Equations (3.13) and (3.14) can be combined to give a more explicit expression for the angular dependence (Colliex and Mory, 1984):

$$\frac{d\sigma_i}{d\Omega} = \frac{4\gamma^2 Z}{a_0^2 k_0^4}\frac{1}{(\theta^2 + \bar{\theta}_E^2)^2}\left\{1 - \left[\frac{\theta_0^4}{(\theta^2 + \theta_E^2 + \theta_0^2)^2}\right]\right\} \tag{3.15}$$

where $\theta_0 = (k_0 r_0)^{-1}$ as in the corresponding formula for elastic scattering, Eq. (3.5). Taking $r_0 = a_0 Z^{-1/3}$ and $\bar{E} \simeq 37$ eV (Isaacson, 1977) leads to the estimates $\bar{\theta}_E \simeq 0.2$ mrad and $\theta_0 \simeq 20$ mrad for a carbon specimen and incident energy $E_0 = 100$ keV.

In the angular range $\bar{\theta}_E < \theta < \theta_0$, which contains a large part of the scattering, $d\sigma/d\Omega$ is roughly proportional to θ^{-2}, whereas above θ_0 it falls off as θ^{-4}; see Fig. 3.7. The differential cross section therefore approximates to a Lorentzian function with an angular width $\bar{\theta}_E$ and a cutoff at $\theta = \theta_0$.

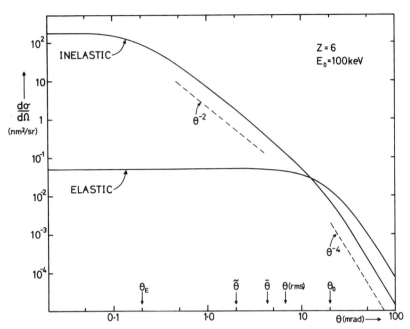

Figure 3.7. Angular dependence of the differential cross sections for elastic and inelastic scattering of 100-keV electrons from a carbon atom, calculated using the Lenz model, Eqs. (3.5), (3.7), and (3.15). Shown along the horizontal axis are (from left to right) the characteristic, median, mean, root-mean-square, and effective-cutoff angles for total inelastic scattering, evaluated using Eqs. (3.53)–(3.56).

On the basis of Bethe theory (Section 3.2.2), cutoff would occur at a mean Bethe-ridge angle $\bar{\theta}_r \approx \sqrt{(2\bar{\theta}_E)}$. In fact, these two angles are generally quite close to one another; for carbon and 100-keV incident electrons, $\theta_0 \approx \bar{\theta}_r \approx 20$ mrad. Using this value as a cutoff angle in Eq. (3.53) and Eq. (3.56), the mean and median angles of inelastic scattering can be estimated as $\bar{\theta} \simeq 20\bar{\theta}_E \simeq 4$ mrad and $\tilde{\theta} \simeq 10\bar{\theta}_E \simeq 2$ mrad for carbon and 100-keV incident electrons. The inelastic scattering is therefore concentrated into considerably smaller angles than the elastic scattering; see Fig. 3.7.

Integrating Eq. (3.15) up to a scattering angle β gives the integral cross section

$$\sigma_i(\beta) \approx \frac{8\pi\gamma^2 Z^{1/3}}{k_0^2} \ln\left[\frac{(\beta^2 + \theta_E^2)(\theta_0^2 + \theta_E^2)}{\theta_E^2(\beta^2 + \theta_0^2 + \theta_E^2)}\right] \qquad (3.16)$$

Extending the integration to all scattering angles, the total inelastic cross section is

$$\sigma_i \approx 16\pi\gamma^2 Z^{1/3}\ln(\theta_0/\theta_E) \approx 8\pi\gamma^2 Z^{1/3}\ln(2/\theta_E) \qquad (3.17)$$

where the Bethe-ridge angle $(2\theta_E)^{1/2}$ has been substituted for the effective cutoff angle θ_0 (Colliex and Mory, 1984). Comparison of Eqs. (3.8) and (3.17) indicates that

$$\sigma_i/\sigma_e \approx 2 \ln(2/\overline{\theta}_E)/Z = C/Z \qquad (3.18)$$

where the coefficient C is only weakly dependent on atomic number Z and incident energy E_0. Atomic calculations (Inokuti *et al.*, 1981) suggest that (for $Z < 40$) $\overline{E}$ varies by no more than a factor of 3 with atomic number; a typical value is $\overline{E} \simeq 40$ eV, giving $C \simeq 17$ for 50 keV and $C \simeq 18$ for 100-keV electrons. Experimental measurements on solids agree surprisingly well with these predictions; see Fig. 3.8. This simple Z-dependence of the inelastic/elastic scattering ratio has been used in the interpretation of scanning transmission ratio images obtained from very thin specimens (Crewe *et al.*, 1975; Carlemalm and Kellenberger, 1982).

More sophisticated atomic calculations differ from Eq. (3.17) in predicting a pronounced effect of the outer-shell structure on the value of σ_i. Instead of a simple power-law Z-dependence, the inelastic cross section reaches minimum values for the compact, closed-shell (rare gas) atoms and maxima for the alkali metals and alkaline earths, which have weakly bound

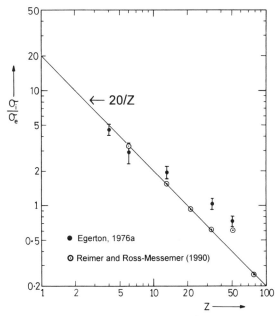

Figure 3.8. Measured values of inelastic/elastic scattering for 80-keV electrons, as a function of atomic number of the target. The solid line represents Eq. (3.18) with $C = 20$.

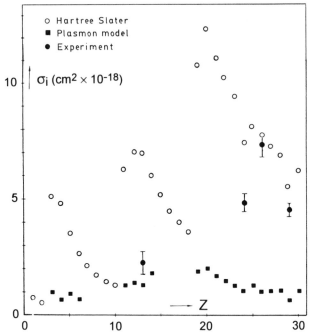

Figure 3.9. Total–inelastic cross section σ_i for 80-keV electrons, as a function of atomic number Z. Open circles show the results of atomic calculations (Inokuti *et al.*, 1975). The squares represent free-electron–plasmon cross sections, Eqs. (3.41) and (3.57), taking into account only the valence electrons. The experimental points for Al, Cr, Fe, and Cu (Munoz, 1983) include scattering up to 9 mrad.

outer electrons; see Fig. 3.9. Closer inspection reveals that in the latter case a large contribution comes from discrete (nonionizing) excitation of the valence shell (Inokuti *et al.*, 1975).

Closely related to the total inelastic cross section is the electron stopping power S, defined by (Inokuti, 1971)

$$S = \frac{dE}{dz} = n_a \overline{E} \sigma_i \qquad (3.19)$$

where E represents energy loss, z represents distance traveled through the specimen, $\overline{E}$ is the mean energy loss per inelastic collision, and n_a is the number of atoms per unit volume of the specimen. Atomic calculations (Inokuti *et al.*, 1981) show that $\overline{E}$ has a periodic Z dependence which largely compensates that of σ_i, giving S a relatively weak dependence on atomic number.

In low-Z elements, *inner* atomic shells contribute relatively little to σ_i, but they have an appreciable influence on the stopping power because the energy losses involved are comparatively large.

3.2.2. Bethe Theory

In order to describe more precisely the inelastic scattering of electrons by an atom (including the dependence of scattered intensity on energy loss), the behavior of each atomic electron must be specified in terms of transition from an initial state of wavefunction ψ_0 to a final state of wavefunction ψ_n. Using the first Born approximation, the differential cross section for the transition is (Inokuti, 1971)

$$\frac{d\sigma_n}{d\Omega} = \left(\frac{m_0}{2\pi\hbar^2}\right)^2 \frac{k_1}{k_0} \left| \int V(r)\psi_0\psi_n^* \exp(i\mathbf{q} \cdot \mathbf{r}) \, d\tau \right|^2 \qquad (3.20)$$

In Eq. (3.20), $\mathbf{k}_0$ and $\mathbf{k}_1$ are wavevectors of the fast electron before and after scattering, $\hbar\mathbf{q} = \hbar(\mathbf{k}_0 - \mathbf{k}_1)$ is the momentum transferred to the atom, $\mathbf{r}$ represents the coordinate of the fast electron, $V(r)$ is the potential responsible for the interaction, and the asterisk denotes complex conjugation of the wavefunction; the integration is over all volume elements $d\tau$ within the atom.

Below about 300 keV incident energy, the interaction potential representing electrostatic forces between an incident electron and an atom can be written as

$$V(r) = \frac{Ze^2}{4\pi\varepsilon_0 r} - \frac{1}{4\pi\varepsilon_0} \sum_{j=1}^{Z} \frac{e^2}{|\mathbf{r} - \mathbf{r}_j|} \qquad (3.21)$$

Although generally referred to as a potential, $V(r)$ is actually the negative of the potential *energy* of the fast electron and is related to the true potential ϕ by $V = e\phi$.

The first term in Eq. (3.21) represents Coulomb attraction by the nucleus (charge $= Ze$); the second term is a sum of the repulsive effects of each atomic electron, coordinate r_j. Because the initial- and final-state wavefunctions are orthogonal, the nuclear contribution integrates to zero in Eq. (3.20), so whereas the elastic cross section reflects both nuclear and electronic contributions to the potential, inelastic scattering involves only interaction with the atomic electrons. Because the latter are comparable in mass to the incident electron, inelastic scattering results in appreciable energy transfer. Combining Eqs. (3.20) and (3.21), the differential cross section can be written in the form

$$\frac{d\sigma_n}{d\Omega} = \left(\frac{4\gamma^2}{a_0^2 q^4}\right) \frac{k_1}{k_0} |\varepsilon_n(q)|^2 \qquad (3.22)$$

where the first term in parentheses is the Rutherford cross section for scattering from a single free electron, obtained by setting $Z = 1$ in Eq. (3.3). The second term (k_1/k_0) is very close to unity provided the energy loss is much less than the incident energy. The final term in Eq. (3.22), known as the *inelastic form factor* or *dynamical structure factor*, is the square of the absolute value of a transition-matrix element defined by

$$\varepsilon_n = \int \psi_n^* \sum_j \exp(i\mathbf{q} \cdot \mathbf{r}_j)\psi_0 \, d\tau = \langle \psi_n | \sum_j \exp(i\mathbf{q} \cdot \mathbf{r}_j) | \psi_0 \rangle \qquad (3.23)$$

Like the *elastic* form factor of Eq. (3.2), $|\varepsilon_n(q)|^2$ is a dimensionless factor which modifies the Rutherford scattering which would take place if the atomic electrons were free; it is a property of the target atom and is independent of the incident-electron velocity.

A closely related quantity is the *generalized oscillator strength* (GOS), defined (Inokuti, 1971) as

$$f_n(q) = \frac{E_n}{R} \frac{|\varepsilon_n(q)|^2}{(qa_0)^2} \qquad (3.24)$$

where $R = (m_0 e^4/2)/(4\pi\varepsilon_0\hbar)^2 = 13.6$ eV is the Rydberg energy and E_n is the energy loss associated with the transition. The differential cross section can therefore be written in the form

$$\frac{d\sigma_n}{d\Omega} = \frac{4\gamma^2 R}{E_n q^2} \frac{k_1}{k_0} f_n(q) \qquad (3.25)$$

where $(k_1/k_0) \simeq 1 - 2E_n(m_0 v^2)^{-1}$ can usually be taken as unity. In the limit $q \to 0$, the generalized oscillator strength $f_n(q)$ reduces to the dipole oscillator strength f_n which characterizes the response of an atom to incident photons (optical absorption).

In many cases (e.g., ionizing transitions to a "continuum") the energy-loss spectrum is a continuous rather than discrete function of the energy loss E, making it more convenient to define a GOS per unit excitation energy (i.e., per unit energy loss); $df(q, E)/dE$. The angular and energy dependence of scattering are then specified by a double-differential cross section:

$$\frac{d^2\sigma}{d\Omega dE} = \frac{4\gamma^2 R}{Eq^2} \frac{k_1}{k_0} \frac{df}{dE}(q, E) \qquad (3.26)$$

To obtain explicitly the angular distribution of inelastic scattering, the scattering vector q must be related to the scattering angle θ. For $\theta \ll 1$ rad and $E \ll E_0$, where E_0 is the incident-beam energy, it is a good approximation to take $k_1/k_0 = 1$ and to write

$$q^2 = k_0^2(\theta^2 + \theta_E^2) \qquad (3.27)$$

where the *characteristic angle* is defined by*

$$\theta_E = \frac{E}{\gamma m_0 v^2} = \frac{E}{(E_0 + m_0 c^2)(v/c)^2} \tag{3.28}$$

where v is the velocity of the incident electron. The resulting equation is

$$\frac{d^2\sigma}{d\Omega dE} \approx \frac{4\gamma^2 R}{E k_0^2}\left(\frac{1}{\theta^2 + \theta_E^2}\right)\frac{df}{dE} = \frac{8a_0^2 R^2}{E m_0 v^2}\left(\frac{1}{\theta^2 + \theta_E^2}\right)\frac{df}{dE} \tag{3.29}$$

At low scattering angles, the main angular dependence in Eq. (3.29) comes from the Lorentzian $(\theta^2 + \theta_E^2)^{-1}$ factor, θ_E representing the "half-width" of the angular distribution. This regime of small scattering angle and relatively low energy loss, where df/dE is almost constant (independent of q and θ), is known as the *dipole region*.

3.2.3. Dielectric Formulation

The equations given in Section 3.2.2 are most easily applied to single atoms or to gaseous targets, in the sense that the required generalized oscillator strength can be calculated (as a function of q and E) using an atomic model. Nevertheless, Bethe theory is also useful for describing the inelastic scattering which takes place in a solid, particularly from inner atomic shells. Outer-shell scattering is complicated by the fact that the valence-electron wave functions are modified by chemical bonding; in addition, collective effects are important, involving many atoms. An alternative approach is to describe the interaction of a transmitted electron with the entire solid in terms of a *dielectric response function* $\varepsilon(q, \omega)$. Although the latter can be calculated from first principles in only a few idealized cases, the same response function describes the interaction of photons with a solid, so this formalism allows energy-loss data to be correlated and compared with the results of optical measurements.

Ritchie (1957) has derived an expression for the electron-scattering power of an infinite medium. The transmitted electron, having coordinate **r** and moving with a velocity **v** in the z-direction, is represented as a point charge $-e\delta(\mathbf{r} - \mathbf{v}t)$ which generates within the medium a spatially dependent, time-dependent electrostatic potential $\phi(\mathbf{r}, t)$ satisfying Poisson's equation:

$$\varepsilon_0 \varepsilon(\mathbf{q}, \omega)\nabla^2\phi(\mathbf{r}, t) = e\delta(\mathbf{r}, t)$$

The stopping power (dE/dz) is equal to the force on the electron in the direction of motion, which is also the electronic charge multiplied by the

---

* For $E_0 < 150$ keV, $\gamma m_0 v^2 \cong 2E_0$ (see Appendix E) and $\theta_E = E/2E_0$ to 10% accuracy.

potential gradient in the z-direction. Using Fourier transforms, one can show that

$$\frac{dE}{dz} = \frac{2\hbar^2}{\pi a_0 m_0 v^2} \int\int \frac{q_y \omega \, \text{Im}[-1/\varepsilon(q, \omega)]}{q_y^2 + (\omega/v)^2} \, dq_y \, d\omega \qquad (3.30)$$

where the angular frequency ω is equivalent to $E/\hbar$ and q_y is the component of the scattering vector in a direction perpendicular to $\mathbf{v}$. The imaginary part of $[-1/\varepsilon(q, \omega)]$ is known as the *energy-loss function* and provides a complete description of the response of the medium through which the fast electron is traveling. The stopping power can be related to the double-differential cross section (per atom) for inelastic scattering by

$$\frac{dE}{dz} = \int\int n_a E \frac{d^2\sigma}{d\Omega dE} \, d\Omega dE \qquad (3.31)$$

where n_a represents the number of atoms per unit volume of the medium. By writing $dq_y \simeq k_0 \theta$ and $d\Omega \simeq 2\pi\theta \, d\theta$, Eqs. (3.30) and (3.31) give

$$\frac{d^2\sigma}{d\Omega dE} \approx \frac{\text{Im}[-1/\varepsilon(q, E)]}{\pi^2 a_0 m_0 v^2 n_a} \left(\frac{1}{\theta^2 + \theta_E^2}\right) \qquad (3.32)$$

where $\theta_E = E/(\gamma m_0 v^2)$ as before. Equation (3.32) contains the same Lorentzian angular dependence and the same v^{-2} factor as the corresponding Bethe equation, Eq. (3.29). Comparison of these two equations indicates that the Bethe and dielectric expressions are equivalent if

$$\frac{df}{dE}(q, E) = \frac{2E}{\pi E_a^2} \text{Im}\left[\frac{-1}{\varepsilon(q, E)}\right] \qquad (3.33)$$

where $E_a^2 = \hbar^2 n_a e^2/(\varepsilon_0 m_0)$, E_a being a "plasmon energy" corresponding to one free electron per atom (see Section 3.3.1).

In the small-angle (dipole) region, $\varepsilon(q, E)$ varies little with q and can be replaced by the optical value $\varepsilon(0, E)$, which is the relative permittivity of the specimen at an angular frequency $\omega = E/\hbar$. An energy-loss spectrum which has been recorded using a reasonably small collection angle can therefore be compared directly with optical data. Such a comparison involves a Kramers–Kronig transformation to obtain $\text{Re}[1/\varepsilon(0, E)]$, leading to the energy dependence of the real and imaginary parts (ε_1 and ε_2) of $\varepsilon(0, E)$, as discussed in Section 4.2. At large energy loss, ε_2 is small and ε_1 tends to unity; $\text{Im}(-1/\varepsilon) = \varepsilon_2/(\varepsilon_1^2 + \varepsilon_2^2)^{1/2}$ becomes proportional to ε_2 and (apart from a factor of E^{-3}) the energy-loss spectrum is proportional to the x-ray absorption spectrum.

The *optical* permittivity is a transverse property of the medium, in the sense that the electric field of an electromagnetic wave displaces electrons in a direction perpendicular to the direction of propagation, the electron density remaining unchanged. On the other hand, an incident electron produces a longitudinal displacement and a variation of electron density. The transverse and longitudinal dielectric functions are precisely equal only in the random phase approximation (see Section 3.3.1) or at sufficiently small q (Nozières and Pines, 1959); nevertheless, there is no evidence for a significant difference between them, as indicated by the close similarity of $Im(-1/\varepsilon)$ obtained from both optical and energy-loss measurements on a variety of materials (Daniels et al., 1970).

3.2.4. Solid-State Effects

If Bethe theory is applied to inelastic scattering of electrons in a *solid* target, one might expect the generalized oscillator strength (GOS) to differ from that calculated for a single atom, owing to the effect of chemical bonding on the wave functions and the existence of collective excitations (Pines, 1963). These effects change mainly the *energy dependence* of df/dE and of the scattered intensity; the angular dependence of inelastic scattering remains Lorentzian (with the same half-width θ_E), at least for small E and small θ.

Likewise, changes in the total inelastic cross section σ_i are limited to a modest factor (generally ≤ 3) because the GOS is constrained by the Bethe f-sum rule (Bethe, 1930):

$$\int \frac{df}{dE} dE = z \tag{3.34}$$

The integral in Eq. (3.34) is over all energy loss E (at constant q) and should be taken to include a sum over excitations to "discrete" final states, which in many atoms make a major contribution to the total cross section. If the sum rule is applied to a whole atom, z is equal to the total number Z of atomic electrons and Eq. (3.34) is exact. If applied to a single atomic shell, z can be taken as the number of electrons in that shell, but the rule is only approximate, since the summation should include a (usually small) negative contribution from "downward" transitions to shells of higher binding energy, which are in practice forbidden by the Pauli exclusion principle (Pines, 1963; Schnatterly, 1979).

Using Eq. (3.33), the Bethe sum rule can also be expressed in terms of the energy-loss function:

$$\int Im\left[\frac{-1}{\varepsilon(E)}\right] E \, dE = \frac{\pi \hbar^2 z n_a e^2}{2\varepsilon_0 m_0} = \frac{\pi}{2} E_p^2 \tag{3.35}$$

where $E_p = \hbar^2(ne^2/\varepsilon_0 m_0)^{1/2}$, is a "plasmon energy" corresponding to the number of electrons, n per unit volume, which contribute to inelastic scattering within the range of integration.

According to Eq. (3.29), the differential cross section (for $\theta \gg \theta_E$) is proportional to $E^{-1}df/dE$, so although df/dE is constrained by Eq. (3.34), the cross section σ_i (integrated over all energy loss and scattering angle) will decrease if contributions to the oscillator strength are shifted towards higher energy loss. This upward shift applies to most solids since the collective excitation of valence electrons generally involves energy losses which are higher than the average energy of *atomic* valence-shell transitions. As a rough estimate of the latter, one might consider the first ionization energy (Inokuti, personal communication); most often, the measured valence-peak energy loss is above this value, as shown in Fig. 3.10, so the valence-electron contribution to σ_i is smaller in the case of a solid. The resulting decrease in σ_i should be more marked for light elements where the valence shell accounts for a larger fraction of the atomic electrons and therefore makes a larger percentage contribution to the cross section (Inokuti *et al.*, 1981). One fairly extreme example is aluminum, where the inelastic cross section per atom is a factor of about 3 lower in the solid (Fig. 3.9), in rough

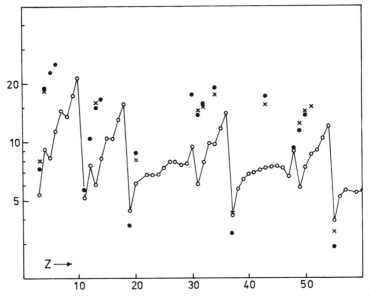

Figure 3.10. First-ionization energy E_i (open circles) and measured plasmon energy E_p (filled circles) as a function of atomic number. Where E_p exceeds E_i, an atomic model can be expected to overestimate the total amount of inelastic scattering. The crosses represent the plasmon energy calculated on a free-electron model.

agreement with the ratio of the plasmon energy (15 eV) and the atomic ionization threshold (6 eV).

Some solid compounds show collective effects which, although not negligible, are less pronounced so that inelastic scattering from the valence electrons retains much of its atomic character. As a rough approximation, one might then imagine each atom to make an independent contribution to the scattering cross section. The effect of chemical bonding is to remove valence electrons from electropositive atoms (e.g., Na, Ca) and reduce their scattering power, whereas electronegative atoms (O, Cl, etc.) have their electron complement and scattering power increased. On this basis, the periodic component of the Z-dependence of σ_i (Fig. 3.9), which is related to the occupancy of the outermost atomic shell, should be reduced in the case of solids. In any event, what is measured experimentally is the sum of the scattering from all atoms (anions and cations); if the reductions and increases in scattering power referred to above are equal in magnitude, the total scattering power will be given simply by a sum of the scattering powers calculated on an atomic model. This additivity principle (when applied to the stopping power) is known as Bragg's rule and is believed to hold to within $\approx 5\%$ accuracy (Zeiss *et al.*, 1977) except for the contribution from hydrogen, which is usually small anyway. It provides some justification for the use of atomic cross sections to calculate the stopping power and range of electrons in solids (Berger and Seltzer, 1982).

Fano (1960) has suggested that the magnitude of collective effects depends on the value of the dimensionless parameter:

$$u_F(E) = \left(\frac{\hbar^2 n e^2}{\varepsilon_0 m_0}\right) \frac{df}{d(E^2)} = \left(\frac{\hbar^2 n e^2}{2\varepsilon_0 m_0}\right) \frac{1}{E} \frac{df}{dE} \tag{3.36}$$

where n is the number of electrons per unit volume (with binding energies less than E) which can contribute to the scattering at an energy E. Collective effects can be neglected if $u_F \ll 1$, but are of importance where u_F *approaches or exceeds* unity within a particular region of the loss spectrum (Inokuti, 1979). Comparison with Eq. (3.33) shows that $u_F = \text{Im}[-1/\varepsilon(E)]/\pi$, so the criterion for neglecting collective excitations becomes

$$\text{Im}[-1/\varepsilon(E)] \ll \pi \tag{3.37}$$

Equation (3.37) provides a more convenient criterion for assessing the importance of collective effects, since any energy-loss spectrum which has been measured up to a sufficiently high energy loss can be normalized, using Eq. (3.35) or (4.27), to give the energy-loss function $\text{Im}[-1/\varepsilon]$, as described in Section 4.2. A survey of experimental data indicates that $\text{Im}[-1/\varepsilon]$ rises to values in the range 3 to 4 for InSb, GaAs, and GaSb (materials which support well-defined plasma oscillations), reaches 2.2 in diamond, and attains values close to unity in the case of Cu, Ag, Pd, and Au, where plasma oscillations are strongly damped (Daniels *et al.*, 1970). Organic solids are similar to the transition metals in the sense that their energy-loss function generally reaches values close to unity for energy losses around 20 eV (Isaacson, 1972a), implying that both atomic transitions and collective effects contribute to their low-loss spectra.

Ehrenreich and Philipp (1962) have proposed more definitive criteria for the occurrence of collective effects (plasma resonance), based on the energy dependence of the real and imaginary parts (ε_1 and ε_2) of the permittivity. According to these criteria, liquids such as glycerol and water, as well as solids such as aluminum, silver, silicon, and diamond, all respond in a way which is at least partly collective (Ritchie *et al.*, 1989).

3.3. Excitation of Outer-Shell Electrons

Most of the inelastic collisions of a fast electron arise from interaction with electrons in outer atomic shells and result in an energy loss of less than 100 eV. In a solid, the major contribution comes from valence electrons (referred to as conduction electrons in a metal), although in some materials (e.g., transition metals and their compounds) underlying shells of low binding energy contribute appreciable intensity in the 0–100 eV range.

We begin this section by considering plasmon excitation, an important process in most solids and one which exhibits several features not predicted by atomic models.

3.3.1. Volume Plasmons

The valence electrons in a solid can be thought of as a set of coupled oscillators which interact with each other and with a transmitted electron via electrostatic forces. In the simplest situation, the valence electrons behave essentially as free particles (although constrained by Fermi–Dirac statistics) and constitute a "free-electron gas," also known as a "Fermi sea" or "jellium." Interaction with the ion-core lattice is assumed to be a minor perturbation which can be incorporated phenomenologically by using an effective mass m for the electrons, different from their rest mass m_0, and by introducing a *damping constant* Γ or its reciprocal τ, as in the Drude theory of electrical conduction in metals. The behavior of the electron gas is described in terms of a dielectric function, just as in Drude theory. In response to an applied electric field, such as that produced by a transmitted charged particle, a collective oscillation of the electron density occurs at a characteristic angular frequency ω_p and this resonant motion would be self-sustaining if there were no damping from the atomic lattice.

Jellium Model

The displacement $\mathbf{x}$ of a "quasifree" electron (effective mass m) due to a local electric field $\mathbf{E}$ must satisfy the equation of motion

$$m\ddot{\mathbf{x}} + m\Gamma\dot{\mathbf{x}} = -e\mathbf{E} \tag{3.38}$$

For an oscillatory field: $\mathbf{E} = \mathbf{E}\exp(-i\omega t)$, Eq. (3.38) has the solution

$$\mathbf{x} = (e\mathbf{E}/m)(\omega^2 + i\Gamma\omega)^{-1} \tag{3.39}$$

The displacement $\mathbf{x}$ gives rise to a polarization $\mathbf{P} = -en\mathbf{x} = \varepsilon_0\chi\mathbf{E}$, where χ is the electronic susceptibility and n is the number of electrons per unit volume. The relative permittivity or "dielectric function" $\varepsilon(\omega) = 1 + \chi$ is then given by

$$\varepsilon(\omega) = \varepsilon_1 + i\varepsilon_2 = 1 - \frac{\omega_p^2}{\omega^2 + \Gamma^2} + \frac{i\Gamma\omega_p^2}{\omega(\gamma^2 + \Gamma^2)} \tag{3.40}$$

Here ω is the angular frequency (rad s^{-1}) of forced oscillation and ω_p is the natural or resonance frequency for plasma oscillation, given by

$$\omega_p = [ne^2/(\varepsilon_0 m)]^{1/2} \tag{3.41}$$

The *energy-loss function* is defined as

$$\mathrm{Im}\left[\frac{-1}{\varepsilon(\omega)}\right] = \frac{\varepsilon_2}{\varepsilon_1^2 + \varepsilon_2^2} = \frac{\omega\Gamma\omega_p^2}{(\omega^2 - \omega_p^2)^2 + (\omega\Gamma)^2} \tag{3.42}$$

A transmitted electron represents a sudden impulse of applied electric field, containing all angular frequencies (Fourier components). Setting up a plasma oscillation of the outer-shell electrons at an angular frequency ω_p is equivalent to creating a "pseudoparticle" of energy $E_p = \hbar\omega_p$, called the *plasmon* (Pines, 1963). As shown in Eq. (3.32), the energy dependence of the inelastically scattered intensity (the energy-loss spectrum) is proportional to $\mathrm{Im}[-1/\varepsilon(E)]$, in which the variable $E = \hbar\omega$ represents energy loss. It is therefore convenient to write Eq. (3.42) in the form

$$\mathrm{Im}\left[\frac{-1}{\varepsilon(E)}\right] = \frac{E_p^2(E\hbar/\tau)}{(E^2 - E_p^2)^2 + (E\hbar/\tau)^2} = \frac{E(\Delta E_p)E_p^2}{(E^2 - E_p^2)^2 + (E\,\Delta E_p)^2} \tag{3.43}$$

where E_p is known as the plasmon energy and $\tau = 1/\Gamma$ is a relaxation time. The energy-loss function $\mathrm{Im}(-1/\varepsilon)$ has a full width at half-maximum (FWHM) given by $\Delta E_p = \hbar\Gamma = \hbar/\tau$ and reaches a maximum value of $\omega_p\tau$ at an energy loss given by

$$E_{\max} = [(E_p)^2 - (\Delta E_p/2)^2]^{1/2} \tag{3.43a}$$

For a material such as aluminum, where the plasmon resonance is sharp ($\Delta E_p = 0.5$ eV), the maximum is within 0.002 eV of E_p, whereas, for the broad resonance found in carbon, Eq. (3.43a) implies that E_{max} is shifted to lower energy by about 2.1 eV. From Eq. (3.40), the energy at which $\varepsilon_1(E)$ passes through zero with positive slope (see Fig. 3.11) is

$$E(\varepsilon_1 = 0) = [(E_p)^2 - (\Delta E_p)^2]^{1/2} \tag{3.43b}$$

This zero crossing is sometimes taken as evidence of a well-defined collective response in the solid under investigation.

Although based on a simplified model, Eq. (3.43) corresponds well with the observed line shape of the valence-loss peak in materials which show a simple plasmon behavior, such as silicon and germanium (Hinz and Raether, 1979). Even with $m = m_0$, Eq. (3.41) provides fairly accurate values for the energy of the main peak in the energy-loss spectrum of many solids, taking n as the density of outer-shell electrons; see Fig. 3.12 and Table 3.2.

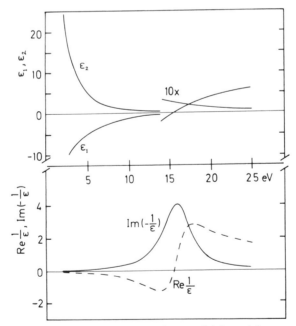

Figure 3.11. Real and imaginary parts of the relative permittivity and the energy-loss function $Im(-1/\varepsilon)$, calculated using a free-electron (jellium) model with $E_p = 15$ eV and $\hbar/\tau = 4$ eV (Raether, 1980). From *Excitations of Plasmons and Interband Transitions by Electrons*, pp. 84–157, © Springer-Verlag, by permission.

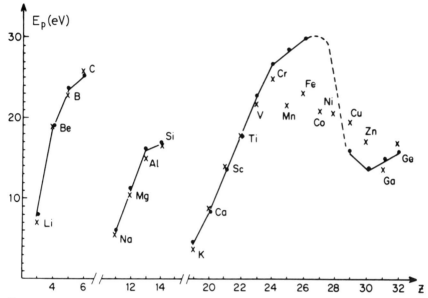

Figure 3.12. Energy of the low-loss peak, as determined experimentally (crosses) and as calculated using the free-electron formula, Eq. (3.41), with $m = m_0$ (Colliex, 1984a).

The relaxation time τ represents the time for plasma oscillations to decay in energy by a factor $\exp(-1) = 0.37$. The number of oscillations which occur within this time is $\omega_p \tau / (2\pi) = E_p / (2\pi \Delta E_p)$. Using experimental values of E_p and ΔE_p, this number turns out to be 4.6 for aluminum, 2.3 for sodium, 0.7 for silicon, and 0.4 for diamond, so in reality the plasma oscillations are highly damped, to a degree which depends on the band structure of the material (see Section 3.3.2).

Table 3.2. Plasmon Energy E_p of Several "Free-Electron" Materials Calculated Using a Free-Electron Model, Experimentally Determined Values of E_p, and Plasmon Linewidth ΔE_p (from Raether, 1980), Characteristic Angle θ_E, Cutoff Angle θ_c [Estimated Using Eq. (3.50) and Eq. (3.51)], and Calculated Mean Free Path [Eq. (3.58)] for 100-keV Incident Electrons

Material	E_p (calc) (eV)	E_p (expt) (eV)	ΔE_p (eV)	θ_E (mrad)	θ_c (mrad)	λ_p (calc) (nm)
Li	8.0	7.1	2.3	0.039	5.3	233
Be	18.4	18.7	4.8	0.102	7.1	102
Al	15.8	15.0	0.53	0.082	7.7	119
Si	16.6	16.5	3.7	0.090	6.5	115
K	4.3	3.7	0.3	0.020	4.7	402

For completely free electrons ($m = m_0$ and $\Gamma \to 0$), $\mathrm{Im}(-1/\varepsilon)$ approximates to a delta function: $(\pi/2)E_p\delta(E - E_p)$. Substitution in Eq. (3.32) and integration over energy loss then gives

$$\frac{d\sigma}{d\Omega} \approx \frac{E_p}{2\pi a_0 m_0 v^2 n_a}\left(\frac{1}{\theta^2 + \theta_E^2}\right) \tag{3.44}$$

where $\theta_E = E_p/(\gamma m_0 v^2)(\approx E_p/2E_0$ for $E_0 \le 100$ keV). Even for $\Gamma \ne 0$, Eq. (3.44) is generally a good approximation; θ_E then represents a "mean" characteristic angle corresponding to the peak or average energy loss E_p.

The Plasmon Wake

When a stationary charged particle is placed in a conducting medium, electrostatic forces cause the electron density to readjust around the particle in a spherically symmetric manner (screening by a "correlation hole"), thereby reducing the extent of the long-range Coulomb field and minimizing the potential energy. When the particle is moving at a speed v, an additional effect occurs as illustrated in Fig. 3.13 (Echenique et al., 1979). Behind the particle, the potential and electron density oscillate at the plasmon frequency (ω_p rad/s), corresponding to spatial oscillation with a wavelength $\lambda_w \approx 2\pi v/\omega_p$. These oscillations also spread laterally, defining a cone of semiangle $\alpha \approx v_F/v$, where v_F is the Fermi velocity in the medium; for a 100-keV electron, this cone is narrow ($\alpha < 1°$), as indicated by the different length scales for the radial and longitudinal distances in Fig. 3.13.

Garcia de Abajo and Echenique (1992) have shown that formation and destruction of the wake occur within distances of approximately $\pi\omega_p/2v = \lambda_w/4$ of the entrance and exit surfaces of the specimen. Consequently the probability of bulk plasmon generation is reduced (begrenzungs effect) if the specimen thickness is less than this value; see also Fig. 3.25.

Plasmon Dispersion; The Lindhard Model

Equation (3.43) describes the energy dependence of the loss spectrum but applies only to small scattering vectors q (dipole region). The jellium model was first extended to higher q by Lindhard (1954), assuming Fermi statistics and using the random-phase approximation (Sturm, 1982), which amounts to neglecting spin exchange and correlation effects arising from Coulomb interaction between the oscillating electrons. The Lindhard model leads to fairly complicated expressions for $\varepsilon(q, E)$ and $\mathrm{Im}(-1/\varepsilon)$ (Tung and Ritchie, 1977; Schnatterly, 1979). In the limit $\Gamma \to 0$, corresponding to

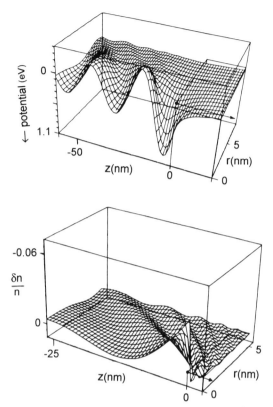

Figure 3.13. (a) Response of a medium (plasmon energy = 25 eV) to a moving charged particle, depicted in terms of the scalar potential calculated by Echenique, Ritchie, and Brandt (1979), with axes relabeled to correspond to the case of a 100-keV electron. Oscillations along the z-axis (direction of electron travel) represent the *wake potential*, which gives rise to plasmon excitation. (b) Corresponding fractional change in electron density. This figure also reveals bow waves which start ahead of the particle and extend laterally as a paraboloidal pattern; they arise from small-impact-parameter colllisions which generate single-electron excitations within the solid. For 100-keV electrons, the bow waves would be more closely spaced than shown since their wavelength scales inversely with particle speed.

completely free electrons, the plasmon energy $E_p(q)$ at which ε_1 passes through zero is given by the equations

$$E_p(q) = E_p + \alpha(\hbar^2/m_0)q^2 \tag{3.45}$$

$$\alpha = (3/5)E_F/E_p \tag{3.46}$$

where E_F is the Fermi energy. Equation (3.45) is the *dispersion relation* for the plasmon, α being the dispersion coefficient. The increase in plasmon

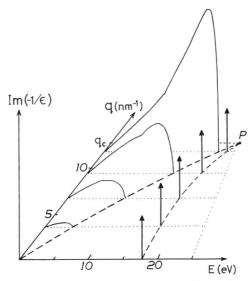

Figure 3.14. Energy-loss function Im($-1/\varepsilon$) computed for silicon using the Lindhard model (Walther and Cohen, 1972). Vertical arrows represent the volume plasmons. The plasmon dispersion curve enters the region of kinematically allowed single-electron excitation at point P, which defines the cutoff wavevector q_c. For $q \gg q_c$, the single electron peak is reduced in intensity and is known as a Bethe ridge.

energy with increasing q (i.e., increasing scattering angle) can be seen in Fig. 3.14, where the plasmon peaks are represented by delta functions (vertical arrows) since damping has been neglected.

The Lindhard model can be extended to include plasmon damping (Mermin, 1970; Gibbons *et al.*, 1976) and insulating materials whose electron distribution is characterized by an energy gap (Levine and Louie, 1982). It is also possible to avoid the random-phase approximation (RPA) and include electron correlation, as first done by Nozières and Pines (1959), who obtained a dispersion relation similar to Eq. (3.45) but with the dispersion coefficient given by

$$\alpha = \frac{3E_F}{5E_p}\left[1 - \left(\frac{E_p}{4E_F}\right)^2\right]\tag{3.47}$$

In the case of aluminum, α is reduced by 11% from its RPA value (0.45), giving improved agreement with most measurements: for example, $\alpha = 0.38 \pm 0.02$ (Batson and Silcox, 1983).

The q-dependence is sometimes used to test the character of a valence-loss peak (Crecelius *et al.*, 1983). If the measured value of the dispersion coefficient is comparable to the RPA value given by Eq. (3.46), collective behavior is suspected; if α is close to zero, an interband transition may be involved.

Unless the energy-loss spectrum is recorded using a sufficiently small collection aperture (semiangle $\ll \theta_E^{1/2}$), contributions from different values of q cause a slight broadening and upward shift of the plasmon peak.

Critical Wave Vector

Above a certain wavevector q_c, the plasma oscillations in a "free-electron gas" are very heavily damped because it becomes possible for a plasmon to transfer all of its energy to a single electron, which can then dissipate the energy by undergoing an interband transition. Such an event must satisfy the usual conservation rules; if an energy E and momentum $\hbar\mathbf{q}$ are to be transferred to an electron of mass m_0 which initially has a momentum $\hbar\mathbf{q}_i$, conservation of both energy and momentum requires

$$E = (\hbar^2/2m_0)(\mathbf{q} + \mathbf{q}_i)^2 - (\hbar^2/2m_0)\mathbf{q}_i^2 = (\hbar^2/2m_0)(q^2 + 2\mathbf{q}\cdot\mathbf{q}_i) \quad (3.48)$$

The *minimum* value of q which satisfies Eq. (3.48) corresponds to the situation where $\mathbf{q}_i$ is parallel to $\mathbf{q}$ and as large as possible, namely, $q_i = q_F$, where q_F is the Fermi wavevector. Denoting this minimum value of q as q_c and substituting for $E = E_p(q)$ using Eq. (3.45), we obtain

$$E_p + \alpha(\hbar^2/m_0)q_c^2 = (\hbar^2/2m_0)(q_c^2 + 2q_c q_F) \quad (3.49)$$

If the dispersion coefficient α is not greatly different from 0.5, the quadratic terms on both sides of Eq. (3.49) almost cancel, and to a rough approximation

$$q_c \simeq m_0 E_p/(\hbar^2 q_F) = E_p/(\hbar v_F) \quad (3.50)$$

where v_F is the Fermi velocity. Equation (3.50) is equivalent to $\omega_p/q \simeq v_F$; in other words, energy transfer becomes possible when the phase velocity of the plasmon falls to a value close to the velocity of electrons at the Fermi surface. More precisely, q_c is defined by intersection of the curves representing Eq. (3.45) and Eq. (3.48) with $q_i = q_F$, as indicated by point P in Fig. 3.14.

A jellium model therefore predicts that the inelastic scattering due to plasmon excitation should fall abruptly to zero above a critical (or cutoff) angle θ_c which is related to the critical wave vector q_c by

$$q_c \simeq k_0(\theta_c^2 + \theta_E^2) \simeq k_0\theta_c \quad (3.51)$$

A more sophisticated calculation based on Hartree–Fock wave functions (Ferrell, 1957) predicts a gradual cutoff described by a function $G(q, q_c)$ which falls (from unity) to zero at $q = 0.74 q_F$, in somewhat better agreement

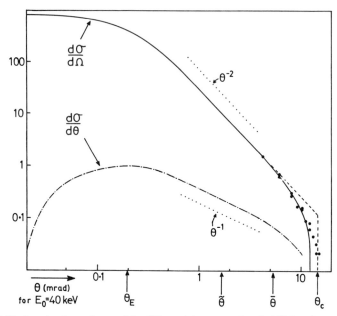

Figure 3.15. Angular dependence of the differential cross section $d\sigma/d\Omega$ for plasmon scattering as calculated by Ferrell (solid curve) and using a sharp cutoff approximation (dashed line). Experimental data of Schmüser (1964) for aluminum and 40-keV incident electrons is indicated by the solid circles. Also shown are the characteristic, median, mean, and cutoff angles, calculated using Eq. (3.51)–(3.56).

with experimental data (see Fig. 3.15). In fact, experimental evidence suggests that inelastic scattering is partly collective in nature at wavevectors considerably above q_c (Batson and Silcox, 1983).

Mean, Root-Mean-Square, and Median Scattering Angles

The *mean* scattering angle $\bar{\theta}$ associated with plasmon scattering can be defined as

$$\bar{\theta} = \int \theta \left(\frac{d\sigma}{d\Omega} \right) d\Omega \bigg/ \int \frac{d\sigma}{d\Omega} \, d\Omega \tag{3.52}$$

where the integrations are over all solid angle Ω. Assuming that the differential cross section has a Lorentzian angular dependence with an abrupt cutoff at $\theta = \theta_c$, and using the approximation $d\Omega = 2\pi(\sin \theta) \, d\theta \cong (2\pi\theta) \, d\theta$ (since $\theta_c \ll 1$), we obtain

$$\bar{\theta} = \int_0^{\theta_c} \frac{\theta^2 \, d\theta}{\theta^2 + \theta_E^2} \bigg/ \int_0^{\theta_c} \frac{\theta \, d\theta}{\theta^2 + \theta_E^2} = \frac{\theta_c - \theta_E \arctan(\theta_c/\theta_E)}{\frac{1}{2}\ln[1 + (\theta_c/\theta_E)^2]} \tag{3.53}$$

Similarly, a mean-square angle can be evaluated as

$$\langle \theta^2 \rangle = \theta_c^2/\ln(1 + \theta_c^2/\theta_E^2) - \theta_E^2 \tag{3.54}$$

The root-mean-square angle $\theta(\text{rms})$ is the square root of $\langle \theta^2 \rangle$ and is of use in describing the angular distribution of *multiple* scattering.

A *median* scattering angle $\tilde{\theta}$ can also be defined, such that half of the scattering occurs at angles less than $\tilde{\theta}$:

$$\int_0^{\theta=\tilde{\theta}} \frac{d\sigma}{d\Omega} d\Omega \bigg/ \int_0^{\theta=\theta_c} \frac{d\sigma}{d\Omega} d\Omega = \frac{1}{2} \tag{3.55}$$

Making the low-angle approximation as above, we obtain

$$\tilde{\theta} = \theta_E(\theta_c/\theta_E - 1)^{1/2} \approx (\theta_E \theta_c)^{1/2} \tag{3.56}$$

For 100-keV incident electrons, $\theta_c/\theta_E \simeq v/v_F \simeq 100$ in a typical material (see Table 3.2), giving $\bar{\theta} \simeq 22\theta_E$ and $\tilde{\theta} \simeq 10\theta_E$. These average scattering angles are at least an order of magnitude larger than θ_E, as a result of the $2\pi \sin \theta$ weighting factor which relates $d\Omega$ and $d\theta$, combined with the wide "tails" of the Lorentzian angular distribution compared to a Gaussian function of equal half-width.

Besides being the half-width of the differential cross section $d\sigma/d\Omega$ (which represents the amount of scattering per unit *solid* angle), θ_E is the *most probable* scattering angle, corresponding to the maximum in $d\sigma/d\theta$; see Fig. 3.15. It is also the appropriate angle to use when calculating the chromatic-aberration limit to spatial resolution in an energy-filtered image of a thin specimen (Egerton and Wong, 1995).

Equations (3.53)–(3.56) can be applied approximately to the case of single-electron excitation (including inner-shell ionization edges) by taking the Bethe-ridge angle $\theta_r \approx (E/E_0)^{1/2} \approx (2\theta_E)^{1/2}$ as a cutoff angle (see Section 3.6.1), and to the total-inelastic scattering by setting $\theta_c \approx \theta_0 = 1/(k_0 r_0)$; see Section 3.2.1.

Plasmon Cross Section and Mean Free Path

Integration of Eq. (3.44) up to an angle β which is less than θ_c gives the *integral cross section* per atom (or per molecule):

$$\sigma_p(\beta) = \frac{E_p \ln(1 + \beta^2/\theta_E^2)}{2n_a a_0 m_0 v^2} \approx \frac{E_p \ln(\beta/\theta_E)}{n_a a_0 m_0 v^2} \tag{3.57}$$

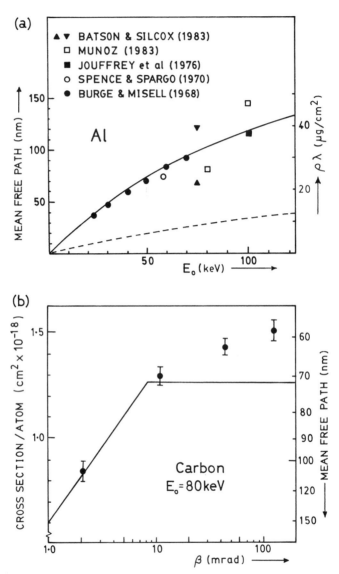

Figure 3.16. (a) Mean free path for valence-electron excitation in aluminum, measured as a function of incident-electron energy. Batson and Silcox (1983) give two values: the upper triangle represents the estimated plasmon mean free path (collective component only), whereas the lower value includes single-electron transitions up to 100 eV. The solid cruve represents the free-electron plasmon model: Eq. (3.58) with $\beta = \theta_c$. The dashed curve shows the totally inelastic mean free path calculation on the basis of a single-atom model (Inokuti *et al.*, 1975). (b) Mean free path for valence excitation in amorphous carbon, as a function of collection semiangle β. The measured values (Egerton, 1975) include energy losses up to 50 eV; the solid line represents the plasmon model, Eq. (3.58), with an abrupt cutoff at $\theta_c = 8.5$ mrad.

where n_a is the number of atoms (or molecules) per unit volume and the approximation is for $\beta \gg \theta_E$. An inverse measure of the amount of scattering below the angle β is the mean free path $\lambda_p(\beta) = [n_a \sigma_p(\beta)]^{-1}$, given by

$$\lambda_p(\beta) = \frac{2a_0 m_0 v^2}{E_p \ln(1 + \beta^2/\theta_E^2)} \approx \frac{a_0}{\gamma \theta_E \ln(\beta/\theta_E)} \tag{3.58}$$

These free-electron formulas give reasonably accurate values for "free-electron" metals such as aluminum (see Fig. 3.16) but apply less well to transition metals (Fig. 3.9), where single-electron transitions considerably modify the energy-loss spectrum.

Assuming a sharp cutoff of intensity at $\theta = \theta_c$, estimates of the *total* plasmon cross section and mean free path are obtained by substituting $\beta = \theta_c$ in Eqs. (3.57) and (3.58). For 100-keV incident electrons, λ is often around 100 nm (see Table 3.2, page 157).

3.3.2. Single-Electron Excitation

As discussed in the preceding section, the plasmon model gives a good description of the valence-loss spectrum of materials such as Na, Al, and Mg where motion of the conduction electrons is relatively unaffected by the crystal lattice. The plasmon peaks are particularly dramatic in the case of alkali metals, where E_p falls below the ionization threshold (Fig. 3.10), giving low plasmon damping.

In all materials, however, there exists an alternative mechanism of energy loss, involving the transfer of energy from a transmitted electron to a single atomic electron within the specimen. This second mechanism can be regarded as competing with plasmon excitation in the sense that the *total* oscillator strength per atom must satisfy the Bethe sum rule, Eq. (3.34). The possible effects of single-electron excitation include the addition of fine structure to the energy-loss spectrum and a broadening and/or shift of the plasmon peak, as we now discuss.

Free-Electron Model

In Section 3.3.1, Eq. (3.48) referred to the transfer of energy from a plasmon to a single atomic electron, but this same equation applies equally well to the case where the energy E is supplied directly from a fast electron. By inspecting Eq. (3.48) it can be seen that, for a given value of q, the maximum energy transfer $E(\text{max})$ occurs when $\mathbf{q}_i$ is parallel to $\mathbf{q}$ and as large as possible (i.e., $q_i = q_F$) so that

$$E(\text{max}) = (\hbar^2/2m_0)(q^2 + 2qq_F) \tag{3.59}$$

The minimum energy loss $E(\text{min})$ corresponds to the situation where $\mathbf{q}_i$ is antiparallel to $\mathbf{q}$ and equal to q_F, giving

$$E(\text{min}) = (\hbar^2/2m_0)(q^2 - 2qq_F) \tag{3.60}$$

Within the region of q and E defined by Eqs. (3.59) and (3.60) (the shaded area in Fig. 3.17), energy loss by "single-electron excitation" is *kinematically allowed* in the free-electron approximation. The Lindhard model (Section 3.3.1) predicts the probability of such transitions and shows (Fig. 3.14) that they occur within the expected region, but mainly at higher values of q. At large q, $\text{Im}[-1/\varepsilon]$ becomes peaked around $E = (\hbar^2/2m_0)q^2$, as predicted by Bethe theory (Fig. 3.36).

Making use of particle concepts, large q corresponds to a collision with small impact parameter b; if the incident elecctron passes sufficiently close to an atomic electron, the latter can receive enough energy to be excited to a higher energy state (single-electron transition). In contrast, atomic electrons further from the path of the fast electron may respond collectively and share the transferred energy.

The relation between collective and single-particle effects is further illustrated in the plasmon-pole (or "single-mode") model as developed by Ritchie and Howie (1977). In their treatment, two additional terms occur in the denominator of Eq. (3.43), resulting in a dispersion relation

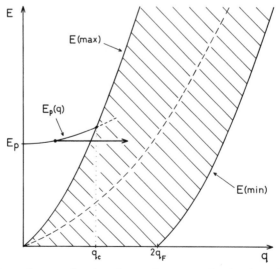

Figure 3.17. Energy loss as a function of scattering vector, showing the region, defined by Eqs. (3.59) and (3.60), over which single-electron excitation is allowed according to the jellium model. Also shown is the plasmon dispersion curve $E_p(q)$. The horizontal arrow indicates momentum transfer from the lattice, resulting in damping of a plasmon.

$$E_q^2 \equiv [E_p(q)]^2 = E_p^2 + (3/5)(q/q_c)^2 E_p^2 + \hbar^4 q^4/4m_0^2 \qquad (3.61)$$

which for small q reduces to the plasmon dispersion relation, Eqs. (3.45) and (3.46), and at large q to the energy-momentum relation ($E = \hbar^2 q^2/2m_0$) for an isolated electron (dashed line in Fig. 3.17). An expression for the energy-loss function can also be derived; neglecting plasmon damping, the resulting differential cross section takes the form (Ritchie and Howie, 1977)

$$\frac{d\sigma}{d\Omega} = \frac{m_0 e^2 E_p^2 (v^2 - 2E_p/m_0)^{1/2}}{2\pi\hbar^4 v E_q q^2 n_a} \approx \frac{e^2 E_p^2}{2\pi\hbar^2 v^2 E_q^2}\left(\frac{1}{\theta^2 + \theta_q^2}\right) \qquad (3.62)$$

with $\theta_q = E_q/\gamma m_0 v^2$. Equation (3.62) becomes equivalent to the plasmon formula [Eq. (3.44)] at small scattering angles and to the Rutherford cross section for scattering from a free electron [Eq. (3.3) with $Z = 1$] at large q, where the momentum is absorbed mainly by a single electron whose energy then considerably exceeds its binding energy.

The Effect of Band Structure

The free-electron plasmon model is a good approximation in many cases (Fig. 3.12), but in most materials the lattice has a significant effect on electron motion, as reflected in the band structure of the solid and the nonspherical nature of the Fermi surface. Single-electron transitions can then occur outside the shaded region of Fig. 3.17, for example at low q, the necessary momentum being supplied by the lattice.

The transition rate is determined by details of the band structure; where the single-electron component in a loss spectrum is sufficiently high (e.g., transition metals) the E-dependence may show a characteristic fine structure. In semiconductors and insulators, this structure reflects a "joint density of states" (JDS) between the valence and conduction bands. Peaks in the JDS occur where branches representing the initial and final states on the energy-momentum diagram are approximately parallel (Bell and Liang, 1976).

Damping of Plasma Oscillations

As remarked in Section 3.3.1, plasma resonance in solids is highly damped and the main cause of this damping is believed to be the transfer of energy to single-electron transitions (i.e., the creation of electron-hole pairs). On a free-electron model, such coupling satisfies the requirements of energy and momentum conservation only if the magnitude of the plasmon wavevector q exceeds the critical value q_c. In a real solid, however, momentum can be supplied by the lattice (in units of a reciprocal lattice vector, i.e., an Umklapp process) or by phonons, enabling the energy transfer to occur at lower values of q, as indicated schematically by the horizontal arrow in Fig. 3.17. The energy of a resulting electron-hole pair is subsequently released as heat (phonon production) or electromagnetic radiation (cathodoluminescence).

In fine-grained polycrystalline materials, grain boundaries may act as an additional source of damping for low-q (long-wavelength) plasmons, resulting in a broadening of the energy-loss peak at small scattering angles (Festenberg, 1967).

Shift of Plasmon Peaks

The single-electron and plasmon processes are also coupled in the sense that the intensity observed in the energy-loss spectrum is not simply a sum of two independent processes. Interband transitions (occurring at an energy E_i) can be incorporated into Drude theory by adding (to the original number n of electrons per unit volume) n_i "bound" electrons which behave as simple-harmonic oscillators with a characteristic angular frequency ω_i and damping constant Γ_i. The bound electrons provide a contribution χ_i to the susceptibility (and to the dielectric function $\varepsilon = 1 + \chi$) given by (Raether, 1980)

$$\chi_i(\omega) = (n_i e^2/\varepsilon_0 m_0)(\omega_i^2 - \omega^2 + i\omega\Gamma_i)^{-1} \tag{3.63}$$

The new plasmon energy E_p^i, defined by $\varepsilon_1(\omega) = 0$, becomes (for small Γ)

$$E_p^i \cong E_p[1 + \chi(\omega_p)]^{-1/2} \tag{3.64}$$

where $E_p = \hbar(ne^2/\varepsilon_0 m_0)^{1/2}$ would be the plasmon energy if all the electrons were free ($n_i = 0$).

If interband transitions occur at an energy *below* the plasmon energy ($\omega_i < \omega_p$), χ_i is negative in the vicinity of $\omega = \omega_p$ and the plasmon energy

Table 3.3. Experimental Energy E_p (expt) of the Main Peak in the Energy-Loss Spectra of Several Semiconductors and Insulators, Compared with the Plasmon-Resonance Energy E_p (calc) Given by Eq. (3.41), where n is the Number of Outer-Shell (Valence) Electrons per unit Volume. Values in eV, from Raether (1980)

Material	E_p (expt)	E_p (calc)
Diamond	34	31
Si	16.5	16.6
Ge	16.0	15.6
InSb	12.9	12.7
GaAs	15.8	15.7
NaCl	15.5	15.7

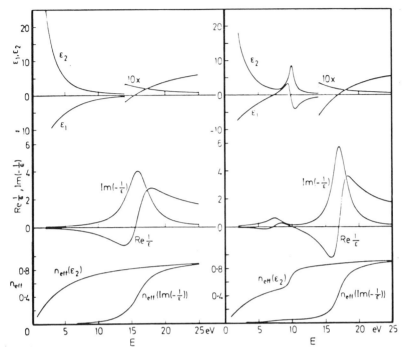

Figure 3.18. Dielectric properties of a free-electron gas with $E_p = 16$ eV and $\Delta E_p = h/\tau = 4$ eV. On the right-hand side, interband transitions have been added at $E_i = 10$ eV (Daniels et al., 1970). From *Springer Tracts in Modern Physics*, Vol. 54, pp. 78–135, © Springer-Verlag, by permission.

is likely to be increased above the free-electron value:

$$(E_p^i)^2 \cong [n_i/n]E_p^2 + E_i^2 \tag{3.65}$$

In an insulator or semiconductor, $n_i \cong n$ (since practically all of the electrons have a binding energy at least equal to the band gap E_g), giving

$$(E_p^i)^2 \cong E_p^2 + E_g^2 \tag{3.66}$$

For all semiconductors and some insulators, $E_g^2 \ll E_p^2$ so $E_p^i \cong E_p$; as a result, the plasmon energy is given fairly well by a free-electron formula,* as illustrated in Table 3.3. Even in the alkali halides, whose spectra show considerable fine structure due to excitons, the energy of the main loss peak is not far from the free-electron value, assuming eight "free" electrons per molecule (Raether, 1980); see page 432.

---

* There are in fact two opposing effects; low-energy interband transitions also increase the effective mass of the valence electrons (Pines, 1963), thereby decreasing the plasma frequency. This may further explain why the free-electcron formula is often a good approximation.

A simple-minded explanation for the occurrence of plasmon effects in insulators in that atomic electrons which receive energy (from a fast electron) in excess of their binding energy E_i are "released" to take part in collective oscillation. On this view, materials whose band gap exceeds E_p should not show evidence of plasma resonance, one possible example being solid neon (Daniels and Krüger, 1971).

In the case of a metal, interband transitions may cause a second resonance if the value of n_i is sufficiently large. For energies just below E_i, $\varepsilon_1(E)$ is forced positive (see Fig. 3.18) and therefore crosses zero with positive slope (the condition for a plasma resonance) at two different energies. Such behavior is observed in silver, resulting in a (highly damped) resonance at 3.8 eV as well as the "free-electron" resonance at 6.5 eV (see Fig. 3.19). In the case of copper and gold, interband transitions lead to a similar fluctuation in ε_1 (Fig. 3.19), but they occur at too low an energy to cause ε_1 to cross zero and there is only one resonance point (at 5 eV in gold).

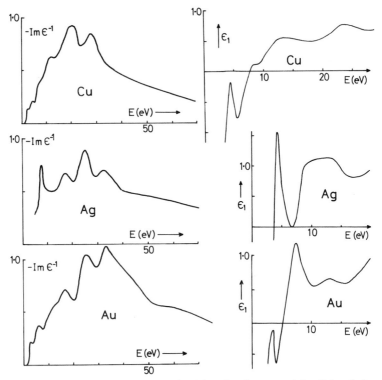

Figure 3.19. The energy loss function $\mathrm{Im}(-1/\varepsilon)$ and real part ε_1 of the dielectric function, derived from transmission energy-loss spectroscopy on polycrystalline films of silver, gold, and copper (Daniels *et al.*, 1970). From *Springer Tracts in Modern Physics*, Vol. 54, pp. 78–135, © Springer-Verlag, by permission.

If interband transitions take place at an energy which is *higher* than E_p, Eqs. (3.63) and (3.64) indicate that $\chi_i(E_p)$ is positive. Polarization of the bound electrons reduces the restoring force on the displaced free electrons and the resonance energy is reduced below the free-electron value.

A general rule is that, where two characteristic energies occur together in the same system, the energy-loss peaks are forced apart. This is illustrated by the case of graphite, which contains one π-electron and three (more strongly bound) σ-electrons per carbon atom. If isolated, these two groups of electrons would have free-electron resonance energies of 12.6 and 22 eV, whereas the observed peaks in the loss spectrum occur at about 7 and 27 eV (Liang and Cundy, 1969).

3.3.3. Excitons

In insulators and semiconductors, it is possible to excite electrons from the valence band to a Rydberg series of states which lie just below the bottom of the conduction band, resulting in an energy loss E_x given by

$$E_x = E_g - E_b/n^2 \tag{3.67}$$

where E_g is the energy gap, E_b is the exciton binding energy, and n is an integer. The resulting excitation can be regarded as an electron and a valence-band hole which are bound to each other to form a quasiparticle known as the exciton.

Although the majority of cases fall between the two extremes, two basic types of exciton can be distinguished. In the *Wannier* (or Mott) exciton, the electron-hole pair is weakly bound ($E_b < 1$ eV) and the radius of the "orbiting" electron is larger than the interatomic spacing. The radius and binding energy can be estimated using hydrogenic formulas: $E_b = e^2/(8\pi\varepsilon\varepsilon_0 r) = m_0 e^4/(8\varepsilon^2\varepsilon_0^2 h^2 n^2)$, where ε is the relative permittivity at the orbiting frequency (usually the light-optical value). The electron and hole travel together through the lattice with momentum $\hbar q$. Such excitons exist in high-permittivity semiconductors (e.g., Cu_2O, CdS) but a high-resolution spectrometer system would be needed to detect the associated energy losses.

Frenkel excitons are strongly bound and relatively compact, the radius of the electron orbit being less than the interatomic spacing. These are essentially excited states of a single atom and in some solids may be mobile via a hopping mechanism. For the alkali halides, in which there is probably some Wannier and some Frenkel character, the binding energy amounts to several eV, tending to be lower on the anion site.

Energy-loss peaks due to transitions to exciton states are observed below the "plasmon" resonance peak in alkali halides (Creuzburg, 1966; Keil, 1968), in rare-gas solids (Daniels and Krüger, 1971) and in molecular

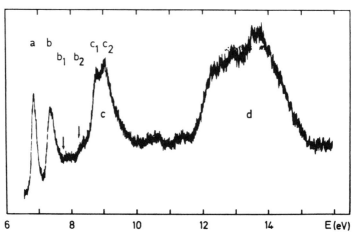

Figure 3.20. Energy-loss spectrum of KBr at a temperature of 80 K (Keil, 1968). Peaks a, b, and c are due to excitons; peak d probably represents a plasma resonance.

crystals such as anthracene. The peaks can be labeled according to the transition point in the Brillouin zone and often show a doublet structure arising from spin-orbit splitting (Fig. 3.20).

The energy E_x of an exciton peak should obey a dispersion relation:

$$E_x(q) = E_x(0) + \hbar^2 q^2 / 2m^* \tag{3.68}$$

where m^* is the effective mass of the exciton. However, measurements on alkali halides show dispersions of less than 1 eV (Creuzburg, 1966) suggesting that $m^* \gg m_0$ (Raether, 1980).

3.3.4. Radiation Losses

If the velocity v of an electron exceeds (for a particular frequency) the speed of light in the material through which it is moving, the electron loses energy by emitting Čerenkov radiation at that frequency. The photon velocity can be written as $c/n = c/\sqrt{\varepsilon_1}$, where n and ε_1 are the refractive index and relative permittivity of the medium, so the Čerenkov condition is satisfied when

$$\varepsilon_1(E) > c^2/v^2 \tag{3.69}$$

In an insulator, ε_1 is positive at low photon energies and may considerably exceed unity. In diamond, for example, $\varepsilon_1 > 6$ for 3 eV $< E <$ 10 eV, so Čerenkov radiation is generated by electrons whose incident energy is 50 keV or higher, resulting in an additional "radiation peak" in the corresponding range of the energy-loss spectrum (see Fig. 3.21a). The photons are

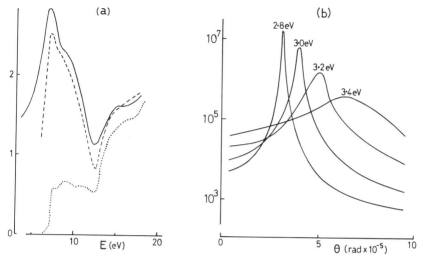

Figure 3.21. (a) Solid line: low-loss spectrum of a 262-nm diamond sample, recorded using 55-keV electrons (Festenberg, 1969). The dashed curve is the intensity using the relativistic theory of Kröger (1968); the dotted curve is the intensity calculated without taking into account retardation. (b) Calculated angular dependence of the radiation-loss intensity for a 210-nm GaP specimen, 50-keV incident electrons, and four values of energy loss (Festenberg, 1969).

emitted in a hollow cone of semiangle $\phi = \cos^{-1}(cv^{-1}\varepsilon_1^{-1/2})$ but are only detectable if the specimen is tilted to avoid total internal reflection.

Kröger (1968) has developed relativistic formulas for the differential cross section, including the retardation effects which are responsible for Čerenkov emission. For specimens which are not too thin, so that surface effects can be neglected, Eq. (3.32) becomes (Festenberg and Kröger, 1968)

$$\frac{d^2\sigma}{d\Omega\,dE} = \frac{\mathrm{Im}(-1/\varepsilon)}{\pi^2 a_0 m_0 v^2 n_a} \frac{\theta^2 + \theta_E^2[(\varepsilon_1 v^2/c^2 - 1)^2 + \varepsilon_2^2 v^4/c^4]}{[\theta^2 - \theta_E^2(\varepsilon_1 v^2/c^2 - 1)]^2 + \theta_E^4 \varepsilon_2^2 v^4/c^4} \qquad (3.70)$$

in which the Lorentzian angular term is replaced by a more complicated function whose "resonance" denominator decreases to a small value (for small ε_2) at an angle $\theta_p = \theta_E(\varepsilon_1 v^2/c^2 - 1)$. The resulting angular distribution of inelastic scattering peaks sharply at small angles (<0.1 mrad) due to the radiation loss. The calculated peak position and width (as a function of energy loss; see Fig. 3.21b) are in agreement with experimental work (Chen *et al.*, 1975). Because the value of θ_p is very small, the radiation-loss electrons will pass through an on-axis collection aperture of typical size and contribute to the energy-loss spectrum recorded in the TEM.

Equation (3.70) indicates that the angular distribution departs significantly from the Lorentzian form (indicating appreciable retardation effects) within the energy range for which $\varepsilon_1 v^2/c^2 > 0.5$ (Festenberg and Kröger, 1968), a less restrictive condition than Eq. (3.69). The retardation-loss condition can also be fulfilled at relatively high energy loss (where, in both conductors and insulators, ε_1 tends to unity) if v^2/c^2 exceeds approximately 0.5, in other words when the incident energy is greater than about 200 keV (see Appendix A).

Energy is also lost by radiation when an electron crosses a boundary where the relative permittivity changes. This "transition radiation" results not from the change of velocity but from a change in electric-field strength surrounding the electron (Frank, 1966). Polarized photons are emitted with energies up to approximately $0.5 \, \hbar\omega(1 - v^2/c^2)^{-1/2}$ (Garibyan, 1960), but the probability of this process appears to be of the order of 0.1%.

3.3.5. Surface Plasmons

Analogous to the bulk or volume plasmons which propagate inside a solid, there exist longitudinal waves of charge density which travel along a surface and are known as surface plasmons. The electrostatic potential at the surface (if it is planar) is of the form $\cos(qx - \omega t) \exp(-q|z|)$, where q and ω are the wavevector and angular frequency of oscillation and t represents time. The charge density at the surface is proportional to $\cos (qx - \omega t)\delta(z)$ and electrostatic boundary conditions lead to the condition

$$\varepsilon_a(\omega) + \varepsilon_b(\omega) = 0 \qquad (3.71)$$

where ε_a and ε_b are the relative permittivities on either side of the boundary. Equation (3.71) defines the angular frequency ω_s of the surface plasmon.

Free-Electron Approximation

The simplest situation corresponds to a single vacuum/metal interface where the metal has negligible damping ($\Gamma \to 0$). Then $\varepsilon_a = 1$ and $\varepsilon_b = 1 - \omega_p^2/\omega^2$, where $\omega_p = E_p/\hbar$ is the bulk-plasmon frequency in the metal. Substitution into Eq. (3.71) gives the energy E_s of the surface-plasmon peak in the energy-loss spectrum:

$$E_s = \hbar\omega_s = \hbar\omega_p/\sqrt{2} = E_p/\sqrt{2} \qquad (3.72)$$

A slightly more general case is a dielectric/metal boundary where the permittivity of the *dielectric* has a positive real part ε_1 and a much smaller imaginary part ε_2 for frequencies close to ω_s. Again assuming negligible damping in the metal, Eq. (3.72) becomes

$$E_s = E_p/(1 + \varepsilon_1)^{1/2} \qquad (3.73)$$

and the energy width of the resonance peak is

$$\Delta E_s = \hbar/\tau = E_s \varepsilon_2 (1 + \varepsilon_1)^{-3/2} \qquad (3.74)$$

Stern and Ferrell (1960) have calculated that a rather thin oxide coating (typically 4 nm) is sufficient to lower E_s from $E_p/\sqrt{2}$ to the value given by Eq. (3.74), and experiments done under conditions of controlled oxidation support this conclusion (Powell and Swan, 1960).

Equation (3.74) illustrates the fact that surface-loss peaks generally occur at lower energy than their volume counterparts and are usually observable only below 10 eV in the energy-loss spectrum. In the case of an interface between two metals, Eq. (3.71) leads to a surface-plasmon energy $[(E_a^2 + E_b^2)/2]^{1/2}$, but this situation is more complicated since the excitation is not necessarily confined to the boundary region (Jewsbury and Summerside, 1980).

The intensity of surface-plasmon scattering is characterized not by a differential cross section (per atom of the specimen) but by a differential "probability" of scattering per unit solid angle, given for a free-electron metal (Stern and Ferrell, 1960) by

$$\frac{dP_s}{d\Omega} = \frac{\hbar}{\pi a_0 m_0 v}\left(\frac{2}{1 + \varepsilon_1}\right)\frac{\theta \theta_E}{(\theta^2 + \theta_E^2)^2} f(\theta, \theta_i, \psi) \qquad (3.75)$$

$$f(\theta, \theta_i, \psi) = \left[\frac{1 + (\theta_E/\theta)^2}{\cos^2 \theta_i} - (\tan \theta_i \cos \psi + \theta_E/\theta)^2\right]^{1/2} \qquad (3.76)$$

where θ is the angle of scattering, θ_i is the angle between the incident electron and the normal to the surface, and ψ is the angle between planes (perpendicular to the surface) which contain the incident- and scattered-electron wave vectors. In Eq. (3.76), θ and θ_i can be positive or negative and θ_E is equal to $E_s/\gamma m_0 v^2$.

The angular distribution of scattering is shown in Fig. 3.22. For normal incidence, $f(\theta, \theta_i, \psi) = 1$ and (unlike the case of volume plasmons) the scattered intensity is zero in the forward direction ($\theta = 0$), a result of the fact that there is then no component of momentum transfer along the boundary plane. The intensity rises rapidly to a maximum at $\theta = \pm \theta_E/\sqrt{3}$, so the minimum around $\theta = 0$ is not observable with a typical TEM collection aperture and incident-beam divergence.

For $|\theta| \gg \theta_E$, the intensity falls proportional to θ^{-3} rather than θ^{-2}, as in the case of bulk losses. This means that a small collection aperture, displaced off-axis by a few θ_E, will tend to exclude surface contributions to the loss spectrum (Liu, 1988).

For nonnormal incidence, $f(\theta, \theta_i, \psi) \neq f(-\theta, \theta_i, \psi)$, leading to an asymmetrical angular distribution which has a higher maximum intensity

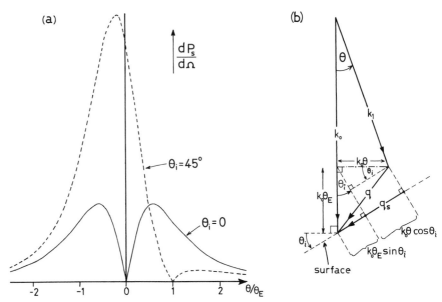

Figure 3.22. (a) Angular distribution of surface scattering; note the asymmetry and the higher integrated intensity in the case of a tilted specimen ($\theta_i = 45°$). (b) Vector diagram for surface scattering, illustrating the relationship between q and q_s for the case where θ and θ_i have the same sign. The scattering angle is assumed to lie within the plane of incidence ($\psi = 0$).

as a result of the $\cos^2 \theta_i$ denominator in Eq. (3.76). This asymmetry has been verified experimentally (Kunz, 1964; Schmüser, 1964). The zero in $dP_s/d\Omega$ again corresponds to the case where the momentum transfer $\hbar\mathbf{q}$ occurs in a direction perpendicular to the surface.

The total probability for surface-plasmon excitation at a single vacuum interface is obtained by integrating Eq. (3.75) over all θ. For normal incidence ($\theta_i = 0$) the result is

$$P_s = \frac{\pi\hbar}{a_0 m_0 v (1 + \varepsilon_1)} = \frac{e^2}{4\varepsilon_0 \hbar v (1 + \varepsilon_1)} \tag{3.77}$$

At 100-keV incident energy, P_s is 0.021 for $\varepsilon_1 = 1$ (vacuum/metal interface) and 0.011 for $\varepsilon_1 = 3$ (typical of many oxides). Taking into account *both* surfaces, the probability of surface-plasmon excitation in an oxidized aluminum sample is therefore about 2% and the corresponding loss peak (at just over 7 eV) will be clearly visible only in rather thin samples, where the inelastic scattering due to bulk processes is weak. However, if the specimen is tilted away from normal incidence, P_s is increased as a result of the $\cos \theta_i$ term in Eq. (3.76).

Dielectric Formulation for Surface Losses

The free-electron approximation $\varepsilon_b = 1 - \omega_p^2/\omega^2$ can be avoided by characterizing the materials (conductors or insulators) on *both* sides of the boundary by frequency-dependent permittivities ε_a and ε_b. Dielectric theory then provides an expression for the differential "probability" of surface scattering at a single interface (Raether, 1980, p. 146):

$$\frac{d^2P_s}{d\Omega\, dE} = \frac{k_0^2|q_s|}{\pi^2 a_0 m_0 v^2 q^4 \cos\theta_i} \operatorname{Im}\left[\frac{(\varepsilon_a - \varepsilon_b)^2}{\varepsilon_a \varepsilon_b(\varepsilon_a + \varepsilon_b)}\right] \tag{3.78}$$

assuming that the electron remains in the plane of incidence ($\psi = 0$). In the small-angle approximation, $q^2 = k_0^2(\theta^2 + \theta_E^2)$ and, as shown in Fig. 3.22,

$$q_s = k_0\theta\cos\theta_i + k_0\theta_E\sin\theta_i \tag{3.79}$$

where q_s is the wavevector of the surface plasmon, equal to the component of the scattering vector which lies along the surface.* *One* of the terms in Eq. (3.79) is negative if θ and θ_i are of opposite sign. Note that Eq. (3.78) and Eq. (3.71) are symmetric in ε_a and ε_b, so the direction of travel of the incident electron is unimportant. The width of the plasmon peak is determined by the imaginary parts of the permittivities on both sides of the boundary. Near the resonance condition, $\varepsilon_a \simeq -\varepsilon_b$, $(\varepsilon_a - \varepsilon_b)^2 \simeq -4\varepsilon_a\varepsilon_b$ and Eq. (3.78) can be simplified to

$$\frac{d^2P_s}{d\Omega\, dE} = \frac{2k_0^2|q_s|}{\pi^2 a_0 m_0 v^2 q^4 \cos\theta_i} \operatorname{Im}\left(\frac{-2}{\varepsilon_a + \varepsilon_b}\right) \tag{3.80}$$

Very Thin Specimens

The surface plasmons excited on each surface of a specimen of thickness t are essentially independent of each other provided that

$$q_s t \simeq k_0\theta t \gg 1 \tag{3.81}$$

For 100-keV incident electrons ($k_0 = 1700$ nm^{-1}) and $\theta \simeq \theta_E/\sqrt{3} \simeq 0.1$ mrad (the most probable angle of surface scattering; see Fig. 3.22), Eq. (3.81) implies $t \gg 10$ nm. If this condition is not fulfilled, the electrostatic fields originating from both surfaces overlap and the plasmons interact with one another (see Fig. 3.23). In the case of a free-electron metal bounded by similar dielectrics ($\varepsilon_a = \varepsilon_c = \varepsilon$; $\varepsilon_b = 1 - \omega_p^2/\omega^2$) the resonance is split into

---

* The component of momentum perpendicular to the surface is absorbed by the lattice.

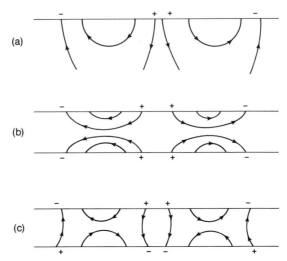

Figure 3.23. Electric field lines associated with surface plasmons excited in a bulk sample (a) and in a thin film (b and c). The plasmon frequency is higher in the symmetric mode (b) than in the asymmetric one (c).

two modes, the frequency of each being q-dependent and given approximately, for large q_s, by (Ritchie, 1957)

$$\omega_s = \omega_p \left[\frac{1 \pm \exp(-q_s t)}{1 + \varepsilon} \right]^{1/2} \tag{3.82}$$

The symmetric mode, where like charges face one another (Fig. 3.23b), corresponds to the higher angular frequency. For small q_s, Eq. (3.82) does not apply; relativistic constraints (Kröger, 1968) cause ω to lie below* the photon line ($\omega = cq_s$) on the dispersion diagram (Fig. 3.24). This dispersion behavior has been verified experimentally (Pettit *et al.*, 1975).

The differential probability for surface excitation at *both* surfaces of a film of thickness t, in the case of normal incidence and neglecting retardation effects, can be expressed (Raether, 1967) as

$$\frac{d^2 P_s}{d\Omega \, dE} = \frac{2\hbar}{\pi^2 \gamma a_0 m_0^2 v^3} \frac{\theta}{(\theta^2 + \theta_E^2)^2} \, \mathrm{Im}\left[\frac{(\varepsilon_a - \varepsilon_b)^2}{\varepsilon_a^2 \varepsilon_b} R_c \right] \tag{3.83}$$

* There are, in fact, *radiative* surface plasmons which lie above this line, but they are less easily observed in the energy-loss spectrum because for small scattering angles their energy is the same as that of the volume plasmons (see Fig. 3.24).

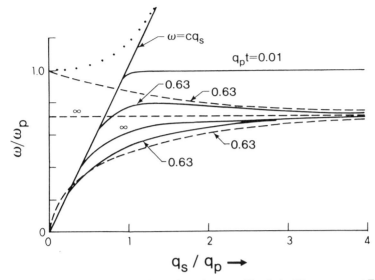

Figure 3.24. Dispersion diagram for surface plasmons. The dashed lines represent Eq. (3.82); the solid curves were calculated taking into account retardation (Raether, 1980), for film thicknesses given by $q_p t = 0.01$, 0.63, and ∞, where $q_p = \omega_p/c$. For $\theta_i = 0$, the horizontal axis is approximately proportional to scattering angle θ of the fast electron, since $q_s/q_p \approx (v/c)^{-1}(\theta/\theta_E)$. The dispersion relation of a radiative surface plasmon is shown schematically (dotted curve).

where

$$R_c = \frac{\varepsilon_a \sin^2(tE/2\hbar v)}{\varepsilon_b + \varepsilon_z \tanh(q_s t/2)} + \frac{\varepsilon_a \cos^2(tE/2\hbar v)}{\varepsilon_b + \varepsilon_a \coth(q_s t/2)} \qquad (3.84)$$

A secondary effect of the surface excitation is to reduce the intensity of the *bulk*-plasmon peak. This happens because the energy-loss function in Eq. (3.83) has a negative value at $E \approx \hbar\omega_p$. For $t > v/\omega_p$, the calculated reduction ΔP_v in the probability P_v of volume-plasmon excitation is just equal to the probability of surface-plasmon generation at a *single* surface (see Fig. 3.25), which is typically about 1% and small compared to P_v. However, this negative surface contribution occurs mainly at small angles ($\theta \approx \theta_E$), owing to the relatively narrow angular distribution of the surface-loss intensity. Its effect is therefore greater if the energy-loss spectrum is recorded with a small angular-collection aperture. For an effective aperture of 0.2 mrad and 50-keV electrons transmitted through a film of aluminum, Raether (1967) has reported a reduction in the volume-loss intensity of 8% at $t = 100$ nm increasing to over 40% at $t = 10$ nm.

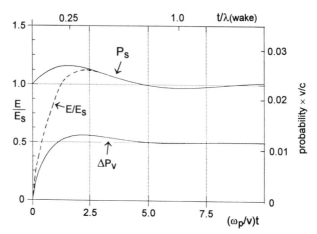

Figure 3.25. Thickness dependence of the probability P_s of surface-plasmon excitation and the associated energy loss (in units of $\hbar\omega_p / \sqrt{2}$: dashed curve) for normally incident electrons of speed v, calculated on the basis of a free-electron model for a specimen with two clean and parallel surfaces (Ritchie, 1957). The lower curve shows the reduction ΔP_v in the probability of volume-plasmon excitation, the so-called begrenzungs effect. This effect has also been calculated for the case of a small sphere (Echenique *et al.*, 1987).

A relativistic version of Eq. (3.83) has been given by Kröger (1968); relativistic effects modify the dispersion relation of the surface plasmon (particularly at low q_s) and result in slightly lower resonance energies which are in closer agreement with experiment.

Integrating Eq. (3.83) over energy loss, Ritchie (1957) investigated how the total probability P_s of surface-plasmon loss (for a free-electron metal) varies with sample thickness. For $t \gg v/\omega_p$ (=7.2 nm for E_0 = 100 keV, E_p = 15 eV), P_s becomes independent of thickness and tends asymptotically to $e^2/(4\varepsilon_0\hbar v)$, twice the value given by Eq. (3.77) since the specimen has two surfaces. For $t < 5(v/\omega_p)$, the surface-loss probability increases slightly (see Fig. 3.25), corresponding to an increase in the asymmetric (ω_-) surface mode. But as shown by Eq. (3.82), the energy loss associated with this mode tends to zero as $t \to 0$, so the total energy loss due to surface-plasmon excitation falls towards zero for very thin films, as indicated by the dashed curve in Fig. 3.25.

3.3.6. Surface-Reflection Spectra

As an alternative to measuring transmitted electrons, energy-loss spectra can be recorded from electrons which have been reflected from the surface of a specimen. The depth of penetration of the electrons (perpendicular to the surface) depends on the primary-beam energy and the angle of incidence θ_i (see Fig. 3.27). For moderate angles, both bulk- and surface-loss peaks occur in the reflection spectrum; at a glancing angle ($\theta_i > 80°$) the penetration depth is small and only surface peaks are observed, particularly if the spectrum has been recorded from specularly reflected electrons (angle of reflection = angle of incidence) and the incident energy is not too high (Powell, 1968).

In the case of a crystalline specimen, the reflected intensity is strong when the angle between the incident beam and the surface is equal to a Bragg angle for atomic planes which *lie parallel to the surface*, the *specular Bragg condition*. The intensity is further increased by adjusting the crystal orientation so that a resonance parabola seen in the reflection diffraction pattern (the equivalent of a Kikuchi line in transmission diffraction) intersects a Bragg-reflection spot, such as the (440) reflection for a {110} GaAs surface, giving a *surface-resonance condition*. The penetration depth of the electrons is then only a few monolayers, but the electron wave travels a short distance (typically of the order of 100 nm) parallel to the surface before being reflected (Wang and Egerton, 1988).

The ratio P_s of the integrated surface-plasmon intensity, relative to the zero-loss intensity in a specular Bragg-reflected beam, has been calculated on the basis of both classical and quantum-mechanical theory (Lucas and Sunjic, 1971; Evans and Mills, 1972). For a clean surface ($\varepsilon_a = 1$) and assuming negligible penetration of the electrons,

$$P_s = e^2/(8\varepsilon_0 \hbar v \cos \theta_i) \tag{3.85}$$

which is identical to the formula for normal transmission through a single interface [Eq. (3.77)] except that the incident velocity v is replaced by its component $v(\cos \theta_i)$ normal to the surface. In the case of measurements made at glancing incidence, this normal component is small and P_s may approach or exceed unity.* Surface peaks then dominate the energy-loss spectrum (see Fig. 3.26); bulk plasmons are observed only for lower values of θ_i or (owing to the broader angular distribution of volume scattering) when recording the spectrum at inelastic-scattering angles θ away from a specular beam (Schilling, 1976; Powell, 1968). An equivalent explanation for the increase in surface loss as $\theta_i \to 90°$ is that the incident electron spends a longer time in the vicinity of the surface (Raether, 1980).

Schilling and Raether (1973) have reported energy *gains* of $\hbar\omega_s$ in energy-loss spectra of 10-keV electrons reflected from a liquid-indium surface at $\theta_i \simeq 88.5°$. Such processes are measurable only when the incident-beam current is so high that the time interval between the arrival of the

---

* Strictly speaking, P_s is a "scattering parameter" analogous to t/λ in Section 3.3.6, but approximates a single-scattering probability if much less than unity.

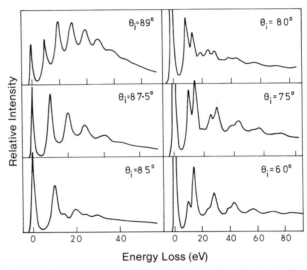

Figure 3.26. Plasmon-loss spectra recorded by reflection of 8-keV electrons from the surface of liquid aluminum (Powell, 1968). At glancing incidence, the spectrum is dominated by plural scattering from surface plasmons, but as the angle of incidence is reduced, volume-plasmon peaks occur at multiples of 15 eV.

electrons is comparable with the surface-plasmon relaxation time τ. Even under the somewhat-optimized conditions used in this experiment, the probability of energy gain was only about 0.2%.

Since the electron microscope allows a reflection diffraction pattern to be observed and indexed, the value of the incident angle θ_i can be obtained from the Bragg angle θ_B of each reflected beam, provided allowance is made for refraction of the electron close to the surface (see Fig. 3.27). The refraction effect depends on the mean inner potential ϕ_0 of the specimen; using relativistic mechanics to calculate the acceleration of the incident electron towards the surface, one obtains

$$\cos^2 \theta_i = \sin^2 \theta_s = \sin^2 \theta_B - \frac{2(e\phi_0)(1 - v^2/c^2)^{3/2}}{m_0 c^2 (v^2/c^2)} \qquad (3.86)$$

Since $\cos \theta_i = \sin \theta_s \simeq \theta_s$, where θ_s is the angle between the incident beam and the surface (measured outside the specimen), Eq. (3.86) leads to the relation $P_s \theta_s = $ const, which has been verified experimentally (Powell, 1968; Schilling, 1976; Krivanek et al., 1983). For a given Bragg reflection, the value of P_s calculated from Eqs. (3.85) and (3.86) tends to increase with increasing incident energy; for the symmetric (333) reflection from silicon, for example, P_s is 1.06 at 20 keV and 1.36 and 80 keV. Experimental values (Krivanek et al., 1983) are somewhat higher (1.4 and 1.8), probably because

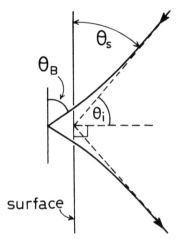

Figure 3.27. Reflection of an electron from the surface of a crystal, showing schematically the difference between the angles θ_s and θ_B, due to refraction.

Eq. (3.85) neglects surface-plasmon excitation during the brief period when the electron penetrates inside the crystal.

Reflection energy-loss spectroscopy is a more surface-sensitive technique when the incident energy is low, partly because the electron penetration depth is smaller (the extinction distance for elastic scattering is proportional to electron velocity v) and because the electrostatic field of a surface plasmon extends into the solid to a depth of approximately $1/q_s \approx (k_0\theta_E)^{-1} = v/\omega_p$ (for $\theta_i \approx \pi/2$), which is proportional to the incident velocity.

Nonpenetrating Incident Beam (Aloof Excitation)

An alternative type of experiment is where the primary electrons pass close to a surface but remain outside, such as when a finely focused electron beam is directed parallel to a face of a cubic MgO crystal (Marks, 1982). In the case of a metal, bulk plasmons are not excited and surface excitations can be studied alone. Classical, nonrelativistic theory gives for the excitation probability (Howie, 1983)

$$\frac{dP_s(x, E)}{dE} = \frac{2z}{\pi m_0 a_0 v^2} K_0(2\omega x/v) \, \mathrm{Im}\left[\frac{\varepsilon(E) - 1}{\varepsilon(E) + 1}\right] \tag{3.87}$$

where z is the length of the electron path parallel to the surface (a distance x away), $K_0(2\omega x/v)$ is a modified Bessel function and $\varepsilon(E)$ is the dielectric function (complex permittivity) of the specimen, which is a function of the energy loss $E = \hbar\omega$. Equation (3.87) can be generalized to deal with the case where the specimen surface is curved (Batson, 1982; Wheatley *et al.*, 1984) or where the incident electron executes a parabolic trajectory as a result of a negative potential applied to the sample (Ballu *et al.*, 1976). A limited range of impact parameter x can be selected by scanning the incident beam and using a gating circuit to switch the spectrometer signal on or off in response to the output of a dark-field detector (Wheatley *et al.*, 1984).

3.3.7. Surface Modes in Small Particles

In the case of an isolated spherical particle (relative permittivity = ε_b) surrounded by a medium of permittivity ε_a, the surface-resonance condition is modified from Eq. (3.71) to become

$$\varepsilon_a + [l/(1 + l)]\varepsilon_b = 0 \tag{3.88}$$

where l is an integer. For a free-electron metal, Eq. (3.88) gives for the surface-resonance frequency

$$\omega_s = \omega_p[1 + \varepsilon_a(l + 1)/l]^{-1/2} \tag{3.89}$$

where ε_a is the relative permittivity of the immediate surroundings (e.g., vacuum or oxide layer). The lowest frequency corresponds to $l = 1$ (dipole mode) and predominates in very small spheres, of radius $r < 10$ nm. As the radius increases, the energy-loss intensity shifts to higher-order modes and the resonance frequency increases asymptotically towards the value given by Eq. (3.73) for a flat surface. This behavior has been verified by experiments on metal spheres (Fujimoto and Komaki, 1968; Achèche *et al.*, 1986), colloidal silver and gold particles embedded in gelatin (Kreibig and Zacharias, 1970), and irradiation-induced precipitates of sodium and potassium in alkali halides (Creuzburg, 1966). The probability of exciting a given mode, averaged over all possible trajectories of the fast electron, is of the form (Fujimoto and Komaki, 1968)

$$\frac{dP_s(l)}{d\omega} = \frac{8\hbar r}{a_0 m_0 v^2 q^4}\left(\frac{\omega_s}{\omega_p}\right)^2 \frac{\omega\omega_s^2\Gamma}{(\omega^2 - \omega_s^2)^2 + \omega^2\Gamma^2} \frac{(2l + 1)^3}{l} \int_{\omega r/v}^{\infty} \frac{[J_l(z)]^2}{z^3}\,dz \tag{3.90}$$

where $\Gamma = 1/\tau$ is the damping constant of the metal and J_l is a spherical Bessel function.

If the spheres touch one another, additional surface modes are excited (Batson, 1982). Dielectric theory has been used to predict the additional peaks which occur when spherical metal particles are attached to a substrate (Wang and Cowley, 1987; Ouyang and Isaacson, 1989; Zabala and Rivacoba, 1991).

For the case of small spheres embedded in a medium, Howie and Walsh (1991) have proposed an *effective energy-loss function* $\text{Im}(-1/\varepsilon)_{\text{eff}}$ which is geometrically averaged over different segments of a typical electron trajectory. They show that this function predicts the observed spectrum of irradiated AlF_3 (containing small Al particles) more successfully than effective medium theories, which give formulas for an *effective permittivity* ε_{eff}. Measurement of $\text{Im}(-1/\varepsilon)_{\text{eff}}$ might yield the average size and volume fraction of fine precipitates or point-defect clusters in specimens whose structure is too fine in scale or too complex to permit direct imaging.

Cavities in a metal or dielectric also have characteristic resonance frequencies, given for the case of a spherical void ($\varepsilon = 1$) in a metal by

$$\omega_s = \omega_p \, [(m + 1)/(2m + 1)]^{1/2} \tag{3.91}$$

so that the frqeuency now decreases towards $\omega_p/\sqrt{2}$ as the integer m increases from zero (Raether, 1980). As an example, helium-filled "voids" in Al/Li alloy ($\hbar\omega_p = 15$ eV) appear bright in the image formed from 11-eV loss electrons (Henoc and Henry, 1970).

3.4. Single, Plural, and Multiple Scattering

If the specimen is very thin, the probability of scattering (inelastic or elastic) is low and the probability of more than one scattering event within the specimen is negligible. Provided the energy resolution is sufficiently good, the intensity $J^1(E)$ in the energy-loss spectrum then approximates to a *single-scattering distribution* (SSD) or single-scattering profile $S(E)$:

$$J^1(E) \approx S(E) = I_0 n_a t (d\sigma/dE) + I_0 (dP_s/dE) \tag{3.92}$$

where I_0 is the zero-loss intensity, approximately equal (because of the low scattering probability) to the total area I_t under the loss spectrum; n_a is the number of atoms (or molecules) per unit volume of the specimen, and t is the specimen thickness within the irradiated area. The energy-differential cross section per atom (or molecule) $d\sigma/dE$ is obtained by integrating the double-differential cross section $d^2\sigma/d\Omega \, dE$ given by Eq. (3.29) or Eq. (3.32) up to a scattering angle equal to the collection semiangle β used when acquiring the spectrum. The last term in Eq. (3.92) represents the intensity arising from surface-mode scattering. Integration over energy loss E gives the total single-scattering intensity:

$$I_1(\beta) = I_0 n_a t \sigma(\beta) + I_0 P_s(\beta) \tag{3.93}$$

where $\sigma(\beta)$ is an integral cross section, given by Eq. (3.57) for the case of volume-plasmon excitation.

3.4.1. Poisson's Law

If inelastic scattering can be viewed in terms of collisions which are *independent* events, their occurrence should obey Poisson statistics: the probability that a transmitted electron suffers n collisions is $P_n = (1/n!)m^n \exp(-m)$, where m is the *mean* number of collisions incurred by an electron which travels through the specimen. For convenience, we set $m = t/\lambda$, where λ is the average distance between collisions, the *mean free path* for inelastic scattering. Sometimes t/λ is referred to as the *scattering parameter* of the specimen. P_n is represented in the energy-loss spectrum by the ratio of the energy-integrated intensity I_n of n-fold scattering, divided by the *total* integrated intensity:

$$P_n = I_n/I_t = (1/n!)(t/\lambda)^n \exp(-t/\lambda) \qquad (3.94)$$

The variation of P_n with scattering parameter is shown in Fig. 3.28. For a given order n of scattering, the intensity is highest when $t/\lambda = n$. In the case of the unscattered ($n = 0$) component (zero-loss peak), the intensity is a maximum at $t = 0$ and decreases exponentially with specimen thickness. For $n = 0$, Eq. (3.94) gives

$$t/\lambda = \ln(I_n/I_t) \qquad (3.95)$$

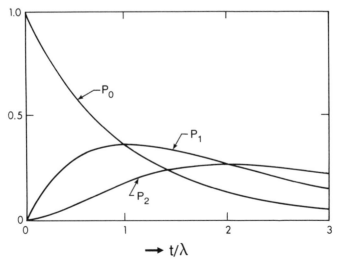

Figure 3.28. Probability of no inelastic scattering (P_0), of single scattering (P_1) and of double scattering (P_2), as a function of scattering parameter (t/λ).

The following comments relate to Eq. (3.94), which is known as Poisson's law.

(1) Angles of scattering are assumed to be small, so that the distance which electrons travel through the specimen is identical for the different orders n of scattering.

(2) If several energy-loss processes (each characterized by a different mean free path λ_i) occur within the energy range over which the spectral intensity is integrated in Eq. (3.93) and Eq. (3.94), the effective scattering parameter is

$$t/\lambda = \sum_i t_i/\lambda_i \qquad (3.96)$$

If the electron passes through several layers, t_i represents the thickness of a layer i; for a single-layer specimen, each t_i is equal to the specimen thickness. Equation (3.96) reflects the fact that the scattering probabilities are additive.

(3) Although Eq. (3.94) refers to bulk processes, surface-mode scattering can be included by adding the surface-loss probability P_s (see Section 3.3.5) to the scattering parameter t/λ. For normal incidence, P_s is sufficiently small ($<5\%$) that second- and higher-order surface scattering is negligible, but for reflection at grazing incidence P_s can exceed unity and multiple surface-plasmon peaks dominate the spectrum (see Fig. 3.26). For this situation, the validity of Poisson statistics has been confirmed experimentally (Schilling, 1976).

(4) Equation (3.94) is exact only if the specimen is of uniform thickness in the area from which the spectrum is recorded. The effect of thinner regions can be visualized by imagining a hole to occur within the analyzed area; electrons passing through the hole contribute to the zero-loss intensity I_0 but not to other orders of scattering. Breakdown of Eq. (3.94) leads to inaccuracy in the removal of plural scattering by Fourier-log deconvolution, as discussed in Chapter 4.

(5) The use of Poisson statistics is justified if all scattering events contribute to the measured intensities I_n. However, the energy-loss spectrum is often recorded with an angle-limiting aperture which accepts only a fraction $F_n(\beta)$ of the electrons of a given order. In this case, Eq. (3.94) retains its validity *only* if the aperture factors obey the relationship

$$F_n(\beta) = [F_1(\beta)]^n \qquad (3.97)$$

so that substitution into Eq. (3.94) gives

$$I_n(\beta)/I_t(\beta) = F_n(\beta)(1/n!)(t/\lambda)^n \exp(-t/\lambda) \qquad (3.98)$$
$$= (1/n!)(F_1 t/\lambda)^n \exp(-t/\lambda) = (1/n!)[t/\lambda(\beta)]^n \exp(-t/\lambda)$$

where $\lambda(\beta) = \lambda/F_1(\beta)$. Note that the exponential factor (representing depletion of intensity from the unscattered component) involves the mean free path λ for scattering through *all* angles, whereas the *relative* intensities of the different orders of scattering depend on an angle-dependent mean free path $\lambda(\beta)$ which is related to the integral cross section for scattering within the aperture:

$$\lambda(\beta) = n_a t \sigma(\beta) \qquad (3.99)$$

If $d\sigma/d\Omega$ is a Lorentzian function with an abrupt cutoff at a scattering angle θ_c,

$$F_1(\beta) = \frac{\lambda}{\lambda(\beta)} = \frac{\sigma(\beta)}{\sigma} = \frac{\ln(1 + \beta^2/\theta_E^2)}{\ln(1 + \theta_c^2/\theta_E^2)} \qquad (3.100)$$

As a result of this logarithmic dependence on the collection semiangle, $\lambda(\beta)$ is somewhat longer than the mean free path λ for scattering through all angles.

To justify the validity of Eq. (3.97) and Eq. (3.98), we must examine the angular distribution of plural inelastic scattering.

3.4.2. Angular Distribution of Plural Inelastic Scattering

In the case of double scattering, the intensity per unit solid angle $dJ^2/d\Omega$ is a two-dimensional convolution of the single-scattering angular distribution. Using polar coordinates to represent the radial component θ and the azimuthal component φ of scattering angle, this convolution can be represented (Fig. 3.29a) as

$$\frac{dJ^2(\theta)}{d\Omega} \propto \int \left[\frac{d\sigma(\theta_2)}{d\Omega}\right] \left[\frac{d\sigma(\theta_1)}{d\Omega_1}\right] d\Omega_1 \qquad (3.101)$$

where the integration is over all solid angles Ω_1. For a Lorentzian $d\sigma/d\Omega$ with *no cutoff*, the integration can be represented analytically:

$$dJ^2/d\Omega \propto \theta^{-1}\{\ln[1 - u^2 + uw - u/w] - \ln[(1 - 2u^2/v + w^2u^2/v^2)^{1/2}$$
$$+ wu/v - u/w]\} \qquad (3.102)$$

where $u = \theta/\theta_E$, $v = 1 + \pi^2/\theta_E^2$, and $w = (4 + \theta^2/\theta_E^2)^{1/2}$. For $\theta_E \ll \theta \ll \pi$, Eq. (3.101) becomes

$$dJ^2/d\Omega \propto \theta^{-2} \ln(\theta/\theta_E) \qquad (3.103)$$

so at higher scattering angles the intensity falls off a little more slowly than the θ^{-2} dependence of single scattering (Fig. 3.29c).

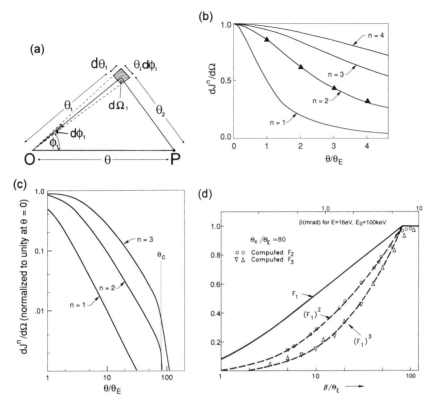

Figure 3.29. (a) Geometry of double scattering, from O to P, showing scattering angles (assumed small, such that $\sin \theta \approx \theta$) projected onto a plane perpendicular to the optic axis. The individual scattering angles are θ_1 and θ_2; the total deflection after double scattering is θ. (b) Scattering per unit solid angle for the first four orders of scattering; triangles represent double scattering calculated from an analytical formula, Eq. (3.102). (c) Scattering per unit angle, calculated up to higher scattering angle and assuming an abrupt cutoff at $\theta_c = 80\theta_E$. (d) Fraction F_n of inelastic scattering (of order n) collected by an aperture of semiangle β. The data points show F_2 and F_3 calculated using different algorithms; dashed curves represent the square and cube of F_1 (Egerton and Wang, 1990). From *Ultramicroscopy*, R. F. Egerton and Z. L. Wang, Plural scattering deconvolution of electron energy-loss spectra recorded with an angle-limiting aperture, © 1990, pp. 137–148, with permission from Elsevier Science B.V., Amsterdam, The Netherlands.

A *truncated* Lorentzian angular distribution of single scattering can be specified by introducing a function $H(\theta)$ which changes from 1 to 0 as θ passes through θ_c, so that Eq. (3.101) becomes

$$\frac{dJ^2(\theta)}{d\Omega} \propto \int_{\theta_1=0}^{\pi} \int_{\varphi_1=0}^{2\pi} \left[\frac{H(\theta_1)}{\theta_1^2 + \theta_E^2}\right] (\theta \, d\varphi_1 \, d\theta_1) \left[\frac{H(\theta_2)}{\theta_2^2 + \theta_E^2}\right] \qquad (3.104)$$

where $\theta_2^2 = \theta^2 + \theta_1^2 - 2\theta\theta_1 \cos(\varphi_1)$, from application of the cosine rule to the vector triangle in Fig. 3.29a. Equation (3.104) can be evaluated numerically, although considerable computing time is needed to achieve good accuracy. Exploiting the fact that the inelastic scattering has axial symmetry, the double integral may be replaced by a single integral involving a Bessel function, the so-called Hankel transform (Bracewell, 1978), with a reduction in computing time (Johnson and Isaacson, 1988; Reimer, 1989).

Extending Eq. (3.104), the intensity of n-fold scattering can be computed as an n-fold convolution of $d\sigma/d\Omega$. Calculated angular distributions are shown in Fig. 3.29b and Fig. 3.29c. Relative to the half-width θ_E of single scattering, the half-widths of the double, triple, and quadruple scattering distributions are increased by factors of 2.6, 5.1, and 7.5, respectively. Since some plural-scattering intensity extends up to an angle $n\theta_c$, the "cutoff" at $\theta = \theta_c$ becomes more gradual as n increases.

To calculate the attenuation factor F_n, the angular distribution of n-fold scattering must be integrated up to an angle β and divided by the integral over all scattering angles. Results for $n = 2$ and $n = 3$ are shown in Fig. 3.29d, in comparison to the square and cube of F_1 (dashed curves). Although discrepancies are observable close to θ_c and at small angles, Eq. (3.97) is found to be accurate to within 3% for $\beta > 15\theta_E$ (Egerton and Wang, 1990; Su et $al.$, 1992). Equation (3.97) has also been inferred from the results of Monte Carlo calculations (Jouffrey et $al.$, 1989) and has been verified experimentally from deconvolution of plasmon-loss spectra; see page 254 and page 256.

For double scattering ($n = 2$), the relation $F_2 = (F_1)^2$ can be proven mathematically if we assume $\beta \gg \theta_E$, since in this case

$$dJ^2/d\Omega \propto \theta^{-1} dJ^2/d\theta \propto \theta^{-1}(dF_2/d\theta) \tag{3.105}$$
$$\propto \theta^{-1} d[\ln^2(\theta/\theta_E)]/d\theta \propto \theta^{-2} \ln(\theta/\theta_E)$$

which is of the same form as Eq. (3.103). Consequently, Eq. (3.97) appears to be a property of the θ^{-2} tail of the Lorentzian single-scattering angular distribution.

3.4.3. Influence of Elastic Scattering

So far, our discussion of angular distributions and plural-scattering probabilities has made no explicit reference to elastic scattering, even though the probability of such scattering is comparable to that of inelastic scattering; see page 145. In general, the angular width of inelastic scattering is less than that of elastic scattering (see Section 3.2) and the collection semiangle β used in spectroscopy is often less than the characteristic angle

θ_0 of elastic scattering or (for a crystalline specimen) the scattering angle $2\theta_B$ of lowest-order diffraction spots. Under these conditions, the total intensity $I_t(\beta)$ in the spectrum is reduced to a value which is considerably below the incident-beam intensity I, particularly for thick specimens. Although this reduction of intensity is the most important effect of elastic scattering on the energy-loss spectrum, we will now consider the effect on the relative intensities of the different orders of inelastic scattering recorded by a spectrometer.

In an *amorphous* material, elastic and inelastic scattering are independent and both are governed by Poisson statistics, so the joint probability of m elastic and n inelastic events is

$$P(m, n) = (x_e/m!)(t/\lambda_e)^m (x_i/n!)(t/\lambda_i)^n \qquad (3.106)$$

where $x_e = \exp(-t/\lambda_e)$ and $x_i = \exp(-t/\lambda_i)$, λ_e and λ_i being the mean free paths for elastic and inelastic scattering through all angles. If a fraction $F(m, n)$ of these electrons passes through a collection aperture of semi-angle β, the recorded zero-loss component $I_0(\beta)$ for an incident-beam intensity I is given by

$$I_0(\beta)/I = \sum_0^\infty P(m, 0)F(m, 0) = x_e x_i + x_e x_i \sum_1^\infty (1/m!)(t/\lambda_e)^m F_m^e(\beta) \qquad (3.107)$$

Here, $x_e x_i$ represents the unscattered electrons and $F_m^e(\beta)$ is the fraction of m-fold *purely elastic* scattering which passes through the aperture. Calculations (Wong and Egerton, 1995) based on the Lenz model suggest that the m-fold elastic-scattering distribution approximates to a broadened single-scattering angular distribution with θ_0 replaced by $(0.7 + 0.5m)\theta_0$, so that Eq. (3.6) becomes, for $m > 1$:

$$F_m^e(\beta) \approx [1 + (0.7 + 0.5m)^2 \theta_0^2/\beta^2]^{-1} \qquad (3.108)$$

The *inelastic* intensity $I_i(\beta)$ transmitted through the aperture, integrated over all orders of scattering, is given by

$$I_i(\beta)/I = \sum_{m=0}^\infty \sum_{n=1}^\infty P(n, m)F(n, m) \qquad (3.109)$$

$$= x_e x_i \sum_{n=1}^\infty (1/n!)(t/\lambda_i)^n F_n^i(\beta) + \sum_{m=1}^\infty \sum_{n=1}^\infty P(n, m)F(m, n)$$

$F_n^i(\beta)$ is the fraction of *inelastically* scattered electrons which pass through the aperture, previously denoted $F_n(\beta)$. The final term in Eq. (3.109) represents electrons which have been scattered both elastically and inelastically. Calculations (Wong and Egerton, 1995) of the angular distribution of this "mixed" scattering, based on Lenz-model single-scattering distributions, indicate that the corresponding aperture function can be approximated by a simple product:

$$F(m, n) \approx F_{m}^{e}(\beta)F_{n}^{i}(\beta) \qquad (3.110)$$

Equation (3.110) probably results from the fact that the elastic and inelastic scattering have very different angular widths. Now Eq. (3.108) can be rewritten, making use of Eq. (3.107) and Eq. (3.97), as

$$I_i(\beta)/I = x_e x_i \sum_{n=1}^{\infty} (1/n!)(t/\lambda_i)^n F_n^i(\beta)[1 + \sum_{m=1}^{\infty} (t/\lambda_e)^m F_m^e(\beta)] \qquad (3.111)$$
$$= [I_0(\beta)/I][\exp(F_1^i t/\lambda_i) - 1]$$

Writing the total intensity recorded through the aperture as $I_t(\beta) = I_0(\beta) + I_i(\beta)$ and defining $t/\lambda_i(\beta) = F_1^i(t/\lambda_i)$ as before, Eq. (3.111) becomes

$$t/\lambda_i(\beta) = \ln[I_t(\beta)/I_0(\beta)] \qquad (3.112)$$

which is the same as Eq. (3.95), derived previously without considering elastic scattering. This equation, which is useful for measuring specimen thickness (Section 5.1), can therefore be justified mathematically on the basis of angular distributions of elastic and inelastic scattering, which are found to be good approximations in amorphous materials (Wong and Egerton, 1995). Experimentally, Eq. (3.112) has been verified to within 10% for $t/\lambda < 5$ in amorphous, polycrystalline, and single-crystal specimens (Hosoi et al., 1981; Leapman et al., 1984a).

Elastic scattering has a greater influence on the recorded intensity of *inner-shell* inelastic scattering (whose angular width is often comparable to that of elastic scattering), as discussed on page 268. In crystalline specimens, additional effects occur as a result of channeling; see page 138.

3.4.4. Multiple Scattering

For relatively thick specimens ($t/\lambda > 5$), individual peaks may not be visible in the loss spectrum; multiple outer- and inner-shell processes combine to produce a Landau distribution (Whelan, 1976; Reimer, 1989) which is broadly peaked around an energy loss of some hundreds of eV; see Fig. 3.30. The position of the maximum (the most probable energy loss) is roughly proportional to specimen thickness (Perez et al., 1977; Whitlock and Sprague, 1982) and is *very approximately* $(t/\lambda)E_p$, where E_p is the energy of the main peak in the single-scattering distribution. This multiple scattering behavior is sometimes called *straggling*.

When the number n of events is large, the angular distribution of scattering tends towards a Gaussian function, a consequence of the central limit theorem (Jackson, 1975). The mean-square angular deflection is n

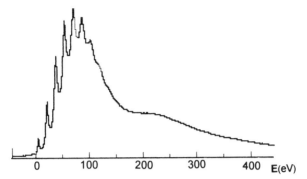

Figure 3.30. Energy-loss spectrum of a thick region of crystalline silicon ($t/\lambda = 4.5$), showing multiple plasmon peaks superimposed on a Landau background and followed by a broadened silicon L-edge.

$\langle\theta^2\rangle$, where $\langle\theta^2\rangle$ is the mean-square angle for single scattering, given by Eq. (3.54) in the case of a truncated Lorentzian angular distribution.

3.4.5. Coherent Double-Plasmon Excitation

A fast electron traveling through a solid can in principle lose a characteristic amount of energy equal to $2\hbar\omega_p$. In accordance with Eq. (3.94), the intensity at an energy loss $E = 2\hbar\omega_p$ should therefore be

$$I_{2p} = I_0[(t/\lambda_p)^2/2 + t/\lambda_{2p}] \tag{3.113}$$

where the first term represents the incoherent production of two plasmons in separate scattering events, in accordance with Poisson's law, while the second term represents *coherent double-plasmon* excitation, characterized by a mean free path λ_{2p}. As seen from Eq. (3.113), the coherent contribution would be fractionally greater in the case of very thin specimens. It should be visible directly if the energy-loss spectrum is deconvoluted to remove the incoherent plural scattering.

Based on a free-electron model, Ashley and Ritchie (1970) deduced that the relative probability P_{rel} of the double process is proportional to the fifth power of the cutoff wavevector q_c. Taking $q_c \approx \omega_p/v_F$ their formula becomes

$$P_{rel} = \lambda_p/\lambda_{2p} \approx 0.013r_s^2 \tag{3.114}$$

where r_s is the radius of a sphere containing one free electron, divided by the Bohr radius a_0. The lowest free-electron density (corresponding to $r_s = 5.7$) occurs in cesium, giving $P_{rel} = 0.34$. For aluminum, Eq. (3.114) yields $r_s = 2.0$ and $P_{rel} = 0.04$; by calculating a many-body Hamiltonian, Srivastava *et al.* (1982) obtained $P_{rel} = 0.024$.

Experimental determinations for aluminum have produced disparate values: 0.135 (Spence and Spargo, 1971), 0.07 (Batson and Silcox, 1983), <0.03 (Egerton and Wang, 1990), and ≤0.005 (Schattschneider *et al.*, 1987). The reason for these discrepancies is unknown; Schattschneider (1988) points out that small holes in the specimen or a variation in thickness would cause an overestimate of P_{rel} obtained via Fourier-log deconvolution. Although the double-plasmon process is of interest in terms of nonlinear physics and plasmon–electron coupling, its apparent low probability suggests that it can normally be neglected in quantitative analysis of the low-loss spectrum.

3.5. The Spectral Background to Inner-Shell Edges

Each ionization edge in the energy-loss spectrum is superimposed on a downward-sloping background which arises from other energy-loss processes and which may have to be subtracted when carrying out elemental analysis or interpreting core-loss fine structure. Since the background intensity is often comparable to or larger than the core-loss intensity, accurate subtraction of the background is essential, as discussed in Chapter 4. It is therefore desirable to minimize the background intensity, which involves understanding the energy-loss mechanisms giving rise to the background. In this section, we discuss possible contributions to the background in the case of a very thin specimen, then consider the effect of plural scattering, which becomes important as the sample thickness increases.

3.5.1. Valence-Electron Scattering

For energy losses below 50 eV, inelastic scattering from outer-shell electrons is largely a collective process in the majority of solids. A "plasmon" peak usually occurs in the range 10–30 eV, above which the intensity falls monotonically with increasing energy loss. Integration of Eq. (3.32) up to a collection semiangle β large compared to the characteristic angle θ_E gives

$$d\sigma/dE \propto \text{Im}(-1/\varepsilon) \, \ln(\beta/\theta_E) \qquad (3.115)$$

The Drude expression for $\text{Im}(-1/\varepsilon)$, Eq. (3.43), is proportional to E^{-3} for large E, so the fall-off of intensity within the plasmon "tail" should vary roughly as E^{-3}, ignoring the logarithm term in Eq. (3.115).

At large energy loss, however, the scattering is likely to have a single-electron character (see Section 3.3.2) and is more appropriately described by Bethe theory, using the same equations as applied to inner-shell excitation in Section 3.6. In particular, $d\sigma/dE$ would be expected to have a power-law energy dependence, as in Eq. (3.154), with the exponent of the order of 4 or 5 for small values of β. Such behavior was confirmed by measurements of the valence-electron scattering from thin films of carbon, in the energy range 100–280 eV (Egerton, 1975; Maher *et al.*, 1979).

Since the binding energy of a valence electron is small compared to the energy losses under consideration, a large proportion of the background intensity will occur at large scattering angles, in the form of a Bethe ridge (Fig. 3.36). Again, experimental data on carbon support this; see Fig. 3.31. In contrast, the inner-shell electrons have large binding energies and the characteristic energy losses are forward-peaked with an approximately Lorentzian angular distribution. Therefore, for the same energy loss, the core-loss intensity is concentrated into smaller scattering angles than is the background. Using a small collection aperture to record the energy-loss spectrum therefore enhances the edge/background ratio; see Fig. 3.32. However, the small aperture results in a weak core-loss signal, giving rise to a relatively high shot-noise component in the spectral data and a low signal/

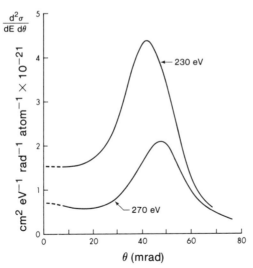

Figure 3.31. Angular dependence of the valence-electron scattering per unit scattering angle, recorded using a thin carbon specimen and 80-keV incident electrons (Egerton, 1975). Both curves correspond to energy losses below the *K*-ionization edge ($E_K = 284$ eV).

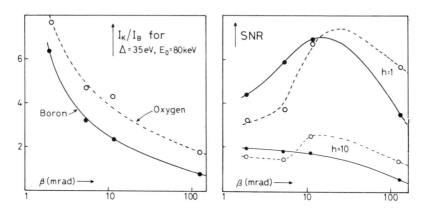

Figure 3.32. Signal/background ratio (I_K/I_b) and signal/noise ratio (SNR) as a function of collection semiangle, measured for the boron and oxygen K-edges in B_2O_3 (Egerton *et al.*, 1976). The parameter h is a factor which occurs in the formula for SNR (see Section 4.4.4) and which depends on the widths of the background-fitting and extrapolation regions.

noise ratio (SNR). Consequently, the SNR is an optimum at some *intermediate* value of collection semiangle, typically around 10 mrad (see Fig. 3.32).

3.5.2. Tails of Core-Loss Edges

In addition to the valence electrons, inner-shell electrons of lower binding energy may contribute intensity to the background underlying an ionization edge. If the preceding edge is prominent and not much lower in binding energy, the angular distribution of the background will be forward-peaked and of comparable width to that of the edge being analyzed. The advantage of a small collection angle (in terms of signal/background ratio) will then be less than in the case where the background arises mainly from valence-electron excitation. Where the two edges are well separated in energy, however, the angular distributions may be sufficiently dissimilar to allow a significant increase in signal/background ratio with decreasing collection angle, as in the case of the oxygen K-edge in B_2O_3 (Fig. 3.32).

3.5.3. Bremsstrahlung Energy Losses

When a transmitted electron undergoes centripetal acceleration in the nuclear field of an atom, it loses energy in the form of electromagnetic radiation (Bremsstrahlen). Although "*coherent* bremsstrahlung" peaks can be recorded from crystalline specimens in certain circumstances (Spence et al., 1983; Reese et al., 1984), the energy spectrum of the emitted photons usually forms a *continuous* background to the characteristic peaks observed in an x-ray emission spectrum.

The differential cross section for bremsstrahlung scattering into angles less than β can be written as (Rossouw and Whelan, 1979)

$$d\sigma/dE = CE^{-1}Z^2(v/c)^{-2} \ln[1 + (\beta/\theta_E)^2] \qquad (3.116)$$

where $C = 1.55 \times 10^{-31}$ m² per atom. Equation (3.116) neglects screening of the nuclear field by the atomic electrons, but is sufficient to show that (for energy losses below 4 keV) the bremsstrahlung background is small in comparison with that arising from electronic excitation (Isaacson and Johnson, 1975; Rossouw and Whelan, 1979).

3.5.4. Plural Scattering

Within the low-loss region ($E < 100$ eV), plural valence-electron or plasmon scattering contributes significant intensity unless the specimen thickness is much less than the plasmon mean free path (of the order of 100 nm for 100-keV electrons; see Section 3.3.1). At an energy loss of several hundred eV, however, multiple scattering involving only plasmon events makes a negligible contribution, since the required number n of scattering events is large and the probability P_n becomes vanishingly small as a result of the $n!$ denominator in Eq. (3.94). For example, a multiple-plasmon loss of $10E_p$ requires (on the average) 10 successive scattering events, giving $P_n < 10^{-6}$ for a sample thickness equal to the plasmon mean free path.

Similarly, it can be shown that the probability of two or more inelastic events of *comparable* energy loss is negligible when the total loss is greater than 100 eV. For example, if the single-scattering probability $P(E)$ is of the form AE^{-r}, the probability of two similar events (each of energy loss $E/2$) is $2^{2r}[P(E)]^2$, which at high energy loss is small compared to $P(E)$ because of the rapid fall-off in the differential cross section.

However, the probability of two *dissimilar* energy losses can be appreciable, as illustrated by use of the following simplified model. The low-loss spectrum is represented by a series of sharp peaks at multiples of the plasmon energy E_p, the area under each being given by Poisson statistics (Stephens, 1980). The energy dependence of the single-scattering background (arising from inner-shell or valence single-electron excitation) is taken to be $J^1(E) = AE^{-r}$, with A and r constants. Provided that scattering events are independent, the joint probability of several events is the product of the individual probabilities and the intensity at an energy loss E, due to one single-electron and n plasmon events, is

$$J^{1+n}(E) = A(E - nE_p)^{-r}(t/\lambda)^n \exp(-t/\lambda)/n! \qquad (3.117)$$

where λ is the plasmon mean free path. This equation allows the contributions from different orders of scattering to be compared for different values of t/λ; see Fig. 3.33.

Plural-scattering contributions to the ionization edge can be evaluated in a similar way (Fig. 3.33). Since the double (core-loss + plasmon) scattering is delayed until an energy loss $E = E_k + E_p$, the core-loss intensity just above the threshold E_k represents only *single* core-loss scattering. If the jump ratio (JR) of the edge is defined as the height of this *initial* rise divided by intensity of the preceding background, JR is clearly a decreasing function of specimen thickness. An alternative measure of the visibility of an edge

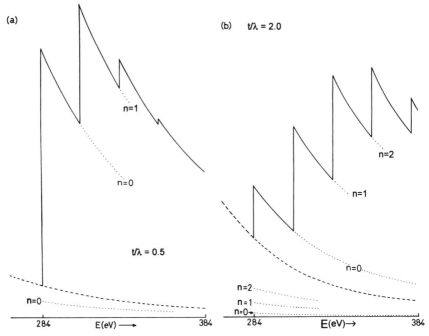

Figure 3.33. Contribution of plasmon scattering (up to order n) to the carbon K-edge (solid lines) and to its preceding background (dashed curve), calculated as a convolution of power-law background and edge profiles with an idealized low-loss spectrum (a series of delta functions at 25-eV intervals). In this simple model, the contribution of each successive order to the edge profile is visible as a sharp step in intensity; in practice, these steps are rounded and barely visible in some materials. The specimen thickness is (a) one half and (b) twice the total-inelastic mean free path.

is its signal/background ratio (SBR), defined as the ratio of core-loss intensity to background intensity integrated over equal energy intervals Δ immediately above and below the threshold. SBR decreases with thickness less rapidly than JR because some plural scattering (up to approximately Δ/E_p) is included in the core-loss integral. In fact, SBR would be independent of specimen thickness, for sufficiently large Δ. In practice, the fact that plural scattering causes the core-loss intensity to be spread over a larger energy range makes extrapolation of the background and the measurement of core-loss integral more difficult in thicker specimens.

Despite the approximations involved, Eq. (3.117) agrees quite well with measurements of SBR for low-energy ionization edges; see Fig. 3.34a. In the case of higher-energy edges (Fig. 3.34b), the agreement is less satisfactory, partly because of a spectrometer contribution to the background (Section 2.4.1). Although extremely thin specimens give the highest signal/background ratio, the core-loss signal itself is very weak, resulting in a high fractional noise content. As a result, the signal/noise ratio (SNR) may be

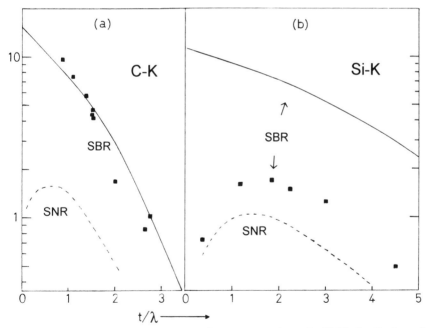

Figure 3.34. Signal/background ratio (SBR) and signal/noise ratio (SNR) for K-edges of (a) elemental carbon and (b) pure silicon, as a function of specimen thickness. Solid curves represent Eq. (3.117) and squares are measurements for 10-mrad collection semiangle, 100-keV incident energy, and 100-eV integration windows. SNR (in arbitary units) was calculated as $I_k/(I_k+hI_b)^{1/2}$, as discussed in Section 4.4.4.

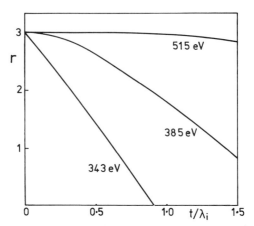

Figure 3.35. Change in slope parameter r characterizing the energy dependence (AE^{-r}) of a carbon loss spectrum, calculated as a function of specimen thickness t (Leapman and Swyt, 1983). The four chosen values of energy loss would immediately precede minor edges due to calcium, nitrogen, and oxygen in the loss spectrum of a biological sample. The slope parameter is assumed to be constant and equal to three in the absence of plural scattering $(t \to 0)$. From *Microbeam Analysis—1983*, p. 165, © San Francisco Press, Inc., by permission.

optimum at some intermediate thickness, as illustrated in Fig. 3.34b. SNR determines the visibility of an edge and the minimum detectable concentration in elemental analysis; see page 354.

A more general method of computing plural-scattering contributions is by self-convolution of the single-scattering energy distribution. Using this approach, Leapman and Swyt (1983) have shown that plural scattering usually causes the "background" exponent r to decrease with increasing sample thickness; see Fig. 3.35. The change in r is particularly large for energy losses just above a major ionization edge (e.g., in the range 300–400 eV for a carbonaceous sample) and is attributable to a change in overall shape of the major edge as a result of "mixed" scattering (see Section 3.7.3). The pre-edge background and jump ratio can also be calculated by Monte Carlo methods (Jouffrey *et al.*, 1985), allowing the precise effects of the collection aperture to be included.

3.6. Atomic Theory of Inner-Shell Excitation

Inner-shell electrons have relatively large binding energies; the associated energy losses are typically some hundreds of eV, corresponding to the x-ray region of the electromagnetic spectrum. As a result of this strong binding to the nucleus, collective effects are generally unimportant. Inner-shell excitation can therefore be described to a good approximation using single-atom models.

3.6.1. Generalized Oscillator Strength

The key quantity in Bethe theory (Section 3.2.2) is the generalized oscillator strength (GOS) which describes the response of an atom when a given energy and momentum are supplied from an external source (e.g., by collision of a fast electron). In order to calculate the GOS, it is necessary to know the initial- and final-state wavefunctions of the inner-shell electron. Calculations are based on several different methods.

The Hydrogenic Model

The simplest way of estimating the GOS, and the first to be developed (Bethe, 1930), is based on wave mechanics of the hydrogen atom. This approach is of interest in energy-loss spectroscopy because it provides realistic values of K-shell ionization cross sections with only a modest amount of computing, enabling cross sections needed for quantitative elemental analysis to be calculated "on line." The ease of application arises from the fact that analytical expressions are available for the wavefunctions of the hydrogen atom, obtained from an exact solution of the Schrödinger equation:

$$(-\hbar^2/2m_0)\nabla^2\psi - (e^2/4\pi\varepsilon_0 r)\psi = E_t\psi \qquad (3.118)$$

where E_t is the "net" (kinetic + electrostatic) energy of the atomic electron.

To make Eq. (3.118) applicable to an inner-shell electron within an atom of atomic number Z, the electrostatic term must be modified to take into account the actual nuclear charge Ze and screening of the nuclear field by the remaining $(Z - 1)$ electrons. Following Slater (1930), an "effective" nuclear charge Z_se is used in the Schrödinger equation. In the case of K-shell excitation, the second $1s$ electron screens the nucleus and reduces its effective charge by approximately $0.3e$, giving $Z_s = Z - 0.3$. For L-shell excitation: $Z_s = Z - (2 \times 0.85) - (7 \times 0.35)$, allowing for the screening effect of the two K-shell and seven remaining L-shell electrons. "Outer" electrons (i.e., those whose principal quantum number is higher than that of the initial-state wavefunction) are assumed to form a spherical shell of charge whose effect is to reduce the inner-shell binding energy by an amount E_s, so that the *observed* threshold energy for inner-shell ionization is

$$E_k = Z_s^2 R - E_s \qquad (3.119)$$

where $R = 13.6$ eV is the Rydberg energy. The Schrödinger equation of the atom is therefore

$$(-\hbar^2/2m_0)\nabla^2\psi - (Z_s e^2/4\pi\varepsilon_0 r)\psi + E_s\psi = E_t\psi \qquad (3.120)$$

The net energy E_t of the excited electron is related to its binding energy E_k and the energy E lost by the transmitted electron:

$$E_t = E - E_k \qquad (3.121)$$

Substituting Eq. (3.119) and Eq. (3.121) into Eq. (3.120) gives

$$(-\hbar^2/2m_0)\nabla^2\psi - (Z_s e^2/4\pi\varepsilon_0 r)\psi = (E - Z_s^2 R)\psi \qquad (3.122)$$

This is the Schrödinger equation for a "hydrogenic equivalent" atom with nuclear charge $Z_s e$ and no outer shells. Since the wavefunctions remain hydrogenic (or "Coulombic") in form, standard methods can be used to solve for the wavefunction ψ and for the GOS.

For convenience of notation, one can define dimensionless variables Q' and k_H which are related (respectively) to the scattering vector q and the energy loss E of the fast electron:

$$Q' = (qa_0/Z_s)^2 \qquad (3.123)$$

$$k_H^2 = E/(Z_s^2 R) - 1 \qquad (3.124)$$

The generalized oscillator strength (GOS) per atom is then given, for $E > Z_s^2 R$ and for K-shell ionization (Bethe, 1930; Madison and Merzbacher, 1975), by

$$\frac{df_K}{dE} = \frac{256E(Q' + k_H^2/3 + 1/3)\exp(-2\beta'/k_H)}{Z_s^4 R^2[(Q' - k_H^2 + 1)^2 + 4k_H^2]^3[1 - \exp(-2\pi/k_H)]} \qquad (3.125)$$

where β' is the value of $\arctan[2k_H/(Q' - k_H^2 + 1)]$ which lies within the range 0 to π. For energy losses in the range $E_k < E < Z_s^2 R$ (i.e., imaginary k_H, corresponding to transitions to discrete states *in the hydrogenic-equivalent atom*) an appropriate expression for the K-shell GOS is (Egerton, 1979)

$$\frac{df_K}{dE} = \frac{256E(Q' + k_H^2/3 + 1/3)\exp(y)}{Z_s^4 R^2[(Q' - k_H^2 + 1)^2 + 4k_H^2]^3} \qquad (3.126)$$

where

$$y = -(-k_H^2)^{-1/2}\log_e\left[\frac{Q' + 1 - k_H^2 + 2(-k_H^2)^{1/2}}{Q' + 1 - k_H^2 - 2(-k_H^2)^{1/2}}\right] \qquad (3.127)$$

Corresponding formulas have been derived for L-shell GOS (Walske, 1956; Choi *et al.*, 1973) and for M-shell ionization (Choi, 1973).

Hartree–Slater Method

Accurate wavefunctions have now been computed for most atoms by iterative solution of the Schrödinger equation with a self-consistent atomic potential. The Hartree–Slater (HS or HFS) method represents a simplification of the Hartree–Fock (HF) procedure, by assuming a central (spherically symmetric) field within the atom. The resulting wavefunctions are close to those obtained using the HF method but require much less computing. The radial component $\phi_0(r)$ of the ground-state wavefunction has been tabulated by Herman and Skillman (1963). For calculation of the GOS, the final-state radial function ϕ_n is obtained by solving the radial Schödinger equation for a net (continuum) energy E_t:

$$\left[\frac{\hbar^2}{2m_0} \frac{d^2}{dr^2} - V(r) - \frac{l'(l'+1)\hbar^2}{2m_0 r^2} + E_t \right] \phi_n(r) = 0 \qquad (3.128)$$

where l' is the angular-momentum quantum number of the final (continuum) state.

Using a central-field model, it is not possible to provide an exact treatment of electron exchange, but an approximate correction can be made by assuming an exchange potential of the form (Slater, 1951)

$$V^x = -6[(3/8\pi)\rho(r)]^{-1/3} \qquad (3.129)$$

where $\rho(r)$ is the spherically averaged charge density within the atom.

The transition-matrix element, defined by Eq. (3.23), can be written (Manson, 1972) as

$$|\varepsilon_{nl}|^2 = \sum_{l'} (2l'+1) \sum_{\lambda} (2\lambda+1) \left[\int_0^\infty \phi_0(r) J_\lambda(qr) \phi_n(r) \, dr \right]^2 \left| \begin{pmatrix} l' & \lambda & l \\ 0 & 0 & 0 \end{pmatrix} \right|^2 \qquad (3.130)$$

where the operator $\exp(i\mathbf{q} \cdot \mathbf{r})$ has been expanded in terms of spherical Bessel functions $J_\lambda(qr)$ and the integration over angular coordinates is represented by a Wigner $(3-j)$ matrix. The GOS is obtained from Eq. (3.24), summing over all important *partial waves* corresponding to different values of l'. Such calculations have been carried out by McGuire (1971), Manson (1972), Scofield (1978), Leapman *et al.* (1980), and Rez (1982, 1989); the results will be discussed later in this section, and in Section 3.7.1.

E- and q-Dependence of the GOS

The generalized oscillator strength is a function of both the energy E and the momentum $\hbar q$ supplied to the atom, and is most conveniently portrayed on a two-dimensional plot known as a *Bethe surface*, an example

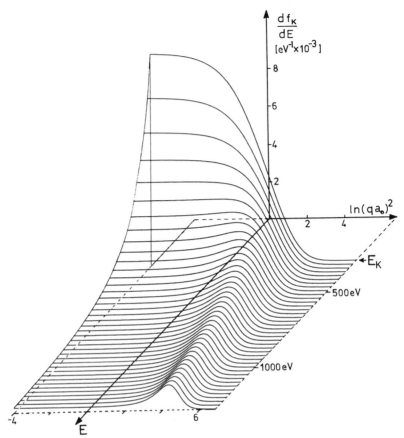

Figure 3.36. Bethe surface for K-shell ionization of carbon, calculated using a hydrogenic model (Egerton, 1979). The generalized oscillator strength is zero for energy loss E below the ionization threshold E_K. The horizontal coordinate is related to scattering angle.

of which is shown in Fig. 3.36. The individual curves in this figure represent *qualitatively* the *angular* dependence of inner-shell scattering, since the double-differential cross section $d^2\sigma/d\Omega\, dE$ is proportional to $E^{-1}q^{-2}\, df/dE$, as in Eq. (3.26), while q^2 increases approximately with the square of the scattering angle [Eq. (3.27)]. For an energy loss not much larger than the inner-shell binding energy E_k, the angular distribution is forward-peaked (maximum intensity at $\theta = 0$, $q = q_{min} \cong k\theta_E$) and corresponds to the *dipole region* of scattering. On a particle model, this low-angle scattering represents "soft" collisions with relatively large impact parameter.

At large energy loss, the scattering becomes concentrated into a *Bethe ridge* (Fig. 3.36) centered around a value of q which satisfies

$$(qa_0)^2 = E/R + E^2/(2m_0c^2R) \approx E/R \qquad (3.131)$$

for which the equivalent scattering angle θ_r is given (Williams *et al.*, 1984) by

$$\sin^2 \theta_r = (E/E_0)[1 + (E_0 - E)/(2m_0c^2)]^{-1} \qquad (3.132)$$

or $\theta_r \cong (E/E_0)^{1/2} \cong (2\theta_E)^{1/2}$ for small θ and nonrelativistic incident electrons. This high-angle scattering corresponds to "hard" collisions with small impact parameter, where the interaction involves mainly the electrostatic field of a single inner-shell electron and is largely independent of the nucleus. In fact, the $E - q$ relation represented by Eq. (3.131) is simply that for Rutherford scattering by a free, stationary electron; the nonzero width of the Bethe ridge reflects the effect of nuclear binding or (equivalently) the nonzero kinetic energy of the inner-shell electron.

The energy dependence of the GOS is obtained by taking cross sections through the Bethe surface at constant q. In particular, planes corresponding to very small values of q (left-hand boundary of Fig. 3.36) give the inner-shell contribution $df_k(0, E)/dE$ to the *optical* oscillator strength per unit energy $df(0, E)/dE$, which is proportional to the *photoabsorption cross section* σ_0:

$$df(0, E)/dE = df_k(0, E)/dE + (df/dE)' = \sigma_0/C \qquad (3.133)$$

where $(df/dE)'$ represents the "background" contribution from outer shells of lower binding energy and $C = 1.097 \times 10^{-20}$ m^2 eV (Fano and Cooper, 1968). Experimental values of photoabsorption cross section have been tabulated (Hubbell, 1971; Veigele, 1973) and these data can be used to test the results of single-atom calculations of the GOS.

Such a comparison is shown in Figs. 3.37 and 3.38. For K-shell ionization, a hydrogenic calculation predicts quite accurately the overall shape of the absorption edge and the absolute value of the photoabsorption cross section. In the case of L-shells, the hydrogenic model gives too large an intensity just above the absorption threshold (particularly for the lighter elements) and too low a value at high energies. This discrepancy arises from the oversimplified treatment of screening in the hydrogenic model, where the effective nuclear charge Z_s is taken to be independent of the atomic coordinate r. In reality, energy losses just above the threshold involve interaction further from the nucleus where the effective charge is smaller (because of outer-shell screening), giving an oscillator strength lower than the hydrogenic value. Conversely, energy losses much larger than E_k correspond to close collisions for which Z_s approaches the full nuclear charge,

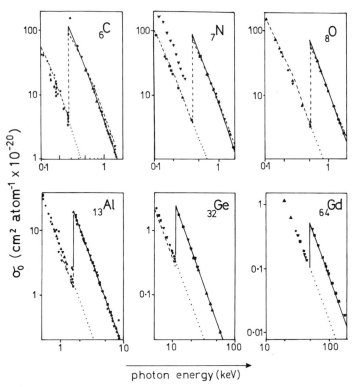

photon energy (keV)

Figure 3.37. K-shell x-ray absorption edges of several elements. Dotted lines represent the extrapolated outer-shell contribution and solid lines are the photoabsorption cross section calculated using a hydrogenic model (Egerton, 1979). The experimental points are taken from Hubbell (1971) and the dashed lines are based on Hartree-Slater calculations of McGuire (1971).

resulting in an oscillator strength slightly higher than the hydrogenic prediction. Also, for low-Z elements, the L_{23} absorption edge has a rounded shape (Fig. 3.38) due to the influence of the "centrifugal" term $\hbar^2 l'(l' + 1)/2m_0 r^2$ in the effective potential [see Eq. (3.128)]. To be useful for L-shells, the hydrogenic model requires an energy-dependent correction chosen to match the observed edge shape of each element (Egerton, 1981a). The correction is even larger in the case of M-shells (Luo and Zeitler, 1991).

The Hartree-Slater model takes proper account of screening and gives a good prediction of the edge shape in many elements (Fig. 3.38). In other cases (e.g., L_{23}-edges of transition metals) agreement is worse, owing to the fact that the calculations deal usually only with ionizing transitions to the continuum and neglect excitation to discrete (bound) states just above the absorption threshold (Section 3.7.1). Moreover, a free-atom

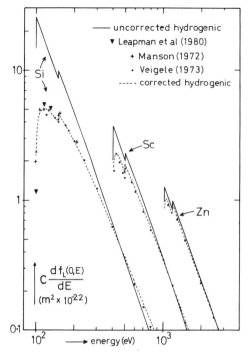

Figure 3.38. *L*-shell photoabsorption cross section, as predicted by hydrogenic calculations (Egerton, 1981a) and by the Hartree-Slater model (Manson, 1972; Leapman *et al.*, 1980). Experimental data points are also shown (Veigele, 1973).

model cannot predict the solid-state fine structure which becomes prominent close to the threshold (Section 3.8).

At large q, a constant-q section of the Bethe surface intersects the Bethe ridge (Fig. 3.36). As a result, the energy-loss spectrum of large-angle scattering contains a broad peak which is analogous to a Compton profile and whose shape reflects the momentum distribution of the atomic electrons; see page 378.

If df/dE is known as a function of q and E, the angular and energy dependence of scattering can be calculated from Eq. (3.26), provided the relationship between q and the scattering angle θ is also known.

3.6.2. Kinematics of Scattering

In the case of elastic scattering, conservation of momentum leads to a simple relation between the magnitude q of the scattering vector and the scattering angle θ (see Fig. 3.2), a given value of q corresponding to a single value of θ. In the case of inelastic scattering, the value of q depends on both the scattering angle and the energy loss.* The relationship between q and θ is derived by applying the conservation of both momentum and energy to the collision. Since a 100-keV electron has a velocity more than half that of the speed of light and a relativistic mass 20% higher than its rest mass, it is necessary to use relativistic kinematics to derive the required relationship.

Conservation of Energy

The total energy W of an incident electron ($=$kinetic energy E_0 + rest energy m_0c^2) is given by

$$W = \gamma m_0 c^2 \tag{3.134}$$

where $\gamma = (1 - v^2/c^2)^{-1/2}$. The incident momentum is

$$p = \gamma m_0 v = \hbar k_0 \tag{3.135}$$

Combining Eq. (3.134) and Eq. (3.135), we obtain

$$W = [(m_0c^2) + p^2c^2]^{1/2} = [(m_0c^2)^2 + \hbar^2 k_0^2 c^2]^{1/2} \tag{3.136}$$

Conservation of energy indicates that

$$W - E = W' = [(m_0c^2)^2 + \hbar^2 k_1^2 c^2]^{1/2} \tag{3.137}$$

where W' is the total energy of the scattered electron and E is the energy loss. Using Eq. (3.136) in Eq. (3.137) leads to an equation relating the change in magnitude of the fast-electron wavevector to the energy loss:†

$$k_1^2 = k_0^2 - 2E[m_0^2/\hbar^4 + k_0^2/(\hbar c)^2]^{1/2} + E^2/(\hbar c)^2 \tag{3.138}$$
$$= k_0^2 - 2\gamma m_0 E/\hbar^2 + E^2/(\hbar c)^2$$

Note that this relationship is independent of the scattering angle.

For numerical calculations, it is convenient to convert each wavevector into a dimensionless quantity by multiplying by the Bohr radius a_0. Making

---

* This indicates an additional degree of internal freedom which on a classical (particle) model of scattering corresponds to the interaction between the incident and atomic electrons taking place at different points within the electron orbit.

† For Rutherford scattering from a free electron (where $E = \hbar^2 q^2/2m_0$) and low incident energies (such that $E_0 = \hbar k_0^2/2m_0$), Eq. (3.138) becomes $k_1^2 = k_0^2 - q^2$, indicating that the angle between $\mathbf{q}$ and $\mathbf{k}_1$ is 90°. In this case, q goes to zero for $\theta \to 0$ (see Fig. 3.39), as implied by Eqs. (3.131) and (3.132).

use of the equality $Ra_0^2 = \hbar^2/2m_0$, where R is the Rydberg energy, Eq. (3.138) becomes

$$(k_1a_0)^2 = (k_0a_0)^2 - (E/R)[\gamma - E/(2m_0c^2)] \qquad (3.139)$$

The E^2 term in Equation (3.139) is insignificant for most inelastic collisions. The value of $(k_0a_0)^2$ is obtained from the incident kinetic energy E_0 or from the "relativistically corrected" energy $T = m_0v^2/2$:

$$(k_0a_0)^2 = (E_0/R)(1 + E_0/2m_0c^2) = (T/R)/(1 - 2T/m_0c^2) \qquad (3.140)$$

Conservation of Momentum

Momentum conservation is incorporated by applying the cosine rule to the vector triangle (Fig. 3.39), giving

$$q^2 = k_0^2 + k_1^2 - 2k_0k_1 \cos \theta \qquad (3.141)$$

Differentiation of this equation gives (for constant E and E_0)

$$d(q)^2 = 2k_0k_1 \sin \theta \, d\theta = (k_0k_1/\pi) \, d\Omega \qquad (3.142)$$

Substituting Eq. (3.138) into Eq. (3.141) then gives

$$(qa_0)^2 = \frac{2T\gamma^2}{R}\left[1 - \left(1 - \frac{E}{\gamma T} + \frac{E^2}{2\gamma^2 Tm_0c^2}\right)^{1/2} \cos \theta\right] - \frac{E\gamma}{R} + \frac{E^2}{2Rm_0c^2} \qquad (3.143)$$

Equation (3.143) can in principle be used to compute qa_0 for any value of θ, but for small θ this procedure would require high-precision arithmetic, since evaluation of the brackets in Eq. (3.143) involves substracting almost identical numbers. For $\theta = 0$, corresponding to $q = q_{min} = k_0 - k_1$, binomial expansion of the square root in Eq. (3.143) indicates that terms up to second order in E cancel, giving

$$(qa_0)^2_{min} = E^2/4RT + E^3/(8\gamma^3RT^2) + \text{higher-order terms} \qquad (3.144)$$

For $\gamma^{-3}E/T \ll 1$ (which applies to practically all collisions), only the E^2 term is of importance and Eq. (3.144) can be written in the form

$$q_{min} \cong k_0\theta_E \qquad (3.145)$$

where $\theta_E = E/(2\gamma T) = E/(\gamma m_0v^2)$ is the "characteristic" scattering angle.

For a nonzero scattering angle, it is convenient to evaluate the corresponding value of q from

$$\begin{aligned}(qa_0)^2 &= (k_0a_0 - k_1a_0)^2 + 2(k_0a_0)(k_1a_0)(1 - \cos \theta) \\ &= (qa_0)^2_{min} + 4(k_0a_0)(k_1a_0) \sin^2(\theta/2) \\ &\cong (qa_0)^2_{min} + 4\gamma^2(T/R) \sin^2(\theta/2)\end{aligned} \qquad (3.146)$$

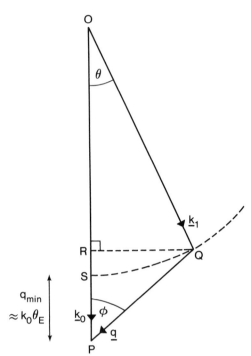

Figure 3.39. Vector triangle for inelastic scattering. The dashed circle represents the locus of point Q which defines the different values of q and θ possible for a given value of k_1, equivalent to a given energy loss; see Eq. (3.138). For $E \ll E_0$ and small θ, $RP \approx SP \approx k_0\theta_E$ and $RQ \approx k_1\theta \approx k_0\theta$; applying Pythagoras' rule to the triangle PQR then leads to Eq. (3.147).

For $\theta \ll 1$ rad, Eq. (3.146) is equivalent to

$$q^2 \cong q_{min}^2 + 4k_0^2(\theta/2)^2 \cong k_0^2(\theta^2 + \theta_E^2) \qquad (3.147)$$

Equation (3.147) is valid outside the dipole region, provided $\sin\theta \cong \theta$, and is relativistically correct provided θ_E is defined as $E/pv = E/2\gamma T$ rather than as $E/2E_0$.

3.6.3. Ionization Cross Sections

For small scattering angles ($\theta \ll 1$ rad), the energy-differential cross section can be obtained by integrating Eq. (3.29) up to an appropriate collection angle β:

$$\frac{d\sigma}{dE} \cong \frac{4R\hbar^2}{Em_0^2v^2} \int_0^\beta \frac{df(q, E)}{dE} 2\pi\theta(\theta^2 + \theta_E^2)^{-1}\, d\theta \qquad (3.148)$$

Within the dipole region of scattering, where $(qa_0)^2 < 1$ (equivalent to $\beta < 10$ mrad at the carbon K-edge, for 100-keV incident electrons) the GOS is approximately constant and equal to the optical value $df(0, E)/dE$, so Eq. (3.148) becomes

$$\frac{d\sigma}{dE} = \frac{4\pi a_0^2 R^2}{ET} \frac{df(0, E)}{dE} \ln[1 + (\beta/\theta_E)^2] \tag{3.149}$$

To evaluate Eq. (3.148) outside the dipole region, df/dE can be computed for each angle θ, related to q by Eq. (3.141) or Eq. (3.147).

Alternatively, one can use q or qa_0 as the variable of integration. From Eq. (3.26) and Eq. (3.142), we have

$$\frac{d\sigma}{dE} = \frac{4\pi \gamma^2 R}{Ek_0^2} \int \frac{df(q, E)}{dE} \frac{d(q^2)}{q^2} \tag{3.150}$$

$$= 4\pi a_0^2 \left(\frac{E}{R}\right)^{-1} \left(\frac{T}{R}\right)^{-1} \int \frac{df(q, E)}{dE} d[\ln(qa_0)^2] \tag{3.151}$$

where $T = m_0 v^2/2$, $R = \hbar^2/(2m_0 a_0^2) = 13.6$ eV, and the limits of integration are, from Eq. (3.144) and Eq. (3.146)

$$(qa_0)^2_{\min} \cong E^2/(4RT) \tag{3.152}$$

$$(qa_0)^2_{\max} \cong (qa_0)^2_{\min} + 4\gamma^2(T/R) \sin^2(\beta/2) \tag{3.153}$$

Integration over a logarithmic grid, as implied by Eq. (3.151), is convenient for numerical (computer) evaluation, because in the dipole region $d\sigma/dE$ peaks sharply at small angles but varies much more slowly at larger θ. From Eq. (3.151) one can see that the energy-differential cross section is proportional to the area under a constant-E section through the Bethe surface between $(qa_0)_{\min}$ and $(qa_0)_{\max}$.*

A computation of $d\sigma/dE$ is shown in Fig. 3.40. Logarithmic axes are used in order to illustrate the approximate behavior:

$$d\sigma/dE \propto E^{-s} \tag{3.154}$$

where s is the downward slope in Fig. 3.40 and is constant over a limited range of energy loss. The value of s is seen to depend on the size of the collection aperture, an effect which has been confirmed experimentally (Maher *et al.*, 1979).

For large β, such that most of the inner-shell scattering contributes to the loss spectrum, s is typically about 3 at the ionization edge ($E = E_k$), decreasing towards 2.0 with increasing energy loss. The asymptotic E^{-2}

---

* The *volume* under the Bethe surface is a measure of the electron stopping power.

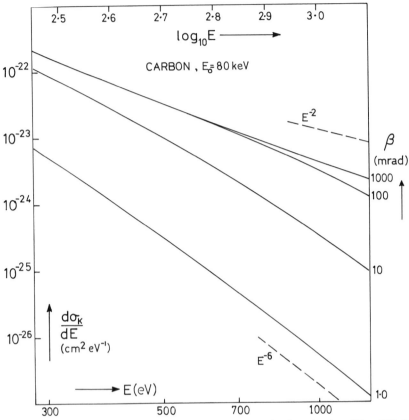

Figure 3.40. Energy-differential cross section for K-shell ionization of carbon ($E_K = 284$ eV) calculated for different collection semiangles β using hydrogenic wavefunctions (Egerton, 1979). $d\sigma_K/dE$ represents the K-loss intensity, after subtraction of the pre-edge background.

behavior is due to the fact that for $E \gg E_k$ practically all the scattering lies within the Bethe ridge and approximates to Rutherford scattering from a free electron, for which $d\sigma/dE \propto q^{-4} \propto E^{-2}$.

For small β, s *increases* with increasing energy loss, the largest value (just over 6) corresponding to large E and very small β. Equation (3.148) gives $d\sigma/dE \propto E^{-1}\theta_E^{-2} df(0, E)/dE \propto E^{-3} df(0, E)/dE$ for very small β, while $df(0, E)/dE \propto E^{-3.5}$ for K-shell excitation and $E \to \infty$ (Rau and Fano, 1967); therefore an asymptotic $E^{-6.5}$ is to be expected.

For thin specimens in which plural scattering is negligible, the inner-shell contribution to the energy loss spectrum (recorded with a collection semiangle β) is the single-scattering intensity $J_k^1(\beta, E)$ given by

$$J_k^1(\beta, E) = NI_0 \, d\sigma/dE \qquad (3.155)$$

where N is the number of atoms per unit specimen *area* contributing to the ionization edge and I_0 is the integrated zero-loss intensity.

Partial Cross Section

For quantitative element analysis, the inner-shell intensity is integrated over an energy range of width Δ beyond an ionization edge. Assuming a thin specimen (negligible plural scattering) the integrated intensity is

$$I_k^1(\beta, \Delta) = N I_0 \, \sigma_k(\beta, \Delta) \tag{3.156}$$

where the "partial" cross section $\sigma_k(\beta, \Delta)$ is defined by

$$\sigma_k(\beta, \Delta) = \int_{E_k}^{E_k+\Delta} \frac{d\sigma}{dE} \, dE \tag{3.157}$$

For numerical integration of $d\sigma/dE$, use can be made of the power-law behavior, Eq. (3.154), to reduce the required number of energy increments to less than 10, as illustrated on page 422.

Figure 3.41 shows the calculated angular dependence of K-shell partial cross sections for first-row (second-period) elements. The cross sections saturate at large values of β (i.e., above the Bethe-ridge angle θ_r) due to the fall-off in df/dE outside the dipole region. The median scattering angle (for energy losses in the range E_k to $E_k + \Delta$) corresponds to a partial cross section equal to one half the saturation value, and is typically $5\bar{\theta}_E$, where $\bar{\theta}_E = (E_k + \Delta/2)/(2\gamma T)$. Figure 3.41 shows that whereas the saturation cross sections decrease with increasing incident energy, the low-angle values increase. This behavior arises from the fact that a small collection aperture accepts a greater fraction of the scattering when the incident energy is high and the angular distribution is therefore more sharply peaked in the forward direction.

Integral and Total Cross Sections

For a very large integration range Δ, the partial cross section becomes equivalent to the "integral" cross section $\sigma_k(\beta)$ for inner-shell scattering into angles up to β and all permitted values of energy loss. The integral cross section can be evaluated by choosing the upper limit of integration in Eq. (3.157) such that contributions from higher energy losses are negligible. For small β, taking an upper limit equal to $3E_k$ gives less than 1% error due to higher losses, but for large β the limit must be set higher because of contributions from the Bethe ridge.

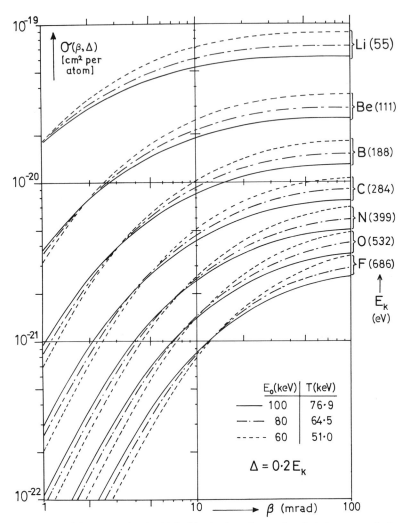

Figure 3.41. Partial cross section for K-shell ionization of second-period elements, calculated for an integration window Δ equal to one fifth of the edge energy, assuming hydrogenic wavefunctions and nonrelativistic kinematics (Egerton, 1979).

For β less than the Bethe-ridge angle $\theta_r \cong (2\theta_E)^{1/2}$, the integral cross section can be predicted with moderate accuracy (Egerton, 1979) by using a formula analogous to Eq. (3.149), namely,

$$\sigma_k(\beta) \simeq 4\pi a_0^2 (R/T)(R/\bar{E})f_k \ln[1 + (\beta/\bar{\theta}_E)^2] \qquad (3.158)$$

where the mean energy loss $\bar{E}$ is defined by

$$\overline{E} = \int_0^\infty E\left(\frac{d\sigma}{dE}\right) dE \bigg/ \int_0^\infty \frac{d\sigma}{dE} dE \qquad (3.159)$$

and $\overline{\theta}_E = \overline{E}/2\gamma T$. The quantity f_k in Eq. (3.158) is the dipole oscillator strength, which for K-shell ionization is given approximately by $f_k \approx 2.1 - Z/27$ and typically $\overline{E} \approx 1.5E_k$.

By setting $\beta = \pi$, the integral cross section becomes equal to the total cross section σ_k for inelastic scattering from shell k. An approximate expression for σ_k is the "Bethe asymptotic cross section" (Bethe, 1930):

$$\sigma_k \cong 4\pi a_0^2 N_k b_k (R/T)(R/E_k) \ln(c_k T/E_k) \qquad (3.160)$$

where N_k is the number of electrons in shell k (2, 8, and 18 for K-, L-, and M-shells), while b_k ($\approx f_k/N_k$) and c_k ($\approx 4E_k/\overline{E}$) are factors which can be obtained by calculation or from experimental measurements of cross section (Inokuti, 1971; Powell, 1976). If experimental values are available for different incident energies, a plot of $T\sigma_k$ against $\ln T$ (known as a Fano plot) should yield a straight line, according to Eq. (3.160). Linearity of the Fano plot is sometimes used as a test of the reliability of the measured cross sections, or (alternatively) of the applicability of Bethe theory (for example at low incident energies). The slope and intercept of the plot give the values of b_k and c_k. At incident-electron energies above about 200 keV, Bethe theory must be modified to take account of retardation effects and a modified form of the Fano plot is required (Appendix A).

3.7. Appearance of Inner-Shell Edges

In this section, we consider first the overall shape of ionization edges, as deduced from atomic calculations (Leapman et al., 1980; Rez, 1982), photoabsorption data (Hubbell, 1971; Veigele, 1973) and libraries of EELS data (Zaluzec, 1981; Ahn and Krivanek, 1983). We concentrate on edges within the energy range 50–2000 eV which are readily observable by EELS. A table of edge shapes and edge energies is given in Appendix D, together with a table showing the relationship between the quantum-mechanical and spectroscopy notations for inner-shell excitation.

3.7.1. Basic Edge Shapes

Because the wave functions of the core electrons change relatively little when atoms aggregate to form a solid, an atomic model provides a useful indication of the general shape of inner-shell edges. Following Manson (1972), Leapman et al. (1980) calculated differential cross sections for

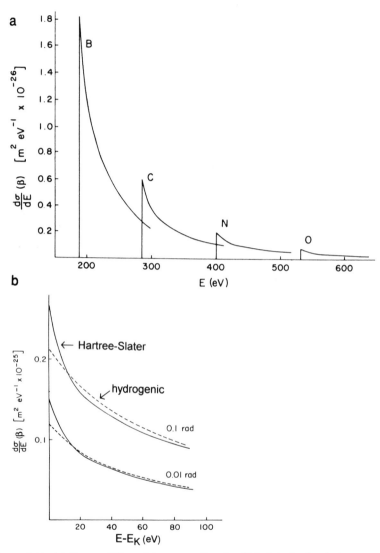

Figure 3.42. (a) Energy-differential cross section for K-shell ionization in boron, carbon, nitrogen, and oxygen, calculated for 80-keV incident electrons and 3-mrad collection semiangle using the Hartree-Slater method. (b) Comparison of Hartree-Slater and hydrogenic cross sections for the carbon K-edge, taking $E_0 = 80$ keV and collection semiangles of 10 mrad and 100 mrad (Leapman et al., 1982).

K-, L-, and M-shell ionization, using Bethe theory and the Hartree–Slater central-field model (Section 3.6.1). Their results for K-shell edges are shown in Fig. 3.42a, where the vertical axis represents the core-loss intensity after background subtraction and in the absence of plural scattering and instrumental broadening. Although the vertical scale in Fig. 3.42 refers to an incident energy of 80 keV and a collection semiangle β of 3 mrad, the K-edges retain their characteristic "sawtooth" shape for different values of E_0 and β; see Fig. 3.42b.

As seen in Fig. 3.43, experimentally determined K-shell edges conform to this same overall shape but with the addition of some pronounced fine structure. The K-ionization edges remain basically sawtooth-shaped for third-period elements (Na to Cl), although their intensities are lower, resulting in a relatively high noise content in the experimental data.

Calculations of L_{23} edges (due to excitation of $2p$ electrons) in third-period elements (Na to Cl) show that they have a more rounded profile; see Fig. 3.44. The intensity exhibits a *delayed maximum* 10–20 eV above the ionization threshold, resulting from the $l'(l + 1)$ term in the radial Schrödinger equation, Eq. (3.128), which causes a maximum to appear in the effective atomic potential. At energies just above the ionization threshold, this "centrifugal barrier" prevents overlap between the initial ($2p$) and final-state wavefunctions, particularly for final states with a large angular-momentum quantum number l'. Measured L_{23} edges display this delayed maximum (Ahn and Krivanek, 1983), although excitonic effects tend to sharpen the edge, particularly in the case of insulating materials (Section 3.8.5).

Fourth-period elements give rise to quite distinctive L-edges. Atomic calculations predict that the L_{23} edges of K, Ca, and Sc will be sharply peaked at the ionization threshold because of a "resonance" effect: the dipole selection rule favors transitions to final states with d-character ($l' = 2$) and the continuum $3d$ wavefunction is sufficiently compact to fit mostly within the centrifugal barrier, resulting in strong overlap with the core-level ($2p$) wavefunction and a large oscillator strength at threshold (Leapman *et al.*, 1980). These sharp threshold peaks are known as *white lines* since they were first observed in x-ray absorption spectra where the high absorption peak at the ionization threshold resulted in almost no blackening on a photographic plate. The white lines are mainly absent in the calculated L_{23}-edge profiles of transition-metal atoms (Fig. 3.44) because the calculations neglect excitation to bound states (i.e., discrete unoccupied $3d$ levels). In a solid, however, these atomic levels form a narrow energy band with a high density of vacant d-states, leading to the strong threshold peaks observed experimentally; see Fig. 3.45.

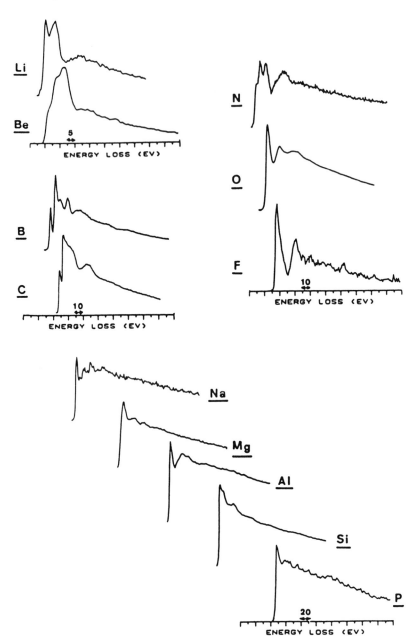

Figure 3.43. *K*-ionization edges (after background subtraction) measured by EELS with 120-keV incident electrons and collection semiangles in the range 3–15 mrad (Zaluzec, 1982).

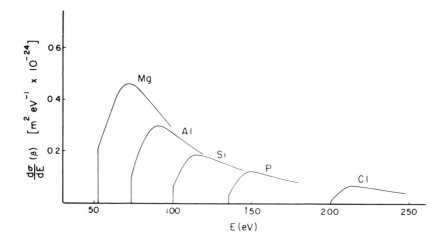

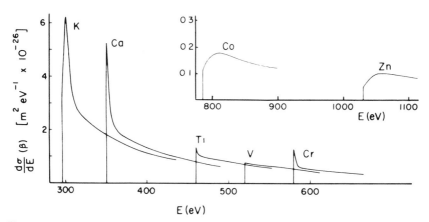

Figure 3.44. Hartree-Slater calculations of L_{23} edges, for 80-keV incident electrons and a collection semiangle of 10 mrad (Leapman *et al.*, 1980).

Spin–orbit splitting causes the magnitude of the L_2 binding energy to be slightly higher than that of the L_3 level. Consequently, two threshold peaks can be observed, whose separation increases with increasing atomic number (Fig. 3.45). The ratio of intensities of the L_3 and L_2 white lines has been found to deviate from the "statistical" value (2.0) based on the relative occupancy of the initial-state levels (Leapman *et al.*, 1982). The explanation may involve spin coupling between the core hole and the final state (Barth *et al.*, 1983).

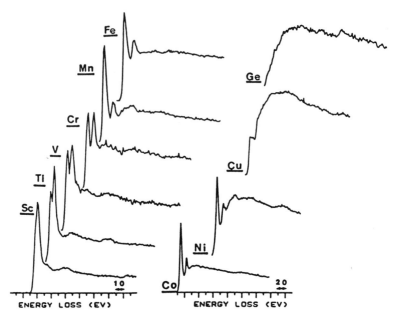

Figure 3.45. *L*-edges of fourth-period elements measured using 120-keV incident electrons and a collection semiangle of 5.7 mrad (Zaluzec, 1982). The background intensity before each edge has been extrapolated and subtracted, as described in Chapter 4.

In the case of metallic copper, the d-band is full and threshold peaks are absent, but in compounds such as CuO electrons are drawn away from the copper atom, leading to empty d-levels and sharp L_2 and L_3 threshold peaks; see Fig. 3.46. Fourth-period elements of higher atomic number (Zn to Br) have full d-shells and the L_{23} edges delayed maxima, as in the case of Ge (Fig. 3.45).

Hartree–Slater calculations of L_1 edges indicate that they have a saw-tooth shape (like K-edges) but relatively low intensity. They are usually observable as a small step on the falling background of the preceding L_{23} edge; see Fig. 3.45.

M_{45} edges are prominent for fifth-period elements, and appear with the intensity maximum delayed by 50 to 100 eV beyond the threshold (Fig. 3.47) because the centrifugal potential suppresses the optically preferred $3d \rightarrow 4f$ transitions just above threshold (Manson and Cooper, 1968). Within the 6th period, between Cs ($Z = 55$) and Yb ($Z = 70$), white-line peaks occur at the threshold due to a high density of unfilled f-states (Fig. 3.48a). The M_4–M_5 splitting and the M_5/M_4 intensity ratio increase with the atomic number (Brown *et al.*, 1984; Colliex *et al.*, 1985). Above $Z = 71$, the M_4

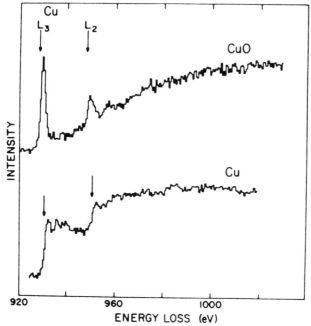

Figure 3.46. Cu L_{23} edges in metallic copper and in cupric oxide, measured using 75-keV incident electrons scattered up to $\beta = 2$ mrad (Leapman *et al.*, 1982).

and M_5 edges occur as rounded steps (Ahn and Krivanek, 1983), making them harder to recognize.

M_{23} edges of elements near the beginning of the fourth period (K to Ti) occur below 40 eV, superimposed on a rapidly falling valence-electron background which makes them appear more like plasmon peaks than as typical edges. M_{23} edges of the elements V to Zn are fairly sharp and resemble K-edges (Hofer and Wilhelm, 1993); M_1 edges are weak and are rarely observed in energy-loss spectra.

N_{67}, O_{23}, and O_{45} edges have been recorded for some of the heavier elements (Ahn and Krivanek, 1983). In thorium and uranium, the O_{45} edges are prominent as a double-peak structure (spin–orbit splitting $\simeq 10$ eV) between 80- and 120-eV energy loss; see Fig. 3.48b.

3.7.2. Dipole Selection Rule

Particularly when a small collection aperture is employed, the transitions which appear prominently in energy-loss spectra are those for which

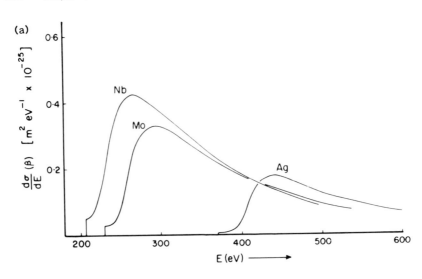

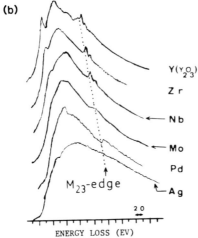

Figure 3.47. (a) M_{45} edges for $E_0 = 80$ keV and $\beta = 10$ mrad, according to Hartree-Slater calculations (Leapman *et al.*, 1980). (b) M_{45} edges of fifth-period elements measured with $E_0 = 120$ keV and $\beta = 5.7$ mrad (Zaluzec, 1982).

the dipole selection rule applies, as in the case of x-ray absorption spectra. From Eq. (3.145), the momentum exchange is approximately $\hbar q_{min} \cong \hbar k_0 \theta_E = E/v$ for small scattering angles, whereas the momentum exchange upon absorption of a photon of the same energy E is $\hbar q(\text{photon}) = E/c$. The ratio of momentum exchange in the two cases is therefore

$$q_{min}/q(\text{photon}) \simeq c/v \tag{3.161}$$

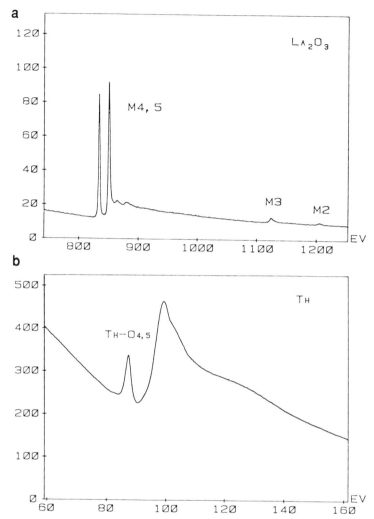

Figure 3.48. (a) Energy loss spectrum of a lanthanum oxide thin film, recorded with 200-keV electrons and a collection semiangle of 100 mrad, and showing M_4 and M_5 white lines and dipole-forbidden M_2 and M_3 edges (Ahn and Krivanek, 1983). (b) O_{45} edge of thorium recorded using 120-keV electrons and an acceptance semiangle of 100 mrad (Ahn and Krivanek, 1983).

and is less than two for incident energies above 80 keV. Therefore the optical selection rule $\Delta l = \pm 1$ applies approximately to the energy-loss spectrum. This dipole rule accounts for the prominence of L_{23} edges ($2p \rightarrow 3d$ transitions) in transition metals and their compounds (where a high density of d-states occurs just above the Fermi level) and of M_{45} edges in the lanthanides (which have a high density of unfilled $4f$ states).

In the case of a large collection aperture, however, the momentum transfer can be several times $\hbar q_{min}$ (the median scattering angle for inner-shell excitation is typically $5\theta_E$; see Fig. 3.41) and dipole-forbidden transitions are sometimes observed (see page 232). For example, sharp M_2 and M_3 peaks are seen in the spectrum of lanthanum oxide (Fig. 3.48a), representing $\Delta l = 2$ transitions from the $3p$ core level to a high density of unfilled $4f$ states. These peaks practically disappear when a small (1.6 mrad) collection angle is used (Ahn and Krivanek, 1983).

3.7.3. Effect of Plural Scattering

So far, we have assumed that the sample is very thin ($t/\lambda \leq 0.3$, where λ is the mean free path for all inelastic scattering) so that the probability of a transmitted electron being scattered inelastically by valence electrons (as well as exciting an inner shell) is small. In thicker samples, this condition no longer applies and a broad double-scattering peak appears at an energy loss of approximately $E_k + E_p$, where E_p is the energy of the "plasmon" peak observed in the low-loss region. In thicker specimens, higher-order satellite peaks merge with the double-scattering peak to produce a broad

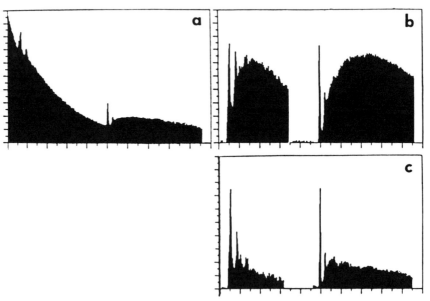

Figure 3.49. (a) Energy-loss spectrum (500–1200 eV) from a thick ($t/\lambda \sim 1.5$) specimen of nickel oxide. (b) Oxygen-K and nickel-L edges, following background removal. (c) Edge profiles following Fourier-ratio deconvolution, with plural scattering removed (Zaluzec, 1983).

hump beyond the edge, completely transforming its shape and obliterating any fine structure; see Fig. 3.33 and Fig. 3.49. This behavior illustrates the desirability of using very thin specimens for the identification of ionization edges and analysis of fine structure. Within limits, however, such plural or "mixed" scattering can be removed from the spectrum by deconvolution (Section 4.3).

3.7.4. Chemical Shifts in Threshold Energy

When atoms come together to form a molecule or a solid, the outer-shell wavefunctions are considerably altered, becoming molecular orbitals or Bloch functions; their energy levels then reflect the overall chemical or crystallographic structure. Although core-level wavefunctions are altered to a much smaller extent, the core-level energies may change by several eV depending on the chemical environment or "effective charge" of the corresponding atom.

Core-level binding energies can be measured directly by *x-ray photoelectron spectroscopy* (XPS; formerly known as ESCA). In this technique, a bulk specimen is illuminated with monochromatic x-rays and an electron spectrometer used to measure the kinetic energies of photoelectrons which have escaped into the surrounding vacuum. The final state of the electron transition therefore lies in the "continuum" far above the vacuum level, and is practically independent of the specimen. In the case of a compound, any increase in binding energy of a core level, relative to its value in the pure (solid) element, is called a *chemical shift.* For metallic core levels in oxides (and most other compounds) the XPS chemical shift is positive because oxidation removes valence-electron charge from the metal atom, reducing the screening of its nuclear field and deepening the potential well around the nucleus.

The ionization-edge threshold energies observed in EELS or in x-ray absorption spectroscopy (XAS) represent the *difference* in energy between a core-level initial state and the lowest-energy final state of an excited electron; the corresponding chemical shifts in threshold energy are more complicated than in XPS because the lowest-energy final state lies *below* the vacuum level and its energy depends on the valence-electron configuration. For example, going from a conducting phase (such as graphite) to an insulator (diamond) introduces an energy-band gap, raising the first-available empty state by several eV and increasing the ionization threshold energy (Fig. 1.4). On the other hand, the edge threshold in many ionic insulators corresponds to excitation to bound exciton states within the energy gap, reducing the chemical shift by an amount equal to the exciton binding energy.

The situation is further complicated by a many-body effect known as *relaxation*. When a positively charged "core hole" is created by inner-shell excitation, nearby electron orbitals are pulled inwards, reducing the magnitude of the measured binding energy by an amount equal to the "relaxation energy." In XPS, where the excited electron leaves the solid, relaxation energies are some tens of eV. In EELS or XAS, however, a core electron which receives energy just slightly in excess of the threshold value remains in the vicinity of the core hole and the screening effect of its negative charge reduces the relaxation energy. In a metal, conduction electrons provide additional screening which is absent in an insulating compound, so although relaxation effects may be less in EELS than in XPS, differences in relaxation energy between a metal and its compounds could have an appreciable influence on the chemical shift (Leapman *et al.*, 1982).

Measured EELS chemical shifts of metal L_3 edges in transition-metal oxides are typically 1 or 2 eV, and can be either positive or negative (Leapman *et al.*, 1982). The shifts of K-absorption edges in the same compounds are all positive and in the range 0.7–10.8 eV (Grunes, 1983). XAS chemical shifts, which should be equivalent to those registered by EELS, have been studied more extensively. The absorption edge of the metal atom

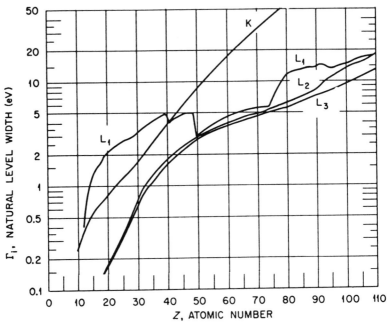

Figure 3.50. Natural widths of K and L levels as a function of atomic number (Krause and Oliver, 1979).

in a compound is usually shifted to higher absorption energy (compared to the metallic element) and these positive chemical shifts range up to 20 eV (for $KMnO_4$) in the case of K-edges of transition metals.

Because transition-series elements can take more than one valency, there exist mixed-valency compounds (e.g., Fe_3O_4, Mn_3O_4) containing differently charged ions of the same species. Since the chemical shift increases with increasing oxidation state, a double or multiple edge may be observed. In chromite spinel, for example, the L_3 and the L_2 white lines are each split by about 2 eV because of the presence of both Cr^{2+} and Cr^{3+} ions. Since the two ions occupy different sites (tetrahedral and octahedral) within the unit cell, the observed splitting will include a contribution (estimated as 0.7 eV) arising from the different site symmetry (Taftø and Krivanek, 1982b).

Site-dependent chemical shifts may also occur in organic compounds in which carbon atoms are present at chemically dissimilar points within a molecule. For example, the carbon K-edge in nucleic acid bases has a pronounced fine structure (page 394) which has been interpreted as being the result of several edges chemically shifted relative to one another because of the different effective charges on the carbon atoms (Isaacson, 1972b).

3.8. Near-Edge Fine Structure (ELNES)

Core-loss spectra recorded from solid specimens show a pronounced fine structure, taking the form of peaks or oscillations in intensity within 50 eV of the ionization threshold. Most of this structure reflects the influence of atoms surrounding the excited atom and requires a solid-state explanation. The basic principles, which are treated in greater detail in several review articles (Brydson, 1991; Sawatzky, 1991; Rez, 1992; Rez et al., 1995), are summarized below.

3.8.1. Densities-of-States Interpretation

Variations in the single-scattering intensity $J_k^1(E)$ can be related to the band structure of the solid in which scattering occurs. The theory is greatly simplified by making the *one-electron approximation*: excitation of an inner-shell electron is assumed to have no effect on the other atomic electrons. According to the Fermi golden rule of quantum mechanics (Manson, 1978), the transition rate is then proportional to a product of the density of final states $N(E)$ and an atomic transition matrix $M(E)$:

$$J_k^1(E) \propto d\sigma/dE \propto |M(E)|^2 N(E) \qquad (3.162)$$

$M(E)$ represents the overall shape of the edge, as discussed in Section 3.7.1 and is determined by *atomic* physics, whereas $N(E)$ depends on the chemical and crystallographic environment of the excited atom. To a *first approximation*, $M(E)$ can be assumed to be a slowly varying function of energy loss E, so that variations in $J_k(E)$ represent the energy dependence of the densities of states (DOS) above the Fermi level. However, the following comments apply.

(1) Transitions can only take place to electron states which are empty; like x-ray absorption spectroscopy, EELS gives information on the density of *unoccupied* states above the Fermi level.

(2) Because the core-level states are highly localized, $N(E)$ is a *local* density of states (LDOS) at the site of the excited atom (Heine, 1980). As a result, there can be appreciable differences in fine structure between edges representing different elements in the same compound; see Fig. 3.51. Even in the case of a single element, the fine structure may be different at sites of different symmetry, as demonstrated by Tafto (1984) for Al in sillimanite.

(3) The strength of the matrix-element term is governed by the dipole selection rule: $\Delta l = \pm 1$, with $\Delta l = 1$ transitions predominating (see Sections 3.7.2 and 3.8.2). As a result, the observed DOS is a *symmetry-projected* density of states. Thus, modulations in K-edge intensity ($1s$ initial state) reflect mainly the density of $2p$ final states. Similarly, modulations in the L_2- and L_3-intensity ($2p$ initial states) are dominated by $3d$ final states, except where $p \rightarrow d$ transitions are hindered by a centrifugal barrier (for example, $p \rightarrow s$ transitions are observed close to silicon L_{23} threshold). As a result of this selection rule, a dissimilar structure can be expected in the K- and L-edges of the same element in the same specimen.

(4) $N(E)$ is in principle a *joint* density of states, the energy dependence of the final-state density being convolved with that of the core level. The core-level width Γ_i is given approximately by the uncertainty relation $\Gamma_i \tau_h \approx \hbar$, where the lifetime τ_h of the core hole is determined by the speed of the deexcitation mechanism (mainly Auger emission in light elements) and the value of Γ_i depends mainly on the threshold energy of the edge; see Fig. 3.50.

The measured ELNES is also broadened by the instrumental energy resolution ΔE; to allow for this broadening and that due to the initial-state width, calculated densities of states can be convolved with a Gaussian or Lorentzian function of width $(\Gamma_i^2 + \Delta E^2)^{1/2}$. In the case of L_{23} or M_{45} edges, two initial states are present with different energy (see page 219). The measured ELNES therefore consists of two shifted DOS distributions, but this effect can be removed by Fourier-ratio deconvolution, with the low-loss region replaced by two delta functions whose strengths are suitably adjusted (Leapman *et al.*, 1982); see Appendix B.4.

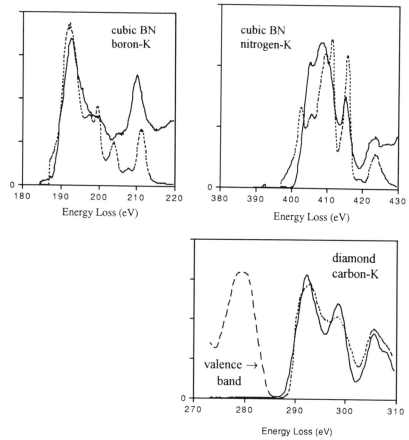

Figure 3.51. Dashed curves: near-edge fine structures calculated for diamond and cubic boron nitride using pseudo-atomic-orbital band theory (Weng *et al.*, 1989). Solid lines: core-loss intensity measured by P. Fallon (P. Rez, personal communication).

A further source of energy broadening arises from the lifetime τ_f of the *final* state. Inelastic scattering limits the mean free path λ_i of the *ejected* electron to a few nanometers for kinetic energies E_{kin} less than 50 eV (Fig. 3.57). Using a free-electron approximation ($E_{kin} = m_0 v^2/2$), the energy broadening is $\Gamma_f \approx \hbar/\tau_f = \hbar v/\lambda_i = (\hbar/\lambda_i)(2E_{kin}/m_0)^{1/2}$. Since λ_i varies inversely with E_{kin} below 50 eV (see Fig. 3.57), Γ_f increases with excitation energy above the threshold and the observed DOS structure is progressively damped with increasing energy loss. As a result, the observed variations in $N(E)$ are weak in the EXELFS region (Section 3.9).

Band-structure calculations which predict the electrical properties of a solid give the *total* densities of states and provide only approximate correlation with measured ELNES. For a more accurate description, the total DOS must be resolved into the correct angular-momentum component at the appropriate atomic site. Pseudopotential methods have been adapted to this requirement and have been used to calculate ELNES for diamond, SiC, and Be_2C (Weng et al., 1989). These more recent calculations show good agreement between DOS peaks and experimental data, both with respect to peak position and relative intensity; see Fig. 3.51. The augmented plane wave (APW) method has provided realistic near-edge structures of transition-metal compounds (Muller et al., 1982; Blaha and Schwarz, 1983).

3.8.2. Validity of the Dipole Approximation

As indicated by Eq. (3.162), energy-loss fine structure represents the density of final states *modulated* by an atomic transition matrix. If the initial state is a closed shell, the many-electron matrix element of Eq. (3.23) can be replaced by a *single-electron* matrix element, defined by

$$M(\mathbf{q}, E) = \int \psi_f^* \exp(i\,\mathbf{q} \cdot \mathbf{r}) \psi_i \, d\tau \qquad (3.163)$$

where ψ_i and ψ_f are the initial- and final-state single-electron wavefunctions and the integration is over all volume τ surrounding the initial state. Expanding the operator as

$$\exp(i\,\mathbf{q} \cdot \mathbf{r}) = 1 + i(\mathbf{q} \cdot \mathbf{r}) + \text{higher-order terms} \qquad (3.164)$$

enables the integral in Eq. (3.163) to be split into three components. The first of these, arising from the unity term in Eq. (3.164), is zero because ψ_i and ψ_f are orthogonal wavefunctions. The second integral containing $(\mathbf{q} \cdot \mathbf{r})$ is zero if ψ_i and ψ_f have the *same symmetry* about the center of the excited atom $(r = 0)$ such that their product is even; $\mathbf{q} \cdot \mathbf{r}$ itself is an *odd* function and the two halves of the integral then cancel. But if ψ_i is an *s*-state (even symmetry) and ψ_f is a *p*-state (odd symmetry), the integral is nonzero and transitions are observed. This forms the basis of the dipole selection rule, according to which the observed $N(E)$ is a symmetry-projected density of states.

For the dipole rule to be valid, the higher-order terms in Eq. (3.163) must be negligible; otherwise a third integral (representing dipole-forbidden transitions) will modify the energy dependence of the fine structure. From the above argument, the dipole condition is defined by the requirement $\mathbf{q} \cdot \mathbf{r} \ll 1$ for all r, equivalent to $q \ll q_d = 1/r_c$, where r_c is the radius of the core state (defining the spatial region in which most of the transitions occur). The hydrogenic model gives $r_c \approx a_0/Z^*$, where Z^* is the effective nuclear charge.

For K-shells, $Z^* \approx Z - 0.3$ (see Section 3.6.1); for carbon K-shell excitation by 100-keV electrons, dipole conditions should prevail for $\theta \ll \theta_d = Z^*/a_0 k_0 = 67$ mrad, a condition fulfilled for most of the transitions since the median angle of scattering is around 10 mrad (Fig. 3.41). In agreement with this estimate, atomic calculations indicate that nondipole contributions are less than 10% of the total for $q < 45$ nm^{-1}, equivalent to $\theta < 23$ mrad for 100-keV electrons (Fig. 3.52). A small spectrometer collection aperture (centered about the optic axis) can therefore ensure that nondipole effects are unimportant. Saldin and Yao (1990) argue that dipole conditions hold only over an energy range ε_{max} above the excitation threshold, but $\varepsilon_{max} \approx 33$ eV for $Z = 3$, increasing to ≈ 270 eV for $Z = 8$. Therefore, dipole conditions should apply to the ELNES of elements heavier than Li and to the EXELFS region for oxygen and heavier elements, for the incident energies used in transmission spectroscopy.

For L_{23} edges, atomic calculations (Saldin and Ueda, 1992) give $q_d a_0 \approx Z^*/9$ with $Z^* = Z - 4.5$, so for silicon and 100-keV incident electrons $\theta_d \approx 11$ mrad. Solid-state calculations for Si (Ma *et al.*, 1990) have suggested

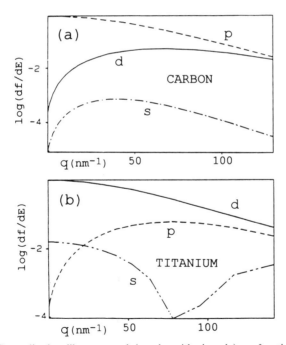

Figure 3.52. Generalized oscillator strength (on a logarithmic scale) as a function of wavenumber for transitions to s, p, and d final states (5 eV above the edge threshold), calculated for (a) carbon K-shell ($1s$ initial state) and (b) titanium L_3-shell ($2p^{3/2}$ initial state) excitation (Rez, 1989).

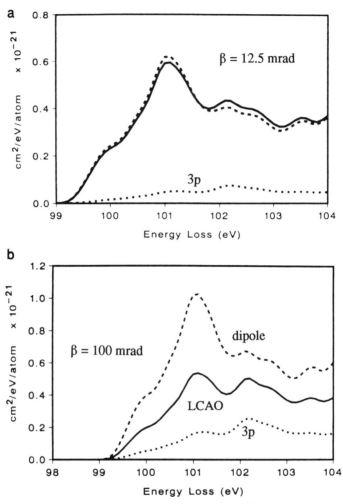

Figure 3.53. Silicon L_{23} differential cross section for 100-keV incident electrons and acceptance angles of (a) 12.5 mrad and (b) 100 mrad. Solid lines are results of LCAO calculations, not using the dipole approximation; dotted lines represent the contribution from $2p \rightarrow 3p$ transitions. Dashed lines represent the dipole approximation (Ma *et al.*, 1990).

that nondipole effects are indeed small (within 5 eV of the threshold) for 12.5-mrad collection semiangle (Fig. 3.53) but are substantial for a large collection aperture, where monopole $2p \rightarrow 3p$ transitions make a substantial contribution (Fig. 3.53b). Monopole transitions have been observed at the Si–L_{23} edge of certain minerals and have been attributed to the low crystal symmetry which induces mixing of *p*- and *d*-orbitals (Brydson *et al.*, 1992a).

A high density of dipole-forbidden states just above the Fermi level may also lead to observable monopole peaks, but mainly in spectra recorded with a displaced collection aperture where the momentum transfer is large (Auerhammer and Rez, 1989). The dipole approximation appears justified for all *M*-edges, at incident energies above 10 keV and with an axial collection aperture (Ueda and Saldin, 1992).

3.8.3. Molecular Orbital Theory

An alternative explanation of ELNES is in terms of molecular orbital (MO) theory (Glen and Dodd, 1968): the local band structure is approximated as a linear combination of atomic orbitals (LCAO) of the excited atom and its immediate neighbors. A simple example is graphite, in which the four valence electrons of each carbon atom are sp^2 hybridized, resulting in three strong σ bonds to nearest neighbors within each atomic layer; the remaining *p*-electron contributes to a delocalized π orbital. The corresponding antibonding orbitals are denoted σ^* and π^*; they are the empty states into which core electrons can be excited, giving rise to distinct peaks in the *K*-edge spectrum (see page 8 and page 387).

In organic materials, the presence of delocalized or unsaturated bonding again gives rise to sharp π^* peaks at an edge threshold. In compounds containing carbon atoms with different effective charge, such as the nucleic acid bases, several peaks are observable and have been interpreted in terms of chemical shifts (Isaacson, 1972a,b). Molecular-orbital concepts have been useful in the interpretation of the fine structure of edges recorded from minerals (Krishnan, 1990; McComb *et al.*, 1992).

3.8.4. Multiple-Scattering (XANES) Theory

A different approach to the interpretation of ELNES makes use of concepts developed to explain x-ray absorption near-edge structure (XANES, also referred to as NEXAFS). This is an extension of EXAFS theory, taking into account *multiple* (plural) elastic scattering of the ejected core electron. Multiple scattering is important in the near-edge region, where backscattering occurs in a larger volume of the specimen as a result of the longer inelastic mean free path at low kinetic energies of the ejected electron (page 241). Nevertheless, the result of the calculations is a property of the *local* environment of the excited atom. The surrounding atoms are divided into shells and the scattering calculated by including a successively larger number of shells, analogous to successive atomic layers in low-energy electron diffraction (LEED) theory. Documented programs for performing such calculations are available (Durham *et al.*, 1982; Vvedensky *et al.*, 1986).

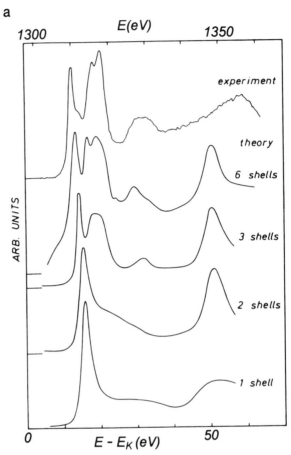

Figure 3.54. Multiple-scattering calculations of the fine structure at (a) the magnesium K-edge and (b) the oxygen K-edge in MgO, shown as a function of the number of shells used in the calculations (Lindner *et al.*, 1986). The spectra at the top are EELS measurements with background subtracted and plural scattering removed by deconvolution.

Usually a moderate number of coordination shells is sufficient, indicating that the scattering is contained mostly within a 1-nm-diameter radius; see Fig. 3.54.

In the case of MgO, the light-metal cations scatter weakly and do not contribute appreciably to the fine structure. The peak labeled C in Fig. 3.54b arises from single scattering from oxygen nearest neighbors and therefore appears when only two shells are used in the calculations. Peak B represents single scattering from second-nearest oxygen atoms and emerges when four

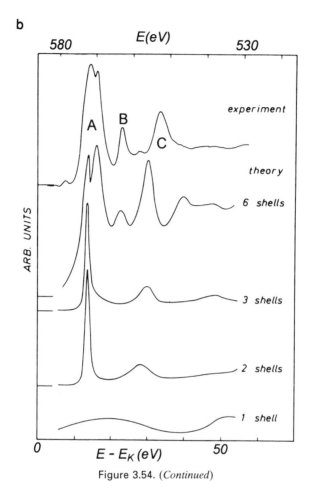

Figure 3.54. (*Continued*)

shells are included. Peak A is believed to arise from *plural* scattering among oxygen nearest neighbors (Rez *et al.*, 1995).

An effect which is found to be important for higher-order shells is focusing of the ejected-electron wave by intermediate shells, in situations where atoms are radially aligned (Lee and Pendry, 1975). A related effect arises from the centrifugal barrier created by first-neighbor atoms, which acts on high-angular-momentum components of the emitted wave, confining it locally for energies just above the edge threshold. This *shape-resonance* effect has been used to interpret absorption spectra of diatomic gases (Dehmer and Dill, 1977) and transition-metal complexes (Kutzler *et al.*, 1980). The resonance value of the ejected-electron wavenumber k obeys the relationship kR = constant, where R is the bond length (Bianconi, 1983; Bianconi et al., 1983a), resulting in the resonance energy (above threshold) being proportional to $1/R^2$. A similar behavior is apparent from band-structure calculations of transition-metal elements (Muller et al., 1982): for metals having the same crystal structure, the energies of DOS peaks are proportional to $1/a^2$, where a is the lattice constant.

Although formally equivalent to a densities-of-states interpretation of ELNES (Colliex et al., 1985), multiple-scattering (MS) calculations are done in *real space*. It is therefore possible to treat disordered systems or complicated molecules such as hemoglobin (Durham, 1983) and calcium-containing proteins (Bianconi et al., 1983b) for which band-structure calculations would not be feasible.

3.8.5. Core Excitons

Because the core hole (inner-shell vacancy) perturbs the final state of the transition, one-electron band-structure theory is not an exact description of the ELNES features observed. One solution is to generalize the concept of density of states $N(E)$ to include temporary bound states formed by interaction between the excited core electron and the core hole, known as *core excitons*. Since their effective radius may be larger or smaller than atomic dimensions, they are analogous to the Wannier or Frenkel excitons observed in the low-loss region of the spectrum (Section 3.3.3). The exciton energies can be estimated by use of the $Z + 1$ or optical alchemy approximation (Hjalmarson et al., 1980) in which the potential used in DOS or MS calculations is that of the next-highest atom in the periodic table. This is equivalent to assuming that the core hole increases (by one unit) the effective charge seen by outer electrons.

In an insulator, core-exciton levels lie within the energy gap between valence and conduction bands and may give rise to one or more peaks below the threshold for ionization to extended states (Pantelides, 1975). Such peaks can be identified as excitonic if band-structure calculations are available on an absolute energy scale and if the energy-loss axis has been accurately calibrated (Grunes et al., 1982). Even when peaks are not visible, excitonic effects may sharpen the ionization threshold, modify the fine structure, or shift the threshold to lower energy loss, as proposed for graphite (Mele and Ritsko, 1979) and boron nitride (Leapman et al., 1983). These effects may be somewhat different in EELS, compared to x-ray absorption spectroscopy, if the excited atom relaxes before the fast electron exits the exciton radius (Batson and Bruley, 1991).

In ionic compounds, the exciton is more strongly bound at the cation than at the anion site (Pantelides, 1975; Hjalmarson, 1980), giving rise to further differences in near-edge structure at the respective ionization edges.

In the case of a metal, the effect of the core hole is screened within a short distance ($\approx$0.1 nm), but electron–hole interaction may still modify the shape of an ionization edge: if the initial state is s-like, the edge may become more rounded; if p-like, it may be sharpened slightly (Mahan, 1975).

3.8.6. Multiplet and Crystal-Field Splitting

The core hole created by inner-shell ionization has an angular momentum which can couple with the net angular momentum of any partially filled shells within the excited atom. Such coupling is strongest when the

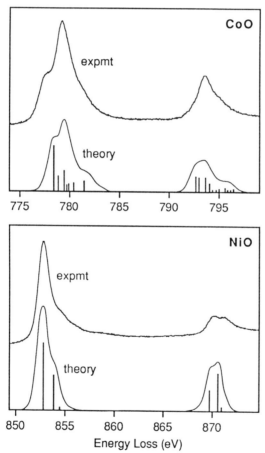

Figure 3.55. Fine structure of the L_2 and L_3 white lines in CoO and NiO, measured by EELS (Krivanek and Paterson, 1990) and calculated by Yamaguchi et al. (1982). The vertical bars represent the calculated states and their relative intensities, from which the smooth curves were derived by convolution with an instrumental resolution function.

hole is created within the partially filled shell itself, as in the case of N_{45} ionization of sixth-period elements from Cs ($Z = 55$) to Tm ($Z = 69$). Both spin and orbital momentum are involved, leading to an elaborate fine structure (Sugar, 1972). A similar effect is observed in the M-shell excitation of third-period elements (Davis and Feldkamp, 1976). Where the core hole is created in a complete shell, separate peaks may not be resolved but the coupling can lead to additional broadening of the fine structure. Although multiplet splitting is basically an atomic effect, its effect is modified by the chemical and crystallographic environment of the excited atom.

High-resolution EELS of transition-metal oxides (Leapman *et al.*, 1982; Krivanek and Paterson, 1990) reveals that the L_2 and the L_3 white-line peaks are each split into two components, due to solid-state as well as atomic effects; see Fig. 3.55. Because of the nonspherical electrostatic field of the oxygen ions surrounding the transition-metal atom, its $3d$ states are split in energy. Similar *crystal-field* splittings are observed by photoelectron spectroscopy (Novakov and Hollander, 1968) and by x-ray absorption spectroscopy. The multiplet structure can be calculated for different values of the crystal-field parameter (Groot *et al.*, 1990); comparison with experiment could give information about the character of the bonding in minerals (Garvie *et al.*, 1994).

3.9. Extended Energy-Loss Fine Structure (EXELFS)

Although the ionization-edge fine structure decreases in amplitude with increasing energy loss, oscillations of intensity are detectable over a range of several hundred eV if no other ionization edges follow within this region. This "extended" fine structure was first observed (as "EXAFS") in x-ray absorption spectra and interpreted as a densities-of-states phenomenon, involving diffraction of the ejected core electron due to the long-range order of the solid. However, quite strong EXAFS modulations are obtained from amorphous samples and the effect is now recognized to be a measure of the short-range order, involving mainly scattering from nearest-neighbor atoms.

If released with a kinetic energy of 50 eV or more, the ejected core electron behaves much like a free electron, the densities of states $N(E)$ in Eq. (3.162) approximating a smooth function proportional to $(E - E_k)^{1/2}$ (Stern, 1974). However, weak oscillations in $J_k^1(E)$ can arise from interference between the outgoing spherical wave (representing the ejected electron) and reflected waves which arise from elastic backscattering of the electron from neighboring atoms; see Fig. 3.56. This interference perturbs the final-state wavefunction in the core region of the central atom and

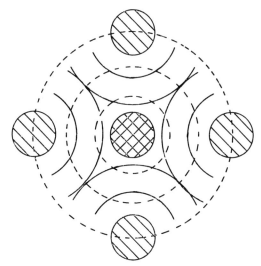

Figure 3.56. Pictorial representation of the electron interference which gives rise to EXELFS. Wavefronts of the outgoing wave (representing a core electron ejected from the central atom) are represented by dashed circles. The solid arcs represent waves back-scattered from nearest-neighbor atoms.

therefore modulates $N(E)$. The interference can be destructive or constructive, depending on the return path length $2r_j$ (where r_j is the radial distance to the jth shell of backscattering atoms) and the wavelength λ of the ejected electron. Since the velocity of the ejected electron is low compared to the speed of light, the magnitude k of the wavevector of the ejected electron is given by classical mechanics:

$$k = 2\pi/\lambda \cong [2m_0(E - E_k)]^{1/2}/\hbar \qquad (3.165)$$

where E is the energy transfer (from an incident electron or an x-ray photon) and E_k the threshold energy of the ionization edge. With increasing energy loss E, the interference is therefore alternately constructive and destructive, giving maxima and minima in the scattering cross section.

Let us assume that instrumental resolution is not a limiting factor and that the effects of plural inelastic scattering of the transmitted electron are negligible or have been removed by deconvolution to yield the single-scattering distribution $J_k^1(E)$, as described in Chapter 4. The oscillatory part of the intensity (following the ionization edge) can then be represented in a normalized form:

$$\chi(E) = [J_k^1(E) - A(E)]/A(E) \qquad (3.166)$$

where $A(E)$ is the energy-loss intensity which would be observed in the absence of backscattering, and which could in principle be calculated using a single-atom model. Using Eq. (3.165), the oscillatory component can also be written as a function $\chi(k)$ of the ejected-electron wavevector. As in an x-ray absorption experiment, the main contribution to $J_k'(E)$ is from dipole scattering (Section 3.7.1), so standard EXAFS theory can be used to interpret the behavior of $\chi(k)$.

Approximating the ejected-electron wavefunction at the backscattering atom by a plane wave and assuming that multiple backscattering can be neglected, EXAFS theory gives (Sayers *et al.*, 1971)

$$\chi(k) = \sum_j \frac{N_j}{r_j^2} \frac{f_j(k)}{k} \exp(-2r_j/\lambda_i) \exp(-2\sigma_j^2 k^2) \sin[2kr_j + \phi(k)] \qquad (3.167)$$

The summation in Eq. (3.167) is over successive shells of neighboring atoms, the radius of a particular shell being r_j. The largest contribution comes from the nearest neighbors ($j = 1$), unless these are very light atoms with a low scattering power (e.g., hydrogen). N_j is the number of atoms in shell j and either N_j itself or N_j/r_j^2 (as a function of r) is known as the *radial distribution function* (RDF) of the atoms surrounding the ionized atom. In the case of a perfect single crystal, the RDF would consist of a series of delta functions corresponding to discrete values of shell radius.

In Eq. (3.167), $f_j(k)$ is the *backscattering amplitude* or form factor for elastic scattering through an angle of π rad; $f_j(k)$ has units of length and can be calculated (as a function of k) knowing the atomic number Z of the backscattering element. Results of such calculations have been tabulated by Teo and Lee (1979). For lower-Z elements, screening of the nuclear field can be neglected, so the backscattering approximates to Rutherford scattering for which $f(k) \propto k^{-2}$, as shown by Eq. (3.1) and Eq. (3.3).

The damping term $\exp(-2r_j/\lambda_i)$ occurs in Eq. (3.167) because of inelastic scattering of the ejected electron along its outward and return path, which changes the value of k and thereby weakens the interference, so the inelastic scattering is sometimes referred to as absorption. Instead of incorporating a damping term explicitly in Eq. (3.167), absorption can be included by making k into a complex quantity whose imaginary part represents the inelastic scattering (Lee and Pendry, 1975). Absorption arises from both electron–electron and electron–phonon collisions and in reality the inelastic mean free path is a function of k; see Fig. 3.57. Because the mean free path is generally less than 1 nm for an electron energy of the order of 100 eV, inelastic scattering provides one limit to the range of shell radii which can contribute to the RDF. Another limit arises from the lifetime τ_h of the core hole.

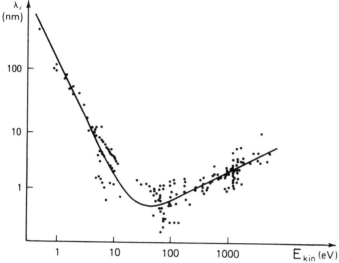

Figure 3.57. Mean free path for inelastic scattering of an electron, as a function of its energy above the Fermi level (Seah and Dench, 1979). The solid curve represents a least-squares fit to the experimental data, measured from a variety of materials.

The Gaussian term $\exp(-2\sigma_j^2 k^2)$ in Eq. (3.167) is the Fourier transform of a radial-broadening function which represents broadening of the RDF due to thermal, zero-point, and static disorders. The disorder parameter σ_j differs from the Debye–Waller parameter u_j used in diffraction theory (see Section 3.1.5) because only *radial* components of the *relative* motion between the central (ionized) and backscattering atoms are of concern. In a single crystal, where atomic motion is highly correlated, σ_j can be considerably less than u_j. The value of σ_j depends on the atomic number of the backscattering atom and on the type of bonding; in an anisotropic material such as graphite, it will also depend on the direction of the vector $\mathbf{r}_j$.

The remaining term in Eq. (3.167), $\sin[2kr_j + \phi_j(k)]$, determines the interference condition. The phase difference between the outgoing and reflected waves consists of a path-length term $2\pi(2r_j/\lambda) = 2kr_j$ and a term $\phi_j(k)$ which accounts for the phase change of the electron wave after traveling through the field of the emitting and backscattering atoms. This phase change can be split into components $\phi_a(k)$ and $\phi_b(k)$ which arise from the emitting and backscattering atom, respectively, and which can both be calculated using atomic wavefunctions, incorporating an effective potential to account for exchange and correlation (Teo and Lee, 1979). In accordance with the dipole selection rule, the emitted wave is expected to have p-symmetry in the case of K-shell ionization (Fig. 3.58) and mainly d-symme-

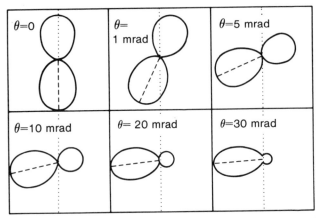

Figure 3.58. Angular distribution of the ejected-electron intensity per unit solid angle, for K-shell ionization in carbon at an energy loss $E = 385$ eV, various scattering angles, and $E_0 = 100$ keV (Maslen and Rossouw, 1983). The dashed line represents the direction of the scattering vector $\mathbf{q}$ of the fast electron; the path of the fast electron is shown by the dotted line.

try in the case of an L_{23} edge. The phase-shift component ϕ_b differs in these two cases.

Measurement and analysis of EXELFS allows interatomic distances r_j (principally those corresponding to first-neighbor atoms) to be measured, as in the case of EXAFS studies. By carefully selecting the scattering angle of the transmitted electron and the specimen orientation, it may be possible to determine bond lengths in a specified direction. In the case of K-shell ionization, for example, the p-type outgoing wave representing the ejected electron probes the atomic environment *predominantly* in the direction of the scattering vector $\mathbf{q}$ of the fast electron; see Fig. 3.58. In other words, the contribution to the EXELFS modulations from atoms which lie in the direction $\mathbf{r}_j$ (which makes an angle Φ with $\mathbf{q}$) is given to a first approximation (Leapman *et al.*, 1981) by

$$\chi(k) \propto (\mathbf{q} \cdot \mathbf{r}_j)^2 \propto \cos^2 \Phi \qquad (3.168)$$

For very small scattering angles ($\theta \ll \theta_E$), atoms lying along the direction of the incident wavevector $\mathbf{k}_0$ make the major contribution to $\chi(k)$, whereas at large scattering angles atoms lying perpendicular to $\mathbf{k}_0$ contribute the most (see Fig. 3.58). However, the scattered intensity in the second case will be much less. To ensure equal intensities in the two spectra, a small collection aperture can be used to record EXELFS data at scattering angles of $+\Theta$ and $-\Theta$. Choosing Θ equal to the characteristic angle $\theta_E = E/\gamma m_0 v^2$, where E is the energy loss corresponding to the middle of the EXELFS region, results in the angle ϕ between $\mathbf{q}$ and $\mathbf{k}_0$ being $\pm 45°$, as seen from

Fig. 3.39. If the specimen itself is oriented at 45° to the incident beam, the two spectra will record preferentially the interatomic distances in directions perpendicular and parallel to the plane of the specimen, respectively. Orientation-dependent EXELFS spectra have been obtained from test specimens of graphite (Disko, 1981) and boron nitride (Leapman *et al.*, 1981). An orientation dependence has also been observed in the *near-edge* fine structure of boron nitride (Leapman and Silcox, 1979) and can be interpreted in terms of the directionality of chemical bonding.

4

Quantitative Analysis of the Energy-Loss Spectrum

This chapter describes some of the spectral-analysis techniques used to obtain quantitative information about the crystallographic structure and chemical composition of a specimen. Although this information is expressed rather directly in the energy-loss spectrum, plural scattering complicates the data recorded from specimens of typical thickness. It is therefore desirable to remove the effects of plural scattering from the spectrum or at least make allowance for them in the analysis procedure.

We deal first with the *low-loss region*, which can be defined somewhat arbitrarily as energy losses below 100 eV. Within this region, the main energy-loss mechanism involves excitation of outer-shell electrons, in other words the valence or (in a metal) conduction electrons. In many solids, a plasmon model (Section 3.3.1) provides the best description of valence-electron excitation, a process which occurs with relatively high probability because the plasmon mean free path is often comparable with the sample thickness.

4.1. Removal of Plural Scattering from the Low-Loss Region

Deconvolution techniques based on the Fourier transform will be described first, since these are in many ways the most accurate and versatile. Alternative methods which are applicable to the low-loss region are outlined in Section 4.1.2.

4.1.1. Fourier-Log Deconvolution

Assuming independent scattering events, the electron intensity I_n, integrated over energy loss and corresponding to scattering of order n, follows a Poisson distribution:

$$I_n = IP_n = (I/n!)(t/\lambda)^n \exp(-t/\lambda) \tag{4.1}$$

where I is the total integrated intensity (summed over all orders), P_n is the probability of n scattering events within the specimen (thickness t), ! denotes a factorial, and λ is the mean free path for inelastic scattering.*

The case $n = 0$ corresponds to the absence of inelastic scattering and is represented in the energy-loss spectrum by the zero-loss peak:

$$Z(E) = I_0 R(E) \tag{4.2}$$

where the resolution function (or instrument response) $R(E)$ has unit area and a full width at half-maximum (FWHM) equal to the experimental energy resolution ΔE.

Single scattering corresponds to $n = 1$ and is characterized by an intensity distribution $S(E)$. From Eq. (4.1),

$$\int S(E) \, dE = I_1 = I(t/\lambda) \exp(-t/\lambda) = I_0(t/\lambda) \tag{4.3}$$

Owing to the limited energy resolution ΔE, single scattering occurs within the experimental spectrum $J(E)$ as a broadened distribution $J^1(E)$ given by

$$J^1(E) = R(E) * S(E) \equiv \int_{-\infty}^{\infty} R(E - E')S(E') \, dE' \tag{4.4}$$

where $*$ denotes a convolution over energy loss, as defined by Eq. (4.4).

Double scattering has an energy dependence of the form $S(E) * S(E)$. However, the area under this self-convolution function is $(I_1)^2 = (I_0 t/\lambda)^2$, whereas Eq. (4.1) indicates that the integral I_2 should be $I_0(t/\lambda)^2/2!$. Measured using an ideal spectrometer system, the double-scattering component would therefore be $D(E) = S(E) * S(E)/(2!I_0)$ but, as recorded by the instrument, it is

$$J^2(E) = R(E) * D(E) = R(E) * S(E) * S(E)/(2!I_0) \tag{4.5}$$

---

* Here λ characterizes all inelastic scattering in the energy range over which the intensity is integrated, and is given by Eq. (3.96) in the case where several inelastic processes contribute within this range. As discussed in Section 3.4, Poisson statistics still apply to a spectrum recorded with an angle-limiting aperture, provided an aperture-dependent mean free path is used.

Likewise, the triple-scattering contribution is equal to $T(E) = S(E) * S(E) * S(E)/(3!I_0^2)$, but is recorded as $J^3(E) = R(E) * T(E)$.

The observed spectrum, including the zero-loss peak, can therefore be written in the form

$$J(E) = Z(E) + J^1(E) + J^2(E) + J^3(E) + \cdots$$
$$= R(E) * [I_0\delta(E) + S(E) + D(E) + T(E) + \cdots]$$
$$= Z(E) * [\delta(E) + S(E)/I_0 + S(E) * S(E)/(2!I_0^2)$$
$$+ S(E) * S(E) * S(E)/(3!I_0^3) + \cdots] \qquad (4.6)$$

where $\delta(E)$ is a unit-area delta function.

The *Fourier transform* of $J(E)$ can be defined (Bracewell, 1978; Brigham, 1974) as

$$j(\nu) = \int_{-\infty}^{\infty} J(E)\exp(2\pi i\nu E)\,dE \qquad (4.7)$$

Taking transforms of both sides of Eq. (4.6), the convolutions become products (Bracewell, 1978), giving the equation

$$j(\nu) = z(\nu)\{1 + s(\nu)/I_0 + [s(\nu)]^2/(2!I_0^2) + [s(\nu)]^3/(3!I_0^3) + \cdots\}$$
$$= z(\nu)\exp[s(\nu)/I_0] \qquad (4.8)$$

in which the Fourier transform of each term in Eq. (4.6) is represented by the equivalent lower-case symbol, and is a function of the transform variable (or "frequency") ν whose units are eV^{-1}. Equation (4.8) can be "inverted" by taking the logarithm of both sides (Johnson and Spence, 1974), giving

$$s(\nu) = I_0\ln[j(\nu)/z(\nu)] \qquad (4.9)$$

Noise Problems

One might envisage taking the *inverse* Fourier transform of Eq. (4.9) in order to recover an "ideal" single-scattering distribution, unbroadened by instrumental resolution. However, as discussed by Johnson and Spence (1974) and by Egerton and Crozier (1988), such "complete" deconvolution is feasible only if the spectrum is coarsely sampled or measured with very high precision. In practice, $J(E)$ contains noise (due to counting statistics, for example) which extends to high frequencies, corresponding to large values of ν. Although not necessarily a monotonic function of ν, the noise-free component of $j(\nu)$ eventually falls towards zero as ν increases. As a result, the *fractional* noise content in $j(\nu)$ increases with ν, and at high "frequencies" $j(\nu)$ is usually dominated by noise. Since $z(\nu)$ also falls with increasing ν, the high-frequency noise content of $j(\nu)$ is preferentially "amplified" when divided by $z(\nu)$, as in Eq. (4.9), and the inverse transform $S(E)$ is submerged by high-frequency noise. Essentially, Eq. (4.9) fails because we are attempting to simulate the effect of a spectrometer system with perfect energy resolution and to recover fine structure in $J(E)$ which is below the resolution limit.

Fortunately, deconvolution based on Eq. (4.9) can be made to work provided we are content to recover the instrumentally broadened single-scattering distribution $J^1(E)$, with little or no attempt to improve the energy resolution. Several slightly different ways of obtaining $J^1(E)$ are listed below, in order of increasing simplicity.

(a) *Use of a Modification Function.* If Eq. (4.9) is multiplied by a function $g(\nu)$ which has unit area and falls rapidly with increasing ν, the high-frequency values of $\ln(j/z)$ are attenuated and noise amplification is controlled. The inverse transform will correspond to $G(E) * S(E)$, the single-scattering distribution recorded with an instrument whose resolution function is $G(E)$. $G(E)$ is known as a modification or reconvolution function. A sensible choice is the unit-area Gaussian: $G(E) = (\sigma\sqrt{\pi})^{-1} \exp(-E^2/\sigma^2)$ whose FWHM is $W = 2\sigma\sqrt{(\ln 2)} = 1.665\sigma$, in which case the single-scattering distribution (SSD) is obtained as the inverse transform of

$$j^1(\nu) = g(\nu)s(\nu) = I_0\exp(-\pi^2\sigma^2\nu^2) \ln[j(\nu)/z(\nu)] \qquad (4.10)$$

A limited improvement in energy resolution is possible by choosing σ such that $W < \Delta E$, but at the expense of increased noise content. If σ is chosen so that $W = \Delta E$, the inverse transform $J^1(E)$ is the SSD which would be recorded using an instrument having the same energy resolution ΔE but possessing a symmetric (Gaussian) resolution function. Besides removing plural scattering from the measured data, procedure (a) therefore corrects for any distortion of peak shapes caused by an irregularly shaped or skew instrument function.

(b) *Reconvolution by Z(E).* If $G(E) = R(E)$, Eq. (4.10) becomes

$$j^1(\nu) = r(\nu)s(\nu) = z(\nu) \ln[j(\nu)/z(\nu)] \qquad (4.11)$$

Upon taking the inverse transform, Eq. (4.11) gives the single-scattering distribution $J^1(E)$ which would have been recorded from a vanishingly thin specimen. The easiest way of obtaining $z(\nu)$ is from the experimental spectrum $J(E)$, setting channel contents to zero above the zero-loss peak. For thicker specimens, however, it may be more accurate to record $Z(E)$ in a second acquisition with no specimen present; a difference in height between the two zero-loss peaks can be shown to result in artifacts which are confined to the zero-loss region (Johnson and Spence, 1974).

(c) Gaussian Approximation for Z(E). Approximating $z(\nu)$ both out-side and within the logarithm of Eq. (4.11) by a Gaussian function of the same width and area, $j^1(\nu)$ can be obtained from the equation

$$j^1(\nu) = I_0\exp(-\pi^2\sigma^2\nu^2)\{\ln[j(\nu)] - \ln[I_0] + \pi^2\sigma^2\nu^2\} \qquad (4.12)$$

where $\sigma = \Delta E/1.665$. This procedure avoids the need to isolate or remeasure $Z(E)$ and calculate its Fourier transform (Egerton and Crozier, 1988).

(d) Replacement of Z(E) by a Delta Function. If the zero-loss peak $Z(E)$ in the original spectrum is replaced by the function $I_0\delta(E)$ *before* calculating the transform $j^d(\nu)$, one can use the approximation

$$j^1(\nu) \cong I_0\ln[j^d(\nu)/I_0] \qquad (4.13)$$

As in the third method, only one forward and one inverse transform are required and noise amplification is avoided. The use of Eq. (4.13) is equiva-lent to treating the experimental spectrum as if it had been recorded with a spectrometer system of perfect energy resolution; the resulting SSD will differ somewhat from that derived using Eq. (4.11) if $J(E)$ contains sharp peaks, comparable in width to the instrumental resolution; see Fig. 4.1.

Practical Details

The Fourier transforms $j(\nu)$ and $z(\nu)$ are (in general) complex numbers of the form $j_1 + ij_2$ and $z_1 + iz_2$, where $i = (-1)^{1/2}$. Therefore we have

$$\frac{j(\nu)}{z(\nu)} = \frac{j_1z_1 + j_2z_2 + i(j_2z_1 - j_1z_2)}{z_1^2 + z_2^2} = re^{i\theta} \qquad (4.14)$$

$$\ln[j(\nu)/z(\nu)] = \ln r + i\theta \qquad (4.15)$$

where

$$r = \frac{[(j_1z_1 + j_2z_2)^2 + (j_2z_1 - j_1z_2)^2]^{1/2}}{z_1^2 + z_2^2} \qquad (4.16)$$

and

$$\theta = \tan^{-1}\left[\frac{j_2z_1 - j_1z_2}{j_1z_1 + j_2z_2}\right] \qquad (4.17)$$

Equations (4.14)–(4.17) enable Eq. (4.10) and Eq. (4.11) to be evaluated by small computers which do not handle complex functions.

In practice, spectral data are held in a limited number N of "channels," each channel corresponding to electronic storage of a binary number. In

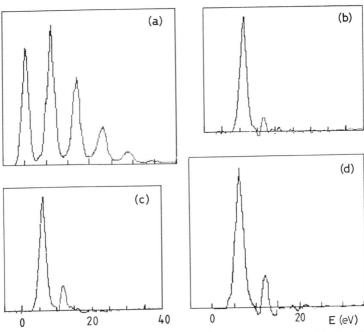

Figure 4.1. (a) Plasmon-loss spectrum of sodium ($t/\lambda = 1.66$) recorded with $E_0 = 80$ keV, $\beta = 4$ mrad (Jones *et al.*, 1984). (b) Spectrum deconvoluted using the "exact" Fourier-log method, Eq. (4.11). The zero-loss peak is absent; the small peak at $E = 2E_p$ may be due to double-plasmon excitation (Section 3.4.5). (c) Zero-loss intensity $Z(E)$ multipled by 1.25 prior to "exact" deconvolution. (d) Original spectrum deconvoluted using the approximate method, Eq. (4.13). (From R. F. Egerton, B. G. Williams, and T. G. Sparrow, Fourier deconvolution of electron energy-loss spectra, *Proc. R. Soc. London* **A398**, 395–404 (1985), by permission of the Royal Society.)

this case $j(\nu)$ is a *discrete* Fourier transform (DFT), defined (Bracewell, 1978) by

$$j(n) = N^{-1} \sum_{m=0}^{m=N-1} J(m) \exp(-2\pi i mn/N) \qquad (4.18)$$

where $J(m)$ is the spectral intensity stored in data channel m (m being linearly related to energy loss) and the integer n replaces ν as the Fourier "frequency."* Because of the sampled nature of the recorded data and its finite energy range, $J(E)$ can be completely represented in the Fourier domain by a limited number of frequencies, not exceeding $n = N - 1$. Moreover, the spectral data $J(E)$ are real (no imaginary part) so that

---

* As an example, the DFT of the unit-area Gaussian is $\exp(-\pi^2 n^2 \sigma^2/N^2)$.

$j_1(-\nu) = j_1(\nu)$, $j_2(-\nu) = -j_2(\nu)$, and $j_2(0) = 0$ (Bracewell, 1978). In practice, the negative frequencies are stored in channels $N/2$ to N, so these relations become $j_1(N-n) = j_1(n)$, $j_2(N-n) = -j_2(n)$, and $j_2(0) = 0$. As a result, only $(N/2 + 1)$ real values and $N/2$ imaginary values of $j(n)$ need be computed and stored, requiring a total of $N + 1$ storage channels for each transform. The requirement becomes just N channels if the zero-frequency value $j_1(0)$, representing the "dc component" of $J(E)$, is discarded (it can be added back at the end, after taking the inverse transform). The fact that the maximum recorded frequency is $n = N/2$ (the Nyquist frequency) means that frequency components in excess of this value ought to be filtered from the data before computing the DFT (Higgins, 1976) in order to prevent spurious high-frequency components appearing in the SSD (known as aliasing). In EELS data, however, this filtering is rarely necessary because frequencies exceeding $N/2$ consist mainly of noise.

Although the limits of integration in Eq. (4.7) extend to infinity, the finite range of the recorded spectrum will have no deleterious effect *provided J(E)* and its derivatives have the same value at $m = 0$ and at $m = N - 1$. In this case, $J(E)$ can be thought of as being part of a *periodic* function whose Fourier *series* contains cosine and sine coefficients which are the real and imaginary parts of $j(n)$. The necessary "continuity condition" is satisfied if $J(E)$ falls almost to zero at both ends of the recorded range. If not, $J(E)$ should be extrapolated smoothly to zero at $m = N - 1$, using (for example) a "cosine-bell" function: $A[1 - \cos r(N - m - 1)]$, where r and A are constants chosen to match the data near the end of the range. Any discontinuity in $J(E)$ creates unwanted high-frequency components which, following deconvolution, give rise to ripples adjacent to any sharp features in the SSD.

In order to record all of the zero-loss peak, the origin of the energy-loss axis must correspond to some nonzero channel number m_0. The result of this displacement of the origin is to multiply $j(n)$ by the factor exp $(2\pi i m_0 n/N)$. However, $Z(E)$ usually has the same origin as $J(E)$, so $z(n)$ gets multiplied by the same factor and the effects cancel in Eq. (4.10). In Eq. (4.11), where $z(\nu)$ also occurs outside the logarithm, the combined effect is to shift the recovered SSD to the right by m_0 channels, so that its origin occurs in channel $m = 2m_0$. To avoid the need for an additional phase-shift term in Eq. (4.12) and Eq. (4.13), $J(E)$ must be shifted so that the center of the zero-loss peak occurs in the first channel ($m = 0$) before computing the transforms. Whenever the data are shifted prior to Fourier processing, the left half of $Z(E)$ must be placed in channels immediately preceding the last one ($m = N - 1$).

The number of data channels used for each spectrum is usually of the form $N = 2^k$, where k is an integer, allowing a *fast-Fourier transform* (FFT) algorithm to be used to evaluate the discrete transform (Brigham, 1974). The number of arithmetic operations involved is then of the order of N $\log_2 N$, rather than N^2 as in a conventional Fourier-transform program (Cochran, 1967), which reduces the computation time by a factor of typically 100 (for $N = 1024$). Short (<50-line) FFT subroutines written in FORTRAN are available in the literature (e.g., Uhrich, 1969; Higgins, 1976) and are used in the Fourier-log program listed in Appendix B.

The zero-loss peak is absent in the inverse transform of $j^1(n)$ but can easily be replaced at the appropriate location, provided an array containing $Z(E)$ is kept throughout the deconvolution. $Z(E)$ is useful in the SSD because it delineates the zero-loss channel and provides an indication of the specimen thickness and the energy resolution.

Thicker Samples

Strictly speaking, Eqs. (4.9)–(4.13) do not specify a unique solution for the single-scattering distribution; it is possible to add any multiple of $2\pi i$ to the right-hand side of Eq. (4.15) and thereby change the SSD without affecting the quantity in square brackets (i.e., the experimental data). This ambiguity will cause problems if, for a particular SSD, the true value of the phase θ (i.e., the imaginary part of the logarithm) lies outside the range (normally $-\pi$ to $+\pi$) generated by a complex-logarithm function, a condition which is liable to occur if the scattering parameter t/λ exceeds π (Spence, 1979). If Eq. (4.17) is used, however, the value of θ will be restricted to the range $-\pi/2$ to $\pi/2$ and trouble may arise when $t/\lambda > \pi/2$.*

One solution to this phase problem (Spence, 1979) is to avoid making use of the imaginary part of the logarithm. This would happen automatically if $S(E)$ were an *even* function (symmetric about $m = N/2$ or $m = 0$), since in this case $s(\nu)$ has no imaginary coefficients (Bracewell, 1978). In practice, $S(E)$ is not even but can always be written as a sum of its even and odd parts: $S(E) = S^+(E) + S^-(E)$. Moreover, *provided* $S(E)$ is zero over one half of its range (see Fig. 4.2), $S(E) = 2S^+(E)$ and it is sufficient to recover $S^+(E)$, thereby avoiding the phase problem. The necessary condition can be satisfied by shifting the spectrum $J(E)$ (before computing its Fourier transform) so that the middle of the zero-loss peak occurs either at the exact center or at the beginning of the data range. In the former case $S(E)$ will be zero for $m < N/2$ (energy *gains* can generally be neglected; Section 3.3.5); in the latter, $J(E)$ must be set equal to zero for $m > N/2$. In either

---

* The range of allowable scattering parameter can be extended to π by adding (to the computed value of θ) a term $\pi \, \text{sgn} \, (j_2 z_1 \text{-} j_1 z_2)$ whenever $j_1 z_1 + j_2 z_2$ is negative.

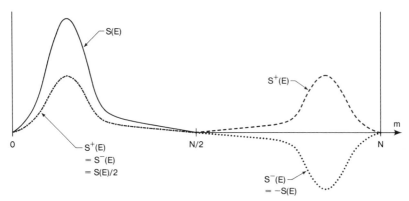

Figure 4.2. Relation between the single-scattering distribution $S(E)$ and its even and odd parts, $S^+(E)$ and $S^-(E)$, for the case where $S(E)$ is zero in the range $N/2 < m < N$.

case, the effects of truncation errors can be avoided by terminating the data smoothly, as discussed earlier.

The Fourier transform $s^+(\nu)$ of $S^+(E)$ is obtained as follows. From Eq. (4.15),

$$\ln[j(\nu)/z(\nu)] = \ln(r) + i\theta = s^+(\nu) + s^-(\nu) \tag{4.19}$$

Since $s^+(\nu)$ is entirely real and $s^-(\nu)$ entirely imaginary,

$$s^+(\nu) = \ln(r) = \ln[|j(\nu)/z(\nu)|] \tag{4.20}$$

Instead of computing the modulus in Eq. (4.20), $\ln(r)$ can be calculated directly from the Fourier coefficients of $J(E)$ and $Z(E)$, using Eq. (4.16). The inverse transform then gives $J^1(E)$ correctly in the region originally occupied by the nonzero $J(E)$ data; a mirror image of $J^1(E)$ appears in the other half of the range. Spence (1979) has shown that a correct SSD can be obtained even from samples with $t/\lambda \approx 10$ using this method.

A disadvantage of this scheme is that it doubles the memory space and computing times required, but these considerations are of little consequence for modern computers. An alternative way of extending deconvolution to thicker specimens is to evaluate θ using Eq. (4.17) and instruct the computer to correct for each discontinuity in the array (Fig. 4.3b) by adding or subtracting π (Egerton and Crozier, 1988). However, this correction becomes problematic if the true change in phase between adjacent coefficients approaches π, as is the case for thicker specimens if the energy range of the original data is too restricted (Su and Schattschneider, 1992a).

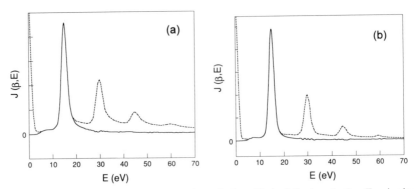

Figure 4.3. Plasmon-loss spectra of aluminum before (dashed line) and after Fourier-log deconvolution, recorded at 120-keV incident energy with (a) no objective-lens aperture ($\beta \approx 1700 \theta_E \approx 120$ mrad) and (b) a 20-μm-diameter objective aperture ($\beta = 54 \theta_E = 3.7$ mrad).

The Effect of Thickness Variations

The analysis so far has assumed that the specimen thickness is uniform over the area from which the spectrum $J(E)$ is recorded. The effect of a variation in thickness can be visualized in an extreme form by imagining part of the sampled are to have zero thickness, corresponding (for example) to a hole in the specimen. Electrons transmitted through the hole will contribute to $Z(E)$ but not to the inelastic intensity. Even if the parameter t represents some average thickness, Eq. (4.1) will not be exact and the SSD derived using the Fourier-log method must be somewhat in error. It appears that double scattering is *undersubtracted* (see Fig. 4.1c), typically by 5% if $Z(E)$ is augmented by 25%.

The effect of a *small* variation Δt in thickness has been calculated by Johnson and Spence (1974). As a fraction of the double-scattering intensity I_2, the residual second-order component ΔI_2 is given by

$$\Delta I_2/I_2 \cong (1/12)(\Delta t/t)^2 \qquad (4.21)$$

This fraction is less than 1% for $\Delta t < 0.35t$.

4.1.2. Misell–Jones and Matrix Methods

In the *absence* of instrumental broadening, the energy-loss spectrum can be written as $I_0 \delta(E) + P(E)$, where the unbroadened plural-scattering distribution $P(E)$ is given by

$$P(E) = S(E) + D(E) + T(E) + \cdots \qquad (4.22)$$

and $p(\nu)$ is its Fourier transform. Using Eq. (4.9) and Eq. (4.22), we obtain

$$s(\nu) = I_0 \ln[1 + p(\nu)/I_0]$$
$$= p(\nu) - [p(\nu)]^2/(2I_0) + [p(\nu)]^3/(3I_0^2) - \cdots \qquad (4.23)$$

Taking inverse transforms of both sides, we obtain

$$S(E) = P(E) - P(E) * P(E)/(2I_0) + P(E) * P(E) * P(E)/3I_0^2) - \cdots \qquad (4.24)$$

The SSD can therefore be computed as a logarithmic series involving *self-convolution* of the original spectrum (Misell and Jones, 1969). Unfortunately, the series converges rapidly only for $p(\nu)/I_0 \ll 1$, that is for $t/\lambda \ll 1$. Also, the method ignores the effect of instrumental broadening, so errors in the SSD can be expected if the measured spectrum $J(E)$ contains features which are comparable in width to the zero-loss peak.

Schattschneider (1983a,b) has developed Eq. (4.24) by replacing each convolution by a Riemann sum which is evaluated as a matrix, so that the procedure is no longer restricted to very thin samples. No truncation errors occur if the data do not fall to zero at both ends of the range, making the matrix procedure attractive for processing a limited range of low-dose data. As in the Misell–Jones procedure, no allowance is made for the energy resolution of the spectrometer system. A FORTRAN program for matrix deconvolution is given in Appendix B.

4.1.3. Deconvolution of Angle-Limited Spectra

The deconvolution methods discussed so far are based on the assumption that intensities recorded in the energy-loss spectrum accurately reflect the relative probabilities of inelastic scattering, which obey Poisson statistics. This assumption is justified if the spectrometer records all of the transmitted electrons. When an angle-limiting aperture precedes the spectrometer, however, the relative intensities of the different orders of scattering are altered.

If the angle-selecting aperture is centered about the optic axis (zero scattering angle), it admits *all* of the unscattered electrons but accepts only a fraction F_n of those which have been inelastically scattered n times, as discussed in Section 3.4. In Eq. (4.9), $z(\nu)$ will be unaffected by the aperture but $j(\nu)$ will be modified. Algebraic analysis (Egerton and Wang, 1990) shows that Fourier-log deconvolution leaves behind a fraction R_2 of the double scattering and a fraction R_3 of the triple scattering, given by

$$R_2 = [F_2 - F_1^2]/F_2, \qquad R_3 = [F_3 - 3F_1F_2 + 2F_1^3]/F_3 \qquad (4.25)$$

Fortunately, Eq. (3.97) indicates that the bracketed terms in Eq. (4.25) are close to zero provided $\beta \gg \theta_E$, where β is the aperture semiangle and θ_E is the characteristic scattering angle at an energy loss E. This condition holds in the low-loss region for typical aperture angles and incident energies above 50 keV: numerical computation of F_2 and F_3, assuming a Lorentzian angular distribution with an abrupt cutoff, shows that less than 3% of the second- and third-order scattering remain after Fourier-log processing, provided $\beta > 10\theta_E$ (Egerton and Wang, 1990; Su et al., 1992), and this prediction has been confirmed experimentally (Fig. 4.3).

If the collection aperture is displaced from the optic axis, as in angular-resolved spectroscopy, Eq. (3.98) is no longer valid and the problem of calculating the SSD becomes more complicated. The convolution integrals must be generalized to include scattering angle θ, treated as a vector with radial and azimuthal components (Misell and Burge 1969). In addition, it may be necessary to deal explicitly with elastic and quasielastic scattering (Bringans and Liang, 1981; Batson and Silcox, 1983) which can appreciably modify the angular distribution of inelastic scattering, particularly in thicker specimens. In principle, the angular distributions of all these scattering processes must be known, although the procedure can be simplified if an energy-loss spectrum which includes all angles of scattering is available (Batson and Silcox, 1983).

For an amorphous or polycrystalline specimen, both the elastic and inelastic scattering are axially symmetric and scattering probabilities (per unit angle) can be written in terms of Hankel transforms, given by Eq. (4.7) with the exponential replaced by a zero-order Bessel function (Johnson and Isaacson, 1988; Reimer, 1989). Su and Schattschneider (1992b) used discrete Hankel transforms to process plasmon-loss spectra recorded from 50nm and 100nm aluminum films at scattering angles up to 13 mrad. Mixed elastic/inelastic scattering gives rise to an undispersed (15 eV) plasmon peak immediately preceding the dispersed one and was successfully removed in the analysis, which required about 1 hour computing time per spectrum.

4.2. Kramers–Kronig Analysis

As shown in Chapter 3, the single-scattering distribution is related to the complex permittivity ε of the specimen and is given (in the absence of instrumental broadening and surface-mode scattering) by

$$J^1(E) \approx S(E) = \frac{2I_0 t}{\pi a_0 m_0 v^2} \mathrm{Im}\left[\frac{-1}{\varepsilon(E)}\right] \int_0^\beta \frac{\theta d\theta}{\theta^2 + \theta_E^2}$$

$$= \frac{I_0 t}{\pi a_0 m_0 v^2} \mathrm{Im}\left[\frac{-1}{\varepsilon(E)}\right] \ln\left[1 + \left(\frac{\beta}{\theta_E}\right)^2\right] \tag{4.26}$$

where I_0 is the zero-loss intensity, t the specimen thickness, v the speed of the incident electron, β the collection semiangle, and $\theta_E = E/(\gamma m_0 v^2)$ is the characteristic scattering angle for an energy loss E. Note that $S(E)$ and the instrumentally broadened intensity $J^1(E)$ are in units of joule^{-1}; a factor of 1.6×10^{-19} is required to convert them to eV^{-1}.

Starting from the single-scattering distribution $J^1(E)$, Kramers–Kronig analysis enables the energy dependence of the real and imaginary parts (ε_1 and ε_2) of the permittivity to be calculated, together with other optical quantities such as the absorption coefficient and reflectivity. Although a typical energy-loss spectrum has poorer energy resolution than that achievable using light-optical spectroscopy, its energy range can be much greater; energy losses equivalent to the visible, ultraviolet, and soft x-ray region may be recorded in the same experiment. Moreover, the energy-loss data are obtainable from microscopic regions of a specimen, which can be characterized in the same instrument using other techniques such as electron diffraction. Published examples of Kramers–Kronig analysis have mostly used specimens of known structure and stoichiometry and the results have formed a useful comparison with (or extension of) existing optical data. However, there exists the possibility of obtaining new data which may be helpful in formulating band structures or in more fully characterizing the specimen (Fink *et al.*, 1983; Turowski and Kelly, 1992).

The first step in the process is to derive the single-scattering distribution $J^1(E)$ from the experimental spectrum $J(E)$, as described in Section 4.1. If the specimen is very thin (less than 10 nm for 100-keV incident electrons), the raw spectrum could be used; however, the spectrum from a very thin sample contains an appreciable surface-loss contribution which may invalidate the use of Eq. (4.26) at low values of E. Some workers minimize this surface contribution by recording the spectrum slightly off-axis, making use of the smaller angular width of the surface losses (Liu, 1988).

4.2.1. Angular Corrections

The next step is to obtain an energy distribution proportional to Im$[-1/\varepsilon]$ by dividing $J^1(E)$ by the logarithmic term in Eq. (4.26). Sometimes this is referred to as an aperture correction, although it is not quite equivalent to simulating the effect of removing the collection aperture, which would require division of $J^1(E)$ by the angular-collection efficiency $\eta = \ln[1 + (\beta/\theta_E)^2]/\ln[1 + (\theta_c/\theta_E)^2]$, where θ_c is an effective cutoff angle (Section 3.3.1). Since $\theta_E \propto E$, applying the angular correction increases the intensity at high energy loss relative to that at low loss.

Equation (4.26) assumes that the angular divergence α of the incident beam is small in comparison with β. If this condition is violated, a further angular correction may be required (Section 4.5). Daniels *et al.* (1970)

give an alternative form of correction which applies when the energy-loss intensity is measured using an off-axis collection aperture.

4.2.2. Extrapolation and Normalization

In order to evaluate subsequent integrals, the data may have to be extrapolated so that $J^1(E)$ falls practically to zero at high energy loss. The form of extrapolation is not critical; an AE^{-r} dependence can be used, where r is estimated from the experimental data or taken as three (as predicted for the "tail" of a plasmon peak by the Drude model; Section 3.3.1).

Unless the values of t, v, and β in Eq. (4.26) are accurately known (Isaacson, 1972a), $\text{Im}[-1/\varepsilon]$ is given an absolute scale by use of a Kramers–Kronig sum rule, obtained by setting $E = 0$ in Eq. (4.28):

$$1 - \text{Re}\left[\frac{1}{\varepsilon(0)}\right] = \frac{2}{\pi} \int_0^\infty \text{Im}\left[\frac{-1}{\varepsilon(E)}\right] \frac{dE}{E} \qquad (4.27)$$

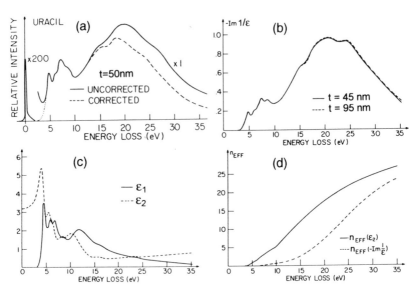

Figure 4.4. (a) Energy-loss spectrum of a 50-nm film of uracil on a 2.5-nm carbon substrate, recorded with 25-keV electrons and $\beta = 0.625$ mrad; the dashed line shows the spectrum after correction for double scattering. (b) Energy-loss function, obtained from films of two different thicknesses. (c) Real and imaginary parts of the dielectric function, derived by Kramers–Kronig analysis. (d) Effective number of electrons (per uracil molecule) as a function of the integration range; the dashed curve was calculated from Eq. (4.32) and the solid curve from Eq. (4.33). (From Isaacson, 1972a.)

Since $\mathrm{Re}[1/\varepsilon(0)] = \varepsilon_1/(\varepsilon_1^2 + \varepsilon_2^2)$, the left-hand side of Eq. (4.27) can be taken as unity for a metal, where both ε_1 and ε_2 become very large for $E \to 0$ (see Fig. 3.11). In the case of an insulator, ε_2 is small for small E and $\mathrm{Re}[1/\varepsilon(0)] \cong 1/\varepsilon_1(0)$, where $\varepsilon(0)$ is the square of the refractive index for visible light. The static permittivity is not appropriate here since the measured spectrum does not extend below $E \simeq 1$ eV because of the limited energy resolution of an electron-microscope system.

Normalization consists of dividing each energy-loss intensity, which is proportional to $\mathrm{Im}[-1/\varepsilon(E)]$ after the plural-scattering and angular corrections, by the corresponding energy loss E and integrating over the whole energy range as in Eq. (4.27). This integral is divided by $(\pi/2)\{1 - \mathrm{Re}[1/\varepsilon(0)]\}$ to yield the proportionality constant $K = I_0 t/(\pi a_0 m_0 v^2)$ and (if the zero-loss intensity and incident energy are known) an estimate of the absolute specimen thickness. The aperture-corrected spectrum is then divided by K to give $\mathrm{Im}[-1/\varepsilon(E)]$; see Fig. 4.4b. If $\mathrm{Re}[1/\varepsilon(0)]$ or the absolute value of K are not known, it may be possible to use Eq. (4.32) or Eq. (4.33) to estimate K, provided the upper limit of the integral can be chosen such that essentially all contributions from a known set of atomic shells are included within the specified energy range.

4.2.3. Derivation of the Dielectric Function

Based on the fact that the dielectric response function is causal (Johnson, 1975) a Kramers–Kronig transformation can be used to derive the function $\mathrm{Re}[1/\varepsilon(E)]$ from $\mathrm{Im}[-1/\varepsilon(E)]$:*

$$\mathrm{Re}\left[\frac{1}{\varepsilon(E)}\right] = 1 - \frac{2}{\pi} P \int_0^\infty \mathrm{Im}\left[\frac{-1}{\varepsilon(E')}\right] \frac{E'\,dE'}{E'^2 - E^2} \qquad (4.28)$$

where P denotes the Cauchy principal part of the integral, avoiding the pole at $E = E'$ (Daniels et al., 1970). In Eq. (4.28), $E'/(E'^2 - E^2)$ acts as a "weighting function," giving prominence to energy losses E' which lie close to E. Values of $\mathrm{Im}[-1/\varepsilon]$ corresponding to $E' < E$ contribute negatively to the integral whereas values corresponding to $E' > E$ make a positive contribution, so $1 - \mathrm{Re}[1/\varepsilon]$ somewhat resembles the differential of $\mathrm{Im}[-1/\varepsilon]$. The principal value of the integral can be obtained by computing $\mathrm{Re}[1/\varepsilon(E)]$ at values of E midway between the $\mathrm{Im}[-1/\varepsilon(E')]$ data points (Johnson, 1972) or by incorporating an analytical expression for the region adjacent to $E = E'$ (Stephens, 1981).

---

* Equation (4.28) applies to only isotropic materials; Daniels et al. (1970) provide equations for the anisotropic case.

The Kramers–Kronig interval can also be evaluated using Fourier-transform techniques (Johnson 1974, 1975), based on the fact that $\mathrm{Re}[1/\varepsilon(E)]-1$ and $\mathrm{Im}[-1/\varepsilon(E)]$ are cosine and sine transforms of the even and odds parts, $p(t)$ and $q(t)$, of the time-dependent dielectric response function: $1/\varepsilon(t) - \delta(t)$. Because a response cannot precede the cause, this function is zero for $t < 0$ so that (as in Fig. 4.2)

$$p(t) = \mathrm{sgn}[q(t)] \tag{4.29}$$

The procedure is therefore to compute $q(t)$ as the sine transform of $\mathrm{Im}[-1/\varepsilon(E)]$, obtain $p(t)$ by reversing the sign of the Fourier coefficients over one half of their range, take the inverse cosine transform, and add unity to obtain $\mathrm{Re}[1/\varepsilon(E)]$. It avoids the need to compute principal parts (there are no infinities on the t-axis) and can be relatively rapid if fast Fourier transforms are used; see Appendix B.

Johnson (1975) has shown that to avoid errors arising from the sampled nature of $\mathrm{Im}[-1/\varepsilon(E)]$, the sharpest peak in this function should contain at least four data points. If this condition is not met, the sine coefficients do not fall to zero at the Nyquist frequency and sign inversion introduces a discontinuity in slope which contributes high-frequency ripple to $\mathrm{Re}[1/\varepsilon(E)]$; see Fig. 4.5. This ripple becomes amplified at low values of E when ε_1 and ε_2 are computed.

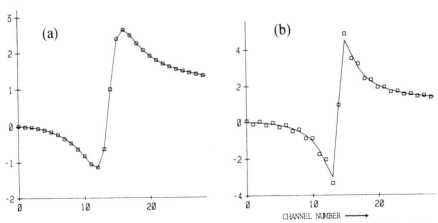

Figure 4.5. $\mathrm{Re}[1/\varepsilon(E)]$ for a free-electron gas, the FWHM of the plasmon peak being (a) four channels and (b) two channels. The solid curve was calculated directly from Drude theory (Section 3.3.1); square data points were derived from the Drude expression for $\mathrm{Im}[-1/\varepsilon(E)]$, using the Fourier method of Kramers–Kronig analysis (Egerton and Crozier, 1988).

After computing $\text{Re}[1/\varepsilon(E)]$, the dielectric function is obtained from

$$\varepsilon(E) = \varepsilon_1(E) + i\varepsilon_2(E) = \frac{\text{Re}[1/\varepsilon(E)] + i\,\text{Im}[-1/\varepsilon(E)]}{\{\text{Re}[1/\varepsilon(E)]\}^2 + \{\text{Im}[-1/\varepsilon(E)]\}^2} \quad (4.30)$$

Equating the real and imaginary parts of Eq. (4.30) gives the separate functions $\varepsilon_1(E)$ and $\varepsilon_2(E)$; see Fig. 4.4c. Other optical quantities can also be calculated, such as the optical absorption coefficient:

$$\mu(E) = (E/\hbar c)[2(\varepsilon_1^2 + \varepsilon_2^2)^{1/2} - 2\varepsilon_1]^{1/2} \quad (4.30a)$$

4.2.4. Corrections for Surface Losses

Equations (4.26)–(4.28) assume that energy losses take place within the interior of the specimen. Particularly if the specimen is very thin, it is desirable to make allowance for surface losses. Assuming a flat and clean (unoxidized) surface and using a free-electron plasmon model of the surface loss, integration of Eq. (3.83) up to a scattering angle β (with $\varepsilon_a = 1$, $\varepsilon_b = \varepsilon = \varepsilon_1 + \varepsilon_2$) gives the single-scattering surface-loss intensity as

$$S_s(E) = I_0 \frac{dP_s}{dE} \quad (4.31)$$

$$= \frac{I_0}{\pi a_0 k_0 T}\left[\frac{\tan^{-1}(\beta/\theta_E)}{\theta_E} - \frac{\beta}{\beta^2 + \theta_E^2}\right]\left[\frac{4\varepsilon_2}{(\varepsilon_1 + 1)^2 + \varepsilon_2^2} - \text{Im}\left(\frac{-1}{\varepsilon}\right)\right]$$

$S_s(E)$ is subtracted from the experimental single-scattering distribution $J_1(E)$ and a new normalization constant K found. Then $\varepsilon(E)$ is recalculated and the whole process repeated if necessary until the result converges (Wehenkel, 1975). At each stage in the procedure, the value of ε_1 at small E (≈ 2 eV) should approximate to the optical value used in applying Eq. (4.27); see Appendix B.3.

4.2.5. Checks on the Data

The existence of the Bethe f-sum rule (Section 3.2.4) gives rise to the concept of an "effective" number of electrons contributing to energy losses *up to* a value E. From Eq. (3.35), one possible definition of the effective number of electrons *per atom* (or per molecule) is

$$n_{\text{eff}}(-\text{Im } \varepsilon^{-1}) = \frac{2\varepsilon_0 m}{\pi\hbar^2 e^2 n_a}\int_0^E E'\,\text{Im}\left[\frac{-1}{\varepsilon}\right] dE' \quad (4.32)$$

where n_a is the number of atoms (or molecules) per unit volume of the sample. Alternatively, Eq. (3.34) can be applied to optical transitions ($q \cong 0$), leading to a second effective number:

$$n_{\text{eff}}(\varepsilon_2) = \int_0^E \frac{df(0, E)}{dE}\, dE = \frac{2\varepsilon_0 m_0}{\pi \hbar^2 e^2 n_a} \int_0^E E'\varepsilon_2(E')\, dE' \qquad (4.33)$$

Because of the $1/E$ weighting factor in the relationship between $d\sigma/dE$ and df/dE, $n_{\text{eff}}(-\text{Im }\varepsilon^{-1})$ remains less than $n_{\text{eff}}(\varepsilon_2)$ at low values of E but the two effective numbers converge towards a plateau value at higher energy loss; see Fig. 4.4d. In favorable cases (elements and simple compounds) this plateau corresponds to a *known* number of electrons per atom, providing a useful check on the derived values of $\varepsilon_2(E)$ and $\text{Im}[-1/\varepsilon(E)]$. In elemental carbon, for example, $n_{\text{eff}} \cong 4$, corresponding to excitation of all the valence electrons, for energies ($\cong 200$ eV) approaching the K-ionization threshold (Hagemann *et al.*, 1974); see page 311. In compounds containing several elements, inner-shell excitation may occur before the valence-electron contribution is exhausted, so the true plateau values are never reached; one can tell if the derived values of n_{eff} are substantially too high but not whether they are too low. This difficulty is removed if the analysis can be carried out up to an energy loss which is several times the largest inner-shell binding energy of the elements involved; the final saturation value should then correspond to the total number of electrons per atom (or molecule).

4.3. Removal of Plural Scattering from Inner-Shell Edges

As noted in Section 3.7.3, plural scattering can drastically alter the observed shape of an inner-shell ionization edge and may have to be removed before near-edge or extended fine structure can be interpreted. This plural scattering takes the form of one or more inelastic collisions with low energy loss (outer-shell excitation) which occur in addition to the core-level ionization. The probability of more than one core ionization can generally be neglected in transmission spectroscopy, where the sample thickness is small compared to the *inner-shell* mean free path. Because the core-loss region typically involves many data points which are equally spaced in energy, fast-Fourier methods are the preferred choice for spectral processing.

4.3.1. Fourier-Log Deconvolution

The deconvolution method due to Johnson and Spence (1974), described in Section 4.1.1, assumes only that scattering events are independent and that the probability of plural scattering is described by Poisson statistics. The technique is therefore capable of removing plural scattering from anywhere within the energy-loss spectrum, including the "mixed" scattering beyond an ionization edge. It involves calculating the Fourier transform of the complete spectrum, from the zero-loss peak up to and beyond the ionization edge(s) of interest. To prevent truncation errors from affecting the SSD in the range of interest, the spectrum is either recorded up to an energy loss well beyond these edges or else extrapolated smoothly towards zero intensity at some high energy loss.

Any gain increment introduced during serial acquisition must be removed from the spectrum before calculating its transform, for example by determining the gain factor G via a linear least-squares fitting of several data points on both sides of the gain change, then multiplying the lower-energy data by G. Similarly, low-loss and core-loss regions which are obtained by separate acquisitions from a parallel-recording spectrometer have to be "spliced" together. The resulting spectrum will often have a large dynamic range (e.g., 10^7), requiring arithmetic of fairly high precision in the Fourier-transform calculations, but the necessary precision is usually available even on a microcomputer (Egerton *et al.*, 1985). The Fourier coefficients are processed using one of the procedures described in Section 4.1.1.

Unlike the Fourier-ratio method to be described in Section 4.3.2, Fourier-log deconvolution removes plural scattering from both the core-loss region and the preceding background. Since the core-loss intensity just above the ionization threshold arises only from *single* inner-shell scattering, the "jump ratio" of an edge is increased by Fourier-log processing, the increase being quite dramatic in the case of moderately thick samples; see Fig. 4.6. In this respect, the deconvoluted spectra are equivalent to those which would have been obtained using a thinner sample or a higher incident energy. However, the noise components arising from the plural scattering remain behind after deconvolution, and statistical errors of background subtraction (Section 4.4.4) remain the same. Therefore deconvolution improves the sensitivity and accuracy of elemental analysis only to the extent that *systematic* errors in background fitting may be reduced, for example if the single-scattering background approximates more closely to a power-law energy dependence (Leapman and Swyt, 1981a).

In addition to increasing the *fractional* noise content of the pre-edge background, Fourier-log deconvolution tends to accentuate small artifacts present in the spectrum as a result of power-supply fluctuations or nonlinearity in the electronics. An example is shown in Fig. 4.6. A gain increment was originally present in the spectrum at 100 eV, corresponding to a change from analog detection to pulse counting during serial recording. In the

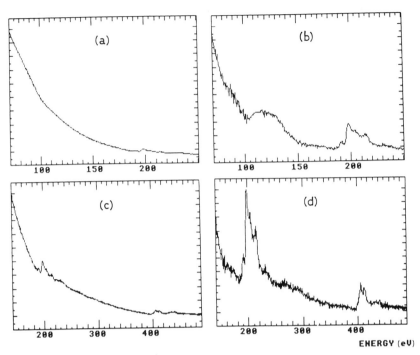

ENERGY (eV)

Figure 4.6. (a) Part of the energy-loss spectrum recorded from a thick region of a boron nitride specimen ($t/\lambda = 1.2$) using 80-keV incident electrons and $\beta = 100$ mrad. (b) The same energy region after Fourier-log deconvolution, showing an artifact generated from the change in slope at 100 eV (the position of the original gain change). (c) An extended energy range, showing the boron and nitrogen K-ionization edges prior to deconvolution. (d) The boron and nitrogen K-edges after Fourier-log deconvolution. (From R. F. Egerton, B. G. Williams, and T. G. Sparrow, Fourier deconvolution of electron energy-loss spectra, *Proc. R. Soc. London* **A398,** 395–404 (1985), by permission of the Royal Society.)

pulse-counting mode, pulse pileup caused partial saturation of the signal, resulting in a change in slope after removal of the gain increment. Deconvolution converts this change in slope into a "hump" extending over several tens of eV, which might be mistaken for an ionization edge. This example is somewhat extreme but illustrates the need for high-quality data prior to deconvolution.

4.3.2. Fourier-Ratio Deconvolution

In this alternative deconvolution technique, the energy-loss spectrum is divided into two regions. The low-loss region, containing the zero-loss peak and energy losses up to typically 100 eV, is used as a deconvolution function or "instrument function" for the second region, which contains the core-loss intensity. Deconvolution is carried out by *dividing* the Fourier transform of the second region by that of the first.

In order to justify such a procedure, we first of all consider "ideal" spectral data, free from instrumental broadening. Let $K^1(E)$ be an ideal core-loss single-scattering distribution, recorded with a spectrometer system having a large acceptance angle and perfect energy resolution. The probability of a transmitted electron suffering an energy loss (between E and $E + dE$) due to core ionization *and no other process* is $K^1(E)\ dE/I$, where I is the total (elastic + inelastic) intensity entering the spectrometer. The probability of no inelastic scattering of any kind is I_0/I, where I_0 is the zero-loss intensity, so the total probability of a *core* loss between E and $E + dE$ is $[K^1(E)dE/I]/(I_0/I) = K^1(E)\ dE/I_0$. The probability that the *total* energy loss falls within the range E to $E + dE$, of which only an amount $E - E'$ is due to inner-shell ionization, is $[K^1(E - E')\ dE/I_0][P(E')dE/I]$, where $P(E)$ is the "ideal" plural scattering intensity as defined in Eq. (4.22). If we define "mixed" scattering to be that which involves one or more low-loss events in addition to an inner-shell excitation, the total probability of such scattering (within the range E to $E + dE$) is

$$\frac{M(E)\ dE}{I} = \int_{E=0}^{E_{max}} \frac{K^1(E - E')\ dE}{I_0} \frac{P(E')\ dE'}{I} = K^1(E) * P(E) \frac{dE}{II_0} \qquad (4.34)$$

where E_{max} is any energy loss exceeding $E - E_k$, E_k being the core-edge threshold energy. The total intensity within the core-loss spectrum (Fig. 4.7) is equal to

$$K(E) = K^1(E) + M(E) = K^1(E) * [\delta(E) + P(E)/I_0] \qquad (4.35)$$

As recorded by a spectrometer system having a resolution function $R(E)$, the core-loss intensity will be

$$J_k(E) = K(E) * R(E) = K^1(E) * \{[I_0\delta(E) + P(E)] * R(E)/I_0\}$$
$$= K^1(E) * J_l(E)/I_0 \qquad (4.36)$$

where $J_l(E)$ is the measured low-loss spectrum, including the zero-loss peak (Fig. 4.7). Taking Fourier transforms of both sides of Eq. (4.36): $j_k(\nu) = k^1(\nu)j_l(\nu)/I_0$, which is inverted to give

$$k^1(\nu) = I_0 j_k(\nu)/j_l(\nu) \qquad (4.37)$$

Equation (4.36) shows that, in principle, the measured core-loss intensity $J_k(E)$ can be corrected for both plural scattering and instrumental broadening by multiplying by I_0 and deconvoluting with respect to the measured low-loss intensity $J_l(E)$, treating the latter as if it were an "instrument

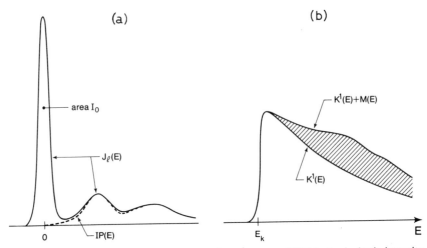

Figure 4.7. (a) Low-loss region of an energy-loss spectrum; IP(E) is the inelastic intensity which would be recorded using a spectrometer with perfect energy resolution. (b) Core-loss region (containing a single ionization edge) with pre-edge background subtracted. The intensities shown are those which would be recorded using an "ideal" spectrometer; the shaded area represents the intensity due to mixed scattering.

function" (Egerton and Whelan, 1974b; Egerton, 1976b). Equation (4.37) illustrates how this deconvolution might be achieved, by taking a ratio of the appropriate Fourier coefficients. However, such "complete" deconvolution is problematic because high-frequency noise components present in $j_k(\nu)$ become "amplified" when divided by $j_l(\nu)$. As discussed in Section 4.1.1, the noise amplification can be avoided in several ways:

(a) By multiplying the ratio of Fourier coefficients by a Gaussian function $\exp(-\pi^2\sigma^2\nu^2)$, we can obtain

$$j_k^1(\nu) = I_0 \exp(-\pi^2\sigma^2\nu^2)j_k(\nu)/j_l(\nu) \tag{4.38}$$

If $\sigma = \Delta E/1.665$, where ΔE is the experimental energy resolution (usually taken as the FWHM of the zero-loss peak), the inverse transform of $j_k^1(\nu)$ will be a core-loss SSD whose energy resolution and noise content are similar to those of the original data. If the zero-loss peak in $J_l(E)$ occurs at data channel $m = m_0$, the inverse transform $J_k^1(E)$ will be shifted to the left of m_0 channels, relative to $J_k(E)$.

(b) If the Fourier transform $r(\nu)$ of the resolution function $R(E)$ is used as the noise-limiting function, we obtain

$$j_k^1(\nu) = r(\nu)k^1(\nu) = z(\nu)j_k(\nu)/j_l(\nu) \tag{4.39}$$

where $z(\nu)$ is the Fourier transform of the zero-loss peak. Assuming that the energy resolution ΔE is independent of energy loss, the inverse transform $J_k^l(E)$ will have an energy resolution identical to that of $J_k(E)$. Equation (4.39) implies $J_k(E) * Z(E) = J_k^l(E) * J_l(E)$, which follows directly from Eq. (4.36) if both sides are convoluted by $R(E)$. Since $J_l(E)$ and $Z(E)$ normally have the same origin on the E-axis, the energy shift associated with method (a) is avoided, but three Fourier transforms must be calculated rather than two.

(c) If the frequency spectra of both the signal $J_k(E)$ and its associated noise $N(E)$ are known approximately, it is possible to choose a noise-rejection function $g(\nu)$ which provides an optimum compromise between noise and resolution in the single-scattering distribution. A function with a sharp cutoff at $\nu = \nu_1$ must be avoided, since it would introduce convolution with $\sin(2\pi\nu_1 E)/(2\pi\nu_1 E)$, resulting in oscillatory artifacts adjacent to any sharp features in the SSD. Ray (1979) showed that a modest increase of energy resolution is possible by using a Wiener-filter function of the form

$$g(\nu) = |j_l(\nu)|^2\{|j_l(\nu)|^2 + N/S\}^{-1} \qquad (4.40)$$

where N/S is an estimated or measured noise/signal ratio in the core-loss region. A greater improvement in resolution might be possible using a maximum-entropy technique (Burch *et al.*, 1983), which produces the smoothest profile lying within specified error limits, subject to other constraints (e.g., that all the output data be positive).

(d) If the resolution function $R(E)$ is narrow compared with peaks in the plural-scattering distribution $P(E)$, the latter may be replaced in Eq. (4.36) by the experimentally measured quantity $P(E) * R(E)$, giving

$$J_k(E) \cong K^l(E) * R(E) * [I_0\delta(E) + P(E) * R(E)]/I_0$$
$$= J_k^l(E) * J_l^d(E)/I_0 \qquad (4.41)$$

so that

$$j_k^l(\nu) \cong I_0 j_k(\nu)/j_l^d(\nu) \qquad (4.42)$$

J_k^d represents the measured low-loss spectrum with the zero-loss peak $Z(E)$ replaced by a delta function of equal area, achieved in practice by summing electron "counts" within $Z(E)$ and placing the sum in the channel corresponding to $E = 0$. Although only approximate, this procedure gives results which are very similar to those obtained using method (b).

Practical Details

Before applying Fourier-ratio deconvolution, the low-loss and core-loss data should be present in computer-memory arrays containing the same number N of channels ($N = 2^k$ where k is an integer). In the case of a parallel-recording spectrometer, these two spectra may have been obtained in separate acquisitions at different integration times or incident-beam intensities. The background to the lowest-energy ionization edge must be removed, as described in Section 4.4, and the intensity extrapolated to zero at the high-E end of the range, making the intensity approximately zero at both ends of the array and thereby satisfying the continuity requirement for a Fourier series; see Section 4.1.1. Likewise, the intensity should approximately zero at both ends of the low-loss spectrum.

If both regions have been recorded with a serial spectrometer in a single scan, a separation point is selected, preceding the first ionization edge and typically at the channel where a gain change was introduced during acquisition. The low-loss data and the background-subtracted core-loss data are transferred to separate arrays of computer-memory array and both extrapolated to zero at the Nth data point.

After computing the necessary Fourier coefficients (see Section 4.1.1 and Appendix B), pairs of coefficients are divided according to the rules of complex division. For example,

$$\frac{j_k(\nu)}{j_l(\nu)} = \frac{j_{k1} + ij_{k2}}{j_{l1} + ij_{l2}} = \frac{j_{k1}j_{l1} + j_{k2}j_{l2}}{j_{l1}^2 + j_{l2}^2} + i\frac{j_{k2}j_{l1} - j_{k1}j_{l2}}{j_{l1}^2 + j_{l2}^2} \tag{4.43}$$

In the case of noise-free data and a specimen of uniform thickness, the Fourier-ratio method can be shown to be equivalent to Fourier-log deconvolution (Swyt and Leapman, 1984). Applied to real data, the two methods give results which are very similar; see Fig. 4.8.

4.3.3. Effect of a Collection Aperture

If F_p, F_k, and F_{pk} are the fractions of plasmon-loss, core-loss, and double (core-loss + one-plasmon) scattering which pass through an angle-limiting collection aperture, the fraction of double scattering remaining after Fourier-log deconvolution is, by analogy with Eq. (4.25),

$$R_{pk} = (F_{pk} - F_p F_k)/F_{pk} \tag{4.44}$$

Algebraic analysis shows that Eq. (4.44) also applies to Fourier-ratio deconvolution.

The angular distribution of double (core+plasmon) scattering can be calculated as an angular convolution of the core-loss and plasmon angular distributions. Taking the latter to be a Lorentzian function with a cutoff at θ_c, R_{pk} is estimated to be less than 4% for $\beta > \theta_c$ (Egerton and Wang, 1990). The plural-scattering intensity left behind after deconvolution is predicted to be appreciable only for small values of β and high-energy edges; see Fig. 4.9.

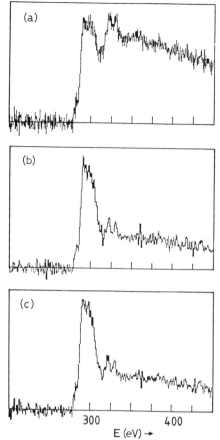

Figure 4.8. (a) Carbon K-edge (after background subtraction) recorded from a thick sample of graphite using 80-keV incident electrons and a collection semiangle $\beta \approx 100$ mrad. (b) Single-scattering K-loss intensity recovered using the Fourier-ratio method. (c) Single-scattering distribution obtained by Fourier-log deconvolution, followed by background subtraction. (From R. F. Egerton, B. G. Williams, and T. G. Sparrow, Fourier deconvolution of electron energy-loss spectra, *Proc. R. Soc. London* **A398,** 395–404 (1985), by permission of the Royal Society.)

4.4. Background Fitting to Ionization Edges

Inner-shell ionization edges occur mostly at energy losses above 100 eV, superimposed upon a monotonically decreasing background which originates from the excitation of atomic electrons of lower binding energy. For quantitative elemental analysis, the core-loss intensity of interest must be isolated by subtracting the background intensity. In this section, we discuss

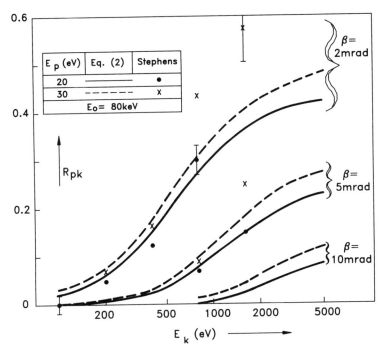

Figure 4.9. Fraction of mixed (plasmon + core-loss) scattering remaining after deconvolution, plotted against edge energy for plasmon energies of 20 eV and 30 eV and three values of collection semiangle (Egerton and Wang, 1990). (From *Ultramicroscopy*, R. F. Egerton and Z. L. Wang, Plural-scattering deconvolution of electron energy-loss spectra recorded with angle-limiting aperture, © 1990, pp. 137–148, with permission from Elsevier Science B.V., Amsterdam, The Netherlands.) The calculations are for an incident energy of 80 keV, to allow direct comparison with the data points based on Monte Carlo simulations of Stephens (1980).

several means of modeling the background, together with the statistical and systematic errors involved in these procedures.

As discussed in Section 3.5, the spectral intensity due to a given energy-loss process has a high-energy tail which approximates to a power-law energy dependence: AE^{-r}. Whereas the coefficient A can vary widely (depending on the incident-beam current, for example), the exponent r is generally in the range 2–6. The value of r usually decreases with increasing specimen thickness (Fig. 3.35), because of plural-scattering contributions to the background; r also decreases if the collection semiangle β is increased, but tends to increase with increasing energy loss, as in Fig. 3.40. Consequently, values of A and r must be measured at each ionization edge. The energy dependence of the background is measured over a "fitting region" immediately preceding the edge, on the assumption that the background intensity has the same energy dependence over at least a limited range beyond the ionization threshold.

Other functions, such as an exponential, polynomial, or log-polynomial, have been used for the E-dependence of the pre-edge background, and have sometimes been found to be preferable to the power-law model. Although they may provide a good match to the experimental intensity within the fitting region, polynomial functions have a tendency to behave wildly when extrapolated to higher energy loss. Such behavior can be avoided by using a "tied" polynomial, forced to pass through a data point far beyond the edge. A similar technique could be used for an exponential or power-law background.

A power-law fit to the experimental background is likely to be improved if the "instrumental" background (introduced, for example, by the detection-system electronics) is subtracted from the spectrum prior to background modeling.

4.4.1. Least-Squares Fitting

A standard technique, which gives good results in the majority of cases, is to match the pre-edge background $J(E)$ to a function $F(E)$ whose parameters (e.g., A and r) minimize the quantity

$$\chi^2 = \sum_i [(J_i - F_i)/\sigma_i]^2 \qquad (4.45)$$

where i is the index of a channel within the fitting region and σ_i represents the statistical error (standard deviation) of the intensity in that channel. For simplicity, σ_i is generally assumed to be constant over the fitting region, in which case the fitting procedure is equivalent to minimizing the mean-square deviation of $J(E)$ from the fitted curve. The statistical formulas required are simpler if some function of $J(E)$ can be fitted to a straight line: $y = a + bx$, for which the least-squares values of the slope and y-intercept are (Bevington, 1969)

$$b = \frac{N\sum x_i y_i - \sum x_i \sum y_i}{N\sum x_i^2 - (\sum x_i)^2} \qquad (4.46)$$

$$a = \sum y_i/N - b \sum x_i/N \qquad (4.47)$$

Here, the summations are over the fitting region, which contains N channels.

In the case of the power-law function $F(E) = AE^{-r}$, *linear* least-squares fitting can be made applicable by taking logarithms of the data coordinates. In other words, $y_i = \log(J_i)$ and $x_i = \log(E_i) = \log[(m - m_0)\delta E]$, where m is the absolute number of a data channel, m_0 is the channel number corresponding to $E = 0$, and δE is the energy-loss increment per channel.* The least-squares values of a and b can be found, using Eq. (4.46) and Eq. (4.47), by running a short program on the microcomputer which controls the data storage. The fitting parameters are then given by $r = -b$ and $\log(A) = a$. The logarithms can be to any base, provided the choice is consistent throughout the program.

As an estimate of the "goodness of fit," the parameter χ^2 can be evaluated using Eq. (4.45), taking $\sigma_i \cong J_i^{1/2}$ (i.e., assuming that electron-beam shot noise determines the uncertainties in the recorded data). More useful is the normalized χ^2 parameter $\chi_n^2 = \chi^2/(N - 2)$, which is less dependent of the number N of channels within the fitting region. Alternatively, a correlation coefficient can be evaluated (Bevington, 1969).

Linear least-squares fitting has been found satisfactory for almost all pre-edge backgrounds (Joy and Maher, 1981a). However, systematic errors can be observed if the number of detected electrons per channel J_i falls to a very low value (<10), a situation which may occur in the case of energy-filtered images (Section 2.6). The fractional uncertainty $\sigma_i/J_i \cong J_i^{-1/2}$ is then large and the error distribution becomes asymmetric, particularly after taking logarithms of the data (Egerton, 1980d). The resulting systematic error in background level can be estimated to be 2% for $J_i = 10$, increasing to 20% for $J_i = 3$.

4.4.2. Two-Area Method

In this alternative method of background fitting, the fitting region is divided into two segments, generally of equal width; the parameters A and r of the power-law background function are found by measuring the respective areas I_1 and I_2 (see Fig. 4.10). If the background decreased *linearly* with energy loss, each area would be given by the parallelogram rule

$$I_1 = (E_3 - E_1)[J(E_1) + J(E_3)]/2 \qquad (4.48)$$

and similarly for I_2. In the case of a power-law background, it turns out to be more accurate to replace the arithmetic average of intensities in Eq. (4.48) by a *geometric* average:

$$I_1 \cong (E_3 - E_1)[J(E_1)J(E_3)]^{1/2} \qquad (4.49)$$

and likewise for I_2, so that

$$\frac{I^1}{I_2} \approx \frac{E_3 - E_1}{E_2 - E_3} \left[\frac{E_1}{E_2}\right]^{-r/2} \qquad (4.50)$$

---

* Note that it is now the quantity $\Sigma[\log(J_i) - \log(F_i)]^2 = \Sigma\, 2|\log(J_i/F_i)|$ which is being minimized.

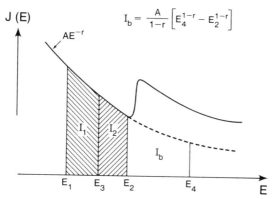

Figure 4.10. Two-area method of background fitting. Values of A, r, and I_b are obtained by measuring the areas I_1 and I_2 under the background just preceding an ionization edge.

If $E_3 = (E_1 + E_2)/2$, Eq. (4.50) becomes $I_1/I_2 \cong (E_1/E_2)^{-r/2}$ and

$$r \cong 2 \log(I_1/I_2)/\log(E_2/E_1) \qquad (4.51)$$

By straightforward integration of $J(E) = AE^{-r}$, Eq. (4.54) can be shown to be exact for $r = 2$. More surprisingly, the formula remains remarkably accurate for higher values of r, the systematic error in the background integral I_b being typically less than 1%, as illustrated in Table 4.1. The factor of 2 in Eq. (4.51) would be absent for narrow and widely spaced energy windows, a situation which is unfavorable in terms of statistical noise.

Table 4.1. Systematic Error Involved in the Two-Area Method
Using a Noise-Free Background[a]

r (exact)	r from Eq. (4.51)	$\dfrac{I_b(\text{exact})}{I_b(I_1 + I_2)}$	$\dfrac{I_b(\text{exact})}{I_b(I_2)}$
2	2.000	1.0000	1.0000
3	3.007	1.0021	1.0014
4	4.019	1.0057	1.0036
5	5.035	1.0109	1.0042

[a] $J(E) = AE^{-r}$ and energies appropriate to a carbon K-ionization edge: $E_1 = 200$ eV, $E_3 = 240$ eV, $E_2 = 280$ eV, and $E_4 = 360$ eV. The last two columns indicate the fractional error in the background integral I_b, calculated using values of A obtained from Eq. (4.52) and (4.53), respectively.

The value of A is obtained from either of the following equations:

$$A = (1 - r)(I_1 + I_2)/(E_2^{1-r} - E_1^{1-r}) \qquad (4.52)$$

$$A = (1 - r)I_2/(E_2^{1-r} - E_3^{1-r}) \qquad (4.53)$$

Having computed A and r, the background contribution I_b beneath the ionization edge can be calculated; see Fig. 4.1.3. Of the two equations for A, Eq. (4.53) will usually result in a more accurate value of I_b; the systematic error is less (see Table 4.1) and, more importantly, the statistical extrapolation error (Sectio 4.4.4) is likely to be smaller since increased weight is given to background channels close to the edge.

Because it involves only a single summation over J_i, the two-area method can be executed faster than a least-squares procedure, a worthwhile consideration if the background fitting must be done a large number of times, as in STEM elemental mapping and spectrum-imaging (Section 2.5.1).

4.4.3. More Sophisticated Methods

Using initial values of A and r derived from the two-area method, a ravine-search program (Bevington, 1969) has been used to provide a better fit to noisy data (Colliex et al., 1981a). This program computes the variances of A and r, and also χ^2 as a test of the significance level of the fit. The computing time can be significant but may be justified if an element is present in low concentration and its ionization edge barely visible above the background noise. Trebbia (1988) has written FORTRAN programs which calculate the background using a maximum-likelihood method, a procedure which avoids bias introduced by the nonlinear transformation in the least-squares method (Pun et al., 1985).

Superior accuracy of background removal is possible if the energy dependence of the core-loss intensity is known. In this case, the fitting procedure can be extended into the core-loss region, as discussed in Section 4.5.4.

4.4.4. Background-Subtraction Errors

In addition to possible systematic errors, any background fitted to noisy data will be subject to a random or statistical error. In the case of peaks superimposed on a smooth background (as in x-ray emission spectra, for example), the background can often be measured on both sides of the peak and its contribution below the peak deduced by *interpolation*. In a core-loss spectrum, the background can be sampled only on the low-energy side of the ionization edge and must be *extrapolated* to higher energies, giving a comparatively large statistical error in the background integral I_b. In the case of linear least-squares fitting, the statistical error can be estimated using standard formulas (Bevington, 1969) for the variances of the parameters a and b in the equation $y = a + bx$. To ensure that the coefficients a and b are statistically independent (and thereby avoid the need to evaluate a covariance term), the origin of the x coordinate must be placed at the *center* of the fitting region; see Fig. 4.11. To illustrate the method of calculation, we first consider the simple case of a linearly decreasing background, for which the background integral is of the form $I_b = n[a + b(m + n)/2]$, where m and n are the number of data channels

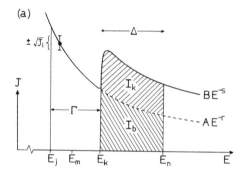

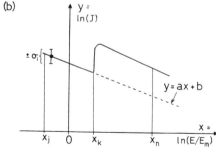

Figure 4.11. (a) Schematic diagram of an ionization edge, defining the widths of the background-fitting and integration regions (Γ and Δ, respectively) and the core-loss and background integrals (I_k and I_b, respectively). (b) The same region of the spectrum, plotted on logarithmic coordinates. The x-axis origin is at the center of the fitting region. One of the data points within the fitting region is shown, together with its standard deviation.

in the fitting region and the integration region, respectively. The variance of I_b can be obtained from the general relation:

$$\text{var}(I_b) = \left(\frac{\partial I_b}{\partial a}\right)^2 \text{var}(a) + \left(\frac{\partial I_b}{\partial b}\right)^2 \text{var}(b) \tag{4.54}$$

Denoting the average standard deviation of the intensity within a single channel in the background-fitting region by σ, the variances of a and b are given by

$$\text{var}(a) = \sigma^2/m \tag{4.55}$$

$$\text{var}(b) \approx \sigma^2 \Big/ \sum_{i=-m/2}^{m/2} i^2 \doteq \frac{\sigma^2}{[i^3/3]_{-m/2}^{m/2}} = \frac{12\sigma^2}{m^3} \tag{4.56}$$

Combining the previous three equations, we obtain

$$\text{var}(I_b) = (\sigma^2 n^2/m)[1 + 3(1 + n/m)^2] \tag{4.57}$$

If σ arises entirely from the counting statistics and if the range of extrapolation is small, so that the electron intensity within the integration region is comparable to that within the fitting region, $\sigma^2 \simeq I_b/n$. In a typical case, these two regions have similar widths ($m \cong n$) and Eq. (4.57) gives $\text{var}(I_b) \cong 13I_b$, where I_b is in units of detected electrons.

The equivalent analysis for a power-law background is similar to the above except that the x-y plot now involves logarithms of the data coordinates and the origin of the x-axis corresponds to an energy loss $E_m = (E_j E_k)^{1/2}$; see Fig. 4.11. The standard deviation on the y-axis is now related to the intensity J in each channel by $\sigma \cong \ln(J - \sqrt{J}) - \ln(J) \cong \sqrt{J}$ and the background integral is given by

$$I_b = \int_{E_k}^{E_n} AE^{-r}\, dE = \frac{E_m e^a}{1 + b}\left(e^{(1+b)x_n} - e^{(1+b)x_k}\right) \tag{4.58}$$

and its variance by

$$\text{var}(I_b) = I_b^2 \text{var}(a) + [C^2 E_m^2 e^{2a}/(1 + b)^4]\,\text{var}(b) \tag{4.59}$$

where

$$C = e^{(1+b)x_n}[(1 + b)x_n - 1] - e^{(1+b)x_k}[(1 + b)x_k - 1]$$

with $a = \ln(AE_m^{-r})$, $b = -r$ and the coordinates x_k and x_n defined in Fig. 4.11.

Experimentally, the inner-shell intensity I_k is obtained by integrating the total intensity between x_k and x_n (to give the integral I_t) and subtracting the background integral I_b. Statistical errors in I_t and in I_b are therefore additive:

$$\text{var}(I_k) = \text{var}(I_t) + \text{var}(I_b) = I_k + I_b + \text{var}(I_b) \qquad (4.60)$$

The last term in Eq. (4.60) represents the background-extrapolation error, which often forms the major part of the uncertainty in I_k.

By treating I_k as the required "signal" and $[\text{var}(I_k)]^{1/2}$ as its statistical uncertainty or "noise," the signal-to-noise ratio can be written as

$$\text{SNR} = I_k[\text{var}(I_k)]^{-1/2} = I_k/(I_k + hI_b)^{1/2} \qquad (4.61)$$

where the dimensionless parameter $h = [I_b + \text{var}(I_b)]/I_b$ represents the factor by which the background-dependent part of $\text{var}(I_k)$ is increased because of the fitting and extrapolation errors.

If the width of the integration region is sufficiently small, the extrapolated background will approximate to a straight line over this region and Eq. (4.57) can be used in estimating the value of h; for example, $h = 14$ if $m = n$. In the more general case, Eq. (4.59) should be used; Fig. 4.12a shows that if a large extrapolation error ($h \gg 1$) is to be avoided, Γ should not be small compared to Δ. Figure 4.12b indicates that as Δ is increased (keeping Γ constant), the signal/noise ratio first increases and then falls slightly. The initial rise is due to the increase in the signal I_k; the subsequent decrease arises from the rapid increase in h as the range of extrapolation is extended.

Berger and Kohl (1993) have considered how statistical and other factors influence the choice of instrumental parameters for elemental mapping. As is always the case in energy-filtered imaging, spatial resolution is of prime importance; the effect of chromatic aberration (Section 2.3.2) puts further constraints on Δ and results in smaller values (typically 20 eV) being used than those which might minimize statistical and systematic errors. In spectroscopy, the statistical uncertainty associated with background extrapolation can be reduced by the use of multiple least-squares (MLS) fitting to calculated or measured core-loss profiles, as discussed in Section 4.5.4.

4.5. Elemental Analysis Using Inner-Shell Edges

Because inner-shell binding energies are separated by tens or hundreds of eV, whereas chemical shifts (Section 3.7.4) amount to only a few eV, identifying ionization edges in the energy-loss spectrum provides a means of qualitative elemental analysis. To make the analysis quantitative, the

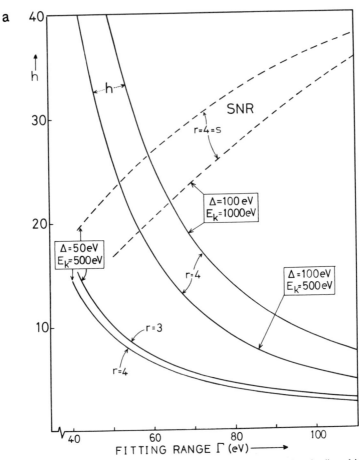

Figure 4.12. (a) Extrapolation parameter *h* and signal/background ratio (in arbitary units) as a function of (a) width Γ of the background-fitting region and (b) width Δ of the integration region (Egerton, 1982a). The calculations assume power-law background ($\propto E^{-r}$) and edge ($\propto E^{-s}$) intensities. Dashed curves show SNR for a weak ionization edge; dotted curves are for a strong edge (equal edge and background intensities). From *Ultramicroscopy*, R. F. Egerton, A revised expression for signal/noise ratio in EELS, © 1982, pp. 387–390, with permission from Elsevier Science B.V., Amsterdam, The Netherlands.

core-loss intensity can be integrated over an appropriate energy window, making allowance for the noncharacteristic background. Such a procedure is more accurate than measuring the *height* of the edge, which is sensitive to near-edge fine structure dependent on the structural and chemical environment of the ionized atom (Section 3.8). If however the element is present at low concentration, it is difficult to obtain sufficient accuracy in the

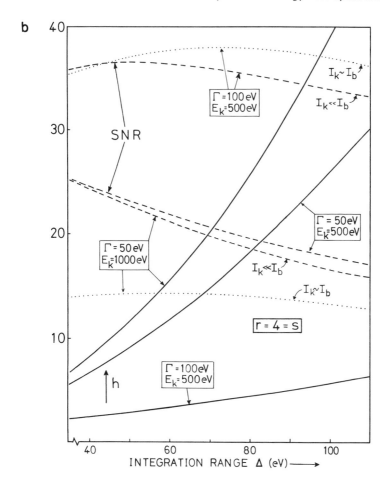

background extrapolation, and in these circumstances it is preferable to fit the experimental spectrum in the core-loss region to an algebraic sum of background and a reference edge, as discussed in Section 4.5.4.

4.5.1. Integration Method

Because incident electrons may be scattered (either once or several times) by elastic, low-loss (plasmon), and core-loss scattering, their angular and energy distributions are quite complicated. However, by making approximations (valid within certain limits of specimen thickness) we can obtain simple formulas which are suitable for routine quantitative analysis.

For the moment, assume that inner-shell excitation is the only form of scattering in the specimen. From Eq. (3.94), the integrated intensity of single scattering from shell k of a selected element, characterized by a mean free path λ_k and a scattering cross section σ_k, would be given by

$$I_k^1 = I_0(t/\lambda_k) = NI_0\sigma_k \qquad (4.62)$$

I_0 represents the unscattered (zero-loss) intensity and N is the areal density (atoms per unit area) of the element, equal to the product of its concentration and the specimen thickness. If we collect the scattering only up to an angle β and integrate its intensity over a limited energy range Δ, the core-loss integral is

$$I_k^1(\beta, \Delta) = NI_0\sigma_k(\beta, \Delta) \qquad (4.63)$$

where $\sigma_k(\beta, \Delta)$ is a "partial" cross section for energy losses within a range Δ of the ionization threshold and for scattering angles up to β, which can be obtained from experiment or calculation (see page 283).

The effect of *elastic* scattering is to cause a certain fraction of the electrons to be intercepted by the angle-limiting aperture. To a first approximation, this fraction is the same for electrons which have caused inner-shell excitation and those which have not, in which case $I_k^1(\beta, \Delta)$ and I_0 are reduced by the same factor and the core-loss integral becomes

$$I_k^1(\beta, \Delta) \approx NI_0(\beta)\, \sigma_k(\beta, \Delta) \qquad (4.64)$$

where $I_0(\beta)$ is the new (observed) zero-loss intensity. Equation (4.64) applies to a core-loss edge from which plural (core-loss + plasmon) scattering has been removed by deconvolution.

If we now permit valence-electron (plasmon) excitation to contribute to the spectrum, its effect is to redistribute intensity towards higher energy loss, away from the zero-loss peak and from the ionization threshold. Not all of this scattering falls within the core-loss energy window, but to a first approximation the fraction which *is* included will be the same as the fraction which falls within an energy window of equal width in the low-loss region. In that case, the core-loss integral (including plural scattering) is given by

$$I_k(\beta, \Delta) \approx NI(\beta, \Delta)\sigma_k(\beta, \Delta) \qquad (4.65)$$

where $I(\beta,\Delta)$ is the low-loss intensity integrated up to an energy loss Δ; see Fig. 4.13.

Although Eq. (4.64) and Eq. (4.65) allow measurement of the absolute areal density N of a given element, an atomic ratio of two elements (a and

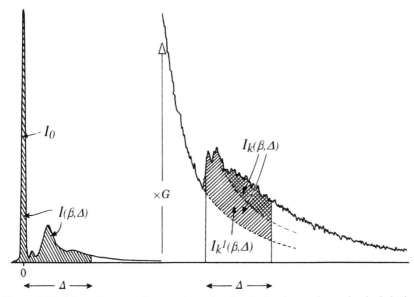

Figure 4.13. Schematic energy-loss spectrum showing the low-loss region and a single ionization edge (intensity scale increased by a factor G). The crosshatched area represents electrons
which have undergone both core-loss and low-loss scattering.

b) is more commonly required. Provided the same integration window Δ
is used for both edges, Eq. (4.65) gives

$$\frac{N_a}{N_b} = \frac{I_{ka}(\beta, \Delta)}{I_{jb}(\beta, \Delta)} \frac{\sigma_{jb}(\beta, \Delta)}{\sigma_{ka}(\beta, \Delta)} \tag{4.66}$$

The shell index can be different for the two edges ($j \neq k$); K-edges are
suitable for very light elements ($Z < 15$) and L- or M-edges for elements
of higher atomic number. However, the approximate treatment of plural
scattering will be more accurate if the two edges have similar shape. If
plural scattering is removed from the spectrum by deconvolution, Eq. (4.64)
leads to

$$\frac{N_a}{N_b} = \frac{I_{ka}^1(\beta, \Delta_a)}{I_{jb}^1(\beta, \Delta_b)} \frac{\sigma_{jb}(\beta, \Delta_b)}{\sigma_{ka}(\beta, \Delta_a)} \tag{4.67}$$

A different energy window ($\Delta_a \neq \Delta_b$) can now be used for each edge, larger
values being more suitable at higher energy loss where the spectrum is
noisier but the edges from different elements are spaced further apart.

Several authors have tested the accuracy of these equations for thin
specimens. Approximating the low-loss spectrum by sharp peaks at multi-

ples of a plasmon energy E_p, Stephens (1980) concluded that the approximate treatment of plural scattering inherent in Eq. (4.65) will lead to an error in N of between 3% and 10% (dependent on E_k) for $t/\lambda_p = 0.5$ and $\Delta/E_p = 5$. The error would be some fraction of this when evaluating elemental ratios. Systematic error arising from the angular approximation inherent in Eq. (4.64) was estimated to be of the order of 1% for a 20-nm amorphous carbon film but considerably larger for polycrystalline or single-crystal specimens if a strong diffraction ring (or spot) occurs just within or just outside the collection aperture (Egerton, 1978a).

As specimen thickness increases, the higher probability of elastic and plural scattering causes Eqs. (4.64)–(4.67) to become less accurate. Elemental ratios given by Eq. (4.66) or Eq. (4.67) are found to change when t/λ exceeds approximately 0.5 (Zaluzec, 1983). This variation has been attributed to the effect of elastic scattering and modeled for amorphous materials of *known* composition, using the angular distribution of elastic scattering given by the Lenz model (Cheng and Egerton, 1993, Su et al., 1995). A computer program has been written to evaluate correction factors for specimens of *unknown* composition, based on additional measurements of the low-loss spectrum at several collection angles (Wong and Egerton, 1995). The correction factor becomes significant at specimen thicknesses above

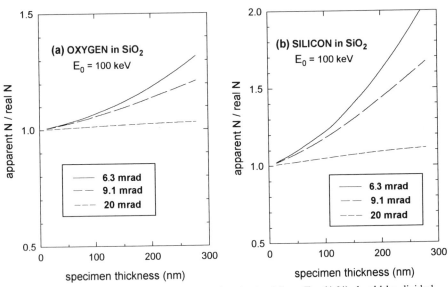

Figure 4.14. Factor by which the areal density obtained from Eq. (4.64) should be divided (to correct for elastic scattering) in the case of (a) the oxygen K-edge and (b) the silicon K-edge in amorphous silicon dioxide (Cheng and Egerton, 1993).

100 nm, particularly for higher edge energy and small collection angle; see Fig. 4.14.

Correction for elastic scattering in crystalline specimens may not be feasible. Accurate treatment would require extensive measurement of intensity in the diffraction plane or a knowledge of the crystal structure, orientation, and thickness of the specimen. The situation in single-crystal specimens is further complicated by the existence of channeling and blocking effects; see page 140 and page 358.

4.5.2. Calculation of Partial Cross Sections

If the core-loss intensity is integrated over an energy window Δ which is wide enough to include most of the fine-structure oscillations, the corresponding cross section $\sigma_k(\beta, \Delta)$ should be little affected by the chemical environment of the excited atom and can therefore be calculated by means of an atomic model. Weng and Rez (1988) have estimated that atomic cross sections are accurate to within 5% for $\Delta > 20$ eV.

The simplest atomic model is based on the hydrogenic approximation, for which the generalized oscillator strength (GOS) is available in analytic form; see Section 3.4.1. Computation is therefore rapid and requires only the atomic number and edge energy as inputs, in addition to the integration range Δ, angular range β and incident-electron energy E_0. FORTRAN programs for the calculation of K- and L-shell cross sections are given in Appendix B.

More sophisticated procedures, such as the Hartree–Slater method, involve more computation and a greater knowledge of atomic properties. However, the resulting GOS can be parameterized as a function of the energy and wavenumber q and the resulting values integrated according to Eq. (3.151) and Eq. (3.157) to yield a cross section. Parameterization can also take account of EELS experiments and (for small q) x-ray absorption measurements. A program giving K-, L-, M-, N-, and O-shell cross sections for small β (as normally employed in elemental analysis) is given in Appendix B.8.

A completely experimental approach to quantification is also possible, the sensitivity factor for a given edge being obtained by measurements on standards (Malis and Titchmarsh, 1986; Hofer, 1987; Hofer et al., 1988). If such measurements are made in the same microscope and under the same experimental conditions as used for analysis of an unknown specimen, this procedure may be relatively insensitive to the effects of chromatic aberration of prespectrometer lenses (Section 2.3.3) and imperfect knowledge of the collection angle and incident-electron energy. It is analogous to the k-factor procedure used in thin-film EDX microanalysis. For wider applicability, an experimentally determined k-factor can be converted to a dipole oscillator strength $f(\Delta)$ which depends on the integration range Δ but is independent of collection angle and incident-electron energy.

4.5.3. Correction for Incident-Beam Convergence

Equations (4.64)–(4.67) are applicable if an angular spread of the incident beam is small in comparison with the collection semiangle β. This condition usually applies to fixed-beam (CTEM) illumination but may be violated if the incident electrons are focused into a very fine probe of large convergence semiangle α. For such a probe, the angular distribution of the core-loss intensity $dI_k/d\Omega$ can be calculated as a vector convolution of the incident-electron intensity $dI/d\Omega$ and the inner-shell scattering $d\sigma_k/d\Omega$. Taking the latter to be a Lorentzian function of width $\theta_E = E/\gamma m_0 v$ (where $E \simeq E_k + \Delta/2$), we obtain

$$\frac{dI_k}{d\Omega} \propto \int_{\theta_0=0}^{\alpha} \int_{\phi=0}^{2\pi} \frac{dI}{d\Omega} \frac{1}{\theta_k^2 + \theta_E^2} \, \theta_0 \, d\theta_0 d\phi \tag{4.68}$$

where $\theta_k^2 = \theta^2 + \theta_0^2 - 2\theta\theta_0 \cos\phi$; see Fig. 4.15. The core-loss intensity passing through a collection aperture of semiangle β is then

$$I_k(\alpha, \beta, \theta_E) = \int_0^\beta \frac{dI_k}{d\Omega} 2\pi\theta \, d\theta \tag{4.69}$$

If the incident intensity per unit solid angle remains *constant* up to a cutoff angle α, the double integral of Eq. (4.68) can be solved analytically (Craven *et al.*, 1981), giving

$$\frac{dI_k}{d\Omega} \propto \ln \left[\frac{\psi^2 + (\psi^4 + 4\theta^2\theta_E^2)^{1/2}}{2\theta_E^2} \right] \tag{4.70}$$

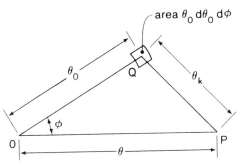

Figure 4.15. Calculation of the core-loss intensity at P (an angular distance θ from the optic axis) due to incident electrons at Q (whose polar coordinates are θ_0 and ϕ). The angle of inelastic scattering is θ_k.

where $\psi^2 = \alpha^2 + \theta_E^2 - \theta^2$. Combining the previous three equations yields

$$F_1 = \frac{I_k(\alpha, \beta, \Delta)}{I_k(0, \beta, \Delta)} = \frac{2/\alpha^2}{\ln[1 + (\beta/\theta_E)^2]} \int_0^\beta \ln\left[\frac{\psi^2 + (\psi^4 + 4\theta^2\theta_E^2)^{1/2}}{2\theta_E^2}\right] \theta\, d\theta$$

(4.71)

F_1 is a factor (<1) representing reduction in the measured core-loss intensity due to incident-beam convergence. Scheinfein and Isaacson (1984) have shown that the integral in Eq. (4.71) can be expressed analytically, and their expression is used to evaluate F_1 in the FORTRAN program CONCOR2 listed in Appendix B. Because F_1 depends to some degree on θ_E (see Fig. 4.16), the convergence correction is *different* for each ionization edge. When using Eq. (4.66) or Eq. (4.67) to obtain an elemental ratio, the effect of beam convergence is included by multiplying the right-hand side by F_{1b}/F_{1a}.

To obtain a correction factor for *absolute* quantification, we must consider the effect of beam convergence on the low-loss intensity. Since $\theta_E/\beta \ll 1$ for valence-electron scattering, the corresponding factor F_1 is close to unity provided $\alpha < \beta$ (see Fig. 4.16). If $\alpha > \beta$, the recorded low-loss intensity is approximately proportional to the area of the convergent-

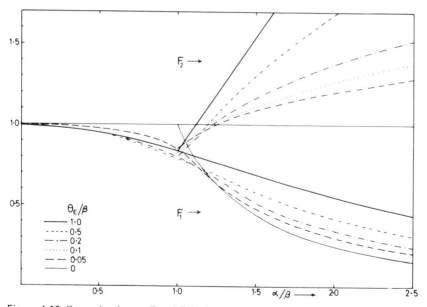

Figure 4.16. Correction factors F_1 and F_2 for incident-beam convergence, plotted as a function of α and for different values of the characteristic scattering angle θ_E. Note that for $\alpha < \beta$ the correction factor ($F_1 = F_2$) first decreases and then increases with increasing θ_E.

beam disk which falls within to the collection aperture, so Eq. (4.65) becomes

$$I_k(\alpha, \beta, \Delta) \approx F_2 N \sigma_k(\beta, \Delta) I(\alpha, \beta, \Delta) \qquad (4.72)$$

where $F_2 \approx F_1$ for $\alpha \leq \beta$ and $F_2 \approx (\alpha/\beta)^2 F_1$ for $\alpha \geq \beta$; see Fig. 4.16.

Alternatively, the effect of incident-beam convergence can be expressed in terms of an *effective collection angle* β^* which differs from β by an amount dependent on α and the edge energy. If defined by $I_k(\alpha, \beta, \Delta) = F_1 I_k(\beta, \Delta) = I_k(\beta^*, \Delta)$, β^* is always less than β; but if defined by Eq. (4.72), $\beta^* > \beta$ for $\alpha > \beta$. Provided the core-loss angular distributions remain Lorentzian up to a scattering angle $\theta = \beta$, both definitions should lead to the same elemental ratio. Kohl (1985) has shown that incident-beam convergence can be taken into account when calculating an *effective cross section* $\sigma_k(\alpha, \beta, \Delta)$, in which case the assumption of a Lorentzian angular dependence is unnecessary. But because the incident-beam intensity $dI/d\Omega$ is assumed to be a rectangular function, any convergence correction is likely to be approximate.

4.5.4. MLS Fitting to Reference Spectra

Because of uncertainties in background extrapolation (Section 4.4.4), the integration method of elemental quantification fails for very noisy data, ionization edges which are weak in relation to the background, or edges which occur in close proximity to each other. In these cases, the situation can be improved if the fitting procedure is extended *across* the ionization threshold, using a multiple-least-squares (MLS) procedure to fit the total spectral intensity $J(E)$ to an expression of the form

$$F(E) = AE^{-r} + B_a S_a(E) + B_b S_b(E) + \cdots \qquad (4.73)$$

The first term represents the background preceding the edge of lowest energy loss; $S_a(E)$, $S_b(E)$, ... represent core-loss reference spectra of the elements of interest and B_a, B_b, ... are coefficients found by minimizing the difference between the left- and right-hand sides of Eq. (4.73), summed over all energies within a region containing the ionization edges (Leapman and Swyt, 1988; Leapman, 1992).

For very thin specimens, or if plural scattering has been removed by deconvolution, the reference spectra could be calculated differential cross sections (Steele *et al.*, 1985), in which case each coefficient B would be simply the product of the zero-loss intensity and the areal density of the appropriate element. More generally, each $S(E)$ can be measured from a standard specimen containing the appropriate element, the atomic ratio of two elements (a and b) being obtained from

$$\frac{N_a}{N_b} = \frac{B_a}{B_b} \frac{I_{ka}(\beta, \Delta)}{I_{kb}(\beta, \Delta)} \frac{\sigma_{kb}(\beta, \Delta)}{\sigma_{ka}(\beta, \Delta)} \qquad (4.74)$$

where $I_{ka}(\beta, \Delta)$ and $I_{kb}(\beta, \Delta)$ are integrals (over some convenient integration range Δ) of the core-loss spectra of the appropriate standards; $\sigma_{ka}(\beta, \Delta)$ and $\sigma_{kb}(\beta, \Delta)$ are the corresponding partial cross sections.

Because the region just above the ionization edge contains prominent fine structure which is sensitive to chemical environment (Section 3.8), this region may have to be excluded from the fitting procedure unless the chemical environment in the reference standard is similar. If the analyzed specimen is appreciably thicker than the standard, it may be necessary to convolve the reference edge with the low-loss region of the analyzed specimen in order to make allowance for plural scattering.

4.5.5. Energy- and Spatial-Difference Techniques

When an energy-loss spectrum is differentiated with respect to energy loss, the slowly varying background to an ionization edge is largely eliminated. First- or second-difference spectra are obtained by digitally filtering conventional spectra (Zaluzec, 1985; Michel *et al.*, 1993) or by using a spectrum-shifting technique with a parallel-recording spectrometer (Section 2.5.5). These spectra can be fitted to energy-difference reference spectra, using MLS fitting without the background term in Eq. (4.73). Because difference spectra are highly sensitive to fine-structure oscillations in intensity, reliable quantification may depend upon the chemical environment being similar in the standard and the unknown material. This condition is more easily met for metals and biological specimens than for ionic and covalent materials (Tencé *et al.*, 1995).

Another way of dealing with the pre-edge background is to record spectra from a region of interest in the specimen (such as an interface or grain boundary) and from a nearby "matrix" region. If the matrix spectrum (scaled if necessary to allow for changes in thickness or diffracting conditions) is subtracted from the original data, the resulting *spatial-difference* spectrum represents changes in ionization-edge intensity due to differences in elemental composition. Because the pre-edge background is largely eliminated, simple integration can provide an estimate of the concentration difference; see Fig. 4.17.

A unique advantage of the spatial-difference method is that, provided the changes in composition are small and the crystal structure is similar in the two locations, systematic variations in background intensity (due, for

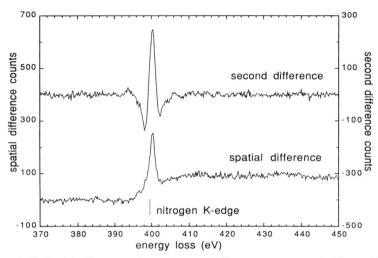

Figure 4.17. Spatial-difference and second-energy-difference spectra recorded from a nitro-gen-containing voidite in diamond (Müllejans and Bruley, 1994). The shape of the *K*-edge (white-line peak followed by a broad continuum) is consistent with molecular nitrogen.

example, to extended fine structures from a preceding edge) are eliminated in the subtraction. Müllejans and Bruley (1994) have discussed other advantages of this technique in terms of signal/noise ratio and provided examples of application; see page 368. A similar spatial-difference procedure can be useful in fine-structure studies (e.g., Bruley and Batson, 1989).

4.6. Analysis of Extended Energy-Loss Fine Structure

As discussed in Section 3.9, the EXELFS modulations which extend up to some hundreds of eV beyond an ionization edge can be analyzed to provide values of interatomic distances. In favorable circumstances, coordination numbers, bond angles, and degree of atomic disorder are also obtainable. This information is of particular value in the case of multielement amorphous materials, where diffraction techniques cannot distinguish the scattering due to different elements.

4.6.1. Fourier-Transform Method of Data Analysis

Following the original EXAFS procedure (Sayers *et al.*, 1971), the radial distribution function (RDF) can be obtained as a Fourier transform of the experimental EXELFS data (Kincaid *et al.*, 1978; Johnson *et al.*,

1981b; Leapman *et al.*, 1981; Stephens and Brown, 1981; Bourdillon *et al.*, 1984). The essential steps involved are as follows.

Background Subtraction and Deconvolution

Unless the specimen is very thin (<10 nm, for 100-keV electrons), plural scattering beyond the edge should be removed by deconvolution. If the Fourier-ratio technique is used (Section 4.3.2), the pre-edge background is subtracted prior to deconvolution; if a Fourier-log method is employed (Section 4.3.1), the background is removed after deconvolution to yield the single-scattering core-loss spectrum J_k^1.

EXELFS analysis is more straightforward if the data is obtained from a K-shell edge, but for atomic numbers greater than 15 the K-loss signal is weak (and therefore noisy) and the L-edge may have to be used (Okamoto *et al.*, 1991). In the case of transition metals, an L_1 edge occurs within the energy range covered by the L_{23} EXELFS and must be removed from the experimental data, for example by subtracting a suitably chosen fraction of the intensity at energies above the L_1 threshold (Leapman *et al.*, 1981). For transition elements beyond Ti ($Z = 22$), the L_2–L_3 splitting exceeds 5 eV, resulting in a "smearing" of the EXELFS; the effect can be eliminated by deconvoluting J_k^1 with a pair of delta functions separated by the appropriate energy and weighted by the ratio 2:1 (Leapman *et al.*, 1982). This deconvolution can be done by division of Fourier coefficients, either before, during, or after the removal of plural scattering.

Isolation of the Oscillatory Component

The oscillatory part $\chi(E)$ of the core-loss intensity is obtained by subtracting from $J_k^1(E)$ a smoothly decaying function $A(E)$, representing the intensity which would have been measured if neighboring atoms were absent. In general, $A(E)$ is not available experimentally and cannot be calculated with sufficient accuracy; it is obtained empirically by fitting a smooth function through J_k^1, as in Fig. 4.18. The function should correctly follow the overall trend of the data but not the EXELFS modulations themselves, otherwise false structure will appear in the RDF at small values of radius r. An odd-order polynomial (Leapman, 1982a) or a cubic spline (Johnson *et al.*, 1981b) has been used, the required fitting programs being generally available on large and medium-sized computers. A power-law function may also be suitable if the exponent is allowed to vary somewhat with energy loss (Stephens and Brown, 1981).

The difference spectrum is normalized by division with $A(E)$ to give

$$\chi(E) = [J_k^1(E) - A(E)]/A(E) \tag{4.75}$$

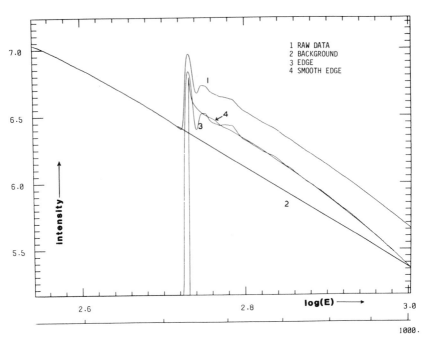

Figure 4.18. (1) Oxygen K-edge of sapphire, recorded using 100-keV incident electrons and collection semiangle $\beta = 16$ mrad. Also shown are (2) the extrapolated background intensity, (3) the core-loss intensity after background subtraction, and (4) a smooth polynomial function $A(E)$ fitted through the core-loss intensity. (A. J. Bourdillon, personal communication.)

The fact that $\chi(E)$ is defined as a ratio of intensities makes it unnecessary to divide by an angular-correction function (Section 4.2).

Scale Conversion

The energy scale of $\chi(E)$ is converted to one of wavenumber k of the ejected electron using Eq. (3.165). If energies are measured in eV and k in nm^{-1}, the formula becomes

$$k = 5.123\sqrt{E_{\text{kin}}} = 5.123(E - E^0)^{1/2} \qquad (4.76)$$

where E_{kin} is the kinetic energy of the ejected inner-shell electron and E^0 is the energy loss corresponding to $E_{\text{kin}} = 0$. E^0 is not precisely equal to the observed threshold energy E_k, since the latter corresponds to the excitation of electrons to the first unoccupied electron level. In a metal, this would be the Fermi level (where $E_{kin} = E_F$) and in the absence of exchange and correlation effects (Stern et al., 1980) one might expect $E^0 = E_k - E_F$. In insulators, however, the initial excitation is often to a bound state (Section 3.8.5), for which $E_{kin} < 0$, leading to $E^0 > E_k$. Because of possible chemical shifts, E^0 is best obtained from the experimental spectrum rather than by calculation. The inflection point at the edge or an energy loss corresponding to half the total rise in intensity (Johnson et al., 1981b) are possible choices for E^0. Unfortunately, an error in E^0 leads to a shift in the RDF peaks; for the boron K-edge in BN, Stephens and Brown (1981) found that the r-values changed by about 5% for a 5-eV change in E^0.

In fact, the most appropriate value of E^0 is related to the choice of energy zero assumed in calculating the phase shifts which are subsequently applied to the data. Lee and Beni (1977) proposed that E^0 be treated as a variable parameter whose value is selected such that peaks in both the imaginary part and the absolute value (modulus) of the Fourier transform of $\chi(k)$ occur at the same radius r. With suitably defined phase shifts (Teo and Lee, 1979), this method of choosing E^0 gave r-values mostly within 1% of known interatomic spacings (up to fifth-nearest neighbors) when applied to EXAFS data from crystalline Ge and Cu (Lee and Beni, 1977) and has since been used fairly widely in EXAFS studies.

Spectral data are usually recorded at equally spaced energy increments but after conversion of $\chi(E)$ to $\chi(k)$, the data points will be unequally spaced. If a fast-Fourier-transform (FFT) algorithm is to be used, the k-increments must be equal and some form of interpolation is needed. For finely spaced data points, linear interpretation is adequate; in the more general case, a sinc function would provide greater accuracy (Bracewell, 1978). A conventional (discrete) Fourier transform takes longer to execute but can use unequally spaced $\chi(k)$ data.

Correction for k-Dependence of Backscattering

According to Eq. (3.167), the RDF is modulated by the term $f_j(k)/k$, where $f_j(k)$ is the backscattering amplitude. The $\chi(k)$ data should therefore be divided by this term. A simple approximation is to take $f_j(k) \propto k^{-2}$, based on the Rutherford scattering formula: Eq. (3.3) with $q = 2k$ (for a scattering angle $\theta = \pi$). As shown in Fig. 4.22, this provides a fairly good approximation for light elements (e.g., C, O) but is not adequate for elements of higher atomic number. In some EXAFS studies, $\chi(k)$ is multiplied by k^n (as in Fig. 4.19b), the value of n being chosen empirically to emphasize either the low-k or high-k data, and thus the contribution of low-Z or high-Z atoms to the backscattered intensity (Rabe et al., 1980).

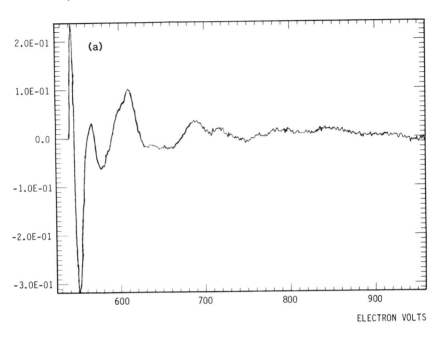

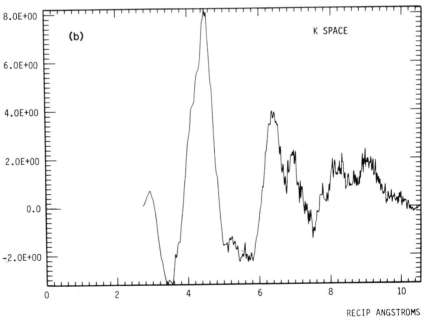

Figure 4.19. (a) $\chi(E)$ and (b) $k^2\chi(k)$, obtained from the data shown in Fig. 4.18 (A. J. Bourdillon, personal communication.)

Truncation of the Data

Before computing the Fourier transform, values of $\chi(k)$ which lie outside a chosen range ($k = k_{min}$ to k_{max}) are removed. The low-k data are omitted because single-scattering EXAFS theory does not apply in the near-edge region and because at low k the phase term $\phi(k)$ becomes nonlinear in k. High-k data are generally excluded because they consist mainly of noise (amplified by multiplying by k^n, as in Fig. 4.19b), which could contribute spurious fine structure to the RDF. The occurrence of another ionization edge at higher energy may also limit the maximum value of k. In typical EXELFS studies (Johnson *et al.*, 1981b; Leapman *et al.*, 1981; Stephens and Brown, 1981), k_{min} lies in the range 20–40 nm^{-1} and k_{max} within the range 60–120 nm^{-1}. If too small a range of k is selected, the RDF peaks are broadened (leading to poor accuracy in the determination of interatomic radii) and accompanied by strong satellite peaks arising from the truncation of the data. These truncation effects can be minimized by using a window function $W(k)$ with smooth edges (Lee and Beni, 1977) or by choosing k_{min} and k_{max} close to zero crossings of $\chi(k)$; when the limits have been suitably chosen, the RDF should be insensitive to the precise values of k_{min} and k_{max}.

Fourier Transformation

The required Fourier transform can be defined as follows:*

$$\bar{X}(r) = \frac{1}{\pi} \int_{-\infty}^{\infty} W(k) \frac{k}{f_i(k)} X(k) \exp(2ikr) \, dk \qquad (4.77)$$

In practice, a discrete Fourier transform is used, so the variable k becomes $\pi m/N$ (see Section 4.1.1) and the limits of integration are $m = 0$ and $m = N$, where N is the number of data points to be transformed. If an FFT algorithm is employed, N must be of the form 2^y, where y is an integer, in which case the $\chi(k)$ data may require extrapolation to values of k larger than k_{max}. A large value of N increases the computing time but gives $\bar{X}$ at more closely spaced intervals of r.

In the Fourier method of EXAFS or EXELFS analysis, interatomic distances are deduced directly from the positions of the peaks in the transform $\bar{X}(r)$. The rationale for this procedure is as follows. Ignore for the moment the effect of the window function and assume that the exponential and Gaussian terms in Eq. (3.167) are unity, corresponding to the case of a perfect crystal with no atomic vibrations and no inelastic scattering of

---

* If the argument of the exponential is written as ikr, or $2\pi ikr$ as in Eq. (4.7), the RDF peaks occur at $r = 2r_j + \phi_1$ and at $r = r_j/\pi + \phi_1/2\pi$, respectively.

the ejected core electron. We must also assume that the phase shift $\phi_j(k)$ can be written in the form

$$\phi_j(k) = \phi_0 + \phi_1 k \tag{4.78}$$

Substitution of Eq. (3.167) and Eq. (4.78) into Eq. (4.77) gives

$$\bar{\chi}(r) = \frac{1}{\pi} \int_{-\infty}^{\infty} \sum_j \frac{N_j}{r_j^2} \sin(2kr_j + \phi_0 + \phi_1 k) \cdot [\cos 2kr - i \sin 2kr]\, dk \tag{4.79}$$

The imaginary part of the Fourier transform is

$$\text{Im}[\bar{\chi}(r)] = -\frac{1}{\pi} \sum_j \frac{N_j}{r_j^2} \int_{-\infty}^{\infty} \{ \sin(2kr) \sin[k(2r_j + \phi_1)]\cos \phi_0 \tag{4.80}$$
$$+ \sin(2kr) \cos[k(2r_j + \phi_1)]\sin \phi_0 \}\, dk$$

and is zero for most values of r, since the (modulated) sinusoid functions average out to zero over a large range of k. However, if r satisfies the condition $2r = 2r_j + \phi_1$,

$$\text{Im}[\bar{\chi}] = -\frac{1}{\pi} \sum_j \frac{N_j}{r_j^2} \cos \phi_0 \int_{-\infty}^{\infty} \sin^2[(2r_j + \phi_1)k]\, dk \tag{4.81}$$
$$= -\frac{1}{2\pi} \sum_j \frac{N_j}{r_j^2} \cos \phi_0$$

$\text{Im}[\bar{\chi}]$ therefore consists of a sequence of delta functions, each of weight proportional to N_j/r_j^2 and located at $r = r_j + \phi_1/2$ ($j = 1,2$, etc., corresponding to successive shells of backscattering atoms). Likewise, $\text{Re}[\bar{\chi}]$ is zero except at $r = r_j + \phi_1/2$, where it takes a value $(1/2\pi) \sin(\phi_0)$. Consequently, the modulus (absolute value) of $\bar{\chi}$ can be written as

$$|\bar{\chi}(r)| = \frac{1}{2\pi} \sum_j \frac{N_j}{r_j^2} (\sin^2 \phi_0 + \cos^2 \phi_0)^{1/2} \delta(r - r_j - \phi_1/2) \tag{4.82}$$
$$= \frac{1}{2\pi} \sum_j \frac{N_j}{r_j^2} \delta(r - r_j - \phi_1/2)$$

and is proportional to the radial distribution function N_j/r_j^2.

Including the Gaussian term of Eq. (3.167) is equivalent to convolving the transform $\bar{\chi}$ with a function of the form $\exp[-r_j^2/(2\sigma_j^2)]$. In other words, the effect of thermal and (in a noncrystalline material) static disorder is to broaden each delta function present in $|\bar{\chi}|$ into a Gaussian peak whose width is proportional to the corresponding disorder parameter σ_j. To the extent that the inelastic mean free path λ_i can be considered to be independent of k, the effect of the exponential term in Eq. (3.167) is simply to attenuate the peaks in $\bar{\chi}$, particularly at larger r_j. Insofar as the window approximates to a rectangular function, its effect will be to convolve each Gaussian peak with a function $\overline{W}$ of the form (Lee and Beni, 1977)

$$\overline{W} = \frac{\sin(2k_{max}r)}{r} - \frac{\sin(2k_{min}r)}{r} \qquad (4.83)$$

Since k_{max} is usually several times k_{min}, the first of these sinc functions is more important at small r, but both terms broaden the peaks in $|\overline{x}|$ and introduce oscillations between the peaks.

In typical EXELFS studies (Johnson et al., 1981b; Leapman et al., 1981; Stephens and Brown, 1981; Bourdillon et al., 1984) the widths (FWHM) of the $\overline{x}(r)$ peaks are typically in the range of 0.02–0.1 nm (see Fig. 4.20), which is considerably larger than the thermal Debye-Waller broadening; at room temperature $\sigma_j < 0.01$ nm in the majority of materials (Stern et al., 1980). These peak widths, which determine the accuracy with which the interatomic radii can be measured, are therefore a reflection of the limited k-range. Fortunately, the sinc functions in Eq. (4.83) are symmetrical

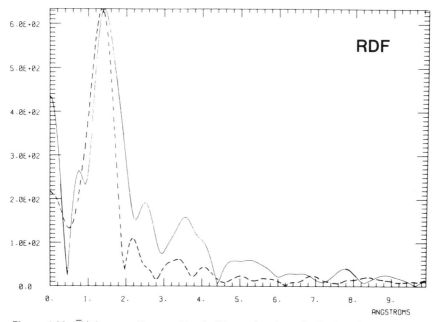

Figure 4.20. $\overline{x}(r)$ for crystalline sapphire (solid curve) and anodically deposited amorphous alumina (broken curve). The first-shell radius is 0.003 nm shorter in the amorphous case, suggesting a mixture of sixfold and fourfold coordination. After applying the phase-shift correction (Teo and Lee, 1979), all peaks are shifted by 0.049 nm to the right. (A. J. Bourdillon, personal communication.)

(about $r = 0$) and do not shift the maxima of $|\bar{\chi}(r)|$ and $\text{Im}[\bar{\chi}(r)]$. However, an error in the choice of E^0 introduces a *nonlinear* term into Eq. (4.78), shifting the maxima in $|\bar{\chi}(r)|$ and $\text{Im}[\bar{\chi}(r)]$ by *unequal* amounts. This forms the basis of the scheme for choosing E^0 by matching the peaks in these two functions (Lee and Beni, 1977).

Correction for Phase Shifts

The final step in the Fourier method is to estimate the linear (in k) component ϕ_1 of the phase shift in order to convert the $\bar{\chi}(r)$ peak positions into interatomic distances. The phase function $\phi_j(k)$ actually contains two contributions: a change $\phi_a(k)$ in phase as the ejected electron first leaves and then returns to the emitting (and "absorbing") atom, and also a phase change $\phi_b(k)$ which occurs upon back-reflection from a particular atomic shell. For K-shell EXELFS, where (as a result of the dipole selection rule) the emitted wave has p-like character:

$$\phi_j(k) = \phi_a^1(k) + \phi_b(k) - \pi \qquad (4.84)$$

where the superscript on $\phi_a(k)$ refers to the angular-momentum quantum number l' of the *final* state (i.e., the emitted wave). The final term $(-\pi)$ in Eq. (4.84) accounts for a factor $(-1)^l$ which should be present before the summation sign in Eq. (3.167), l referring to the angular momentum of the *initial* state. In the case of EXELFS on an L_{23} edge, the emitted wave is expected to be mainly d-like ($l' = 2$) and the phase term takes the form (Teo and Lee, 1979)

$$\phi_j(k) = \phi_a^2(k) + \phi_b(k) \qquad (4.85)$$

Note that $\phi_b(k)$ depends on j and therefore on the atomic number of the backscattering atom, while $\phi_a(k)$ depends on the atomic number of the emitting atom and on the angular-momentum quantum number l' of the emitted wave. The phase term $\phi_j(k)$ is therefore a property of a pair of atoms (the emitting atom being specified) and of the type of inner shell (K, L, etc.) giving rise to the EXELFS. It might also be postulated that $\phi_j(k)$ depends on the chemical environment, but experimental work suggests that this is not the case (Citrin et al., 1976; Lee et al., 1981), at least for $k > 40 \, \text{nm}^{-1}$. In other words, "chemical transferability" of the phase shift can be applied, provided the energy zero E^0 in Eq. (4.75) is chosen consistently.

Among others, Teo and Lee (1979) have therefore carried out "*ab initio*" calculations of $\phi_a(k)$ and $\phi_b(k)$ using atomic wavefunctions and have tabulated these functions for certain values of k and atomic number. Data for intervening elements can be obtained by interpolation. Groundstate wavefunctions were assumed for most of the elements, which could lead

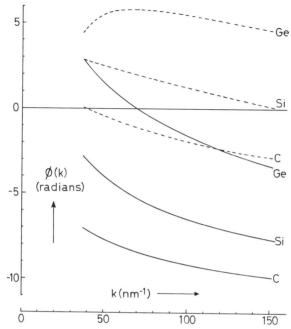

Figure 4.21. Phase shifts ϕ_a (solid curves) and ϕ_b (broken curves) as a function of ejected-electron wavenumber, calculated by Teo and Lee (1979) using Herman-Skillman and Clementi-Roetti wavefunctions.

to a systematic error in r_j if the emitting atom is strongly ionic (Stern, 1974; Teo and Lee, 1979). Some EXAFS workers have preferred to calculate phase shifts using wavefunctions of the higher adjacent element in the Periodic Table (the $Z + 1$ or "optical alchemy" approximation) in order to allow for relaxation of the atom following inner-shell ionization (Section 3.8.5).

The value of ϕ_1 for use in Eq. (4.78) may be taken as the *average* slope of the $\phi_j(k)$ curve in the region k_{min} to k_{max} (Leapman *et al.*, 1981; Johnson *et al.*, 1981b). Since this average slope is negative (Fig. 4.21) and since $r_j = r - \phi_1/2$ at the peak of $\overline{X}(r)$ the r-value corresponding to each $\overline{X}$ peak is *increased* by $|\phi_1|$.

4.6.2. Curve-Fitting Procedure

Because of the width of the RDF peaks computed by Fourier transformation of $\chi(k)$, it would be difficult to accurately distinguish atomic shells which are separated by less than 0.02 nm. An alternative procedure, which has been employed successfully in EXAFS studies, is to use Eq. (3.167) to calculate $\chi(k)$, starting from an assumed model of the atomic structure, and to fit this calculated function to the experimental data by varying the parameters (f_j, N_j, σ_j) of the model.

This approach has several advantages. More exact expressions can be used for the phase function $\phi_j(k)$, by including (for example) k^2 and k^{-3} terms in Eq. (4.78) (Lee et al., 1977). The k-dependence of the backscattering amplitude $f_j(k)$ can be taken into account more precisely, for example by using a Lorentzian function (Teo et al., 1977). This k-dependence can be different for different atomic shells and departs significantly from the k^{-2} approximation in the case of medium- and high-Z elements; see Fig. 4.22. The k-dependence (energy-dependence) of the inelastic mean free path λ_i (Fig. 3.57) can also be included in the analysis. With this more accurate treatment, it has been possible in EXAFS investigations to estimate the coordination number N_j and disorder parameter σ_j as well as interatomic distances.

Particularly in the case of a completely unknown structure, the Fourier transform method (Section 4.6.1) may be used as a basis for selecting the initial parameters of the atomic model. These parameters are then refined by curve fitting to the experimental data, the whole process being an iterative one. Sometimes use is made of the technique known as Fourier filtering, in which the $\chi(k)$ modulation arising from a single atomic shell is generated

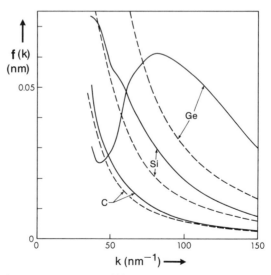

Figure 4.22. Backscattering amplitude $f(k)$ from the atomic calculations of Teo and Lee (solid curves) and from the Rutherford scattering formula: Eq. (3.3) with $\gamma = 1$ and $q = 2k$. Note that $f(k)$ departs from the k^{-2} dependence as the atomic number of the element increases.

by back-transforming a small range of $\bar{X}(r)$, corresponding to a single peak in the RDF (Eisenberger *et al.*, 1978); the parameters of the model are then fitted shell by shell, starting with the shell of smallest radius. However, in many cases of practical interest the Debye–Waller and inelastic terms in Eq. (3.167) damp the small-k oscillations (corresponding to large r_j) to such an extent that only nearest-neighbor separations can be considered reliable.

A further advantage of the curve-fitting procedure is that curvature of the emitted wave (Pettifer and Cox, 1983) and multiple scattering of the ejected electron (Lee and Pendry, 1975) can be taken into account. Equation (3.167) represents a plane-wave approximation and tends to fail for higher-order shells (i.e., at low k) where backscattering takes place further away from the nucleus of the backscattering atom, which therefore "sees" a larger portion of the wavefront. Equation (3.167) also assumes only *single* (elastic) scattering of the ejected electron, a condition which no longer applies to higher-order shells. By removing these restrictions, EXAFS theory has been extended to the near-edge (XANES) region (Durham *et al.*, 1981, 1982; Bianconi, 1983), and this theory also forms a possible basis for analyzing energy-loss near-edge structure (ELNES); see Section 3.8.4.

5

Applications of Energy-Loss Spectroscopy

The aim of this final chapter is to illustrate how the instrumentation, theory, and techniques described earlier are combined to extract practical information from electron-microscope specimens. Brief experimental details are given where necessary to show how the technique has been applied in specific cases. As in previous chapters, we begin with low-loss spectroscopy and energy filtering, proceed to elemental analysis and mapping, and then discuss how structural information about a specimen is obtained by analyzing fine structure of the energy-loss spectrum. The final section shows how many of these techniques have been used to investigate particular problems in materials science and biology. Table 5.1 summarizes the various types of information obtainable by EELS and by alternative high-resolution methods.

5.1. Measurement of Specimen Thickness

It is sometimes necessary to know the local thickness of a TEM specimen, for example when converting areal densities provided by EELS or EDX microanalysis into elemental concentrations or when estimating defect concentration from a TEM image. Several alternative techniques are available for *in situ* thickness measurement. Analysis of a convergent-beam diffraction pattern sometimes achieves 5% accuracy (Castro-Fernandez *et al.*, 1985), but the technique is time consuming and works only for crystalline specimens. Methods based on tilting the specimen and observing the lateral shift of surface features (e.g., contamination spots) are less accurate and may interfere with subsequent microscopy of the same area. Measurement of the bremsstrahlung continuum in an x-ray emission spectrum (Hall, 1979)

Table 5.1. Analytical Data Obtainable by TEM and Other Methods

EELS measurement	Information obtainable	Alternative methods
Low-loss intensity	Local thickness, mass-thickness	CBED, stereoscopy
Plasmon energy	Valence-electron density	
Plasmon peak shift	Alloy composition	CBED, EDXS
Low-loss fine structure	Dielectric function, JDOS	Optical spectroscopy
Low-loss fingerprinting	Phase identification	e^- or x-ray diffraction
Core-loss intensities	Elemental analysis	EDXS, AES
Orientation dependence	Atomic-site location	X-ray ALCHEMI
Near-edge fine structure	Bonding information	XAS (XANES)
Chemical shift of edges	Oxidation state, valency	XPS, XAS
L or M white-line ratio	Valency, magnetic properties	XPS, XAS
Extended fine structure	Interatomic distances	EXAFS, diffraction
Bethe ridge (ECOSS)	Bonding information	γ-ray Compton

can give the mass-thickness of organic specimens to an accuracy of 20% but involves substantial electron dose and possible mass loss (Leapman *et al.*, 1984). Measurement of the amount of elastic scattering from an amorphous specimen yields thickness in terms of an elastic mean free path or in terms of absolute mass-thickness if the chemical composition is known (Langmore and Smith, 1992), but for convenience and accuracy a STEM with annular dark-field detector is required (Langmore *et al.*, 1973).

5.1.1. Log-Ratio Method

The easiest procedure for measuring specimen thickness within a region defined by the incident beam (or an area-selecting aperture) is to record an energy-loss spectrum and use simple integration to compare the area I_0 under the zero-loss peak with the total area I_t under the whole spectrum. From Section 3.4, the thickness t is given by

$$t/\lambda = \ln(I_t/I_0) \tag{5.1}$$

where λ is a total mean free path for all inelastic scattering. As discussed in Section 3.4.1, λ in Eq. (5.1) must be interpreted as an *effective* mean free path $\lambda(\beta)$ if a collection aperture is used to restrict the scattering angles recorded by the spectrometer to a maximum angle β.

Before applying Eq. (5.1), any instrumental background should be subtracted from the spectrum. Particularly for very thin specimens, correct estimation of this background is essential to an accurate thickness measurement. In the case of parallel acquisition, the background corresponds to a dark-current spectrum recorded with exactly the same integration time, shortly before or after the energy-loss data.

Measurement of I_t and I_0 involves a choice of the energies ε, δ, and Δ which define the limits of integration; see Fig. 5.1. The lower limit ($-\varepsilon$) of the zero-loss region can be taken anywhere to the left of the zero-loss peak where the intensity has fallen essentially to zero. The separation point δ for the zero-loss and inelastic regions may be taken as the first minimum in intensity (Fig. 5.1) on the assumption that errors arising from the overlapping tails of the zero-loss and inelastic components approximately cancel. The upper limit Δ should correspond to an energy loss above which the further contributions to I_t are insignificant in relation to the required accuracy. For very thin specimens composed of light elements, $\Delta \approx 100$ eV is sufficient, but for thicker or high-Z specimens, intensity is shifted to higher energy loss (due to contributions from plural scattering and inner shells respectively), requiring a larger value of Δ. The "compute thickness" procedure in the Gatan EL/P software reduces the need for recording a large energy range by extrapolating the spectrum to higher energy loss; in addition, it models the right-hand tail of the zero-loss as an exponential (a reflection of the left-hand tail) in order to deal with the problem of overlap around $E = \delta$.

As shown in Section 3.4.3, Eq. (5.1) is relatively unaffected by elastic scattering; the formula has been shown to give 10% accuracy for t/λ as large as 5 (Hosoi et al., 1981; Leapman et al., 1984b). If $t/\lambda < 0.1$, surface excitations may be significant and would cause an overestimate of thickness (Batson, 1993). For $t/\lambda > 5$, an alternative procedure is available for thickness measurement, based on the peak energy and width of the multiple-scattering distribution (Perez et al., 1977; Whitlock and Sprague, 1982); see Section 3.4.4.

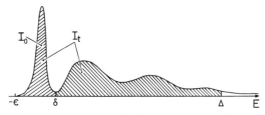

Figure 5.1. The integrals and energies involved in measuring specimen thickness by the log-ratio method.

Measurement of Absolute Thickness

Equation (5.1) provides *relative* specimen thickness t/λ, a quantity which can be useful for assessing the applicability of elemental analysis procedures. To obtain absolute thickness, a value of the total-inelastic mean free path is required. A rough estimate is given by λ (in nm) = $(0.8)E_0$, where E_0 is the incident-electron energy in keV. For 100-keV electrons and $\lambda > 5$ mrad, this estimate is valid within a factor of 2 for typical materials except ice; see Table 5.2.

For materials of known composition, it is possible to calculate a value for the mean free path. However, atomic models such as that of Lenz (1954) yield cross sections which are usually too high (Fig. 3.9), leading to an underestimate of λ; the free-electron plasmon formula, Eq. (3.58), gives

Table 5.2. Values of E_m for Use in Eq. (5.2), Obtained from Energy-Loss Measurements

Material	Type of specimen	Reference	E_m (eV)	100-keV MFP in nm	
				λ(10 mrad)	λ(no ap.)
Al	Single-crystal foil	M&88	17.2	100	82
Al	Polycrystalline film	C90,YE94	16.8	101	83
Al_2O_3	Polycrystalline film	E92	15.9	106	88
Ag	Polycrystalline film	EC87,C90	26.3	71	52
Au	Polycryalline film	EC87,C90	35.9	56	40
Be	Single-crystal foil	M&88	12.4	129	111
BN	Crystalline flake	E81c	17.2	99	84
C	Arc-evaporated film	C90,E92	14.2	116	95
C	C_{60} thin film	E92	14.4	115	94
C	Diamond crystal	E92	19.1	88	74
Cr	Polycrystalline film	E92	25.1	74	57
Cu	Polycrystalline film	C90	30.8	63	47
Fe	Polycrystalline film	EC87,C90	25.0	74	57
(Fe)	306 stainless steel	M&88	23.3	78	61
GaAs	Single crystal	E92	18.2	95	74
Hf	Single-crystal foil	M&88	35.3	57	41
H_2O	Crystalline ice	S&93,E92	6.7	220	200
NiO	Single crystal	M&88	19.8	89	71
Si	Single crystal	EC87	15.0	111	91
SiO_2	Amorphous film	E92	13.8	119	99
Zr	Single-crystal foil	M&88	24.5	75	57

C90 = Crozier (1990); E81c = Egerton (1981c); E92 = Egerton (1992a); EC87 = Egerton and Cheng (1987); M&88 = Malis *et al.* (1988); S&93 = Sun *et al.* (1993); YE94 = Yang and Egerton (1994). The last two columns give mean free paths for 100-keV incident electrons: λ(10 mrad) for β =10 mrad and λ(no ap.) for no angle-limiting aperture, obtained from λ(10 mrad) by making use of the *angular distribution* predicted by Eq. (3.16).

mean free paths which are appropriate for some materials but are generally an overestimate. More realistic mean free paths can be obtained by using scattering theory to parameterize λ in terms of the collection semiangle β, the incident energy E_0 and a mean energy loss E_m which depends on the chemical composition of the specimen:

$$\lambda \approx \frac{106F(E_0/E_m)}{\ln(2\beta E_0/E_m)} \qquad (5.2)$$

In Eq. (5.2), λ is given in nm, β in mrad, E_0 in keV, and E_m in eV; F is a relativistic factor (0.768 for $E_0 = 100$ keV, 0.618 for $E_0 = 200$ keV) defined by

$$F = \frac{1+E_0/1022}{(1+E_0/511)^2} \qquad (5.3)$$

Note that Eq. (5.2) is based on the dipole approximation and is valid only for $\beta \ll (E/E_0)^{1/2}$; in practice, this means collection semiangles up to about 15 mrad at $E_0 = 100$ keV. By recording the low-loss spectrum from a specimen of known thickness, with known β and E_0, λ can be determined from Eq (5.1) and converted to E_m by iterative use of Eq. (5.2). Materials for which this has been done are listed in Table 5.2; the appropriate values of E_m can be used in Eq. (5.2) to calculate the mean free path appropriate to a particular collection angle, for example by use of the BASIC program given in Appendix B.11.

For a specimen which is not listed in Table 5.2 but whose atomic number Z is known, E_m can be obtained from the approximate formula (Malis et al., 1988)

$$E_m \approx 7.6Z^{0.36} \qquad (5.4)$$

This equation is roughly consistent with the Lenz atomic model of inelastic scattering, Eq. (3.16), but makes no allowance for differences in crystal structure or electron density; for example, it would predict the same mean free path for graphite, diamond, and amorphous carbon. In the case of a compound, the Lenz model suggests that an *effective* atomic number for use in Eq. (5.4) can be defined by

$$Z_{\text{eff}} = \frac{\sum_i f_i Z_i^{1.3}}{\sum_i f_i Z_i^{0.3}} \qquad (5.5)$$

where f_i is the atomic fraction of each element of atomic number Z_i. The experimental data on which Eq. (5.4) was based are shown in Fig. 5.2.

Bonney (1990) reported that Eq. (5.4) gave thickness to within 10% when tested on sub-μm vanadium spheres whose thickness was taken to

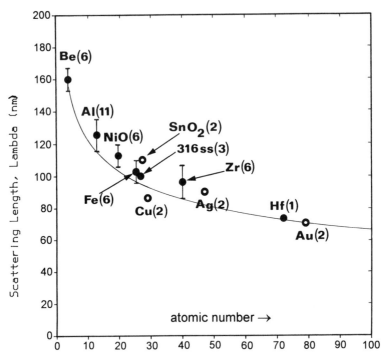

Figure 5.2. Inelastic mean free paths measured for 120-keV incident electrons and an effective collection semiangle of 5.3 mrad (Malis *et al.*, 1988). Solid circles denote measurements on crystalline foils whose thickness was determined from 2-beam CBED fringes; open circles are from thin films whose mass-thickness was determined by weighing. The number of measurements contributing to each data point is indicated in parentheses; the curve represents Eq. (5.2) and Eq. (5.4).

be the same as their diameter. Crozier (1990) found that Eq. (5.2) gave λ to within 15% for $E_0 = 100$ keV and collection semiangles of 5 mrad and 21 mrad, using thin-film standards of C, Al, Fe, Cu, Ag, and Au whose mass-thickness was determined by weighing.

For large collection apertures ($\beta > 20$ mrad for $E_0 = 100$ keV, > 10 mrad at 200 keV), Eq. (5.2) becomes inapplicable; the mean free path actually saturates at a value independent of β, as illustrated in Fig. 3.15. This *total-inelastic* mean free path, appropriate to low-loss spectra recorded *without* an angle-limiting aperture, is given to a reasonable approximation by substituting $\beta = 25$ mrad (15 mrad at 200 keV) in Eq. (5.2). Total-inelastic mean free paths for 100-keV electrons are given in the last column of Table 5.2.

Biological specimens may vary in porosity and are usually characterized in terms of *mass-thickness* ρt, which can be determined from

$$\rho t = \rho \lambda \ln(I_t/I_0) = (1/\sigma')\ln(I_t/I_0) \tag{5.6}$$

where σ' is a cross section *per unit mass.* Calculations (Leapman *et al.*, 1984a,b) based on Thomas–Fermi and Hartree–Fock cross sections and on a dielectric model for mean free path (Ashley and Williams, 1980) suggest that $\rho\lambda$ varies by less than $\pm20\%$ for biological compounds, although the different models gave absolute values of $\rho\lambda$ differing by almost a factor of 2 (e.g., 8.8 $\mu g/cm^2$ to 15 $\mu g/cm^2$ for protein at $E_0 = 100$ keV). More recent measurements and calculations based on the Bethe sum rule indicate $\rho\lambda = 17.2$ $\mu g/cm^2$ for protein at 100-keV beam energy (Sun *et al.*, 1993).

The only data processing involved in the log-ratio method is division of the spectrum into zero-loss and inelastic components, both of which are strong signals and relatively noise-free. Measurements can therefore be performed rapidly, with an electron exposure of no more than 10^{-13} C. Even for organic materials, where structural damage or mass loss may occur at a dose as low as 10^{-3} C/cm² (see page 393), thickness can be measured with a lateral spatial resolution below 100 nm. In this respect, the log-ratio technique is an attractive alternative to the x-ray continuum method (Hall, 1979) which requires electron exposures of 10^{-6} C or more to obtain adequate statistics (Leapman *et al.*, 1984a).

Rez *et al.* (1992) employed the log-ratio procedure with parallel recording to determine the thickness of paraffin crystals with a reported accuracy of 0.4 nm under low-dose (0.003 C/cm²) and low-temperature ($-170°$C) conditions. Leapman *et al.* (1993a) used similar methods to measure 200-nm-diameter areas of protein (crotoxin) crystals and achieved good agreement with thicknesses determined using the STEM annular dark-field signal, after correcting the latter for nonlinearity arising from plural elastic scattering.

Zhao *et al.* (1993) measured $t\lambda$ for a biological thin section by fitting its carbon K-edge to a sum of components derived from K-loss and plasmon-loss single-scattering distributions, both recorded from a pure-carbon film, giving a "carbon-equivalent" thickness. This procedure is convenient to the extent that it does not require recording of the low-loss region of the thin section, but it involves a radiation dose about 100 times higher than that required by the log-ratio method.

5.1.2. Absolute Thickness from the K–K Sum Rule

As described in Section 4.2, Kramers–Kronig analysis of an energy-loss spectrum provides a value of the absolute specimen thickness, as well as energy-dependent dielectric data, without any need to know the chemical composition of the specimen. The method involves extraction of the single-scattering distribution $S(E)$ from the measured spectrum, use of a Kramers–Kronig sum rule to derive the energy-loss function $\mathrm{Im}[-1/\varepsilon(E)]$, and removal of the surface-scattering component of $S(E)$ by iterative computation; see Appendix B.

If specimen thickness is the only requirement, the procedure can be simplified considerably. By combining Eq. (4.27) with Eq. (4.26), we obtain the following expression for specimen thickness:

$$t = \frac{4a_0 F E_0}{I_0\{1 - \text{Re}[1/\varepsilon(0)]\}} \int_0^\infty \frac{S(E)\, dE}{E \ln(1 + \beta^2/\theta_E^2)} \tag{5.7}$$

where $a_0 = 0.0529$ nm, F is the relativistic factor given by Eq. (5.3), and θ_E is the characteristic angle defined by Eq. (3.28). In general, $\text{Re}[1/\varepsilon(0)] = \varepsilon_1/(\varepsilon_1^2 + \varepsilon_2^2)^2$, where ε_1 and ε_2 are the real and imaginary parts of the *optical* permittivity. However, as discussed on page 259, $\text{Re}[1/\varepsilon(0)]$ can be taken as $1/n^2$ for an insulator of refractive index n, and as zero for a metal or semiconductor.

The $1/E$ weighting factor makes the integral of Eq. (5.7) less sensitive to higher orders of scattering. Provided the specimen is not too thick ($t/\lambda < 1.2$), the effect of plural scattering can be approximated by dividing t by a correction factor (Egerton and Cheng, 1987):

$$C = 1 + 0.3(t/\lambda) = 1 + 0.3 \ln(I_t/I_0) \tag{5.8}$$

Kramers–Kronig analysis carried out on thin films of Al, Cr, Cu, Ni, and Au (Egerton and Cheng, 1987) showed that correction for surface-mode scattering correction reduced the thickness estimate by an amount Δt (≈ 8 nm), which was independent of incident energy (in the range 20–100 keV) and collection angle (5–100 mrad). With these approximations, and assuming $\beta^2/\theta_E^2 \gg 1$ over the energy range where the inelastic intensity is significant, Eq. (5.7) can be simplified to:

$$t = \frac{2a_0 T}{CI_0\{1 - n^{-2}\}} \int_0^\infty \frac{J(E)\, dE}{E \ln(\beta/\theta_E)} - \Delta t \tag{5.9}$$

Here $J(E)$ represents the inelastic component of the energy-loss spectrum, including plural scattering but excluding the zero-loss peak. A short computer program to evaluate Eq. (5.9) is available (Egerton and Cheng, 1987).

Because of the $1/E$ weighting in Eq. (5.9), the value of t is particularly sensitive to data at very low energy loss. The procedure used to separate the elastic-scattering peak from the inelastic intensity is therefore important. The simplest procedure is to truncate the spectrum at the first minimum, $E = \delta$ in Fig. 5.1, but this results in an underestimate of t (square data points in Fig. 5.3.) because the contribution below $E = \delta$ (largely surface-mode scattering) is missing. Omitting the surface correction, by setting Δt to zero in Eq. (5.9), compensates for this missing contribution and gives a more realistic thickness value (solid circles in Fig. 5.3, which are within 5%

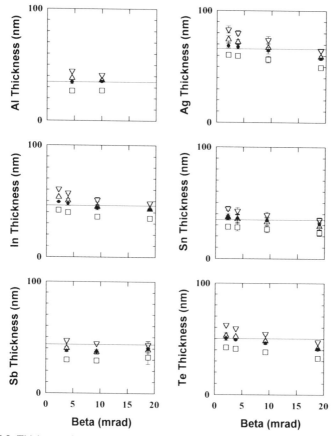

Figure 5.3. Thickness of Al, Ag, In, Sn, Sb, and Te films measured via the Kramers–Kronig sum rule with different treatments of the low-energy limit (Yang and Egerton, 1995). Inverted and upright triangles denote linear and parabolic interpolation (respectively) between the origin and the first minimum. Eliminating data below the minimum gave values represented by the squares and solid circles; in the latter case, the surface term Δt was also set to zero. Horizontal lines represent the film thickness measured by weighing, assuming bulk densities.

of the thickness determined by weighing). Linear or parabolic extrapolation of the inelastic intensity (at $E = \delta$) to zero (at $E = 0$) also gives acceptable results (triangles in Fig. 5.3) and should be applicable over a wider thickness range. A further possibility is to model the tail of the zero-loss peak, as described in Section 5.1.1, and subtract this from the spectrum to yield the inelastic intensity $J(E)$.

The errors involved in these procedures can be minimized by optimizing the energy resolution (through careful focusing of the spectrum and use of a small spectrometer-entrance aperture). In the case of a Gatan

PEELS spectrometer, spectra should be recorded with a high energy dispersion (low eV/channel) to minimize the energy range over which tails on the zero-loss peak (page 102) are significant, even though this results in spectra with a restricted energy range. Because of the $1/E$ weighting in Eq. (5.9), the upper limit of integration can be as low as 100 eV (for sufficiently thin specimens) without incurring significant error.

If t/λ exceeds unity, the plural-scattering correction in Eq. (5.9) becomes a poor approximation. A better procedure would be to use Fourier-log deconvolution to remove the plural-scattering component and employ Eq. (5.9) with C set to unity.

Because the Kramers–Kronig procedure is based on the equivalence of energy-loss and optical data, the spectral intensity $J(E)$ should be dominated by small-angle scattering. In principle, this implies a very small value of β. In practice, collection semiangles up to 15 mrad give acceptable results at $E_0 = 100$ keV (Fig. 5.3); at 200 keV, this condition would become $\beta < 8$ mrad.

5.1.3. Mass-Thickness from the Bethe Sum Rule

As in Section 3.2.2, the single-scattering intensity $S(E)$ can be written in terms of a differential oscillator strength df/dE rather than the energy-loss function $\text{Im}(-1/\varepsilon)$. Combining Eq. (3.33) and Eq. (4.26) gives

$$S(E) = \frac{4\pi I_0 N a_0^2 R^2}{EFE_0} \ln\left(1 + \frac{\beta^2}{\theta_E^2}\right) \frac{df}{dE} \tag{5.10}$$

where N is the *total* number of atoms per unit area and df/dE is the dipole (small-q) oscillator strength. Taking all the E-dependent terms to the right-hand side of this equation, integrating over energy loss and making use of the Bethe sum rule, described by Eq. (3.34), we have

$$\rho t = ANu = \frac{uFE_0}{4\pi a_0^2 R^2 I_0}\left(\frac{A}{Z}\right) \int_0^\infty \frac{ES(E)}{\ln(1 + \beta^2/\theta_E^2)} dE \tag{5.11}$$

here u is the atomic mass unit and ρt is the mass-thickness of the specimen, A and Z being its atomic weight and atomic number. For a compound, A should be replaced by the molecular weight and Z by the total number of electrons per molecule.

The integral in Eq. (5.11) is relatively insensitive to the instrumental energy resolution, allowing $S(E)$ to be taken as the intensity $J^1(E)$ obtained from Fourier-log deconvolution of experimental data. The combined effect of the other terms within the integral is to weight $S(E)$ by a factor which is typically between E and E^2, implying that the spectrum must be measured up to rather high energy loss in order to ensure convergence of the integral.

Convergence is further delayed because inner atomic shells contribute to $S(E)$ only at an energy loss higher than their binding energy. This means that Eq. (5.11) is useful only for specimens composed of light elements, having K-shell binding energies below 1000 eV. However, this category includes most organic and biological materials, which contain mainly carbon, oxygen, and hydrogen. Figure 5.4 shows the E-dependence of Eq. (5.11) for pure carbon, where the integral reaches its saturation value at about 1000 eV.

The need for such an extended energy range means that the spectrum must be recorded with a gain change introduced during serial recording, or sequentially with different integration times in the case of parallel recording. Because the intensity is integrated in Eq. (5.11), good statistics and energy resolution are not important; the high-E data can be quite noisy. Any *detector* background (dark current) should be removed before splicing the two (or more) segments together; it may also be necessary to remove the effect of *spectrometer* background which arises from stray scattering (Crozier and Egerton, 1989) as discussed in Chapter 2. Unless the specimen is extremely thin, Fourier-log deconvolution should be used to remove plural scattering.

For light elements, the ratio A/Z in Eq. (5.11) is close to 2, but a better approximation for most biological materials is to take $A/Z = 1.9$ (Crozier and Egerton, 1989), in which case Eq. (5.11) can be rewritten as

$$\rho t = \frac{BE_0}{I_0} \frac{(1 + E_0/1022)}{(1 + E_0/511)^2} \int_0^\infty \frac{EJ^1(E)}{\ln(1 + \beta^2/\theta_E^2)} \, dE \qquad (5.12)$$

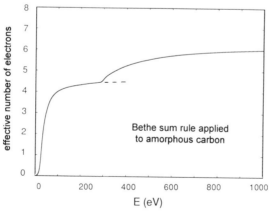

Figure 5.4. Value of the integral in Eq. (5.11), expressed as an effective number of contributing electrons defined by Eq. (4.33). Note that n_{eff} almost saturates as E approaches the K-shell binding energy (284 eV) but approaches its full value (6) only for energy losses of around 1000 eV (Sun *et al.*, 1993). Courtesy of Elsevier Science Publishers.

where $B = 4.88 \times 10^{-11}$ g/cm^2, E is in eV, and E_0 in keV. This equation has been tested on thin films of copper phthalocyanine (ρt up to 30 μg/cm^2, corresponding to $t/\lambda \approx 1.5$) and yielded mass-thickness values within 10% of those determined by weighing (Crozier and Egerton, 1989). The Bethe sum rule has also been used to determine the cross section per unit mass of protein and water (Sun et al., 1993).

The Bethe sum-rule method involves an electron exposure (typically 10^{-7} C for serial recording, 10^{-10} C for parallel recording) which is higher than that needed to apply the log-ratio method ($\approx 10^{-13}$ C) but its potential accuracy is higher. The bremsstrahlung continuum method (Hall, 1979) is useful for thicker specimens ($t > 0.5$ μm) but involves an electron exposure of the order of 10^{-6} C, sufficient to cause significant mass loss in most organic materials if the diameter of the incident beam is less than 1 μm (Leapman et al., 1984a).

5.2. Low-Loss Spectroscopy

The 1–100 eV region of the energy-loss spectrum contains one or more peaks which arise from inelastic scattering by outer-shell electrons. In most materials, the major peak corresponds to collective oscillation of conduction or valence electrons; its energy is closely related to valence-electron density and its width reflects the damping effect of single-electron transitions (Section 3.3.2). In some cases, interband transitions appear directly in the low-loss spectrum as a peak or as fine-structure oscillations superimposed on the plasmon peak. The low-loss spectrum is characteristic of the material present within the electron beam and can in principle be used to identify that material, provided that suitable comparison standards are available and that the spectral data has a sufficiently low noise level.

5.2.1. Phase Identification from Low-Loss Fine Structure

If the specimen contains regions which produce sharp plasmon peaks, such as sodium, aluminum, or magnesium, these materials are readily identified (Sparrow et al., 1983; Jones et al., 1984). Other materials are harder to characterize because their plasmon peaks are broad and tend to occur within a limited range, typically 15–25 eV, but by careful comparison with low-loss spectra recorded from several candidate materials, it is sometimes possible to identify an unknown phase.

This fingerprinting method was used many years ago to identify 25–250 nm precipitates in internally oxidized Si/Ni alloy as amorphous SiO$_2$ (Cundy and Grundy, 1966) and 10–100 nm precipitates in silicon as SiC (Ditchfield and Cullis, 1976). More recently, Evans *et al.* (1991) quantified the depth profile of aluminum in spinel (a possible fusion-reactor material) implanted with 2-MeV Al^+ ions, the depth-dependent low-loss spectrum being fitted to reference spectra of aluminum and spinel; see Fig. 5.5.

As a result of plural scattering, the overall shape of the low-loss spectrum depends on specimen thickness. To avoid errors in the fitting due to differences in thickness between the unknown and reference materials, plural scattering should be removed (e.g., by Fourier-log deconvolution) before spectra are compared. Published libraries of energy-loss spectra (Ahn and Krivanek, 1983; Zaluzec, 1981) allow a *visual* comparison of low-loss spectra with those of commonly occurring elements and compounds. A logical development would be the compilation of single-scattering spectra in digital form, perhaps as the energy-loss function so that simple computer processing could simulate the spectrum at a particular incident energy and collection aperture.

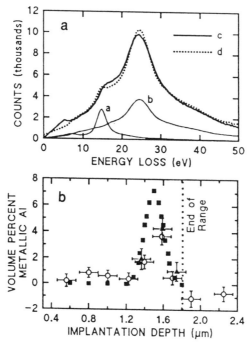

Figure 5.5. (a) Low-loss spectra, with plural scattering removed by Fourier-log deconvolution: a = metallic aluminum, b = undamaged spinel, c = material at an implant depth of ≈1.6 μm, d = best fit from multiple regression analysis, indicating 3.7 ± 0.6 vol.% Al. (b) Profile showing aluminum concentration as a function of depth. From Evans *et al.* (1991), *Microbeam Analysis*—1991, p. 440, © San Francisco Press, Inc., by permission.

Gatts *et al.* (1995) have described the application of neural pattern recognition to the analysis of a series of low-loss spectra recorded across a Si/SiO_2 interface. In the form of a computer program (ART2), the artificial neural network (ANN) sorts the spectra into a small number of groups (*classes*) with similar properties and constructs a *standard* spectrum which is representative of each class. Linear algebra is then used to fit each spectrum to a weighted sum of the standards (representing Si, SiO_2, and an interface loss in this example), the weighting coefficients providing a profile of the specimen. First-difference spectra were used to accentuate changes in peak shape and suppress differences in background level. The value of the ANN method is that it reduces the data set to a small number of standards, enabling the analyst to concentrate on the physical meaning of a few basic patterns, and that it is capable of providing unbiased and quantitative results.

Chen *et al.* (1986) have shown that the *width* of the plasmon peak measured from quasicrystalline Al_6Mn is larger (3.1 eV) than that of the amorphous (2.4 eV) or crystalline phases (2.2 eV), suggesting that the icosahedral material has a distinct electronic band structure which favors the decay of plasmons via interband transitions. On the other hand, Levine *et al.* (1989) found the plasmon widths of icosahedral $Pd_{59}U_{21}/Si_{20}$ and $Al_{75}Cu_{15}V_{10}$ to be the same as those of amorphous materials of the same composition.

Some organic compounds provide a distinctive fine structure in the energy region below the main plasmon peak (Hainfeld and Isaacson, 1978); see Fig 5.6. Even if the structure is not prominent, a careful comparison using multiple least-squares fitting may enable the composition to be measured. Sun *et al.* (1993) used this method to determine the water content of cryosectioned red blood cells as $70 \pm 2\%$, with water and frozen solutions of bovine serum albumin (BSA) as standards. Using a parallel-recording spectrometer, acceptable statistical errors in the MLS fitting were obtained with an electron dose of around 10^{-12} C, allowing measurements on areas down to 100-nm diameter without bubbling or devitrification when the specimen was held at $-160°C$.

Even at 1-eV resolution, different chromophores are distinguishable on the basis of their distinctive low-loss spectrum (Reimer, 1961) and these dyes can be used to selectively stain biological tissue. Electron spectroscopy or energy-selected imaging would then provide a spatial resolution much better than that obtainable with light-optical techniques (Jiang and Ottensmeyer, 1994).

Identification of an unknown phase is more likely to be successful if the spectrometer system offers good energy resolution, so that peak positions can be measured accurately and any fine structure resolved. Using a

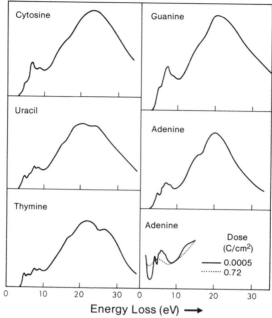

Figure 5.6. Low-loss fine structure in the energy-loss function measured from evaporated thin films of three pyramidines (cytosine, uracil, thymine) and two purines (guanine, adenine). The effect of irradiation is also shown. Data from Johnson (1972) and Isaacson (1972a).

LaB$_6$ source run at reduced temperature, a resolution better than 1 eV is possible; at low emission current, a field-emission tip can provide 0.3-eV resolution, or 0.16 eV with Fourier sharpening (Batson *et al.*, 1992).

An example of the use of a field-emission source combined with a high-resolution electron spectrometer is shown in Fig. 5.7. The large tail of the zero-loss peak has been subtracted from the low-loss data to reveal a steadily rising intensity which can be modeled on the assumption that the joint densities of states (across the energy gap E_g) is proportional to $(E - E_g)^{0.5}$. With the STEM probe located at a misfit dislocation, the effective value of E_g is reduced, the additional low-loss scattering probably arising from electron excitation from filled states (at the dislocation core) to the conduction band. Since these states are related to the structure of the dislocation, it may be possible to use EELS to determine the nature of individual dislocations.

Takeda *et al.* (1994) reported that line-defect self-interstitials in silicon, identified by HREM imaging, give rise to an energy-loss peak at 2.5 eV. Tight-binding calculations showed this peak to be consistent with the presence of eight-membered rings.

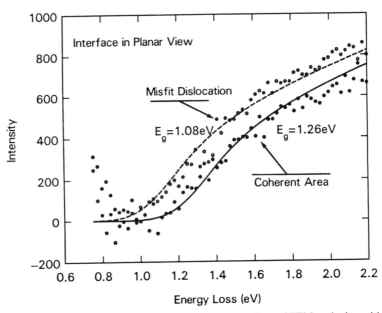

Figure 5.7. Increase in the intensity of inelastic scattering when a STEM probe is positioned on a misfit dislocation in GaAs (Batson *et al.*, 1986).

Resolution in the meV range, with useful incident-beam current and spatial resolution, awaits the development of an efficient monochromating system (Section 2.1.4). This would allow materials of similar elemental composition (such as polymers) to be distinguished on the basis of their chemical bonding.

5.2.2. Measurement of Alloy Composition from Plasmon Energy

If an alloying element is added to a metal, the lattice parameter and/ or valency may alter, leading to a different (outer-shell) electron density and plasmon energy E_p, as discussed in Section 3.3.1. The plasmon energy is also sensitive to changes in effective mass of the electrons, resulting from changes in band structure. The shift in E_p is therefore difficult to calculate but can be determined experimentally for a given alloy system, using calibration samples of known composition. In many cases, the plasmon energy varies with composition x in an approximately linear fashion:

$$E_p(x) \approx E_p(0) + x(dE_p/dx) \tag{5.13}$$

Table 5.3. Shift in Plasmon Energy (expressed in eV) for Aluminum and
Magnesium Alloys $A_{1-x}B_x$, Measured up to
a Fractional Concentration x_{max} of the Alloying Element[a]

A	B	x_{max}	dE_p/dx
Al	Li	0.25	−4.0
Al	Mg	0.08	−4.4
Al/Zn	Mg	0.04	−4.7
Al	Zn	0.3	−0.2
Al	Ge	0.1	+0.1
Al	Ag	0.06	+1.6
Mg	Al	0.09	+5.9

[a]Data from Williams and Edington (1976).

Table 5.3 lists the coefficient dE_p/dx for several Al and Mg alloys. These materials have sharp plasmon peaks, making the small shifts in plasmon energy easier to measure. It is difficult to achieve the required accuracy by locating the position of the maximum in the intensity distribution, but by least-squares fitting to a polynomial (or by dividing the plasmon peak into two equal areas) the mean or median energy can be determined with a *statistical* error given by (Wang *et al.*, 1995b):

$$\delta E = \Delta E_0/(N_0)^{1/2} + \Delta E_p/(N_p)^{1/2} \tag{5.13a}$$

where ΔE_0 and ΔE_p are the widths of the zero-loss and plasmon peaks; N_0 and N_p are the number of primary electrons represented by each peak. The use of parallel recording (to collect the maximum number of electrons) and a LaB_6 or field-emission source (for good energy resolution) helps in obtaining low δE. Hibbert and Eddington (1972) obtained an accuracy of better than 0.1 eV for aluminum and magnesium alloys using photographic recording and a tungsten thermionic source.

To achieve maximum signal/noise ratio in the plasmon peak, the specimen thickness should be equal to the plasmon mean free path (Johnson and Spence, 1974). Ultrathin specimens are therefore neither necessary or desirable. But in thick specimens, the tail of the double-scattering peak displaces the first-plasmon peak towards higher energy loss; to ensure that changes in specimen thickness do not interfere with the measurement, plural scattering should be removed by deconvolution.

Surface oxide or contamination layers also cause a shift in peak position and must be minimized by careful specimen preparation and clean vacuum conditions (e.g., use of liquid-N_2 decontaminator) in the TEM. If measurements are made close to a composition boundary, the latter should be oriented parallel to the incident beam so that spectra are not recorded from overlapping regions. If the specimen is crystalline, strongly diffracting orientations should be avoided (Hibbert and Edington, 1972).

Figure 5.8 shows the abrupt change in E_p at a reaction front in an Al–Li alloy and indicates that the spatial resolution of plasmon-loss micro-analysis can be considerably better than 10 nm, in accord with current ideas on the localization of inelastic scattering (Section 5.5.3). The technique has also been used to demonstrate solute depletion at grain boundaries, to estimate diffusion constants, and to examine solute redistribution in splat-cooled alloys (Williams and Edington, 1976). From processing of spectrum-image data, Hunt and Williams (1991) have constructed plasmon-shift images which directly reflect elemental composition. Tremblay and L'Esper-ance (1994) have used the same technique to measure the volume fraction of Al(Mn,Si,Fe) precipitates in commercial aluminum alloys.

McComb and Howie (1990) employed low-loss analysis to study the dealumination of zeolite catalysts, which are damaged by electron doses beyond about 6 C/cm². The effect of aluminum depletion was mainly to shift the plasmon peak about 1 eV lower in energy. Each spectrum was modeled by fitting the corresponding dielectric function (obtained from K–K analysis) to a Drude expression, Eq. (3.40), but including up to four oscillator contributions with different strengths and frequencies, leading to a value for the overall valence-electron concentration. The results enabled

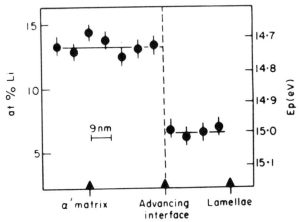

Figure 5.8. Lithium concentration across a type (ii) discontinuous reaction front in aged Al-13at.% Li alloy, measured by plasmon-loss spectroscopy (Williams and Edington, 1976). From D. B. Williams and J. W. Edington, High resolution microanalysis in materials science using electron energy-loss measurements, *Journal of Microscopy* **108**, 113–145, by permission of the Royal Microscopical Society.

a choice to be made between two alternative mechanisms of the dealuminum process.

The plasmon-shift method has also been applied to metal–hydrogen systems, where hydrogen usually introduces an upward shift in the plasmon energy; see Table 5.4. For these systems, the free-electron plasmon formula gives plasmon energies which are generally too low by 1 to 3 eV, but predicts *differences* between the metal and hydride free-electron values which agree quite well with observation. Zaluzec (1992) has pointed out that the *normalized* energy shift $\Delta E_p/x$ decreases with increasing group number of the metal in the periodic table. For Mg, Sc, and Y, hydrogen increases E_p by contributing extra valence electrons; in vanadium (Group V) and FeTi (Pseudogroup VI) lattice expansion apparently counteracts this effect. Other authors (Stephens and Brown, 1980; Thomas, 1981) observed no upward shift in plasmon energy. Instead, the presence of hydrogen in their Ti, V, and Nb specimens introduced weak energy-loss peaks in the 4–7 eV region, presumably by creating a band of states several eV below the conduction band.

Woo and Carpenter (1992) investigated the zirconium hydride system and found the plasmon energy to be higher in δ- and ε-hydrides than in the γ-hydride. This allowed identification of small precipitates in Zr/Nb alloys used in nuclear reactor pressure tubes. Although similar identification was possible by electron diffraction, it was very time consuming. In contrast, the energy-loss spectrum could be acquired and analyzed on-line in a few minutes.

The value of these plasmon-shift studies lies in the fact that dispersed hydrogen cannot be detected by core-loss, WDX, or EDX spectroscopy,

Table 5.4. Plasmon-Peak Energies E_p (in eV) for Metals M and their Hydrides MH_x; the Energy Shift ΔE_p is Compared with the Value $\Delta E_p(\text{FE})$ Calculated from Eq. (3.41)

M	x	E_p(M)	E_p(MH_x)	ΔE_p	ΔE_p(FE)	Reference
Sc	2	14.0	17.2	3.2	3.0	Colliex *et al.*
Y	2	12.5	15.3	2.8	2.8	(1976b)
Tb	2	13.3	15.6	2.3	2.4	
Er	2	14.0	16.8	2.8	2.7	
Mg	2	10.0	14.2	4.2	2.5	Zaluzec *et al.*
Ti	2.0	17.2	20.0	2.8	1.7	(1981)
Zr	1.6	16.6	18.1	1.5	1.5	
V	0.5	22.0	22.0	0	0.6	
Fe/Ti	2	22.0	22.0	0	−0.8	

while quantification of lithium is difficult by core-loss EELS and impossible by EDX methods.

5.2.3. Characterization of Surfaces, Interfaces, and Small Particles

Surface-mode scattering involves an energy loss below that of the volume plasmon peak (see Section 3.3.5). The nature of this scattering depends on the geometry of the interface and the mismatch in the energy-dependent dielectric constant. In the case of a small probe and a single spherical particle, the spectral intensity at a particular energy loss depends on the probe position, and this dependence is different for a silicon sphere and a sphere covered with silicon oxide. Ugarte *et al.* (1992) observed an additional peak in the 3–4 eV region which they could explain only in terms of a thin conducting spherical shell on the outside of the oxide layer, perhaps caused by oxygen depletion by the electron beam.

When the incident beam samples a large number of particles, the inelastic scattering response can be characterized by an effective-medium energy-loss function; see page 185. The low-loss spectrum might then be useful for characterizing fine-scale dispersions (Howie and Walsh, 1991).

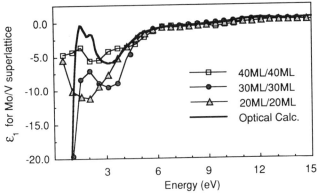

Figure 5.9. Real part ε_1 of the dielectric function for various thicknesses (in monolayers) of the films within a molybdenum/vanadium superlattice, compared to the results of an optical calculation assuming thick layers (Zaluzec, 1992).

If transmission measurements are made with electron beam parallel to an interface, surface-mode contributions are maximized. In the case of a metal multilayer system, differences in scattering are further amplified by performing Kramers–Kronig analysis to extract ε_1; see Fig. 5.9. As the spacing of the layers decreases, the observed structure departs from that calculated from bulk properties of the individual layers, perhaps indicating structural transformation to a strained-layer superlattice (Zaluzec, 1992). Turowski and Kelly (1992) used a field-emission STEM to record low-loss spectra as a function of position across $Al/SiO_2/Si$ field-effect transistor structures. Using Kramers–Kronig analysis, the dielectric function was calculated at each position of the beam, as well as the electronic polarizability $\alpha_e(E)$ which may be a measure of dielectric strength. The maximum polarizability and the energy E_{max} of this maximum were lower near the Al and Si interfaces (Fig. 5.10), suggesting that contact materials reduce the dielectric strength in very thin oxides.

As discussed in Section 3.3.6, energy-loss spectra can be obtained in reflection mode within the TEM. Figure 5.11 shows reflection low-loss spectra from a MgO specimen raised to successively higher temperatures. Above 500 K, the 10 eV peak starts to disappear; reflection K-loss spectra suggest this effect is due to desorption of surface oxygen atoms. Above 1400 K, the main plasmon-peak energy shifts upward; core-loss spectroscopy suggests that a surface layer (a few nm thick) of MgO_2 is formed, which remains after subsequent cooling and exposure to air.

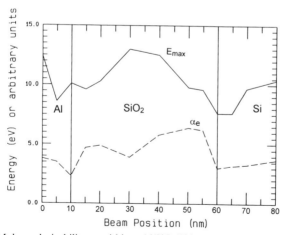

Figure 5.10. Molar polarizability α_e within a $Al/SiO_2/Si$ heterostructure and energy position E_{max} of the maximum in α_e, derived from energy-loss spectroscopy (Turowski and Kelly, 1992).

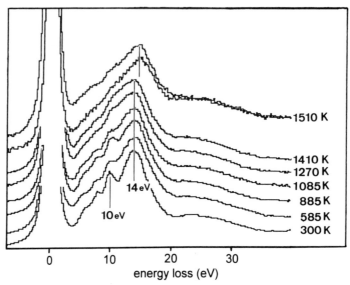

Figure 5.11. Low-loss spectra recorded using the specular (400) reflection from a (100) MgO surface heated to different temperatures (Wang, 1993).

5.3. Energy-Filtered Images and Diffraction Patterns

As discussed in Chapter 2, some electron spectrometers can act as a bandpass energy filter: an energy-selecting slit in the spectrum plane produces an electron image or diffraction pattern from a chosen range of energy loss. This provides information in a convenient form, with greater ease than by spectroscopy at many points on the specimen (or at many scattering angles). For example, a core-loss image can indicate the spatial distribution of an element, at least semiquantitatively, allowing an intelligent choice of the locations where quantitative spectroscopy is required.

As discussed in Section 2.6, energy filtering is possible in a fixed-beam (CTEM) instrument, using either an in-column filter or an imaging spectrometer below the lens column, and in a scanning-transmission (STEM) system. Various modes of operation are discussed in Chapter 2 and (for a CTEM system) by Reimer *et al.* (1988, 1992). A detailed discussion of the contrast mechanisms in energy-filtered images is given by Spence (1988b), Reimer and Ross–Messemer (1989, 1990), Bakenfelder *et al.* (1990), and Reimer (1995). Colliex *et al.* (1989) discuss energy-filtered STEM imaging of thick biological specimens, while energy-filtered diffraction is dealt with by Spence and Zuo (1992).

5.3.1. Zero-Loss Images

By operating the spectrometer in diffraction-coupled mode (dispersed diffraction pattern at the energy-selecting slit) and adjusting the spectrometer excitation or accelerating voltage so that the zero-loss peak passes through the energy-selecting slit, a zero-loss image is produced with greater contrast and/or resolution than the normal (unfiltered) image; see Figs. 5.12 and 5.15. The main factors responsible for such improvement are listed below; their relative importance varies according to the type of specimen and whether a conventional TEM or STEM is involved. Some factors may be identified as affecting the *resolution* and others the *contrast* of an image, although in the case of small-scale image features these two concepts are closely related (Nagata and Hama, 1971).

Chromatic Aberration and Contrast-Reducing Effects

For CTEM imaging of all types of specimen, an energy-selecting slit centered on the zero-loss peak eliminates almost all of the inelastically scattered electrons. In very thin specimens, the inelastic image corresponding to an energy loss E is blurred *relative* to the elastic image by a Lorentzian-shaped point-spread function (Section 2.3.2) of width $2r_E = 2\vartheta_E C_c(E/E_0)$

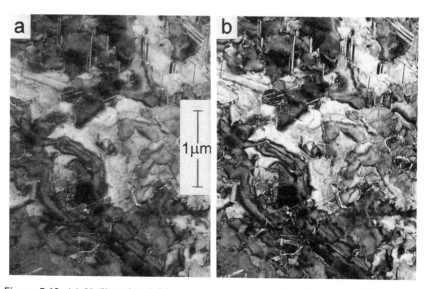

Figure 5.12. (a) Unfiltered and (b) zero-loss micrographs of a 40-nm epitaxial gold film, recorded with a 80-keV electrons and a 10-mrad objective aperture. Energy filtering increased the crystallographic contrast by about a factor of 2 (Egerton, 1976c).

and according to the Rayleigh criterion the image resolution cannot be better than

$$r_i \approx C_c(E/E_0)^2 \tag{5.14}$$

where C_c is the chromatic-aberration coefficient of the objective lens. In the case of 100 keV incident electrons, $C_c = 2$ mm and $E \approx 37$ eV (the *average* energy loss for a thin specimen of carbon), $r_i \approx 0.3$ nm. At *high energy loss,* the characteristic scattering angle ϑ_E may exceed the semiangle β of the angle-limiting aperture (objective aperture in CTEM) and the chromatic radius is given by $\beta C_c(E/E_0)$ rather than by Eq. (5.14); but in thin specimens this situation occurs only for energy losses above 1000 eV . (for $E_0 = 100$ keV) or 2000 eV ($E_0 = 200$ keV), even for a small aperture ($\beta = 5$ mrad).

As the specimen thickness increases, the *fraction* of electrons which are inelastically scattered approaches unity, plural scattering causes the *average* energy loss to increase, and plural elastic/inelastic scattering broadens the inelastic angular distribution, resulting in a chromatic width greater than given by Eq. (5.14). As a result of these three factors, the degradation of resolution by chromatic aberration is more serious in thicker specimens. Zero-loss imaging has therefore been used extensively to improve the images of thick TEM specimens, particularly biological tissue where inelastic scattering is strong relative to elastic scattering (Section 3.2.1).

In the case of high-resolution (phase-contrast) imaging, inelastic scattering is often assumed to produce a structureless background which reduces image contrast, but if the low-loss spectrum contains sharp plasmon peaks the inelastic scattering could produce image artifacts (Krivanek *et al.*, 1990). Energy filtering therefore permits a more quantitative comparison of image intensities with theory (Stobbs and Saxton, 1988). Even after energy filtering, some chromatic aberration remains as a result of the energy spread ΔE of the electron source, as in the case of STEM imaging. In phase-contrast image theory, its effect is described in terms of an envelope function (e.g., Krivanek, 1988).

The virtual elimination of chromatic aberration means that, for examination of thick specimens, energy-filtered microscopy (EFTEM) with 80-keV or 100-keV electrons is an attractive alternative to the use of higher accelerating voltages, where chromatic aberration is reduced in proportion to $1/E_0^2$ according to Eq. (5.14). However, zero-loss filtering reduces the image intensity by a factor of exp (t/λ), where λ is the total-inelastic mean free path, limiting the maximum usable specimen thickness to about 0.5 μm at 80-keV incident energy. Energy filtering combined with higher voltage would take advantage of both factors.

It is convenient to compare zero-loss images with unfiltered images obtained by removing the energy-selecting slit, but such a comparison assumes that the energy filter has imaging properties which are completely achromatic. A safer comparison is with an image produced by a regular (nonfiltering) TEM under similar illumination and imaging conditions.

Staining of biological tissue creates regions containing a high concentration of heavy-metal atoms surrounded by material comprised mainly of light elements (H,C,O). The resulting strong variations in elastic scattering power provide usable contrast. Because the inelastic/elastic scattering ratio is high for light elements, electron scattering in unstained regions is mainly *inelastic* and can therefore be removed by energy filtering, leading to a further improvement in contrast or the option of reducing the concentration of staining agent. Reimer and Ross–Messemer (1989) report that the contrast of large-scale features in OsO_4-stained myelin was increased by a factor of 1.3 after zero-loss filtering.

Because unstained biological specimens give very low contrast, the image is often defocused to create phase contrast. Langmore and Smith (1992) found that zero-loss filtering increased the image contrast from air-dried and frozen-hydrated TMV images by factors between 3 and 4. This improved contrast allows a reduction in the amount of defocusing, allowing better spatial resolution and increased signal/noise ratio or reduced electron dose to the specimen (Schröder *et al.*, 1990).

In the case of crystalline specimens, defects are visible through *diffraction* contrast which arises from variations in the amount of elastic scattering, depending on the local excitation error (deviation of lattice planes from a Bragg-reflecting orientation). The angular spread caused by inelastic scattering creates a variation in excitation error, reducing the contrast of dislocations, planar defects, bend contours, and thickness contours (Metherell, 1967). Consequently, diffraction contrast is improved by zero-loss filtering. Higher incident energy also gives less spread in excitation error; including also the effect of chromatic aberration, Bakenfelder *et al.* (1989) concluded that, for a 500-nm Al film, zero-loss filtering is equivalent (in terms of image quality) to raising the microscope voltage from 80 kV to 200 kV. Making use of zero-loss filtering of 80-keV electrons and of the increased transmission (channeling) which occurs when a crystal is oriented close to a zone axis, Lehmpfuhl *et al.* (1989) obtained clear images of disclocations in gold films as thick as 350 nm.

At high energy loss or large scattering angle, inelastic scattering in crystals is believed to be partly *interbranch* scattering: the character of the electron Bloch wave changes upon scattering and Bragg contrast is not preserved (Hirsch *et al.*, 1977). Although zero-loss filtering removes such scattering, the overall effect is negligible because it contributes negligible intensity. Phonon scattering involves interbranch scattering, but because the energy losses are below 1 eV it is not removed by energy filtering.

5.3.2. Zero-Loss Diffraction Patterns

Energy-filtering of diffraction patterns can be carried out with an imaging filter combined with a conventional TEM; see Section 2.6. A less efficient method is to scan a diffraction pattern across the entrance aperture of a nonimaging spectrometer (Graczyk and Moss, 1969). Because the data recording then takes considerable time, drift of the spectrum can be a problem, although it is alleviated by using a parallel-recording spectrometer and tracking a particular feature such as the zero-loss peak (Cockayne *et al.*, 1991). As an alternative, the array detector of a PEELS system has been rotated through 90° to allow parallel recording of a range of scattering angles at a particular energy loss (Holmestad *et al.*, 1993).

Zero-loss filtering removes the diffuse background arising from the angular distribution of inelastic scattering and allows faint diffraction features to become visible (Midgley *et al.*, 1995). Since the inelastic scattering is strongest at small angles (Fig. 3.7), filtering should be particularly advantageous for low-angle diffraction and would improve analysis of materials with large unit cell or periodic arrays of macromolecules. Filtering has also been used to improve the visibility of reflection diffraction patterns recorded in a TEM (Wang and Cowley, 1994) and to facilitate the accurate measurement of fringes within convergent-beam diffraction disks; see Fig. 5.13.

Removal of inelastic scattering facilitates quantitative analysis of the electron diffraction pattern of an amorphous material. After subtracting a smoothly varying *atomic* scattering factor, Fourier transformation leads to a reduced density function (RDF) whose peak positions provide values of interatomic spacing to an accuracy of typically 5 pm. From the RDF, Liu *et al.* (1988) demonstrated that the first- and second-neighbor distances in amorphous silicon alloys decrease by up to 40 pm upon doping with boron and phosphorus.

Zero-loss filtering of convergent-beam diffraction (CBED) patterns, combined with efficient least-squares modeling, has made it possible to derive lattice parameters of crystalline TEM specimens with 0.1 pm accuracy, sufficient to measure small strains or variations in oxygen deficiency in YBCO superconductors for example (Zuo, 1992). Temperature factors and subtle bonding effects can also be measured (Spence, 1992; Spence and Zuo, 1992). The use of a field-emission electron source allows CBED data to be acquired from specimen areas less than 1 nm in diameter (Xu

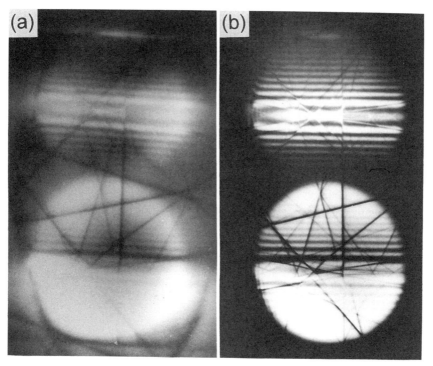

Figure 5.13. Two-beam (220) CBED patterns of silicon: (a) no energy filtering and 1-s exposure, (b) zero-loss filtering and 3-s exposure. From the fringe spacings, the thickness was calculated to be 270 nm (Mayer *et al.*, 1991). From *Proc. Electron Microsc. Soc. Amer.* (1991), p. 787, © San Francisco Press, Inc., by permission.

et al., 1991). Even in the Tanaka LACBED method, in which a small SAD aperture provides a degree of energy filtering, zero-loss filtering can substantially increase the contrast in the central region of the pattern (Burgess *et al.*, 1994).

5.3.3. Low-Loss Images

For materials with sharp plasmon peaks in their low-loss spectrum, an image of relatively high intensity can be formed at the corresponding energy loss and will show the spatial distribution. For example, Be precipitates in an Al alloy appear dark in a 15-eV image (plasmon loss of Al) but brighter than their surroundings in an image recorded at 19 eV, the plasmon loss of beryllium (Castaing, 1975).

Making use of the shift in plasmon energy upon alloying, Williams and Hunt (1992) processed spectrum-image data to display the distribution of Al_3Li precipitates in Al/Li alloys. Tremblay and L'Espérance (1994) used a similar technique to image Al(Mn,Si,Fe) particles in aluminum alloy and deduced the volume fraction of precipitates to be 0.81%.

Inelastic scattering by surface-plasmon excitation provides intensity at energies below the volume-plasmon peak. At this energy loss, small particles show a bright outline because the probability of surface-plasmon excitation is substantially larger for an electron which travels at a glancing angle to the surface (Section 3.3.6). If an unidentified peak is seen in the low-loss spectrum of an inhomogeneous specimen, forming an image at that energy loss will help to determine if it arises from surface-mode scattering.

Some organic dyes (chromophores) have absorption peaks at energies of a few eV, corresponding to visible or UV photons, and can be used as chemically specific stains in light microscopy. By forming an image at the corresponding energy loss, their distribution can be mapped at much higher spatial resolution in an energy-selecting electron microscope (Jiang and Ottensmeyer, 1994).

5.3.4. Z-Ratio Images

Z-ratio contrast in a STEM image is formed by taking a ratio of the high-angle scattering (recorded by an annular detector) and the low-angle scattering (measured through an electron spectrometer which removes the zero-loss beam); see Section 2.6.7. For very thin specimens, the high-angle dark-field signal represents mainly elastic scattering, whereas the spectrometer signal arises from inelastic scattering. Intensity in the ratio image is therefore a measure of the local elastic/inelastic scattering ratio, which is roughly proportional to the local (mean) atomic numnber Z; see Section 3.2.1. The aim is to distinguish differences in elemental composition, while suppressing the effects of varying specimen thickness and fluctuations in incident-beam current. This technique was first used by Crewe et al. (1975) as a way of displaying images of single high-Z atoms on a very thin (<10 nm) carbon substrate, and was subsequently used for imaging small catalyst particles on a crystalline or amorphous support (Treacy et al., 1978; Pennycook, 1981b).

The Z-ratio technique has also been applied to thin sections of biological tissue (Garavito et al., 1982); see Fig. 5.14. If the section thickness is below 50 nm, so that plural scattering is not severe, contrast due to thickness variations (cause by the microtome) largely cancels in the ratio image, allowing small differences in scattering power to be discerned in unstained specimens (Carlemalm and Kellenberger, 1982).

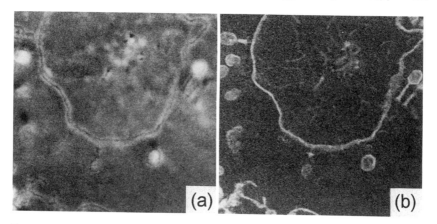

Figure 5.14. T4 bacteriophages adsorbed to *E. coli* in a 30-nm section embedded in 24%-Sn resin (1.1-μm field of view). (a) Annular dark-field image; (b) inelastic/elastic ratio image in which regions of lower atomic number appear bright (Carlemalm *et al.*, 1982).

The Z-contrast image sometimes appears to have better spatial resolution than either the dark-field or inelastic image but this effect occurs because the inelastic image is more blurred than the elastic one (partly due to delocalization of inelastic scattering); upon division of the two intensities, high-frequency components of the dark-field image are preferentially amplified, equivalent to unsharp masking of photographic negatives (Ottens-meyer and Arsenault, 1983). Therefore the ratio image contains no additional information but its fine structure is electronically enhanced. If the contrast in the elastic or inelastic signals is too high, the nonlinear process of division can create image artifacts (Reichelt *et al.*, 1984).

5.3.5. Contrast Tuning and MPL Imaging

Contrast tuning denotes the ability to choose an energy loss (typically in the range 0–200 eV) where contrast is adequate but not so high that the dynamic range of the image cannot be recorded in a single micrograph (Bauer *et al.*, 1987; Wagner, 1990). Dynamic range is often a problem in zero-loss images of thick (e.g., 0.5 μm) sections of biological tissue because stained regions scatter very strongly relative to unstained ones.

Structure-sensitive contrast in biological tissue can be maximized by choosing an energy loss around 250 eV, just below the carbon K-edge, so that the contribution of carbon to the image is minimized. This allows structures containing elements with lower-lying edges (sulfur, phosphorus, or heavy-metal stain) to appear bright in the image, giving a reversed "dark-field" contrast; see Fig. 5.15. Imaging at 260 eV has proved useful for

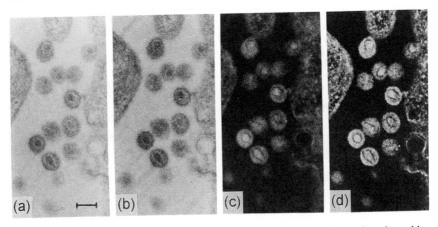

Figure 5.15. A 30-nm section of HIV-1 producing cells, embedded in Epon after glutaralde-hyde and OsO₄ fixation and staining with uranyl acetate. (a) Unfiltered image (the bar measures 100 nm); (b) zero-loss image showing improved contrast; (c) 110-eV image showing reduced contrast; (d) 250-eV image showing structure-sensitive reversed contrast (Özel *et al.*, 1990).

observing microdomain morphology in unstained polymers (Du Chesne *et al.*, 1992).

The *most probable* loss (MPL) is the energy loss at which the spectral intensity is highest. For thin specimens ($t/\lambda < 1$), the zero-loss peak is the most intense, but in thick specimens this peak becomes weak and the MPL corresponds to the broad maximum of the Landau distribution (Fig. 3.30), around 80 eV for 0.5-μm Epon and 270 eV for a 1-μm section (Reimer *et al.*, 1992). An image obtained at this loss will have maximum intensity, a desirable property for accurate focusing and reasonably short recording times so that specimen drift is not a problem. Intensity is also increased by widening the energy-selecting slit, but the spatial resolution may suffer because of higher chromatic aberration.

Pearce–Percy and Cowley (1976) showed that STEM images of thick biological objects can be obtained with near-optimum signal/noise ratio if an electron spectrometer is used to accept all energy losses *below* or *above* the MPL (giving bright-field and dark-field energy-loss images, respectively). Using a cutoff at 150 eV, they obtained 100-keV dark-field images with high contrast from chick fibroblast nuclei about 1 μm thick.

5.3.6. Core-Loss Images and Elemental Mapping

The ability to show the two-dimensional distribution of specific elements makes an imaging filter a powerful tool for practical materials analysis. As discussed in Section 2.6.5, elemental mapping involves recording at least two images, before and after the ionization edge. The simplest procedure is to subtract the pre-edge image from the postedge image; see Fig. 5.16. This *two-window subtraction* procedure works well enough for edges of high jump ratio (as obtained with very thin specimens and high concentrations of the analyzed element) but is unsatisfactory in other circumstances or if quantitative results are required (Leapman and Swyt, 1983). Negative intensities are generated in regions devoid of the selected element, where the energy-loss intensity decreases continuously with energy loss (Crozier, 1995).

An alternative procedure involves *dividing* the postedge image by the pre-edge image (Section 2.6.5), yielding a *jump-ratio image* which is largely insensitive to variations in specimen thickness and diffracting conditions. This is dramatically illustrated in Fig. 5.17, where the diffraction contrast disappears and carbide precipitates (along grain boundaries and within grains) become visible in jump-ratio images recorded at the L_{23} edges of Fe, Cr, and V. For *very* thin specimens, the ratio-image intensity would be proportional to elemental concentration; but, in specimens of typical thickness, plural-scattering background components make the image only a qualitative indication of elemental distribution (Hofer *et al.*, 1995; Crozier, 1995).

For quantitative elemental mapping, it is necessary to record *two* pre-edge images. If the two energy windows are *adjacent* to each other as

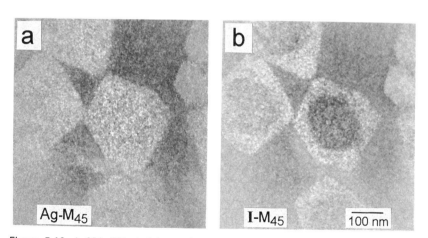

Figure 5.16. A 256×256-pixel image (recorded in 3 s using a Zeiss 902 EFTEM) of a 30-nm microtomed section of a photographic emulsion, showing silver halide microcrystals containing a AgBr core and AgBrI shell (Lavergne *et al.*, 1994). (a) Silver M_{45} map obtained by subtracting 370-eV and 360-eV images. (b) Iodine M_{45} image from subtraction of 615-eV and 580-eV images.

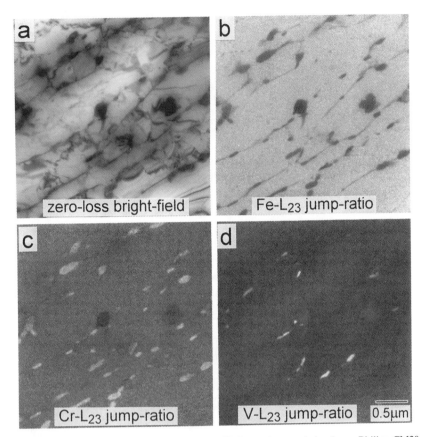

Figure 5.17. Micrographs of electropolished 10%Cr steel, recorded using a Philips CM20 TEM equipped with a Gatan imaging filter (Hofer *et al.*, 1995). (From *Ultramicroscopy*, F. Hofer, P. Warbichler, and W. Grogger. Characterization of nanometre sized precipitates in solids by electron spectroscopic imaging, © 1995, pp. 15–31, with permission from Elsevier Science B.V. Amsterdam, The Netherlands.) (a) Zero-loss bright-field image, whose contrast arises mainly from bend contours and grain boundaries. The succeeding images show jump-ratio maps recorded with 20-eV windows on either side of the L_{23} edges of (b) iron, (c) chromium, and (d) vanadium.

Figure 5.18. Collection of 512×512-element energy-selected images ($E_0 = 200$ keV, $\beta = 7.6$ mrad) of an ion-milled foil of ODS-niobium alloy containing 0.3at.% Ti and 0.3at.% oxygen (Hofer *et al.*, 1995). (From *Ultramicroscopy*, F. Hofer, P. Warbichler, and W. Grogger. Characterization of nanometre sized precipitates in solids by electron spectroscopic imaging, © 1995, pp. 15–31, with permission from Elsevier Science B.V. Amsterdam, The Netherlands.) Images (a) and (b) were recorded with 20-eV windows below the Ti L_{23} edge, (c) with a 20-eV window just above the Ti-edge threshold. (d) Ti elemental map obtained by three-window modeling; (e) Ti-edge jump-ratio map; (f) image representing the ratio of the two pre-edge images.

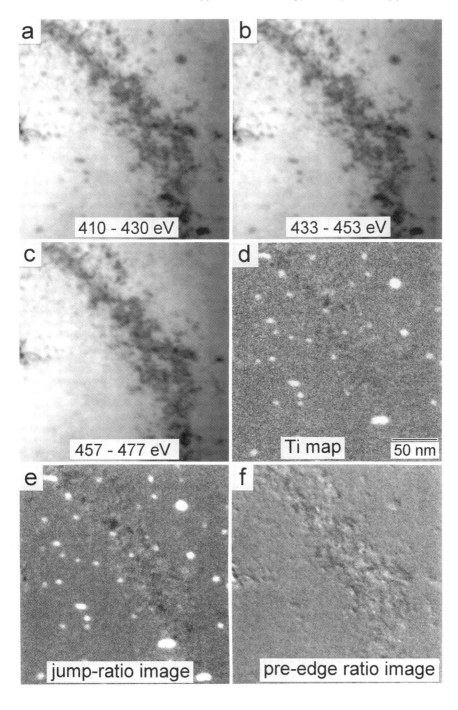

in Fig. 4.10, Eqs. (4.51)–(4.53) can be used to evaluate the background parameters A and r (assuming a background of the form AE^{-r}), from which the background contribution to the postedge image can be calculated. Such calculation is straightforward when the images are acquired electronically into computer memory. An example is shown in Fig. 5.18. The two pre-edge images and the postedge image all look similar, but when they are combined to form an elemental map, titanium oxide precipitates become clearly visible and bend-contour contrast within the foil largely disappears. Diffraction contrast may be further suppressed by dividing by a low-loss image or a zero-loss image (Crozier, 1995). As further confirmation that a jump-ratio or three-window image represents elemental concentration, taking the ratio of the two *pre-edge* images should yield very little contrast. This is the case in Fig. 5.18f except along the bend contour, where intensity fluctuations probably arise from a small change in specimen orientation between acquisition of the two images.

Similar elemental maps of a ceramic composite, consisting of coated SiC fibers embedded in a glass matrix, showed the fibers to consist mostly of carbon and silicon, but with oxygen, boron, magnesium, and aluminum (from the matrix) present in an oxidation layer (Krivanek *et al.*, 1994). The fiber coating was found to contain boron, nitrogen, and carbon, with smaller concentrations of oxygen, magnesium, and silicon; from the fine structure of the images, the coating was deduced to consist of a carbonaceous matrix containing a porous BN structure and small particles of silicon oxide.

As a result of background-extrapolation errors (Section 4.4.4), three-window modeling produces a noisier image than the two-window methods (subtraction *or* division), as seen by comparing images (d) and (e) in Fig. 5.18. To achieve adequate statistics, the recording time can be increased or (in STEM imaging) the number of pixels reduced. A good strategy is to first acquire a jump-ratio image from the area of interest, which requires only a short exposure time. If the results are encouraging, the three-window method can then be used to obtain a more quantitative elemental map.

5.4. Elemental Analysis by Core-Loss Spectroscopy

Elemental analysis is the most frequent electron-microscope application of EELS. As discussed in Chapter 1, EELS can outperform EDX spectroscopy (in terms of elemental sensitivity) in certain cases but generally involves more skill on the part of the operator. In this section, we discuss the data-collection strategies which have been found effective in particular cases, to complement the general description of spectrum-processing techniques in Chapter 4. Techniques which are specific to particular elements are discussed in this Section, while results from particular materials systems are given in Section 5.7. We begin by reviewing some choices of instrument and method which are directly relevant to core-loss spectroscopy.

The relative advantages of serial and parallel acquisition of energy-loss spectra are discussed in a general context in Section 2.4. For elemental analysis, the serial method has no clear advantage; in fact, parallel recording has been shown to offer higher elemental sensitivity (Shuman and Somlyo, 1987; Leapman and Newbury, 1993). A potential problem with parallel recording (particularly important when detecting low concentrations) is the variation in sensitivity across the array, but techniques are available for dealing with this artifact (Section 2.5.5). In the case of a STEM instrument fitted with a parallel-recording spectrometer, the use of spectrum-imaging (Section 2.6.4) is attractive because it allows extensive processing of the data after acquisition, but data-storage and acquisition-time constraints place a practical limit on the number of pixels used. Similar complete information can be recorded by an energy-filtering TEM, although the radiation dose required to achieve the same (time-integrated) signal is higher by a factor equal to the number of images collected; see Section 2.6.6. For many inorganic specimens, this increased dose may be unimportant and may be outweighed by the higher current available in a broad beam, allowing shorter recording times and therefore less drift of the specimen and high voltage.

As discussed in Chapter 2, spectroscopy can be carried out with a conventional TEM operating in either its imaging or diffraction mode. In *image mode*, a region of analysis of known diameter is conveniently selected by means of the spectrometer entrance aperture and, for quantitative analysis, the collection semiangle is known (provided the objective aperture has been calibrated in terms of angle). But particularly for higher-energy edges, chromatic aberration of the microscope imaging lenses prevents the area-selecting procedure from selecting the analyzed area precisely (see Section 2.3.2) and may result in incorrect elemental ratios (Section 2.3.3). The most effective way of avoiding chromatic effects is to change the microscope high voltage by an amount equal to the energy loss being analyzed, a technique which is easiest to implement in the case of serial recording. In *diffraction mode*, the diffraction pattern must be carefully centered about the spectrometer-entrance aperture (close to the center of the viewing screen); the collection angle will depend on the camera length and aperture diameter. Provided the incident probe diameter is small, chromatic aberration is not troublesome.

A small collection angle (5–10 mrad) increases the visibility (signal/background ratio) of an ionization edge (Section 3.5) and is appropriate for lower-energy edges. For higher-energy edges, the problem of low intensity is alleviated by choosing larger β. However, quantitative analysis becomes problematic if strong Bragg spots (or rings) appear just inside or outside the aperture (Egerton, 1978a). In the case of a small probe, the convergence angle α may be as large as 10 mrad, and to avoid considerable loss of signal β should exceed α (Section 4.5.3). As shown by Fig. 4.16, quantitative analysis involves a convergence correction unless $\beta/\alpha > 2$.

The next decision is the area of specimen to be analyzed. Sometimes this is obvious from the TEM image (possibly aided by diffraction) but if the instrumentation allows elemental mapping, a jump-ratio image (Section 5.3.6) can be very useful in choosing a region for detailed analysis. Particularly in the case of quantitative analysis and ionization edges below 200 eV (where plural scattering can greatly increase the pre-edge background), a very thin region of specimen is desirable: ideally $t/\lambda < 0.5$, which implies that the zero-loss peak comprises at least 60% of the counts in the low-loss spectrum (Section 5.1). EELS is usually carried out at the maximum available incident energy, since this is equivalent to using a thinner specimen.

In the case of a crystalline specimen, its orientation relative to the incident beam has an influence on microanalysis. It is advisable to avoid strongly diffracting conditions, such as the Bragg condition for a low-order reflection or where the incident beam is parallel to a low-order zone axis. Choice of a less-diffracting situation increases the collected signal and minimizes quantification errors arising from the effect of channeling on the core-loss cross section (Section 5.6.1) and the contribution of Bragg beams to the core-loss intensity (Section 4.5.1). It may also help in optimizing the spatial resolution of analysis by reducing beam spreading (Section 5.5.2).

Elements of atomic number greater than 12 allow a choice of edge used for elemental analysis. In general, only a major edge (listed in italics in Appendix D) should be used, and those with a threshold energy in the range 100–2000 eV are preferable. Edges which are sawtooth-shaped or peaked at the threshold (denoted h and w in Appendix D) are more easily identified and quantified, especially if the element has a low concentration. Ionization cross sections of K-edges are mostly known to within 10% but the situation for other edges is more variable, as indicated by error estimates in Appendix B.8.

Quantification of the core-loss signal necessitates its separation from the background. The simplest procedure is to model the pre-edge background by linear least-squares fitting (usually to a power-law function of energy loss) and make allowance for this background when integrating the core-loss signal over an energy window (typically 50–200 eV) following the edge; see Sections 4.4.1 and 4.5.1. This procedure becomes difficult where two or more edges are close in energy, where an element is present at low concentration, or at a low-energy edge in a specimen which is not extremely thin. In these cases, the edge signal should be modeled by multiple least-squares (MLS) fitting as a sum of background and edge components; see Section 4.5.4. Usually standard specimens are used to record the edge *shape* but calculated cross sections are used to obtain elemental ratios. Another tactic is to investigate *differences* in concentration by subtracting spectra recorded from different regions of specimen (Section 4.5.5).

To determine elemental ratios, a choice has to be made between a standardless procedure (using cross sections which are calculated or param-eterized; see Appendix B) and a standards-based (k-factor) method. The standardless approach is convenient, but requires that the collection semian-gle be known, at least approximately. The k-factor method involves the use of one or more standard specimens of known composition, covering each element to be analyzed. However, the incident-beam energy and collection angle need not be known, provided the same values are used for the unknown and standard specimens. Some sources of systematic error, such as poorly known cross sections and chromatic-aberration effects, should be absent when using the k-factor method. Standards which have been found useful include the minerals apatite ($Ca_5P_3O_{12}F$) and rhodizite ($K_{46}Cs_{36}Rb_6Na_2Al_{399}Be_{455}B_{1135}Li_2O_{28}$).

Although core-loss spectroscopy can in principle identify any element in the periodic table, some are more easily detected than others. EELS is most commonly used for analyzing elements of low atomic number, which are difficult to quantify by EDX spectroscopy. In the following section, we show how EELS has been employed to detect or measure particular elements.

5.4.1. Measurement of Hydrogen and Helium

Hydrogen in its elemental form is detectable from the presence of an ionization edge. Although the ionization energy is 13.6 eV, this value corresponds to transitions to continuum states of an isolated atom. At slightly lower energy loss, there is a Lyman series of transitions to discrete levels, giving peaks which are not resolved in TEM spectrometer systems: the result is a structureless edge with a maximum at about 12 eV, followed by a gradual decay on the high-loss side (Ahn and Kirvanek, 1983). EELS has been used to detect molecular hydrogen present as bubbles in ion-implanted SiC (Hojou *et al.*, 1992) and in frozen-hydrated biological speci-mens after irradiation within the electron microscope (Leapman and Sun, 1995).

Hydrogen combined with other elements (in a compound or solution) transfers its electrons to the band structure of the whole solid, destroying the energy levels which would give rise to a characteristic ionization edge.

Nevertheless, metallic hydrides have been detected from their low-loss spectra; electrons donated by H atoms usually increase the valence-electron density, shifting the plasmon peak upwards by 1 or 2 eV from that of the metal; see Section 5.2.2.

Hydrogen in an organic compound again influences the low-loss spectrum. Hydrocarbon polymers have their main "plasmon" peak at a *lower* energy than that of amorphous carbon (24 eV) because the presence of hydrogen reduces the mass density. If hydrogen is lost, for example during prolonged electron irradiation, the plasmon energy increases towards that of amorphous carbon (Ditchfield *et al.*, 1973).

Hydrogen in an organic material also increases the inelastic/elastic scattering ratio ν, measurable in a conventional TEM from the total intensity I and zero-loss intensity I_0 in a spectrum recorded *without* an angle-limiting aperture, together with the zero-loss intensity I_u recorded with a small (~ 2 mrad) angle-limiting aperture. Making allowance for plural scattering,

$$\nu = \frac{\ln(I/I_0)}{\ln(I_0/I_u)} \tag{5.15}$$

The specimen should be thick enough to avoid I_0 and I_u being almost equal; otherwise fluctuations in incident-beam current may result in poor accuracy. This type of measurement has been used to monitor the loss of hydrogen from 9,10-diphenyl anthracene as a function of electron dose (Egerton, 1976a).

Helium is produced in the form of nanometer-sized bubbles when 20/25/Nb stainless steel (used as fuel cladding in nuclear reactors) is irradiated with neutrons. By positioning a STEM probe at a bubble and on the nearby metal matrix, McGibbon and Brown (1990) recorded energy-loss spectra which were subtracted to reveal a helium K-ionization edge. Quantification of this edge (using a hydrogenic K-shell cross section) led to an estimate of the He concentration in a 20-nm bubble: 2×10^{28} atoms/m^3, corresponding to a He pressure of 2 kbar (0.2 GPa).

Employing energy-loss spectroscopy with a broad electron beam, Fink (1989) estimated the average pressure in He bubbles formed in Al and Ni by ion implantation. Typically, the excitation threshold was shifted from the free-atom value (21.23 eV) to about 24 eV, probably because of Pauli repulsion between excited-state ($2p$) electrons and the ground-state electrons of neighboring He atoms. Taking this "blue shift" to be proportional to He density, with a calculated proportionality constant of 22×10^{30} m^{3}eV, the He pressure P (derived from an equation of state) was found to be inversely proportional to bubble radius r ($Pr = 90$ kbar nm) for bubbles in aluminum. In the case of implanted Ni, the measured pressure inside the smallest bubbles exceeded 250 kbar, corresponding to a density 10 times larger than liquid helium. The helium was therefore assumed to be present in solid form. Confirmation by electron diffraction is hampered by the small scattering amplitude of He, but electron diffraction peaks have been obtained from bubbles of other rare gases ion-implantated into Al, and were interpreted as indicating epitaxy with the surrounding matrix.

5.4.2. Measurement of Lithium, Beryllium, and Boron

The elements Li, Be, and B give K-ionization edges in the 30–200 eV region, but superimposed on a relatively large background. In thin specimens ($t/\lambda < 0.5$), this background represents the tail of the valence-electron plasmon peak; in thicker ones, plural plasmon scattering dominates. Hofer and Kothleitner (1993) found Fourier-log deconvolution helpful in obtaining an AE^{-r} fit to Li and Be edges recorded from mineral specimens. In this energy regime, it is likely that a K-edge will overlap with L- or M-edges of other elements, complicating quantitative analysis. The problem is alleviated by using a small energy window ($\Delta < 50$ eV) but at the risk of systematic error due to energy-loss near-edge structure (Section 4.5.2). A more satisfactory solution for Li and Be edges (Hunt, 1991; Hofer and Kothleitner, 1993) is to employ MLS fitting to reference edges recorded from simple compounds which have the same coordination, thereby ensuring similar near-edge structure.

Lithium cannot be analyzed with current EDX detectors and is difficult to measure by WDX spectroscopy. Chan and Williams (1985) evaluated EELS as a means of quantitative analysis of Al/Li alloys, which have aerospace applications because of their high strength/weight ratio. To minimize the background at the Li K-edge, very thin (<50 nm) specimens and small collection angles ($\beta < 5$ mrad) were used. The pre-edge background could then be successfully modeled by an AE^{-r} function and extrapolated over a 40-eV interval containing *both* the Li K- and Al L_{23} edges; see Fig. 5.19. By measuring the areas I_1, I_2, and I_3 above background and using hydrogenic (SIGMAK2) cross sections for the appropriate energy regions, the measured Al/Li ratio in Al_3Li δ'-phase precipitates was 3.6 ± 0.6 (expected value 3.0). Using the Al_3Li precipitate as a "k-factor" standard, Sainfort and Guyot measured the atomic percentage of Li in the surrounding matrix as 8.0 ± 1%, consistent with the value of 7% determined from small-angle x-ray scattering.

Strutt and Williams (1993) used EELS to study copper alloys containing up to 2% of Be, in which coherent precipitates confer high strength through age-hardening. They found that the Be/Cu ratio of γ-phase precipitates increased with decreasing ageing temperature, contrary to the expected phase diagram. To perform quantitative measurements with narrow integra-

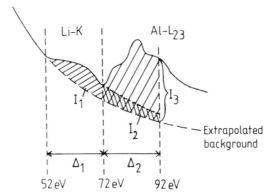

Figure 5.19. Analysis of two overlapping edges, requiring the measurement of integrals I_1, I_2, and I_3 (Chan and Williams, 1985).

tion windows (10 eV), hydrogenic cross sections for $\Delta = 100$ eV were scaled according to edge shapes taken from the Gatan core-loss atlas. The lowest quantifiable concentration of Be was about 10 at.%.

In their analysis of BeO-doped SiC, Liu *et al.* (1991) avoided conventional background fitting by recording a spectrum from pure SiC. After scaling and subtracting the SiC spectrum, weak features at 188 eV suggested a beryllium content of considerably less than 1%. EELS has been employed to detect 10-nm Be grains in lung tissue (Jouffrey *et al.*, 1978).

Boron is easier to quantify because of its higher *K*-edge energy (188 eV). Using a 50-nm-diameter probe in a field-emission STEM and second-difference PEELS recording, Leapman (1992) detected about 1% of boron in silicon. The edge was invisible in the normally acquired PEELS spectrum because of channel-to-channel gain variations and because EXELFS modulations from the preceding Si L_{23} edge persist up to at least 200 eV. About 0.5 at% B in Ni_3Al was detected from PEELS data (Fig. 5.20) in which gain variations were reduced to 0.005% by use of an iterative averaging procedure (Section 2.5.5). A second-difference filter was applied to suppress EXELFS modulations from the preceding Ni *M*- and Al *L*-edges. Good energy resolution helps in distinguishing the intrinsically sharp boron edge from the more gradual EXELFS modulations.

Boride particles (3–5 nm diameter) in silicon have been identified in CTEM core-loss images, confirming previous HREM interpretation in terms of coherent SiB precipitates (Frabboni *et al.*, 1991). Boron-containing compounds are being investigated for use in neutron-capture cancer therapy. Bendayan *et al.* (1989) have used electron spectroscopic imaging (ESI) to show that B-containing biopolymeric conjugates are absorbed intracellularly by colorectal cells.

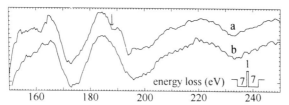

Figure 5.20. Spectra obtained by use of a second-difference filter (see inset, widths in channels) from (a) Ni$_3$Al containing 0.5% B, showing boron K-edge at 188 eV (arrowed), and (b) pure Ni$_3$Al (Boothroyd *et al.*, 1990). From Proc. XIIth Int. Cong. Electron Microsc., p. 81, © San Francisco Press, Inc., by permission.

5.4.3. Measurement of Carbon, Nitrogen, and Oxygen

The absence of beam-induced hydrocarbon contamination is an obvious prerequisite to the unambiguous identification and measurement of carbon. Microscope-induced contamination is reduced by using an ion-getter pump, liquid-nitrogen trap, or cold finger, ensuring a low partial pressure of hydrocarbons in the vicinity of the specimen. Specimen-borne contamination can be minimized by careful attention to cleanliness during specimen preparation and by liquid-nitrogen cooling of the specimen during microscopy, which reduces the migration of hydrocarbons on the surface of the specimen. Surface hydrocarbons are removed or rendered immobile (by polymerization) through mild baking of a specimen, either inside the microscope or before insertion, or by irradiating regions surrounding those to be analyzed with a broad beam in the TEM (by withdrawing the final condenser aperture and defocusing the illumination).

The identification of carbide and nitride precipitates in steel has been an important application of core-loss spectroscopy. Figure 5.21 illustrates how TiC and TiN precipitates may be distinguished more easily from K-loss spectra than from their morphology (their diffraction patterns are also quite similar). The atomic ratios of transition metals within carbides can be estimated from the appropriate L-ionization edges (Fraser, 1978; Baker *et al.*, 1982).

Extraction replicas are a convenient way of isolating small particles for spectroscopy, but the usual replicating materials (carbon or polymers) complicate any analysis for carbon. Garratt–Reed (1981) used a 50-nm coating of evaporated aluminum for extraction and was able to show that the carbon content of presumed V(C, N) precipitates in vanadium HSLA steel was less than his detection limit, about 5 at.%. Silicon extraction replicas, made by RF sputtering in argon, have also been used (Duckworth *et al.*, 1984); one problem which emerged was the loss of carbon from vanadium carbide precipitates when subjected to the small (15 or 1.5 nm)

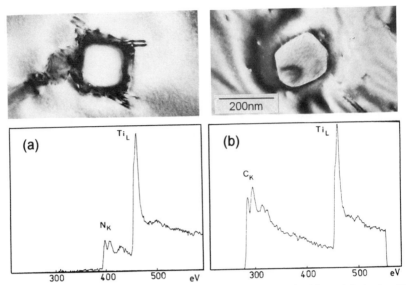

Figure 5.21. Micrographs and core-loss spectra of (a) TiN and (b) TiC precipitates in a Ti-rich phase in stainless steel (Zaluzec, 1980a).

probe in a field-emission STEM. However, use of a parallel-recording spectrometer should provide sufficient sensitivity to allow analysis of VC precipitates down to 1 nm diameter, enabling changes in composition to be monitored during the entire growth sequence of the particles (Craven *et al.*, 1989).

Nitrogen in solution in the γ-phase of duplex stainless steel was measured by Yamada *et al.* (1992) to be 0.26 ± 0.04 wt.%, using a 120-kV TEM in diffraction mode, with a 3-mm spectrometer aperture giving 15-mrad collection semiangle. The nitrogen content of the α-phase was at or below their detection limit, equivalent to 0.2 at.%. This low limit was achieved by iterative averaging of PEELS data, by employing a top-hat filter to form second-differential spectra and by maintaining an energy resolution close to 1 eV with a LaB_6 electron source. Quantification involved the use of narrow (2.5 eV) integration windows around the N and Fe second-differential peaks and a calibration curve of N/Cr intensity ratio against N/Cr concentration (measured using high-nitrogen alloys).

Some types of natural diamond contain octahedral-faceted inclusions, a few nm in size, known as voidites. EELS measurements in a STEM (Bruley and Brown, 1989) showed a sharply peaked *K*-edge at 400 eV, indicating the presence of nitrogen. Analysis of 20 voidites using SIGMAK cross sections gave an average nitrogen concentration which was independent of voidite size and equal to about half the carbon concentration in diamond; see also Section 5.7.1. The shape of the edge was consistent with the presence of N_2 rather than NH_3 (proposed as an explanation for previous lattice images of voidites). Bruley (1992) has established that nitrogen may be present at platelet defects in diamond, but only at a level of the order of a tenth of a monolayer; see page 381.

Using a field-emission STEM with a serial-recording spectrometer, Bourret and Colliex (1982) reported evidence for the segregation of oxygen at dislocation cores in germanium. AE^{-r} fitting to the oxygen edge and extrapolation over 100 eV revealed an oxygen signal $\approx$ 1% of the background. Subsequent HREM imaging of the analyzed dislocations showed that oxygen had been removed by electron irradiation; the characteristic dose was estimated to be 10^4 C/cm^2, corresponding to 40 s of STEM observation with a 10-pA probe scanned over a 2-nm-square area. Future studies of this kind could take advantage of the improved collection efficiency of a parallel-recording spectrometer and second-difference techniques, as employed by Yamada et al. (1992) for example.

Disko et al. (1991) utilized the fact that the $Al-L_{23}$ threshold in Al_2O_3 is shifted upwards in energy by 4.5 eV (relative to Al metal) to distinguish regions of Al–Al and Al–O bonding in oxide-strengthened aluminum alloys formed by cryomilling in liquid nitrogen. This procedure provided an indication of the oxide/metal fraction in different regions of the specimen. The parallel-recording spectrometer was also used to acquire pairs of core-loss spectra shifted by 7 eV; a peak at 400 eV in the ratio (log-derivative spectrum) provided a qualitative but sensitive test for small percentages of nitrogen. Nitrogen quantification was achieved by careful background fitting (Trebbia, 1988) to an AE^{-r} dependence and gave N/O ratios as large as unity in some of the oxide particles.

Oxides and nitrides of silicon have been analyzed by R. W. Carpenter and co-workers, by examining the shape of the Si L-edge. Comparing the L-edge fine structure with that recorded from several candidates, Skiff et al. (1986) showed that oxygen precipitation in Si produces SiO, not SiO_2. From fine structures, and by scaling nitrogen and silicon edges to match those of a standard, they established that precipitates in damaged regions of N^+-ion implanted Si were Si_3N_4. Using the Si L- and O K-edges of SiO_2 as standards, analysis of semi-insulating polycrystalline oxygen-doped silicon (SIPOS) revealed only 15 at.% of oxygen (Catalano et al., 1993), but after annealing at 900°C the SIPOS decomposed into silicon nanocrystals and a matrix containing 36 at.% oxygen. The native oxide formed on silicon at room temperature has been shown to have a composition close to SiO (Kim and Carpenter, 1990). These results suggest that metastable amorphous solid solutions of silicon and oxygen can exist over the whole range from Si to SiO_2 as a single phase.

5.4.4. Measurement of Fluorine and Heavier Elements

Some fluorinated organic compounds have normal biological activity and can be used as molecular markers for specific sites in cells. By forming energy-selected images with the fluorine K-edge, the segregation of (difluoro)serotonin was demonstrated (Costa et al., 1978). Fortunately, compounds in which fluorine is directly attached to an aromatic ring are relatively stable under electron irradiation, sometimes withstanding doses as high as 10^4 C/cm^2 if the specimen is cooled to $-160°C$ (Ciliax et al., 1993).

The distribution of sulfur, phosphorus, and calcium is of great interest in biological systems but the dose required for mapping these elements by x-ray K-emission spectroscopy is often destructive. EELS provides the option of using L-shell ionization, for which the cross section is relatively large, allowing higher detection sensitivity (Fig. 1.13). Due to the low L-edge energies of S and P (135, 165 eV) the spectral background is high, even if the specimen is very thin. In the case of parallel recording, this background is effectively suppressed by the use of first- or second-difference techniques (Section 4.5.5); the sensitivity then depends on the noise level and to some extent on the energy resolution. Detection of Ca, Ti, and transition elements is helped by the fact that these elements display sharp white-line peaks at the ionization threshold (Fig. 3.45). These peaks become amplified in difference spectra, making possible the detection of less than 100 ppm of alkaline earths, transition metals, and lanthanides in a glass test specimen (Leapman and Newbury, 1993).

Based on calculations and experimental results, Wang et al. (1992) have prescribed optimum conditions for detection of phosphorus in biological tissue with a parallel-recording spectrometer, namely $t/\lambda \approx 0.3$ and a 15-eV shift between spectra if first-difference recording is used. EELS was estimated to be 15 times more sensitive than EDX spectroscopy: with 0.5-nm beam current and 100-s recording time, the minimum detectable concentration was calculated as 8.4 mmol/kg ($\approx$ 100 ppm), equivalent to 34 phosphorus atoms in a 15-nm probe.

Electron spectroscopic imaging in a conventional TEM has been used extensively to provide qualitative or semiquantitative elemental maps from biological specimens. Since phosphorus is a constituent of DNA, P–L_{23} images have been used to investigate DNA configurations within 80s ribosomes (Shuman et al., 1982) and chromatin nucleosomes (Ottensmeyer, 1984). The latter contain about 300 atoms of phosphorus; the signal/noise ratio of the corresponding phosphorus signal was estimated to be 30. Köpf–Maier (1990) employed 80-keV energy-selected imaging to analyze the distribution of titanium and phosphorus in human tumors as a function of time after therapeutic doses of titanocene dichloride. The maximum Ti concentration in cell nuclei and nucleoli occurred after 48 hours and was accompanied by an enrichment of phosphorus, confirming that the primary interaction occurs with nucleic acids, particularly DNA.

ESI has also been used to image the distribution of heavier elements such as thorium, cerium, and barium, formed as a cytochemical reaction product in order to detect enzyme activities within a cell (Sorber *et al.*, 1990). In some cases, *K*-edges have been used: for example, to detect aluminum in newt larvae (Böhmer and Rahmann, 1990). Here the advantages of EELS over EDX spectroscopy are less obvious; however, edges in the 1000–3000 eV region can have good signal/background ratios (relatively unaffected by plural scattering) so that specimens as thick as 0.5 μm can be used (Egerton *et al.*, 1991).

5.5. Spatial Resolution and Detection Limits

The spatial resolution obtainable in energy-loss spectroscopy (or in an energy-selected image) depends on several factors, which we now discuss. In the case of core-loss microanalysis, spatial resolution is closely connected with the concept of elemental detection limits, as explained in Section 5.5.4.

5.5.1. Electron-Optical Considerations

In a scanning transmission electron microscope (STEM), the spatial resolution of the image (or of a point analysis) is determined largely by the dimensions of the incident probe. By employing a field-emission source with small emitting area, the source-size contribution to the incident-beam diameter can be made less than 0.2 nm, but spherical and chromatic aberration of the probe-forming lenses increase this value to typically 0.5 nm (Colliex and Mory, 1984). The sampled volume of specimen is also increased by an amount which depends on the convergence semiangle α of the illumination; a point probe focused at the mid-plane of the specimen defines a circle of diameter $t\alpha$ ($\approx$ 1 nm for a focused STEM probe) at the entrance and exit face.

In a thermionic-source TEM, a subnanometer probe can be formed but with much lower current. Its current-density profile contains a sharp central peak surrounded by electron-beam tails which contain an appreciable fraction of the incident current and may extend for many nanometers

(Cliff and Kenway, 1982). The sharp central peak provides STEM images with high-resolution detail but the extended tails can result in contributions to the core-loss signal (for example) considerably removed from the center of the probe. Deconvolution techniques have been used to correct concentration profiles for the effect of the aberration tails, based on a measured or calculated incident-beam profile (Thomas, 1982; Weiss and Carpenter, 1992).

If the TEM is operated in the usual broad-beam imaging mode, a diameter smaller than that of the incident beam can be defined by an area-selecting aperture, either a selected-area diffraction aperture or (with an image on the screen) a spectrometer-entrance aperture. However, the imaging lenses suffer from spherical and chromatic aberration, and the latter can be severe at high energy loss (Section 2.3.2) unless the high voltage is raised by an equal amount or the objective lens refocused (Schenner and Schattschneider, 1994). Spherical aberration cannot be circumvented in this way and can seriously degrade the resolution in thicker specimens if not limited by means of an objective aperture.

5.5.2. Loss of Resolution Due to Elastic Scattering

When an electron beam enters a specimen, it spreads laterally to some degree as a result of electron scattering (mainly elastic scattering, for which the average deflection angle is larger). As depicted in Fig. 1.11, this beam broadening degrades the spatial resolution of x-ray microanalysis; simple models suggest a broadening of 10 nm or more for a 100-nm-thick foil and 100-keV incident electrons (Goldstein *et al.*, 1977). For EELS, the angular divergence of the recorded signal can be limited to some chosen value β by means of a collection aperture which will effectively collimate the inelastic scattering (Collett *et al.*, 1981). The aperture will also tend to exclude electrons present in incident-probe aberration tails, since these electrons travel at larger angles relative to the optic axis, and the EELS signal is unaffected by secondary electrons, which can generate x-rays at considerable distances from the incident beam. The result is somewhat better ultimate spatial resolution than for EDX spectroscopy, as confirmed experimentally (Collett *et al.*, 1984; Titchmarsh, 1989).

For an amorphous specimen, the collimation effect can be estimated from simple geometry, as shown in Fig. 5.22a. If scattering occurs at the top of the foil, the volume of specimen sampled by the recorded electrons (contained with in a cone defined by the collection aperture) is a maximum; if it occurs at the bottom, the volume is zero. Averaging over the thickness of the specimen, the fraction $F_e(r)$ of electrons which have traveled radial distances up to r within the specimen is given by Fig. 5.22b. About 90% of the electrons collected by the aperture are contained within a specimen

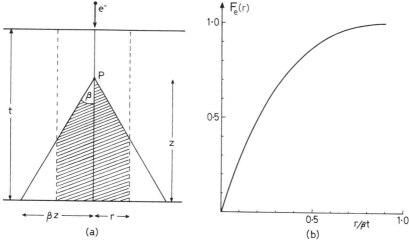

Figure 5.22. (a) Beam broadening due to scattering at a point P, a distance z from the exit surface of the specimen. The shaded region represents the excited volume which lies within a distance r of the optic axis and which gives rise to inelastic scattering within the collection aperture. (b) Fraction $F_e(r)$ of the elastically scattered electrons (scattering angle less than β) which are contained within a radius r of the incident beam axis. This estimate assumes that the angular width of elastic scattering is large compared with the collection semiangle β and the angular width of the inelastic scattering.

volume of diameter $2r \approx \beta t$, which is below 1 nm for a typical aperture and specimen thickness.

In the case of a crystalline specimen, the collection angle can be chosen to exclude Bragg beams, suggesting a radial spread (less than βt) which arises mainly from inelastic scattering and incident-beam convergence. Even in the absence of an aperture, electron channeling and dynamical effects reduce the radial broadening (Browning and Pennycook, 1993). In effect, beam broadening is delayed up to a depth (below the entrance surface) at which s-type Bloch waves (more localized at the atomic centers) are dispersed by inelastic scattering. STEM imaging of atomic columns in a thin crystal is believed to rely on this principle (Pennycook and Jesson, 1991). A practical way of reducing beam spreading is to orient the specimen so as to avoid strongly diffracting conditions (D. Muller, personal communication).

5.5.3. Delocalization of Inelastic Scattering

The width of the interaction potential associated with inelastic scattering imposes an ultimate limit to the spatial resolution obtainable in an energy-selected image or small-probe analysis. The *delocalization* of the inelastic scattering can be defined as the width of the real-space distribution of scattering probability, sometimes called an object function (Pennycook *et al.*, 1995b), or operationally as the blurring of the inelastic image after all instrumental aberrations and elastic effects have been accounted for (Muller and Silcox, 1995).

On a classical (particle) description of scattering, localization is represented by the impact parameter b of the incident electron (Fig. 3.1). Small b implies strong electrostatic interaction and large scattering angle, so the scattering event should appear more localized if observed using an off-axis detector (Howie, 1981; Rossouw and Maslen, 1984) as observed in channeling studies; see Section 5.6.1.

The wave nature of the electron can be incorporated approximately by invoking the uncertainty relation (Howie, 1981; Self and Buseck, 1983) or equivalently by using the Rayleigh criterion for the resolution limit set by diffraction. At high energy loss, where the angular width of inelastic scattering is broad and scattering effectively fills a collection aperture of semiangle β, the resolution is limited to $0.6\lambda/\beta$, as depicted by the horizontal dotted line in Fig. 5.23. At lower energy loss, scattering does not fill the aperture: 50% of the electrons are contained within a median scattering angle $\tilde{\theta}$ which depends on the characteristic scattering angle θ_E and the cutoff angle θ_c of the angular distribution. Taking θ_c as the Bethe-ridge angle $(2\theta_E)^{1/2}$, Eq. (3.56) gives $\tilde{\theta} \approx 1.2(\theta_E)^{3/4}$ and using $\tilde{\theta}$ as an effective angular limit, the corresponding resolution is $0.6\lambda/\tilde{\theta} = 0.5\lambda/\theta_E^{3/4}$. Combining this expression with the diffraction limit imposed by the aperture leads to the following expression for the diameter d_{50} containing 50% of the inelastic intensity:

$$(d_{50})^2 \approx (0.5\lambda/\theta_E^{3/4})^2 + (0.6\lambda/\beta)^2 \tag{5.17}$$

As shown in Fig. 5.23, Eq. (5.17) yields values of d_{50} which are similar to those given by more-sophisticated calculations.

Kohl and Rose (1985) employed quantum-mechanical theory to calculate intensity profiles for the inelastic image of a single atom, using the dipole approximation and ignoring lens aberrations. Schenner and Schattschneider (1994) extended the method to include the effects of spherical aberration, chromatic aberration, and objective-lens defocus, while Muller and Silcox (1995) have investigated the effect of different detector geometries. The calculations involve separate terms for momentum transfer perpendicular and parallel to the optic axis, but the perpendicular component is usually dominant.

Results of these calculations are plotted in Fig. 5.23. Diamonds and inverted triangles represent the full width at half-maximum (FWHM) of a STEM image for incident-convergence and collection semiangles of 10

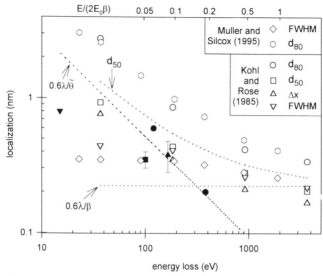

Figure 5.23. Localization width of inelastic scattering for $\beta = 10$ mrad and $E_0 = 100$ keV. Open data points represent calculations by Muller and Silcox (1995) and by Kohl and Rose (1985) of the diameter containing 50% and 80% of the scattering and of the FWHM of the image of a single atom; Δx is the edge resolution (20%–80% intensity) for a straight boundary of independently scattering atoms. Straight lines denote diffraction limits (see text); the dashed curve represents Eq. (5.17). Solid data points are experimental estimates: ▲ Adamson-Sharpe and Ottensmeyer (1981), ▼ Scheinfein *et al.* (1985), ● Shuman *et al.* (1986), and ■ Mory *et al.* (1991). As seen from the top scale, the scattering ought to become more delocalized at higher incident energy.

mrad. This width increases as the energy loss decreases and amounts to about 0.4 nm at an energy loss of 100 eV, in rough agreement with the estimate of Mory *et al.* (1991) based on a statistical analysis of STEM images of uranium atoms, recorded with O_{45}-loss electrons. Circles and hexagons give the diameter d_{80} containing 80% of the intensity of the single-atom image; squares represent the diameter d_{50} containing 50% of the intensity. Because the calculated image takes the form of a broad ring with an extended tail, d_{80} considerably exceeds d_{50} and FWHM values, especially at lower energy loss. At 20-eV loss, d_{80} is a factor of 10 larger than FWHM, indicating that the EELS signal may contain contributions from a considerably larger area of specimen than is apparent from an energy-selected image.

Triangles in Fig. 5.23 represent the edge resolution Δx (between 20% and 80% of the full intensity) calculated by Kohl and Rose for a straight boundary to a region of independently scattering atoms. Below 50 eV, Δx approaches 1 nm, in reasonable agreement with the 1.4-nm resolution of the inelastic signal (recorded as a line scan across the edge of a thin carbon film, mean energy loss $\approx$ 40 eV) using a 43-keV STEM (Isaacson *et al.*, 1974). This experiment has been repeated with 100-keV electrons by Muller and Silcox (1995), who measured the inelastic signal at a known energy loss, as well as the high-angle elastic signal recorded using an annular dark-field (ADF) detector; see Fig. 5.24. At high energy loss (77 eV in Fig. 5.24) the inelastic scattering becomes very localized and decays with distance at the same rate as the elastic signal. Allowing for the width of the electron probe (measured from the ADF signal), it appears that the energy-loss probability decays exponentially with the distance b (impact parameter) between the incident electron trajectory and the edge:

$$P(b,E) \propto \exp(-2b/b_0) \qquad (5.18)$$

where $b_0 = \hbar v/E$ is the impact parameter (introduced by Bohr) beyond which the target electrons are dynamically screened. Batson (1992) has shown that the surface scattering probability for an oxidized aluminum surface obeys the same equation. For $b < b_0$, screening is ineffective and Eq. (5.17) no longer holds; in fact, the scattering should then be described by classical physics, the product $E \cdot P(b,E)$ being equivalent to the *classical* energy loss for interaction between a moving and a stationary bound electron, which is proportional to $1/b^2$ (Jackson, 1975).

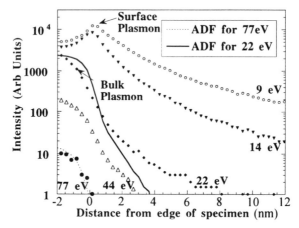

Figure 5.24. Measurements of the inelastic scattering intensity as a function of distance b from the edge of a 3-nm amorphous carbon specimen (Muller and Silcox, 1995). The intensity scale is logarithmic, so the scattering probability $P(b,E)$ falls to very low values at larger energy loss. Note also that P(b,E) achieves a minimum at the edge ($b\approx0$) for $E = 9$ eV and 14 eV, characteristic of surface-plasmon behavior and resulting in enhanced intensity of the bulk-plasmon (22 eV) signal; see Fig. 3.25. From *Ultramicroscopy*, D. A. Muller and J. Silcox, Delocalization in inelastic scattering, © 1995 pp. 195–213, with permission from Elsevier Science B.V., Amsterdam, The Netherlands.

The low FWHM values in Fig. 5.23 indicate that good spatial resolution is possible in a plasmon-loss image, consistent with the rapid fall in the aluminum plasmon peak within 0.8 nm of an abrupt Al/AlF_3 boundary (Scheinfein et al., 1985). The resolution could be further enhanced by use of an off-axis detector (Muller and Silcox, 1995). Note that we are referring here to modulations in the plasmon signal which arise from changes in chemical composition. Particularly in thicker specimens, variations in the amount of plural (elastic + plasmon) scattering can introduce crystallographic contrast in the plasmon-loss image and this contrast has high resolution because of the more localized nature of elastic scattering (Egerton, 1976c; Craven and Colliex, 1977).

Other procedures have been used to estimate localization. Pennycook (1982) invoked the uncertainty relation: $\Delta p \, \Delta x \approx \hbar$, taking Δx as the impact parameter b and $\Delta p = \hbar \Delta k$ as the momentum transfer in an inelastic collision; weighting b^2 with the intensity per unit angle leads to an equation for the root-mean-square impact parameter. Shuman et al. (1986) estimated the width of a single-atom image as the half-width of a contrast transfer function (CTF), obtained as the autocorrelation function of the inelastic scattering amplitude $f(\theta)$. By analogy with elastic scattering, $f(\theta)$ is the square root of the inelastic angular distribution, as given in Eq. (3.1). It is necessary to assume that the imaginary part of $f(\theta)$ is negligible for core losses close to the ionization threshold, where the excited core-loss electron leaves the atom with only a small fraction of the transferred energy. The data points for carbon (imaged using K-loss electrons) and uranium (at the M_{45} edge) are plotted as solid circles in Fig. 5.23. Batson (1992) has related the localization for bulk-plasmon excitation to the plasmon wake (Fig. 3.13) and has deduced the spatial resolution to be better than 0.5 nm.

Energy-selecting microscopes have been used to image small clusters of atoms on a crystalline substrate of different atomic number (e.g., Hashimoto et al., 1992), a goal being to distinguish the atomic number of single atoms. Figure 5.25 shows the calculated intensity in a Be K-loss image of a single Be atom supported by a 7.2-nm gold film (Spence and Lynch, 1982). Because of the low energy loss, the localization length of the Be K-scattering is greater than the first-order lattice spacing of the substrate (in the diffraction plane, the Be inelastic scattering is confined mainly to angles less than the Bragg angle for gold). The image intensity is modulated by lattice fringes of the substrate, arising from Bragg scattering of Be K-loss electrons in the gold. These simulations show that, due to elastic scattering, atoms of atomic number other than that selected by the energy filter may influence the image, unless the selected energy loss is sufficiently high.

The localization of inelastic scattering is also of relevance in channeling experiments (Section 5.6.1). Here the inelastically scattered electrons are

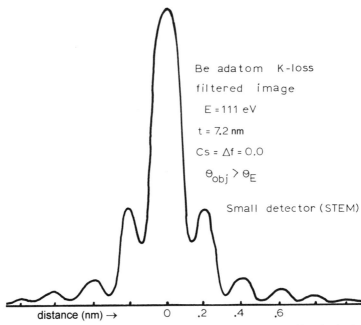

Figure 5.25. Dynamical calculation of a line profile through the Be K-loss lattice image of a single Be adatom on a crystalline gold substrate (Spence and Lynch, 1982). Multiple peaks representing lattice fringes from the gold appear within the inelastic localization range, despite filtering for the Be ionization edge.

recorded in a diffraction plane (at a chosen scattering angle) rather than in a specimen image. The aperture term in Eq. (5.16) is therefore absent and the localization distance attains subatomic dimensions at high energy loss (diagonal dashed line in Fig. 5.23). This condition allows the nonuniform current density which arises from channeling (Section 3.1.4) to appreciably affect the core-loss intensity, dependent on the specimen orientation. Bourdillon *et al.* (1981b) and Bourdillon (1984) found that measured orientation effects were consistent with a root-mean-square impact parameter (equivalent to localization distance) given by

$$b_{RMS} = f(\hbar v/E) \approx (f/2\pi)\lambda/\theta_E \qquad (5.19)$$

with $f \approx 0.5$, as expected from time-dependent perturbation theory (Seaton, 1962).

5.5.4. Statistical Limitations

Because inner-shell ionization cross sections are relatively low, the core-loss signal may represent a rather limited number of inelastically scattered electrons. The useful spatial resolution, detection limits, and accuracy of elemental microanalysis are therefore strongly influenced by statistical considerations. In the analysis below, we estimate the statistical constraints on the detection of a small quantity, N atoms per unit area, of an element in a matrix (or on a support film) which has an areal density of N_t atoms per unit area, such that $N_t \gg N$. For convenience of notation, all intensities are assumed to represent *numbers of recorded electrons*. The radiation dose D received by the specimen (during spectrum acquisition time T) is in units of Coulombs per unit area.

According to Eq. (4.65), the core-loss signal (recorded with a collection semiangle β and integrated over an energy window Δ) is given by

$$I_k \approx NI(\beta,\Delta)\sigma_k(\beta, \Delta) \qquad (5.20)$$

where $\sigma_k(\beta, \Delta)$ is the appropriate core-loss cross section and $I(\beta, \Delta)$ is the total intensity in the low-loss region, integrated up to an energy loss Δ. Assuming that the energy window Δ contains most of the electrons transmitted through the collection aperture and that this aperture is small enough to exclude most of the elastic scattering (mean free path λ_e) we can write

$$I(\beta, \Delta) \approx (I/e)T \exp(-t/\lambda_e) = (\pi/4)d^2(D/e)\exp(-t/\lambda_e) \qquad (5.21)$$

where I is the probe current (in amp), e the electronic charge, and d the probe diameter. We express $I(\beta, \Delta)$ in terms of an electron dose D because, in the absence of instrumental drift, radiation damage provides a fundamental limit to the acquisition time and is certainly the main practical limitation for *organic* specimens. For small Δ, an additional factor of $\exp(-t/\lambda_i)$ is required in Eq. (5.20), λ_i being a mean free path for inelastic scattering (Leapman, 1992).

An equation analogous to Eq. (5.19) can be written for the background intensity beneath the edge (as shown in Fig. 4.11):

$$I_b \approx N_t I(\beta, \Delta)\sigma_b(\beta, \Delta) \qquad (5.22)$$

where $\sigma_b(\beta, \Delta)$ is a cross section for all energy-loss processes which contribute to the background. The core-loss signal/noise ratio measured by an ideal spectrometer is given by Eq. (4.61), but using Eq. (2.33) to make allowance for the detective quantum efficiency (DQE) of the spectrometer, the measured signal/noise ratio will be

$$\text{SNR} = (\text{DQE})^{1/2}I_k/(I_k + hI_b)^{1/2} \approx (\text{DQE})^{1/2} I_k(hI_b)^{-1/2} \qquad (5.23)$$

where h is a factor representing the statistical error associated with background subtraction, typically in the range 5 to 10 (Section 4.4.4). Combining Eqs. (5.20), (5.22), and (5.23), the atomic fraction of the analyzed element is

$$f = \frac{N}{N_t} = \frac{\text{SNR}}{\sigma_k(\beta, \Delta)} \left[\frac{h\sigma_b(\beta, \Delta)}{N_t I_t(\beta, \Delta)} \right]^{1/2} (\text{DQE})^{-1/2} \qquad (5.24)$$

Taking SNR = 3 (corresponding to 98% certainty of detection, according to Gaussian statistics) and using Eq. (5.21), the minimum detectable atomic fraction is

$$f_{\min} \approx \left(\frac{3}{\sigma_k(\beta, \Delta)} \right) \left(\frac{1.1}{d} \right) \left(\frac{h\sigma_b(\beta, \Delta)}{(\text{DQE})(D/e)N_t} \right)^{1/2} \exp\left(\frac{t}{2\lambda_e} \right) \qquad (5.25)$$

Apart from a factor of $(h/\text{DQE})^{1/2}$, Eq. (5.25) is equivalent to Eq. (18) of Isaacson and Johnson (1975) except that $f_{\min}$ is an atomic ratio rather than a minimum *mass* fraction (MMF). For a given radiation dose D, and provided the specimen is sufficiently uniform, a large beam diameter d favors the detection of low atomic concentrations. This is so because the radiation damage is spread over a larger volume of material, permitting a larger beam current or acquisition time. A thermal-emission source (capable of providing a large beam current) may therefore be preferable. On the other hand, the minimum detectable number of atoms (MDN) is given by

$$\text{MDN} = \frac{\pi}{4} d^2 f_{\min} N_t \approx \frac{2.7\, N_t d}{\sigma_k(\beta, \Delta)} \left(\frac{h\sigma_b(\beta, \Delta)}{(\text{DQE})(D/e)} \right)^{1/2} \exp\left(\frac{t}{2\lambda_e} \right) \qquad (5.26)$$

For a given radiation dose D, a *small* probe diameter d is required to detect the minimum number of atoms, which favors the use of a field-emission source.

Figure 5.26 shows the detection limits for calcium (L-loss signal) within a 50-nm-thick carbon matrix, calculated for 100-keV incident electrons and a parallel-recording spectrometer. Single calcium atoms may be detectable with a subnanometer probe, but would involve a high radiation dose; even in the absence of radiolytic processes, $D \approx 10^6$ C/cm^2 is sufficient to remove six layers of carbon atoms by sputtering (Leapman and Andrews, 1992). For larger probe sizes (or larger scanned areas in STEM), the detection of Ca/C ratios down to a few parts per million is predicted, in agreement with measured error limits of about 0.75 mm/kg or 9 ppm (Shuman and Somlyo, 1987; Leapman *et al.*, 1993b). In fact, these experimental estimates were based on second-difference recording (requiring the acquisition of three spectra) and MLS fitting over a narrow energy range (making h close to unity), which introduces a correction factor $(3\sqrt{h})\exp(-t/\lambda_i)$ in Eq. (5.25) and Eq. (5.26). However, this factor is not far from unity.

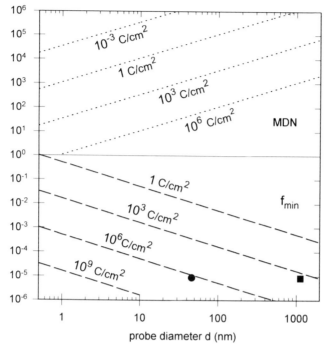

Figure 5.26. Minimum number of calcium atoms (dotted lines) and minimum fraction of calcium (dashed lines) detectable in a 30-nm carbon matrix ($N_t = 2.7 \times 10^{17}$ cm^{-2}), calculated using Eq. (5.26) and Eq. (5.25) for several dose levels of 100-keV incident electrons. The calculations assume DQE = 0.5, $h = 9$, $\lambda_e = 200$ nm, $\beta = 10$ mrad, $\Delta = 50$ eV, and hydrogenic cross sections: $\sigma_L(\beta, \Delta) = 9.9 \times 10^{-21}$ cm^2 and $\sigma_b(\beta, \Delta) = 1.9 \times 10^{-21}$cm^2. The circular data point is from Leapman *et al.* (1993b) and the square point is from Shuman and Somlyo (1987); both points represent experimental data. For scanned-probe analysis, the effective probe diameter is $(4A/\pi)^{1/2}$, A being the scanned area.

Although calcium represents a favorable case, the detection limits for phosphorus are comparable. From energy-selected CTEM images, Bazett–Jones and Ottensmeyer (1981) reported a phosphorus signal/noise ratio of 29 from a nucleosome containing 140 base pairs of DNA (280 phosphorus atoms), equivalent to the detection of 29 atoms at SNR = 3. Using a STEM and parallel-recording spectrometer, Krivanek *et al.* (1991) measured the O_{45} signal from clusters of thorium atoms on a thin carbon film; quantification revealed that the signal originated from only a few atoms. As seen from Eq. (5.25) and Eq. (5.26), f_{min} and MDM will be lowest for elements with a high core-loss cross section $\sigma_k(\beta, \Delta)$—in other words, low-energy edges and edges which are sharply peaked at the threshold. However, edges below 100-eV loss may be less favourable since $\sigma_b(\beta, \Delta)$ becomes large (due to valence-electron excitation) and the background is further increased by plural scattering.

Comparison with EDX Spectroscopy

The characteristic x-ray signal (in terms of the number of detected photons) recorded by an EDX detector, using the same probe current I and the same acquisition time T, is

$$I_x = N(I/e)T\omega_k\sigma_k(\pi, E_0)\eta_x \qquad (5.27)$$

where ω_k is the fluorescence yield and $\sigma_k(\pi, E_0)$ the *total* ionization cross section for shell k; η_x is the collection efficiency of the x-ray detector, including photon absorption in the front contact and dead layers, the detector window, and the specimen. Using Eq. (5.20) and Eq. (5.27), we can compare the EELS and x-ray signals acquired under identical conditions:

$$\frac{I_k}{I_x} = \frac{\sigma_k(\beta, \Delta)}{\omega\sigma_k(\pi, E_0)} \frac{1}{\eta_x} \exp\left(\frac{-t}{\lambda_e}\right) \qquad (5.28)$$

The exponential term (typically 0.3) represents loss of EELS signal as a result of elastic scattering outside the collection aperture. Core-loss intensity is also reduced by the aperture and because only the fraction which lies within an energy range Δ of the ionization threshold is utilized. As a result, the cross-section ratio in Eq. (5.28) is appreciably less than unity; 0.1 would be a typical value. However, the x-ray fluorescence yield, which is close to unity for K lines of heavy elements, is below 0.05 for photon energies below 2000 eV ($\omega_k \approx 0.002$ for carbon-K radiation). In addition, η_k is normally below 0.01 because of the low collection solid angle for x-rays, and can be considerably less for low-energy photons because of absorption in the specimen and at the detector front-end. As a result, the EELS signal usually exceeds the EDX signal, by a modest factor for heavy elements but by a factor of several thousand for a light element such as carbon.

The major advantage of EDX spectroscopy is that the background to characteristic peaks is relatively low. Moreover, the background can be subtracted by interpolation rather than extrapolation, equivalent to $h = 2$ in Eq. (4.61) so that the signal/noise ratio is given by

$$(\text{SNR})_x = I_x/(I_x + 2I_b)^{1/2} \qquad (5.29)$$

For low elemental concentrations, I_b cannot be neglected relative to I_x and a model for the background intensity I_b is required in order to calculate detection limits (Joy and Maher, 1977). As an alternative, Leapman and

Hunt (1991) obtained the SNR and $(SNR)_x$ from χ^2 values produced by MLS fitting to energy-loss and x-ray spectra recorded simultaneously from test specimens (10-nm carbon films containing small concentrations of F, Na, P, Cl, Ca, and Fe). The ratio $SNR/(SNR)_x$ is plotted in Fig. 1.13 and shows that EELS is capable of higher sensitivity for the low-Z elements and for medium-Z elements which have L_{23} edges in the 30–700 eV range.

5.6. Structural Information from EELS

As discussed in Chapters 3 and 4, inelastic scattering in a solid is sensitive to the crystallographic and electronic structure of a specimen as well as to its elemental composition. This structural information is present in the dependence of the inelastic intensity on specimen orientation, in the angular dependence of scattering, and in the form of fine structure in the low-loss or core-loss regions of the energy-loss spectrum.

5.6.1. Orientation Dependence of Ionization Edges

In a crystalline specimen, the electron wavefunction can be written as a sum of Bloch waves, and the probability of inner-shell excitation is proportional to the square of the modulus of this sum (Cherns *et al.*, 1973). The total intensity distribution must possess the periodicity of the lattice and so has nodes and antinodes whose position within each unit cell is a sensitive function of crystal orientation (Section 3.1.4). At high energy loss, where electron scattering from inner shells is localized near the center of each atom (Section 5.5.3), the core-loss intensity will therefore change as the specimen is tilted about an axis perpendicular to the incident beam. Known (from x-ray work) as the Borrmann effect, this represents a potential source of error for EELS elemental analysis in crystalline materials. However, it can also be used constructively to determine the crystallographic site of a particular element, selected by means of its ionization energy.

The orientation of the specimen relative to the incident beam determines the *channeling* condition, which affects the rate of inner-shell ionization and the rate of x-ray production, as discussed in Section 3.1.4. In the ALCHEMI method of atomic site determination, based on *planar channeling* (Spence and Taftø, 1983), the orientation dependence of x-ray emission is measured for two elements which lie on alternate planes (parallel to the incident-beam direction). The crystallographic site of a third element can then be determined by comparing its x-ray orientation dependence with that of the other two. An alternative *axial-channeling* method involves measuring characteristic x-ray signals from a matrix element and an impurity (whose atomic site is unknown) with the incident beam traveling along a low-index zone axis and in a random orientation; the ratio of the fractional changes in signal provides the fraction F of impurity atoms which lie on particular atomic columns of the matrix (Pennycook, 1988). Roussouw *et al.* (1989) have described the application of multivariate statistical procedures to the analysis of spectra recorded under several zone-axis diffraction conditions, in order to reduce the influence of experimental errors.

In the EELS case, primary electrons which have caused inner-shell excitation must traverse the remainder of the specimen before being detected. In doing so, they are again subject to channeling, which affects their probability of escape in a particular direction (an effect known as *blocking*). If this direction is defined by a collection aperture centered about the incident-beam direction, the blocking effect of the crystal on the inelastically scattered electrons *augments* the channeling of the incident beam (equivalent to *double alignment* in particle-channeling experiments) and the orientation dependence observed by EELS can be larger than that seen in x-ray emission spectroscopy. If R is the factor by which an elemental ratio (ratio of the core-loss signals due to two different elements present at nonequivalent crystallographic sites) changes with orientation, one might expect (Taftø and Krivanek, 1982a)

$$R_{\text{EELS}} = (R_x)^2 \qquad (5.30)$$

However, factors related to the localization of inelastic scattering reduce R_{EELS} relative to $(R_x)^2$. Characteristic x-rays may be produced by any energy loss above the ionization threshold E_k; if the core-loss intensity is proportional to E^{-s}, the mean energy loss is

$$\langle E \rangle = \int E^{1-s}\,dE \bigg/ \int E^{-s}\,dE \approx E_k(s-1)/(s-2) \qquad (5.31)$$

For scattering through all angles, s is close to 3 for most energy losses (see Fig. 3.40), giving $\langle E \rangle \approx 2E_k$. On the other hand, the core-loss signal is usually measured at energy losses just above the edge threshold, corresponding to an average loss close to E_k. As seen from Fig. 5.23, this lower mean energy loss implies less localized scattering and therefore a reduced Borrmann effect (Bourdillon *et al.*, 1981b).

In fact, poor localization of the inelastic scattering will reduce both R_{EELS} and R_x in the case of light elements (such as oxygen) with low threshold energies. Pennycook (1988) observed that the fractional change in x-ray signal between axial and random orientations is reduced for photon energies below 5000 eV (by a factor of 0.6 at 1300 eV). Calculations of Spence *et al.* (1988) found the reduction to be less for planar channeling; in fact, Qian *et al.* (1992) observed orientation dependence of the oxygen K-signal by both EELS and EDX spectroscopy, although they emphasize that for quantitative analysis a correction factor to allow for delocalization would be necessary. Rossouw *et al.* (1989) have argued that their statistical procedure makes approximate allowance for the degree of localization.

Fortunately, the orientation dependence in EELS can be increased (at the expense of reduced signal) by collecting electrons deflected through larger angles, where the scattering is more localized. For planar channeling (incident beam far from a major zone axis), the chosen scattering angle can be made arbitrarily large, without affecting the blocking (or channeling) conditions, by displacing the collection aperture in a direction parallel to the appropriate Kikuchi band (Taftø and Krivanek, 1981). It is convenient to use beam-deflector coils to translate the diffraction pattern (visible by lowering the TEM screen) in an appropriate direction relative to a fixed spectrometer-entrance aperture, making use of the observed Kikuchi bands (Taftø and Krivanek, 1982a). Localization is also improved by using lower incident energies (see Fig. 5.23), but thinner specimens may be necessary in order to maintain edge visibility.

An example of these orientation effects is shown in Fig. 5.27. Spectra (a) and (b) were recorded with the collection aperture centered about the zero-order diffraction spot; the ratio of the Mg and Al K-edge intensities varies by only a factor of 1.8 when the specimen orientation is changed so that the Kikuchi line at the edge of the (400) band crosses the collection aperture. In cases (c) and (d), the diffraction pattern has been displaced parallel to the (400) band in order to increase the localization of the inelastic scattering which enters the aperture; as a result, the Al/Mg ratio changes by a factor of 9 as the specimen is tilted through the (400) Bragg condition. If the illumination and detector apertures are placed on *opposite* sides of a Kikuchi line, the channeling and blocking effects largely cancel (Taftø and Krivanek, 1981).

To determine the atomic site of a particular element from planar channeling, the specimen orientation must be carefully chosen. In the case of minerals with the a spinel structure (general formula AB_2O_4), the incident beam should be nearly parallel to (800) planes but away from a principal zone axis. The standing-wave field is then determined mainly by those (800) planes which contain all of the oxygen atoms and two-thirds of the metal atoms on octahedral sites (Taftø *et al.*, 1982). The remaining metal atoms occur in tetrahedral coordination on the intervening (800) planes and are strongly ionized if the incident-beam direction lies just *outside* the (400) Kikuchi band (Fig. 5.27b). Conversely, the octahedral (and oxygen) atoms are strongly excited if the incident beam lies just *inside* the (400) band (Fig. 5.23a). In a *normal* spinel, A atoms are on tetrahedral sites and B atoms

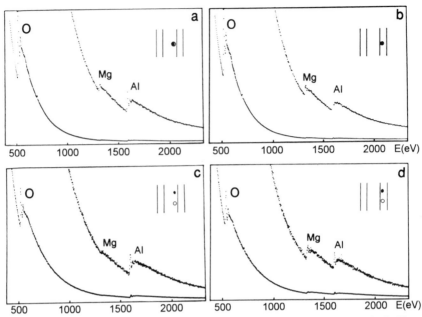

Figure 5.27. Core-loss spectra measured from a normal spinel ($MgAl_2O_4$) with the incident beam nearly parallel to (100) planes. Insets show the location of the incident beam (solid circle) and collection aperture (open circle) relative to the (400) and (800) Kikuchi bands. In (a) and (b), the collection aperture is centered about the illumination axis; the Mg/Al intensity ratio increases by a factor of 1.8 as the crystal is tilted so that the aperture is outside rather than inside the (400) band. In (c) and (d), the collection aperture has been displaced by 10 mrad parallel to the (400) band to increase the localization of core-loss scattering; the Mg/Al ratio increases by a factor of 9 between the two crystal orientations. (From Taftø and Krivanek, 1982a).

on octahedral sites; in an *inverse* spinel the tetrahedral sites are filled by half of the B atoms, the octahedral sites accommodating the A and remaining B atoms. Provided components A and B give rise to clearly observable ionization edges, the orientation dependence of the latter can be used to determine whether a given spinel has the normal or inverse structure, even when both A and B contain a mixture of two different elements (Taftø *et al.*, 1982).

In *mixed-valency* compounds, one of the elements (such as a transition metal) is present as differently charged ions whose ionization edges may be distinguishable as a result of a chemical shift (Section 3.7.4). For example, Fe^{2+} and Fe^{3+} ions in chromite spinel give rise to sharp white-line threshold peaks shifted by about 2 eV. Taftø and Krivanek (1982b) utilized this chemical shift, together with the orientation dependence of the ionization edges, to show that in their sample of chromite spinel all of the Fe^{2+} atoms were on tetrahedral sites and all of the Fe^{3+} atoms on octahedral sites.

Channeling experiments require specimens containing single-crystal regions as large as the incident-beam diameter, usually several nm because the convergence angle of smaller probes will reduce the orientation effect. The specimen thickness should be at least equal to the extinction distance ξ_g in order to provide a pronounced variation in current density within each unit cell; otherwise the orientation dependence will be weak and the method prone to statistical error. The extinction distance is proportional to the incident-electron velocity but also depends on crystal orientation and atomic number (Hirsch *et al.*, 1977; Reimer, 1993). For 100-keV electrons and a strong-channeling direction, $\xi_g \approx 50$ nm for carbon, decreasing to 20 nm for gold. A parallel-recording spectrometer is beneficial in reducing the recording time and/or the noise level of the core-loss data. Except where chemically shifted peaks must be resolved, an energy resolution of a few eV is sufficient and a field-emission source is unnecessary.

5.6.2. Core-Loss Diffraction Patterns

In the last section, we discussed the variation of core-loss intensity as the specimen is tilted, keeping the collection aperture at a fixed angle. We next consider the variation in intensity with scattering angle, for a fixed sample orientation. An amorphous specimen would have an axially symmetric distribution of scattering; at low energy loss the intensity is peaked about the unscattered direction, while for energy losses far above an ionization threshold it takes the form of a diffuse ring, representing a section through the Bethe ridge (Fig. 3.36). In the case of a crystalline specimen, elastic scattering results in a diffraction pattern containing of Bragg spots (or rings, for a polycrystal) and Kikuchi lines or bands. Energy-filtered diffraction patterns can be recorded using a scanning transmission electron microscope, by rocking the incident beam in angle or by using postspecimen deflection (Grigson) coils to scan the pattern across a small-aperture detector, but a more efficient procedure (requiring much shorter recording time and electron dose to the specimen) is to use a conventional TEM fitted with an imaging filter (Section 2.6).

At low energy loss, the diffraction pattern resembles a zero-loss pattern but with the diffraction spots broadened somewhat by the angular width of inelastic scattering. This regime corresponds to the median angle $\tilde{\theta}$ of inelastic scattering being less than the angular separation between Bragg beams, approximately the lowest-order Bragg-reflection angle θ_B. Taking the localization length as $L = 0.6\lambda/\tilde{\theta}$ (Section 5.5.3) and using the Bragg equation: $\lambda = 2d\theta_B$, this condition is equivalent to $L > d$ or (since the interplanar spacing d is comparable to the lattice constant) localization of the inelastic scattering exceeding the unit-cell dimensions. Under these conditions, inelastic scattering does not greatly change the angular distribution of elastic scattering and diffraction contrast is preserved in energy-selected images of defects such as stacking faults and dislocations (Craven *et al.*, 1978).

At higher energy loss, the inelastic scattering becomes highly localized and the mean inelastic-scattering angle exceeds the angular separation the diffracted beams. The Bragg spots therefore disappear from the energy-filtered diffraction pattern, which starts to resemble the Kossel pattern from an iostropic source of electrons inside the crystal or the Kikuchi pattern obtained from a thick specimen; see Fig. 5.28.

By subtracting the diffraction intensities recorded just above and below an ionization edge of a chosen element, a core-loss diffraction pattern can be produced. Spence (1980b, 1981) has carried out dynamical calculations to find out under what conditions this pattern would have the symmetry of the local coordination of a chosen element, rather than the symmetry of the entire crystal. The necessary conditions are that the localization of the core-loss scattering and the inelastic spread $2t\bar{\theta}$ within the specimen thickness t must both be less than unit-cell dimensions. For $E_0 = 100$ keV and an inorganic crystal with small unit cell (0.6 nm), the first condition

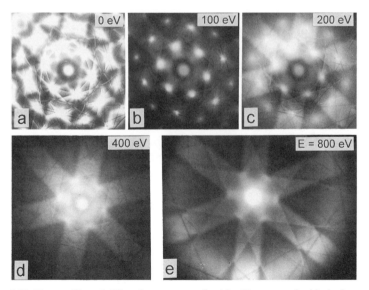

Figure 5.28. Energy-filtered diffraction patterns of a thin silicon crystal with the beam close to a $\langle 111 \rangle$ axis. (a) Zero-loss pattern, overexposed to show the Kikuchi lines resulting from phonon scattering; (b), (c), (d), and (e) inelastic patterns recorded at multiples of 100 eV, showing the broadening and diminution of diffraction spots and the development of Kikuchi bands and the emergence of a diffuse ring representing the Bethe ridge.

requires a core loss E_k above 200 eV; the second implies $t < 50$ nm for $E_k = 200$ eV or $t < 15$ nm for $E_k = 1000$ eV. The requirements are relaxed for larger unit cells. Core-loss diffraction could therefore provide an alternative to channeling and ELNES techniques for determining the atomic site of light atoms in a crystal.

From measurements on $LaAlO_3$, Midgley et al. (1995) concluded that examination of HOLZ intensities in a core-loss CBED pattern might be used to determine which atomic species (or which sublattice in a complex crystal) contributes most to a particular Bloch state, information which would contribute to the solution of unknown crystal structures.

5.6.3. Coordination from ELNES Fingerprinting

In compounds containing coordinate bonding, such as minerals and organic complexes, it is sometimes desirable to know the coordination number and the symmetry of the nearest-neighbor ligands surrounding a metal ion. As discussed in Section 3.8.1, the energy-loss near-edge structure (ELNES) of an ionization edge is an approximate representation of a *local* densities of states at the atom giving rise to the edge. This interpretation is consistent with multiple-scattering (XANES) calculations of the backscattering of the ejected electron (Section 3.8.4), which show that in most cases the scattering is quite localized, involving just a few near-neighbor shells. The localized nature of the near-edge signal is particularly obvious for amorphous materials and molecular structures, where long-range order may be absent, and for crystals with large unit cell, such as complex minerals and molecular crystals.

Sometimes the scattering of the ejected core electron is mostly from a single shell of atoms, where these are strongly scattering species such as the O^{2-} or F^- ions. In effect, the ions form a cage or potential barrier which impedes the escape of the inner-shell electron (Bianconi et al., 1982). The near-edge fine structure then serves as a coordination fingerprint and can yield useful information when applied to mineral specimens (Taftø and Zhu, 1982). In the case of polymers and macromolecules, the XANES structure can provide a fingerprint of the functional groups which act as building blocks for the entire structure (Stohr and Outka, 1987). The fact that the scattering is localized allows molecular orbital calculations to be used as a basis for interpreting and labeling the peaks (Sauer et al., 1993).

ELNES where the excited atom is in planar trigonal coordination (three nearest neighbors lying in a plane) is shown in Fig. 5.29. The carbon K-edge of a carbonate group is characterized by a narrow π^* peak followed (at 10–11 eV separation) by a broader σ^* peak, quite different from the fine structure of carbon in any of its elemental forms. Previous measurements by

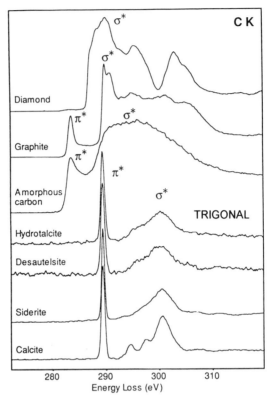

Figure 5.29. Carbon K-edges of minerals containing the carbonate anion, an example of planar trigonal bonding, compared with the K-edges from elemental carbon. (From L. A. Garvie, A. J. Craven, and R. Brydson, Use of electron-energy loss near-edge fine structure in the study of minerals, *American Mineralogist* **79**, 411–425, © 1984 by the Mineralogical Society of America, by permission.) The spectra were recorded with an energy resolution of 0.5 eV and deconvolved to remove plural scattering and peak distortion arising from the asymmetrical energy distribution of the field-emission source.

Hofer and Golub (1987) yielded similar results and the same kind of K-edge fine structure is observed for prismatic (nonplanar) coordination (Brydson, 1991) and for trigonally coordinated boron in the mineral vonsenite (Rowley *et al.*, 1990). Therefore it appears that trigonal coordination frequently gives rise to a fine structure which can be characterized as N10W, where N denotes a narrow peak, W a wide peak, and the number in between is an approximate value of the peak separation in eV.

Situations involving tetrahedral coordination are shown in Fig. 5.30. The Si L-edge of the SiO_4 tetrahedron shows two sharp peaks (separated by ≈ 7 eV) followed by a third prominent broad peak separated about 22

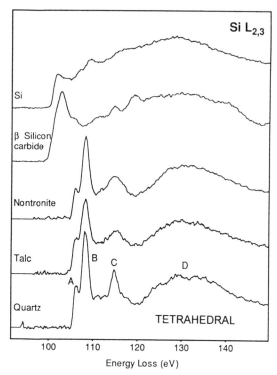

Figure 5.30. Silicon L_{23} edges from three silicates containing SiO_4 tetrahedra, compared with the L_{23} edges of SiC and Si. (From L. A. Garvie, A. J. Craven, and R. Brydson, Use of electron-energy loss near-edge fine structure in the study of minerals, *American Mineralogist* **79**, 411–425, © 1984 by the Mineralogical Society of America, by permission.) The spectra, recorded at 0.5-eV resolution, have been corrected for plural scattering and peak distortion introduced by the field-emission source.

eV from the first peak. When observed at high energy resolution (0.5 eV), the first peak is seen to contain a smaller peak about 1.9 eV lower in energy. A similar peak structure was observed for the chlorate, sulfate, and phosphate anions (Hofer and Golub, 1987; Brydson, 1991) with some differences of detail. Therefore it appears that the structure of the major peaks can be characterized as N7N22W, in the notation of the previous paragraph.

Examples of octahedral coordination are shown in Fig. 5.31. A sharp peak (a) is followed by two broader peaks (b) and (c), displaced by about 7 eV and 19 eV. Fairly similar structures are observed at the K-edges of MgO, where both Mg and O atoms are octahedrally coordinated (Colliex *et al.*, 1985; Lindner *et al.*, 1986) and can be summarized as N7W19W.

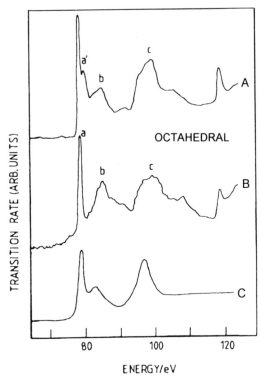

Figure 5.31. Aluminum L_{23} edge in (A) chrysoberyl and (B) rhodizite, together with (C) ICXANES calculations for aluminum octahedrally coordinated to a single shell of oxygen atoms (Brydson *et al.*, 1989).

Several factors complicate this simple concept of a coordination finger-prints. If the symmetry of the coordination is distorted, the near-edge peaks are broadened or split into components (Buffat and Tuilier, 1987; Brydson *et al.*, 1992b). For example, the Si L_{23}-edge of zircon shows five peaks in place of the first two which characterize SiO_4 tetrahedra (McComb *et al.*, 1992b). Atoms outside the first coordination shell may contribute, which probably explains the two additional peaks seen in the calcite carbon-K ELNES (Fig. 5.29). Core-hole effects and crystal-field splitting cause the L_3 and L_2 edges of transition metals to appear as two components when observed at high energy resolution (Krivanek and Paterson, 1990; Krishnan, 1990); careful modeling of these effects should yield detailed information relating to crystal-field strength (Garvie *et al.*, 1994).

In addition, several factors can cause *apparent* differences in structure measured from different specimens or in different laboratories. If a thermi-onic electron source (energy width typically 1–2 eV) is used, the energy resolution may be insufficient to resolve peak splittings such as those visible in Fig. 5.30. Early ELNES data were recorded by serial EELS, with relatively long recording times and increased possibility of degraded resolution due to high-voltage drift and of specimen contamination or radiation damage. If recorded with high energy resolution, peak shapes may be distorted because of the asymmetry of the emission profile, as seen in the zero-loss peak. If the specimen thickness is greater than a few hundred nm, the overall shape of the ionization edge will be modified by plural scattering (Section 3.7.3), increasing the heights of peaks which occur 15 eV or more from the threshold. This distortion can be removed by Fourier deconvolution, which can also correct for the asymmetry of the emission profile; see Appendix B.

It is therefore important when recording near-edge fine structures to aim for high energy resolution (using a field-emission source and parallel recording if possible) and to minimize plural scattering by using thin specimen areas and/or deconvolution techniques so that a fair comparison can be made between spectra and with reference materials. A small collection angle (<15 mrad for 100 keV) simplifies interpretation by ensuring that nondipole transitions are excluded, although these transitions might be used creatively to explore the wavefunction symmetry (Auerhammer and Rez, 1989). A single crystal is not necessary; in fact, the fine structure may be more reproducible if recorded from a polycrystalline area containing several grains, so as to suppress orientation effects (Brydson et al., 1992). In principle, these effects yield additional information about the directionality of bonding (Leapman et al., 1983) but require careful control over the specimen orientation.

Because the initial-state wavefunction is more localized, higher-energy edges may offer more characteristic fingerprints, favoring K-edges rather than the low-energy N_{23} or O_{23} edges whose plasmon-like shape is practically independent of the atomic environment (Colliex et al., 1985). But the $1s$ states for $Z > 14$ and $2p$ states for $Z > 28$ have natural widths which exceed 0.5 eV (Fig. 3.50), so K-edges above 2000 eV and L_{23} edges above 1000 eV may show less fine structure.

ELNES fingerprinting has recently been applied to several materials science problems. By comparing the Al K-edges of ion-thinned specimens of blast-furnace slag cement with those recorded from minerals (orthoclase, pyrope, hydrotalcite) and with multiple-scattering calculations, Brydson et al. (1993) concluded that two phases were present: one with Al atoms substituted for Si at tetrahedral sites, the other (Mg-rich) phase containing Al at octahedral sites. Bruley et al. (1994) recorded Al L_{23} spectra from a diffusion-bonded niobium/sapphire interface (Fig. 5.32). Agreement was obtained with MS calculations, assuming an interfacial monolayer of aluminum atoms tetrahedrally bonded to three oxygen and one Nb atom with

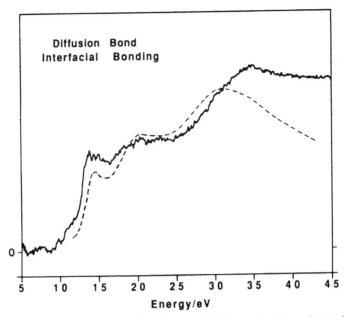

Figure 5.32. Aluminum L_{23} edge recorded using spatial-difference techniques from a diffusion-bonded Nb/Al_2O_3 interface. The dashed curve shows a multiple-scattering simulation of tetrahedral $Al(O_3Nb)$ with two thirds of the Al–Nb bonds shortened to 0.134 nm, as suggested by the Knauss–Mader model (Bruley *et al.*, 1994).

Al–Nb bonds providing the cohesive energy. Interfaces produced by molecular beam epitaxy showed no modification of the Al–L ELNES at the interface, suggesting that the sapphire is terminated by oxygen and that charge transfer from Nb provides the cohesive force. Similar spatially resolved ELNES studies were used to establish the symmetry and coordination number of Al atoms at a 35.2° grain boundary in sapphire (Bruley, 1993).

Another example of the use of ELNES is the work of Rowley *et al.* (1991) on the oxidation of Fe/Cr alloys in superheated steam. The presence of boron increases the oxidation resistance because a thin microcrystalline film of composition $(M_xB_{1-x})_2O_3$, where M = Fe,Cr, or Mn, forms a diffusion barrier. The boron K-edge (Fig. 5.33) exhibits a sharp peak at 194 eV and a broad peak containing two components centered on 199 and 203 eV, which are believed to represent BO_3 and BO_4 borate groupings (from comparison with the K-edges of vonsenite and rhodizite). The proportions of these two components varied as the electron beam was moved across the sample, suggesting that the film consisted of separate phases of MBO_3 and M_3BO_6 (norbergite structure, containing only BO_4 groups) in the ratio

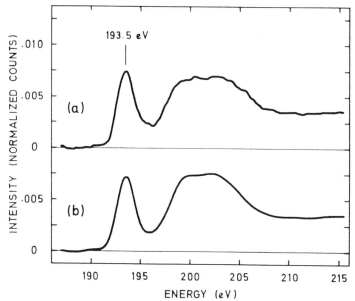

Figure 5.33. (a) Boron-K ELNES of a boron-doped Fe/Cr oxide layer formed on stainless steel in superheated steam. (b) Composite spectrum formed by adding boron K-edges recorded from colemanite (BO_3 fingerprint) and rhodizite (containing BO_4) in the ratio 7:3 (Sauer *et al.*, 1993).

3:1 (Sauer *et al.*, 1993). In the case of heavy boron doping, a small prepeak preceding the 194-eV maximum was taken to indicate the presence of a boride, whose π^* peak is shifted down in energy because of the lower effective positive charge on the boron atom.

Using thin films of silicon alloys, Auchterlonie *et al.* (1989) showed that Si atom nearest neighbors (B, C, N, O, or P) can be distinguished using the fine structure of the Si L_{23} edge, as well as from the energy (chemical shift) of the first ELNES peak. The bonding type in various forms of carbon and carbon alloys can be measured from the π^*/σ^* ratio at the carbon K-edge, as discussed in Section 5.7.1. As discussed in Chapter 3, ELNES fine structure can also serve as a guide to the densities of unoccupied states in a solid and can therefore be used as a check on band-structure calculations.

Bianconi *et al.* (1983a) have predicted that the energy of the broad shape-resonance peak (due to transitions to σ^* states) is proportional to $1/R^2$, where R is the nearest-neighbor distance and the proportionality constant depends on the type of nearest neighbors. This simple rule could be useful (as in XAS) for measuring changes in bond length.

5.6.4. Determination of Valency from White-Line Ratios

As discussed in Section 3.7.1, the L_{23} edges of transition metals and M_{45} edges of rare earths are characterized by two white-line peaks close to the ionization threshold. The energy separation of these peaks reflects the spin-orbit splitting of the initial states in the transition, but their relative intensity need not correspond to the numbers of electrons in the initial states, as would be expected if the matrix element and final densities of states were the same for all electrons. Influence of a spin selection rule results in a white-line intensity ratio R_w which depends on the number of electrons in the final ($3d$ or $4f$) state, and therefore varies with atomic number and (in some cases) oxidation state. White-line measurements are therefore a potential source of information about oxidation state.

Values of R_w are shown in Fig. 5.34. Whereas for the lanthanides this ratio is a monotonic function of final-state occupancy, values for transition metals appear to decrease at higher occupancy, although the experimental results show considerable scatter. Part of this variability may be due to different methods of processing the spectra. For example, Manoubi *et al.* (1990) fitted each white line with a Lorentzian function and the continuum background with an arctangent function but did not make explicit allow-

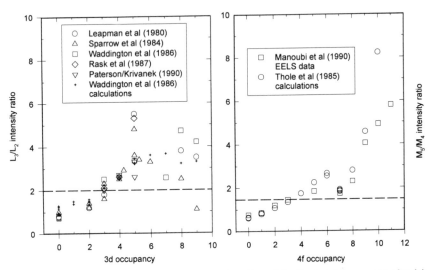

Figure 5.34. White-line intensity ratio R_w as a function of the final-state occupancy for (a) elements and compounds of transition metals (elements Ca to Cu) and (b) lanthanide rare-earth metals and oxides (elements La–Lu). EELS measurements are denoted by open symbols, calculations by crosses. The dashed horizontal lines show the so-called statistical ratio, based on the number of inner-shell electrons (Appendix D).

ance for plural scattering. Sparrow *et al.* (1984) used Fourier-log decon-volution to remove plural scattering but made no allowance for the contin-uum beneath the peaks. Some of the EELS data was obtained with an energy resolution of 4 eV, whereas the EELS atlas data used by Waddington *et al.* (1986) has an energy resolution of 1–2 eV. In addition, Wong (1994) has shown (for the case of nickel and its silicides) that variations in R_w can result from solid-state effects rather than variations in d-state occupancy.

An alternative measurement is the ratio R_c of the total intensity of both white lines in relation to the following continuum (representing transi-tions to delocalized states). One procedure for measuring this ratio is indi-cated in Fig. 5.35; the continuum is integrated over a 50-eV window, starting 50 eV beyond the edge, and linear extrapolation from this window up to the center of each white-line peak allows the continuum background to be subtracted at each peak. With this definition, the white-line/continuum ratio R_c is shown as a function of d-band occupancy n_d in Figs. 5.35b and d, which are based on EELS measurements on metallic films (Pearson *et al.*, 1993). An approximately linear decrease is seen, in accordance with

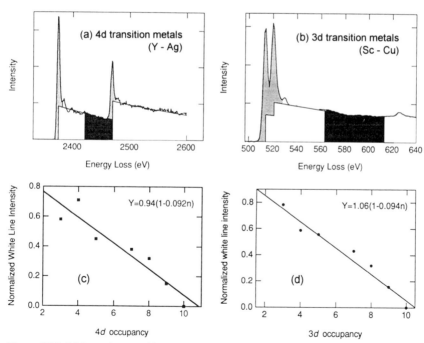

Figure 5.35. (a) Procedure used by Okamoto *et al.* (1992) for measuring white-line/continuum ratio R_c of $4d$ transition metals and (b) its variation with $4d$ occupancy; (c) procedure for measuring R_c of $3d$ transition metals and (d) its variation with $3d$ occupancy.

the fact that the number of unoccupied d-states is $10 - n_d$, where n_d is the d-state occupancy. In fact, the linearity is improved if allowance is made for variations in the transition-matrix element, values of which have been calculated and tabulated (Pearson *et al.*, 1993).

The variation of R_c with n_d has been used to measure charge transfer in disordered and amorphous copper alloys (Pearson *et al.*, 1994). Corrections were made for changes in matrix element and the Cu white-line intensity was measured relative to a normalized L_{23} edge of metallic copper (which has no white lines). The copper atoms were found to lose 0.2 ± 0.06 electrons when alloyed with Ti or Zr, between zero and 0.06 electron when alloyed with Au or Pt, and between zero and 0.09 electron when alloyed with Pt. Electron transfer back to the copper atoms was observed after the alloys became crystalline.

White-line ratios at the iron L_{23} edge have been used to study the conditions for ferromagnetism in amorphous alloys (Morrison *et al.*, 1985). As germanium is added to iron, R_c remains approximately the same, indicating that the d-band occupancy is unaltered. Therefore the gradual loss of ferromagnetism cannot be explained in terms of charge transfer in or out of the $3d$ band. However, R_w does change, indicating an redistribution of electrons between the $d_{5/2}$ and $d_{3/2}$ sub-bands and a change in spin pairing, which may account for the change in magnetic moment. Measurements on a crystalline $Cr_{20}Au_{80}$ alloy showed that the L_3/L_2 white-line ratio increased by a factor of 1.6 compared to pure chromium (see Fig. 5.36), indicating a substantial shift in spin density between $j=5/2$ and $j=3/2$ states which may be the reason for a sevenfold increase in the magnetic moment.

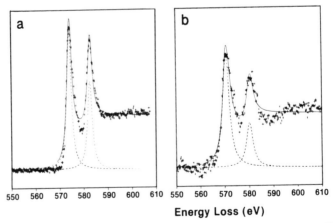

Figure 5.36. L_{23} edge of (a) elemental chromium and (b) chromium in $Cr_{20}Au_{80}$ alloy, showing an increase in L_3/L_2 white-line ratio (Pease *et al.*, 1986). The dashed lines are Lorentzian fits to the white lines.

5.6.5. Use of Chemical Shifts

The shift in threshold energy of an ionization edge can in principle be used to measure charge transfer but, as discussed in Chapter 3, the XAS and EELS chemical shift represents the *net* effect on the initial and final states. Coordination number also has an effect, accounted for in the concept of coordination charge (Brydson *et al.*, 1992b).

A simple example is the shift in energy of the π^* peak from 284 eV in graphite to 288 eV in calcite (Fig. 5.29) as a result of the highly electronegative O atoms surrounding the C atom. Martin *et al.* (1989) observed that the carbon K-edge recorded from a thin film of calcium alkylaryl sulfonate micelles, which consist of a calcium carbonate core surrounded by hydrocarbon molecules, can be represented as a superposition of the K-edges of calcite and graphite. Peaks in the carbon K-edge fine structure of nucleic acid bases were interpreted by Isaacson (1972b) and Johnson (1972) in terms of chemical shifts of the π^* peak due to the different environments of the carbon atoms within each molecule. The shift in peak energy was found to be proportional to the effective charge at each site (Kunzl's law). Auchterlonie *et al.* (1989) showed that for silicon alloys the energy of the first peak at the silicon L-edge was displaced by an amount proportional to the electronegativity of the nearest-neighbor atoms (B, P, C, N, and O); see Fig. 5.37. On the basis of this shift and the near-edge structure, the different alloys could be uniquely identified. Brydson *et al.* (1992) have explained the shape of the oxygen K-edges of the minerals rhodizite, wollastonite, and titanite in terms of the potential at each of the oxygen sites.

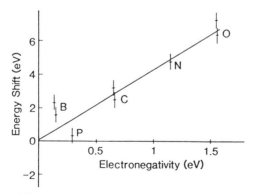

Figure 5.37. Energy shift of the first peak at the silicon L_{23} edge, plotted against the Pauling electronegativity (relative to Si) of ligand atoms (Auchterlonie *et al.*, 1989). From *Ultramicroscopy*, G. J. Auchterlonie, D. R. McKenzie, and D. J. H. Cockayne, Using ELNES with parallel EELS for differentiating between a-Si:X thin films. © 1989, p. 219, Figure 3, with permission from Elsevier Science B.V., Amsterdam, The Netherlands.

5.6.6. Use of Extended Fine Structure

As shown in Section 4.6, a radial distribution function (RDF) specifying interatomic distances relative to a particular element can be derived by Fourier analysis of the extended fine structure (EXELFS), which starts about 50 eV beyond an ionization-edge threshold. This procedure has been tested on various model systems and, after correction for phase shifts, has yielded first- and second-nearest-neighbor distances which agree with x-ray measurements to accuracies between 0.01 nm and 0.001 nm (Johnson *et al.*, 1981b; Kambe *et al.*, 1981; Leapman *et al.*, 1981; Stephens and Brown, 1981; Sevely *et al.*, 1985; Qian *et al.*, 1995).

Higher precision is possible when measuring *changes* in interatomic distance in specimens of similar chemical composition. The RDF of sapphire (α-Al_2O_3) and amorphous (anodized) alumina were compared in Fig. 4.20 and indicate a change in Al–O distance of 0.003 nm. There was no measurable shift in the nearest-neighbor peak after crystallizing the amorphous layer in the electron beam, consistent with the crystallized material being γ- rather than α-alumina (Bourdillon *et al.*, 1984). Aluminum K-edge EXELFS of cross-sectional specimens has been used to investigate the structure of ion-implanted α-Al_2O_3 (Sklad *et al.*, 1992). Implantation at $-185°C$ with 160-keV Fe ions ($4 \times 10^{16} cm^{-2}$) produced a 160-nm-thick amorphous layer which recrystallized epitaxially to α-Al_2O_3 upon annealing in argon at 960°C. Oxygen-edge EXELFS required a restricted energy range because of the presence of an iron L_{23} edge at 708 eV, but yielded consistent results. Implantation with a stoichiometric mixture of Al and O ions gave an amorphous layer which recrystallized into a mixture of γ-Al_2O_3 and epitaxial α-Al_2O_3, as determined from Al K-edge EXELFS; see Fig. 5.38.

Batson and Craven (1979) used a field-emission STEM to acquire K-edge EXELFS from amorphous carbon films, which showed differences in RDF depending on whether the substrate was mica or KCl. These results demonstrated that, even with serial recording, EXELFS can yield structural information from very small areas (below 10 nm diameter). As always, the fundamental limit to spatial resolution is radiation damage, but here the problem is particularly serious because of the need to achieve excellent signal/noise ratio from inner-shell edges.

The situation is improved in the case of lower edge energies; for example, the Si L-edge from SiC was found to be sensitive to the atomic environment up to the sixth coordination shell (Martin and Mansot, 1991). However, low-energy EXELFS are difficult to analyze quantitatively because of the high pre-edge background (not necessarily a power law) and more restricted energy range due to nearby ionization edges. At the other extreme, EXELFS of the titanium K-edge (4966-eV threshold energy) has

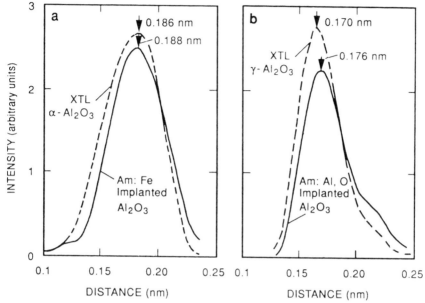

Figure 5.38. Nearest-neighbor peak in the RDF obtained from aluminum K-edge EXELFS of (a) α-Al_2O_3 implanted with 160-keV Fe ions (4×10^{16} cm^{-2}), compared with the crystalline substrate, and (b) a layer implanted with a stoichiometric mixture of Al and O ions, compared with γ-Al_2O_3 produced by annealing for 1 hour in argon (Sklad *et al.*, 1992). Data have been corrected for phase shifts using empirical correction factors.

been measured using 400-keV electrons and parallel recording (Blanche *et al.*, 1993) and yielded results comparable with synchroton-radiation EXAFS. Kaloyeros *et al.* (1988) used boron-edge EXELFS and Ti K-edge EXAFS (from a rotating-anode x-ray tube) in their investigation of the high-temperature stability of amorphous films of titanium diboride, which were made by electron-beam evaporation onto liquid-nitrogen-cooled substrates.

As an example of a biological application, Fig. 5.39 shows RDF's recorded from iron-rich clusters (siderosomes) extracted from lung fluids of a patient suffering from silicosis (Diociaiuti *et al.*, 1995). The RDF obtained from the iron L_{23} edge is similar to that recorded from a hematite standard but the RDF obtained from the oxygen K-edge exhibits a displaced nearest-neighbor peak and a second-nearest-neighbor (O–O) peak reduced in intensity compared to hematite. These discrepancies can be accounted for by presuming that oxygen is also present in a protein coat surrounding the biomineral core, forming short O=C bonds which shift the center of the first peak to lower radius. Diociaiuti and colleagues (1991, 1992a,b) also recorded EXELFS from chromium, copper, and palladium clusters in

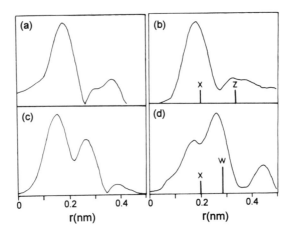

Figure 5.39. Radial distribution functions for (a) iron atoms and (b) oxygen atoms, obtained from EXELFS of alveolar macrophages (siderosomes). The iron and oxygen RDF recorded from hematite are given in (c) and (d), where vertical bars marked X, Z, and W show the expected Fe–O, Fe–Fe, and O–O interatomic distances (Diociaiuti *et al.*, 1995).

order to quantify the increase in nearest-neighbor distance (up to 5%) with decreasing particle size, as well as the effect of oxidation.

EXELFS measurements as a function of temperature have enabled Okamoto *et al.* (1992) to study ordering in undercooled alloys, particularly chemical short-range order which is difficult to measure by techniques such as diffuse x-ray scattering. The temperature dependence of the mean-square relative displacement (MSRD) was used to deduce Einstein and Debye temperatures, good agreement being obtained with force-constant theory and with existing experimental data. Differences in Einstein temperature between ordered and disordered alloys, reflecting differences in vibrational states, are illustrated in Fig. 5.40. Since the EXELFS data can in principle be obtained from small specimen volumes, measurement of a local Debye temperature might characterize the defect density at interfaces or in small precipitates (Disko *et al.*, 1989).

In order to achieve the 0.1% statistical accuracy which is desirable for analyzing EXELFS data, it is necessary to record as many as 10^6 electrons within each resolution element (typically 2 eV). Parallel recording helps to reduce the acquisition time and, therefore, the likelihood of radiation damage and specimen drift. Careful monitoring of relative peak heights

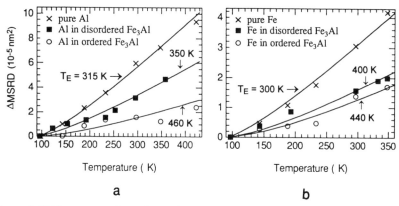

Figure 5.40. Temperature dependence of the nearest-neighbor mean-square relative displacement measured for (a) aluminum and (b) iron atoms in the pure elements and in chemically disordered and ordered Fe_3Al (Okamoto et al., 1992). All values are relative to measurements made at 97 K; solid lines represent the Einstein model, with Einstein temperatures T_E as indicated.

within each readout can alert the operator to damage and changes in specimen thickness within the beam (Qian et al., 1995). High-voltage and spectrometer drift can be corrected by shifting the readouts back into register. Diode-array dark current can be accurately removed by substracting from each spectrum a readout recorded in an equal integration time with the electron beam blocked from the array, and gain variations can be reduced below statistical errors, for example by recording a smoothly decreasing pre-edge region (Section 2.5.5).

Even with parallel recording, EXELFS analysis requires an incident-electron exposure of typically 10^{-4} C. For a 1μm-diameter incident beam, this is equivalent to 1 C/cm^2 dose, enough to destroy the structure of most organic specimens and even some inorganic ones (see Section 5.7.3). Because most damage-producing inelastic collisions involve energy losses outside the EXELFS region, electrons produce more damage than the monochromatic x-rays used in EXAFS studies (Isaacson and Utlaut, 1978). If there exist materials in which structural damage takes place only as a result of inner-shell scattering, this conclusion would not apply (Stern, 1982). Provided radiation damage is not a problem, the electron microscope is competitive with synchrotron sources in the sense that the EXELFS recording time is less than that of EXAFS for ionization edges below 3 keV (Isaacson and Utlaut, 1978; Stern, 1982).

The core-loss intensity is improved by increasing the collection semiangle β, but at the expense of a higher pre-edge background (Section 3.5). If β is too large, nondipole contributions could complicate the EXELFS analysis, but such effects are believed to be small (Leapman et al., 1981; Disko, 1981); see Section 3.8.2. For 100-keV electrons, a semiangle in the range 10–20 mrad should allow dipole theory to be used, while transmitting typically half of the core-loss scattering. This large angle will average out the directional dependence of EXELFS (Section 3.9); to study the directionality of bonding, a much smaller value of β is necessary.

5.6.7. Electron-Compton (ECOSS) Measurements

Electron energy-loss spectra are usually recorded with a collection aperture centered on the optic axis (around the unscattered beam). If this aperture is displaced or the incident beam tilted through a few degrees so that only large-angle scattering is collected, a new spectral feature emerges in the form of a broad peak; see Fig. 5.41. Known as an electron-Compton profile, this peak represents a cross section (at constant scattering vector $\mathbf{q}$) through the Bethe ridge; see Figs. 3.31 and 3.36. It is centered about an energy loss E which satisfies Eq. (3.132), the scattering angle θ_r being determined by the collection-aperture displacement or the tilt of the incident beam. For $\sin \theta_r \ll 1$ and $E \ll E_0$, Eq. (3.132) is equivalent to the relation $E = \theta_r^2 \gamma^2 T$ for Rutherford scattering from a free stationary electron.

The fact that the atomic electrons are not free is shown by the width of the peak, which is a measure of the electron binding energy. Therefore, although the Compton profile contains overlapping contributions from both outer- and inner-shell electrons, the latter give a much broader energy distribution and contribute mainly to the tails of the profile. Conversely, the central region of the peak represents mainly scattering from the bonding (valence) electrons.

The spread of the Compton peak can also be thought of as a Doppler broadening due to the "orbital" velocity or momentum distribution of the atomic electrons, which is closely related to the electron wavefunctions. Quantitative analysis of the data (Williams et al., 1981) is rather similar to the Fourier method of EXELFS analysis; after subtracting the background contribution from the tails of lower-energy processes, the energy scale is converted into one of momentum (or wavevector k) acquired by the atomic electron. The Fourier transform of the resulting profile (Fig. 5.41b) is known as the reciprocal form factor $B(r)$ and is the autocorrelation function of the ground-state atomic wavefunction in a direction specified by the scattering vector $\mathbf{q}$ (i.e., by the azimuthal position of the detector aperture). If the

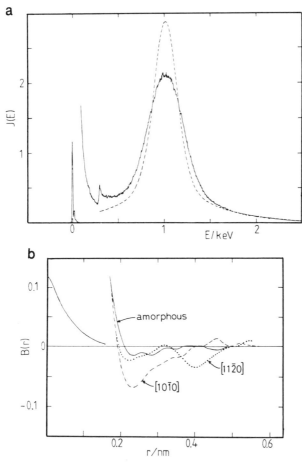

Figure 5.41. (a) Energy-loss spectrum of amorphous carbon, recorded using 120-keV incident electrons and a scattering angle of 100 mrad. Zero-loss and plasmon peaks are visible (because of elastic and multiple scattering), in addition to the carbon K-edge and Compton peak. The dashed curve shows a computed free-atom Compton profile. (b) Reciprocal form factor derived for amorphous carbon (solid line), and for the [10$\bar{1}$0] direction (dashed line) and [11$\bar{2}$0] direction (dotted line) in graphite (Williams and Bourdillon, 1982).

specimen is an insulator or a semiconductor, zero-crossings in $B(r)$ are expected to coincide with the lattice spacings.

The optimum scattering angle for recording the electron-Compton profile from a light element sample (such as graphite) appears to be about 100 mrad for 60–100 keV electrons, resulting in a profile whose center lies at about 1 keV loss. An energy resolution of 10 eV is sufficient, but should be combined with an angular resolution of about 3 mrad (Williams *et al.*,

1984). The signal/background ratio at the Compton peak is maximized by making the sample as thin as possible, showing that the background arises from plural (or multiple) scattering, chiefly large-angle elastic (or phonon) scattering accompanied by one or more small-angle (low-loss) inelastic events. Because of the large width of the Compton peak, accurate subtraction of the background is not easy. To obtain a tolerably low background, the specimen thickness should be less than about 30 nm in the case of 100 keV incident energy and a low-Z element such as carbon. Even thinner samples would be required for higher-Z elements and would result in longer recording times.

Because of the need for both angular and energy discrimination, the collection efficiency of the Compton signal is of the order of 10^{-8} for serial EELS and 10^{-6} for parallel recording. Adequate statistics within the Compton peak involves recording at least 10^6 electrons, requiring an incident exposure $\approx 3 \times 10^{-3}$ C or a dose of 0.4 C/cm^2 for a 10-μm incident beam and parallel recording. Radiation damage and specimen contamination are therefore potential problems, alleviated by cooling the specimen, and iterative gain averaging (Fig. 2.30) has been used to remove nonuniformities of a photodiode-array detector (Jonas and Schattschneider, 1993).

Electron-Compton measurements have been used to show that the bonding in arc-evaporated carbon films is predominantly graphitic (Williams et al., 1983) and has revealed apparent inadequacies in band-structure calculations for graphite (Vasudevan et al., 1984). By using single-crystal silicon specimens with [100] and [111] orientation and placing the collection aperture at selected points in the diffraction pattern, anisotropies in the electron-momentum distribution have been measured and compared with γ-ray experiments and with theory (Jonas et al., 1992; Schattschneider and Exner, 1995).

5.7. Application to Specific Systems

In this final section, we review some interesting applications of EELS in selected areas of materials science and biology. Answers to the appropriate questions have often required low-loss spectroscopy, core-loss spectroscopy, and fine-structure analysis, in addition to TEM imaging, electron or x-ray diffraction and other analytical techniques. The topics chosen are a personal choice; there have been significant applications in other fields such as metallurgy and advanced materials (Okamoto et al., 1992; Zaluzec, 1992), semiconductor devices (Batson, 1992a; Lakner et al., 1992; Batson, 1995; Jäger and Mayer, 1995), catalysts and ceramics (Wang et al., 1987; Bentley, 1992; Brydson et al., 1995).

5.7.1. Carbon-Based Materials

With recent developments in hard coatings and polymers, carbon remains an especially important element. At room temperature, elemental carbon exists in three common forms which can be distinguished from their low-loss and K-edge spectra (Egerton and Whelan, 1974a; Fallon, 1992); see Fig. 1.4.

Diamond

Diamond has the highest demonstrated hardness and thermal conductivity, extremely low electrical conductivity (when pure) and high refractive index (2.4) combined with transparency to visible light. Natural diamond has traditionally been classified into different types on the basis of infrared spectra. Type I diamonds contain an appreciable amount of nitrogen, either segregated (type Ia) or dispersed (type Ib); in the former case, transmission electron microscopy reveals the presence of microscopic (typically 10–100 nm) *platelets* lying on {100} planes.

One model for the structure of these platelets (due to A.R. Lang) calls for the presence of at least two monolayers of nitrogen, which should be detectable by EELS if the platelet is viewed edge-on in a field-emission STEM. Experimental estimates have varied but all of them place the nitrogen content substantially lower than that required by the Lang model. According to recent measurements (Bruley, 1992; Fallon *et al.*, 1995), the nitrogen content can vary between 0.08 and 0.47 monolayer. The nitrogen K-edge appears to be similar in shape to that of carbon, suggesting that the local environment of N and C atoms is similar, as would be the case for nitrogen atoms present as isolated substitutional impurities; see Fig. 5.42a. Additional scattering at an energy loss around 5 eV has been observed at platelets and has been interpreted in terms of localized states arising from partial dislocations (Bursill *et al.*, 1981).

A small rise in intensity occurring about 5 eV before the carbon K-edge (e.g., Fig. 5.29) is found to be more prominent in thin specimens and could be due to excitation to surface states within the band gap of diamond or to π^* levels of a graphitic surface layer which is present on ion-milled or oxygen-thinned specimens. Using spatial-difference EELS, Bruley and Batson (1989) showed that additional intensity appears in the pre-edge region when the electron beam is placed close to a dislocation, and discussed this in terms of excitation to defect or impurity states at the dislocation. Differences were observed between individual dislocations, which may represent differences in dislocation structure. The presence of a monolayer of oxygen at the {111} free surface of diamond has been deduced from the

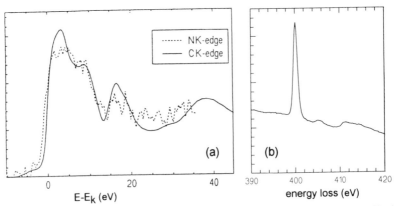

Figure 5.42. (a) Nitrogen K-edge recorded at a platelet, scaled to match the carbon K-edge from nearby diamond. The shapes are basically similar but the N-edge appears to have higher intensity within the range 20 eV to 30 eV from the threshold (Fallon *et al.*, 1995). (b) Nitrogen K-edge recorded from a region of diamond containing a voidite (Luyten *et al.*, 1994).

observation of an oxygen K-edge in reflection-mode energy-loss spectra (Wang and Bentley, 1992).

Some natural diamonds contain octahedral-faceted inclusions, a few nm in size, known as voidites. Bruley and Brown (1989) have shown that some voidites contain nitrogen, a sharp peak at the ionization threshold indicating N_2 rather than NH_3 (previously proposed to explain lattice images). The nitrogen concentration appeared to be independent of voidite size, its average value being equal to about half the carbon concentration in diamond. Despite the high pressure involved, the nitrogen did not appear to be metallic, as evidenced by the lack of additional intensity below 5 eV (where inelastic scattering in diamond is minimal because of the 5.4-eV band gap). No diffraction spots were observed, suggesting that nitrogen is present in an amorphous phase. Luyten *et al.* (1994), however, found moiré fringes and tetragonal-phase diffraction spots at voidites which gave a strong nitrogen signal; see Fig. 5.42b. Such differences might provide information about the geological conditions which have given rise to diamond formation.

A metastable hexagonal form of diamond (lonsdaleite) can be synthesized by shock-wave conversion of graphite. Its plasmon peak occurs at slightly lower energy (32.4 eV) than cubic diamond, possibly due to the presence of lattice defects (Schmid, 1995). Moreover, its plasmon peak is roughly symmetrical, unlike cubic diamond which has a "shoulder" around 23 eV arising from interband transitions.

Very small (0.5–10 nm) crystals of diamond have been found in chondritic meteorites. Their carbon K-edge (Fig. 5.43) shows a prominent feature just below the main absorption threshold, characteristic of transitions to

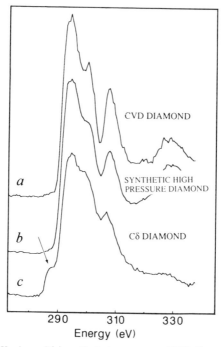

290   310   330
Energy (eV)

Figure 5.43. Carbon K-edges of (a) synthetic low-pressure CVD diamond, (b) synthetic high-pressure diamond, and (c) Cδ component of the Allende CV3 meteorite (Blake *et al.*, 1988). The arrow marks pre-edge structure due to $1s \rightarrow \pi^*$ transitions, characteristic of graphitic bonding.

π^* states in sp^2-bonded carbon and indicating that graphitic or amorphous carbon is present at the surface of each grain. This finding (Blake *et al.*, 1988) gave support to a proposal that the diamond was formed by pressure conversion of graphite during grain–grain collisions in interstellar space.

Thin films of diamond have been grown by chemical vapor deposition (CVD) onto single-crystal silicon substrates, but are usually found to be polycrystalline. From detection and measurement of a π^* peak at 285 eV, Fallon and Brown (1993) concluded that amorphous carbon is present at grain boundaries and at the free surface. From its plasmon energy and low estimated sp^3 fraction, this carbon is believed to be nonhydrogenated; see Fig. 5.44. The lack of epitaxy may result from the presence of a subnanometer layer of amorphous carbon between the substrate and film, visible in STEM images of cross-sectional specimens formed from energy losses just before and just after the diamond threshold (Muller *et al.*, 1993). However, some deposition conditions give areas with oriented growth and absence of an interfacial layer (Tzou *et al.*, 1994).

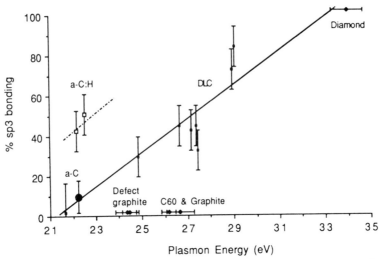

Figure 5.44. Percentage of sp^3 bonding (measured from the C–K $1s$-π^* transition) as a function of plasmon energy (position of the main peak in the low-loss spectrum) for different forms of carbon (Fallon and Brown, 1993). (From *Diamond and Related Materials*, P. J. Fallon and L. M. Brown, Analysis of chemical-vapour-deposited diamond grain boundaries using transmission electron microscopy and parallel electron energy loss spectroscopy in a scanning transmission electron microscope, © 1993, p. 1005, figure 2, with permission from Elsevier Science B.V., Amsterdam, The Netherlands.) The solid line indicates the general trend for nonhydrogenated carbon films. The presence of hydrogen lowers the film density and plasmon energy (dotted line). The large filled circle at the bottom left represents grain-boundary amorphous carbon.

Diamondlike Carbon and Alloys

Transparent films of tetrahedral-amorphous carbon (ta-C, also known as hard carbon or amorphous diamondlike carbon: a-DLC) can be made by laser ablation or by creating a vacuum arc on a graphite cathode, with magnetic filtering to ensure that 20–2000 eV ions are selected from the plasma. They have high hardness and a density about 80% of the diamond value (3.52). Berger *et al.* (1988) determined the type of bonding in these films in terms of the parameter:

$$R = \frac{I_k(\pi^*)}{I_k(\Delta)} \frac{I_l(\Delta)}{I_0} \tag{5.32}$$

where $I_k(\pi^*)/I_k(\Delta)$ is the ratio of K-shell intensity in the pre-edge π^* peak to intensity in the whole edge, integrated over an energy range Δ (at least 50 eV). The factor $I_l(\Delta)/I_0$ (ratio of low-loss and zero-loss intensities) corrects for the fact that $I_k(\pi^*)$ contains only *single* K-shell scattering, whereas $I_K(\Delta)$ includes plural (*K*-shell + plasmon) events; this factor would be omitted for spectra which have been deconvolved to remove plural scattering. The fraction of graphitic bonding was then evaluated as $f = R/R_g$, where R_g is the value of R measured from the spectrum recorded from graphitized carbon (single-crystal graphite is unsuitable because its *K*-edge ratio depends on specimen orientation).

A variation on the above procedure is to measure the intensity within a *narrow* window Δ_1 centered around the π^* peak and over a broader window Δ_2 under the σ^* peak, the atomic fraction X of sp^2 bonding being given by

$$\frac{I_\pi(\Delta_1)}{I_\sigma(\Delta_2)} = k\,\frac{X}{4 - X} \qquad (5.33)$$

where the value of k is again obtained by comparison with graphite ($X=1$). With $\Delta_1 = 2$ eV and $\Delta_2 = 8$ eV, this procedure was found to yield a variability of about 5% and an absolute accuracy of $\pm 13\%$ in the sp^3 fraction when applied to a large number of ta-C films (Bruley *et al.*, 1995).

For ta-C, Eq. (5.32) predicts that 15% of the bonding is sp^2, the remainder being sp^3 (on the assumption that sp^1 bonding is absent). This diamondlike bonding may occur because a graphitic surface layer is subjected to high compressive stress or high local pressure by energetic incident ions. The ta-C films contain no hydrogen, unlike amorphous silicon and germanium films in which hydrogen is required to stabilize the structure. The low-loss spectrum contains a weak peak at 5 eV (characteristic of π electrons) and energy-selected imaging at this energy has shown that ta-C can contain a low density of disk-like inclusions which are mostly sp^2 bonded (Yuan *et al.*, 1992).

By adjusting the ion energy selected by magnetic filtering, the sp^3 fraction and film density of ta-C (measured from the plasmon energy or by RBS) can be varied; plotting sp^3 fraction against plasmon energy, the data lie close to a straight line with bulk diamond as an extrapolation; see Fig. 5.44. Amorphous carbon produced from an arc discharge between pointed carbon rods (the usual method of making TEM support films) also lies on this line, its sp^3 fraction being typically 8%. A variation of Fig. 5.44 is to plot sp^2 fraction against the *inverse square* of the plasmon energy, in which case straight-line behavior is in accordance with the free-electron model (Bruley *et al.*, 1995). Plasmon-loss and π^*-peak measurements have been used to investigate the variation of film density and sp^2 fraction with deposition method (e-beam evaporation, ion sputtering laser ablation) and with temperature and thermal conductivity of the substrate (Cuomo *et al.*, 1991.

By introducing N_2 into the cathodic arc, up to 30% of nitrogen can be incorporated into ta-C, but with gradual loss of sp^3 bonding (and reduction in compressive stress in the film) as the nitrogen content is increased. The fine structures of the nitrogen and carbon K-edges are similar at all compositions, consistent with substitutional replacement of carbon by nitrogen and a similar matrix element and density of states at N and C sites (Davis *et al.*, 1994).

Hard, transparent films of amorphous carbon *containing hydrogen* (a-C:H) are produced by plasma deposition from hydrocarbons. Fink *et al.* (1983) used a Bethe-sum method, based on Eq. (4.32), to measure the sp^2 fraction in these materials as $f = R/R_g$, where

$$R = \frac{n_{\mathrm{eff}}(\pi)}{n_{\mathrm{eff}}(\Delta)} = \frac{\int_0^\delta E \, \mathrm{Im}(-1/\varepsilon) \, dE}{\int_0^\Delta E \, \mathrm{Im}(-1/\varepsilon) \, dE} \tag{5.33}$$

The permitivity function $\varepsilon(E)$ is obtained from Kramers–Kronig analysis (Section 4.2); the energy Δ was taken as 40 eV and δ as the intensity minimum (about 8 eV) between the π-resonance peak (around 6 eV) and the $(\sigma + \pi)$ resonance (around 24 eV). This procedure assumes that inelastic intensity below $E = \delta$ corresponds entirely to excitation of carbon π-electrons (ignoring any contributions from hydrogen), by analogy with graphite where Eq. (5.33) gives the expected value of $R_g = 0.25$, corresponding to one π electron out of a total of four valence electrons per atom (Taft and Philipp, 1965).

For hydrogenated films, Fink *et al.* (1983) obtained $R \approx 0.08$, implying that one third of the bonding is graphitic. Upon annealing at 650°C, the graphitic fraction increased to two thirds and the $(\sigma + \pi)$ plasmon energy decreased by 2 eV. Since annealing removes hydrogen from the films, this decrease in plasmon energy could be partly due to loss of the electrons previously contributed by hydrogen atoms. Upon annealing to 1000°C, the plasmon energy increases, indicating an increase in physical density (Fink, 1989).

Measurement of the energy of the π-plasmon peak as a function of scattering angle gave a dispersion coefficient close to zero, implying that the π electrons undergo single-electron rather than collective excitation. The π-electron states are therefore localized in a-C:H, as in the case of states within the "mobility gap" of amorphous semiconductors. Upon removal of hydrogen by annealing, the π peak becomes dispersive, indicating formation of a band of delocalized states. Fink (1989) interpreted these results in terms of model for a-C:H in which π-bonded clusters are surrounded by a sp^3 matrix.

McKenzie *et al.* (1986) studied the fine structure of the carbon *K*- and Si *L*-edges in a-Si$_{1-x}$C$_x$:H alloys as a function of composition and used the chemical shifts of individual peaks to derive information about compositional and structural disorder. Amorphous carbon/nitrogen alloys (CN$_x$, where $x < 0.8$) have also been studied; the carbon and nitrogen *K*-edges provide a convenient measurement of film composition, while the presence of a strong π^* peak suggests that the material is primarily sp^2 bonded (Chen *et al.*, 1993).

Fullerenes and Nanotubes

In 1985, it was discovered that the soot condensed from carbon vapor contains molecules (known as fullerenes or buckyballs) composed of graphite-like sheets bent into an ellipsoidal or spherical structure. The solid form (fullerite) can be extracted with benzene or by subliming the deposit onto a substrate to create a thin film. The energy-loss spectrum of C$_{60}$ fullerite shows a main ($\sigma + \pi$) plasmon peak at 25.5 eV, a π-resonance peak at 6.4 eV, and several subsidiary peaks (Hansen *et al.*, 1991; Kuzuo *et al.*, 1991). The peaks of other fullerites, such as C$_{70}$, C$_{76}$, and C$_{84}$, are shifted in energy and their fine structure is different (Terauchi *et al.*, 1994; Kuzuo *et al.*, 1994a). The carbon *K*-edge structures also distinguishable and are unlike that of graphite (Fig. 5.45), making EELS a useful tool for analyzing small volumes of these materials.

In 1991, needle-shaped structures (nanotubes or buckytubes, consisting of coaxial cylinders of graphitic sheets) were found in arc-discharge soot. These multishell tubes give π-plasmon peaks at 5.2 or 6.4 eV and a main plasmon peak between 22.0 and 24.5 eV, depending on the tube radius

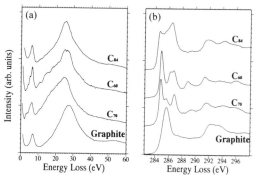

Figure 5.45. (a) Low-loss and (b) *K*-loss spectra of fullerites, compared with graphite (Kuzuo *et al.*, 1994a).

(Kuzuo *et al.*, 1992). Using a catalyst (Fe, Co, or Ni), single-shell tubes can be produced, sometimes in the form of hexagonally packed bundles. The main plasmon peak recorded from such a bundle has a maximum at 20.6 eV, quite close to the value (19.9 eV) given by Eq. (3.41) assuming four electrons per atom and m equal to the free-electron mass (Kuzuo *et al.*, 1994b). Because of the small size of nanotubes, EELS combined with electron microscopy provides a unique tool for examining their electronic structure, which (according to band-structure calculations) may have interesting properties and eventual applications.

5.7.2. Polymers and Biological Specimens

Microtomed thin sections of polymers and biological tissue present problems for TEM imaging and microanalysis because of their radiation sensitivity and the fact that image contrast is often very low. Energy-filtered imaging can be used to increase the contrast or to examine thicker sections, as discussed in Section 5.3. From spectrum-image data, Hunt *et al.* (1995) formed "chemical" maps at 7 eV loss (where double-bonded compounds show a characteristic peak) showing polystyrene-rich regions in unstained sections of polyethylene blend. Ade *et al.* (1992) utilized differences in carbon-K near-edge structure (variations in π^*-peak energies) to image polymer blends and chromosomes in a scanning transmission x-ray microscope (STXM) with 55-nm spatial resolution. The analogous ELNES imaging could be performed in an energy-selecting TEM with greater spatial resolution but with higher radiation dose; see page 377.

Energy-loss spectroscopy of polymers is facilitated by the use of a parallel-recording spectrometer. More *et al.* (1991) used their PEELS system to detect sulfur in a 0.5-μm-square area of polyether sulfone (PES), for which the maximum safe dose (deduced from decay of the 6-eV peak in a time-resolved series of spectra) was estimated to be 0.24 C/cm². Rao *et al.* (1993) were able to detect a 15% increase in carbon concentration in 40-nm-sized regions of ion-implanted polymers, using K-edges together with low-loss spectra (to allow for differences in local thickness).

Biological TEM studies are always strongly dependent on specimen preparation. The ability to prepare ultrathin sections minimizes the unwanted background in energy-loss spectra (Section 3.5) and mass-thickness contributions to core-loss images (Section 2.6.5). For phosphorus L-edge measurements, the optimum specimen thickness has been estimated to be 0.3 times the total-inelastic mean free path (Wang *et al.*, 1992); with 100-keV primary electrons, this would correspond to about 100 nm of dry tissue and 60 nm of hydrated tissue. Rapid freezing techniques reduce the migration or loss of diffusible species, as required for quantitative elemental analysis. Use of a low-temperature stage (drift rate of less than 2 nm/minute if high resolution is required) minimizes the diffusion and loss of elements during microscopy.

Leapman and Ornberg (1988) have evaluated the requirements for elemental analysis of biological specimens. Carbon, nitrogen, and oxygen are the major constituents; their ratio (together with P and S) can be useful for identifying proteins and nucleoties (DNA, ATP, etc.). Using a serial-recording spectrometer and hydrogenic cross sections, ratios of C, N, O, and P (or S) measured in test samples of DNA and insulin agreed within experimental error ($\approx 10\%$) with expected values. Leapman and Ornberg also found that N:O:F ratios measured from fluorohistidine were within 10% of the nominal values, provided the radiation dose was kept below 2 C/cm^2. Fluorine is of potential importance as a label, for example for identifying neurotransmitters in organelles (Section 5.4.4).

Na, K, Mg, Cl, P, and S are typically present at levels between typically 0.03% and 0.6% dry-mass fraction (25–500 millimoles/kg dry weight, equivalent to 5–100 millimoles/kg wet wt., assuming 80% water content). These elements can be analyzed by EDX spectroscopy (Shuman *et al.*, 1976; Fiori *et al.*, 1988), but mass loss and specimen drift limit the spatial resolution to about 50 nm. In the case of EELS, higher sensitivity for S, P, Cl, and Fe is obtainable by choosing *L*-edges, which have higher scattering cross sections. The *L*-edges of sodium and magnesium lie too low in energy while that of potassium overlaps strongly with the carbon *K*-edge, so these three elements are probably better detected by EDX methods (Leapman and Ornberg, 1988).

Calcium is present in high concentration ($\approx 10\%$) in mineralizing bone but otherwise at the millimolar level. At this concentration, a 50-nm-diameter region in a 50-nm-thick specimen contains only about 50 Ca atoms; measuring small changes in concentration therefore requires nearly single-atom sensitivity. Due to the presence of over 10^6 other atoms per Ca atom, this is a more difficult (but more important) problem than the detection of single atoms on a thin carbon support, but one which is being successfully addressed (Shuman and Somlyo, 1987; Leapman *et al.*, 1993b) using parallel-recording spectrometers and MLS processing of the energy-loss spectrum (Section 4.5.4).

Elemental mapping of phosphorus, sulfur, and calcium was pioneered in the energy-filtering CTEM by Ottensmeyer and colleagues and applied towards solving the structure of chromatin nucleosomes and mineralizing cartiliage (Bazett–Jones and Ottensmeyer, 1981; Arsenault and Ottensmeyer, 1983; Ottensmeyer, 1984). The use of very thin specimens minimized plural-scattering and mass-thickness contributions to the image, but for the most part these elemental maps were obtained simply by subtracting a scaled pre-edge image from the postedge image. With a STEM and computer processing, pre-edge modeling can be carried out at each image point, allowing the possibility of more accurate background subtraction during or after acquiring the image; see Sections 2.6.5 and 5.3.6.

Spectrum-imaging offers the possibility of extensive data manipulation after the spectra have been recorded. For example, it allows a procedure known as *segmentation* to be used for measuring small concentrations of elements in particular organelles. If the PEELS system has recorded an extended range of the energy-loss spectrum from each pixel within a STEM image, regions of similar composition can be recognized by examining the K-edges of major constituents (C,N,O). Spectra from one or more of these regions (or arbitrary shape) are summed in order to provide adequate statistics for measuring the average trace-element concentration. Leapman *et al.* (1993) used this technique to measure calcium concentrations (50–100 ppm) in mitochondria and endoplasmic reticulum (see Fig. 5.46) with a

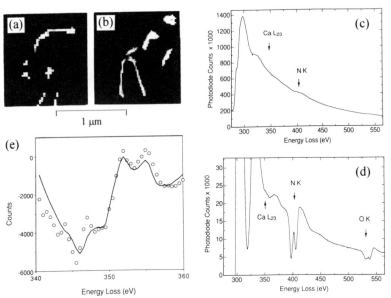

Figure 5.46. (a, b) Regions of endoplasmic reticulum in mouse cerebellar cortex, segmented on the basis of their nitrogen content. (c) Spectrum obtained by summing contributions from both segmented regions. (d) First-difference spectrum, showing a weak calcium L-edge. (e) Multiple-least-squares fit of the calcium L-edge data points to a $CaCl_2$ reference spectrum (Leapman *et al.*, 1993b). From *Ultramicroscopy*, R. D. Leapman, J. A. Hunt, R. A. Buchanan and S. B. Andrews, Measurement of low calcium concentrations in cryosectioned cells by parallel-EELS mapping © 1993, pp. 230–231, Figures 4, 5, and 6, with permission from Elsevier Science B.V., Amsterdam, The Netherlands.

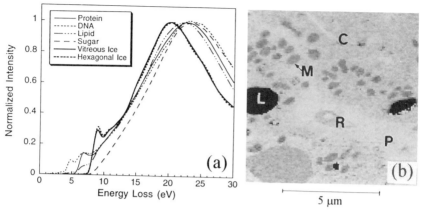

Figure 5.47. (a) Low-loss spectra of protein, DNA, lipid, sugar, and ice. (b) Water map of frozen hydrated liver tissue: L = lipid droplet (zero water content), M = mitochondria (average content 57%), R = erythrocyte (65% water), P = plasma (91% water). (Sun *et al.*, 1995).

precision of better than 20%. A similar technique has been used in EDX spectroscopy; it is somewhat analogous to the use of crystallized proteins when determining the structure of a single molecule by x-ray or electron diffraction.

The low-loss spectra of biologically important substances all exhibit a broad peak around 23 eV, whereas ice shows a peak around 20 eV; see Fig. 5.47a. Sun *et al.* (1995) exploited this difference (and differences in fine structure below 10 eV) to measure the water content within cells, with a precision of around 2% and a spatial resolution of 80 nm. Their procedure involved MLS fitting of spectrum-image data (6–30 eV region) to standard spectra from ice and protein. They also produced maps of water content, showing pronounced differences between mitochondria, cytoplasm, red blood cells, plasma, and lipid components; see Fig. 5.47b. This contrast arises from subtle differences in the spectral shape, rather then differences in elastic or energy-integrated inelastic cross sections, and is therefore not evident in the normal dark-field or bright-field images.

5.7.3. Radiation Damage and Hole Drilling

The energy transfer which occurs during inelastic scattering usually results in valence electrons excited to orbitals of higher energy. When an atom or molecule returns to its ground state, the chemical bonds with its neighbors may be permanently broken (creating an ion or free radical) or may take up a different configuration, resulting in permanent structural damage. This damage places a fundamental limit on the spatial resolution obtainable from TEM images and from electron-beam microanalysis, as discussed in Section 5.5.4. In crystalline specimens, *structural* disorder is seen as a disappearance of lattice fringes and a gradual fading of the spot diffraction pattern (Glaeser, 1975; Zeitler, 1982). The disruption of chemical bonding is observed more directly as a disappearance of fine structure in an optical absorption or energy-loss spectrum (Reimer, 1975; Isaacson, 1977).

Electron irradiation may also result in the removal of atoms from the irradiated area, known as *mass loss*. This process is of concern in elemental analysis by EELS or EDX spectroscopy because some elements are removed more rapidly than others, giving a change in chemical composition. Either of these spectroscopies can be used to monitor the loss of specific elements but EELS is the more sensitive technique for low-Z elements (see page 25) and therefore more suitable for investigating differential mass loss in organic specimens. The dose required for a single measurement is minimized if the spectrum is collected from as large an area of specimen as possible; for energy-loss spectroscopy in TEM image mode, this implies low magnification and a large spectrometer-entrance aperture. If a spectrum is recorded at a time t after the start of irradiation, the accumulated dose is $D=It/A$, where I is the beam current and A the cross-sectional area of the beam. The remaining amount (N atoms/area) of a particular element can be obtained from its ionization edge, making use of Eq. (4.65). If $\ln(N)$ is plotted against D, the initial slope of the data gives the *characteristic dose* D_e (the dose which would cause N to fall to $1/e$ of its initial value, in a simple case). D_e is an *inverse* measure of the radiation sensitivity of the specimen for small fractional depletion of a particular element.

Organic Specimens

Measurements on organic materials indicate that elemental loss usually depends on the accumulated dose and not on the dose rate (i.e., D_e is independent of current density). Table 5.5 lists D_e for selected organic compounds exposed to 100-keV incident electrons. Values for other incident energies may be estimated by assuming D_e to be proportional to the effective incident energy: $T=m_0v^2/2$ (Isaacson, 1977). Not surprisingly, D_e is low for compounds containing unstable groups such as nitrates, and aromatic compounds are generally more stable than aliphatic ones. Replacement of hydrogen by halogen atoms (as in chlorinated phthalocyanine) further reduces the radiation sensitivity, partly due to the increased steric hindrance (cage effect) from surrounding atoms. Fluorine attached directly to an aromatic ring can be remarkably stable, especially at low temperatures (Ciliax *et al.*, 1993).

Table 5.5. Approximate Values of Characteristic Dose D_e for Removal of Specified Elements from Organic Compounds by 100-keV Electrons[a]

Material	Element removed	D_e (C/cm²) 300 K	D_e (C/cm²) 100 K
Nitrocellulose	C	0.07	0.4
(collodion)	N	0.002	0.3
	O	0.007	0.6
Polymethyl methacrylate	C	0.6	1
(PMMA)	O	0.07	0.6
Cu-phthalocyanine	N	0.8	
Cl₁₅Cu-phthalocyanine	Cl	4	>10
Perfluorotetracosane	F	0.2	0.2
Amidinotetrafluorostilbene	F	0.8	>10⁴

[a]From Egerton (1982d) and Ciliax et al. (1993).

As seen in Table 5.5, cooling an organic specimen to 100 K reduces the rate of mass loss, sometimes by a large factor. Cryogenic operation probably does not reduce the number of broken bonds but prevents atoms from leaving the irradiated area by reducing their diffusion rate. EELS measurements confirm that gaseous atoms are released from the irradiated area when the specimen returns to room temperature (Egerton, 1980c). Lamvik et al. (1989) found that mass loss in collodion is further reduced at a temperature of 10 K.

An alternative way of reducing mass loss is to coat the specimen on both sides with a thin ($\approx$10 nm) film of carbon. Probably because each surface film acts as diffusion barrier, mass loss is reduced by a factor of typically 2 to 6 (Egerton et al., 1987). Carbon-contamination films produced in the electron beam (in the presence of hydrocarbons) are believed to have a similar protective effect. There is evidence that encapsulation also helps to preserve crystallinity, perhaps by aiding recombination processes (Fryer and Holland, 1984).

Radiation damage can also be monitored from the low-loss spectrum, with a reduction in the dose required for measurement. From the plasmon-peak intensity, Egerton and Rossouw (1976) measured the rate of hydrocarbon contamination as a function of specimen temperature and found that it became negative (indicating etching by oxygen or water vapor) below $-50°C$. By measuring a shift in the main plasmon energy towards that of amorphous carbon, Ditchfield et al. (1973) observed that polyethylene loses a significant fraction of its hydrogen at doses as low as 10^{-3} C/cm². In addition, they observed the creation of double bonds (due to cross-linking) from the appearance of a π-excitation peak around 6 eV. In the case of polystyrene, an initially visible π peak *decreased* upon irradiation, indicating that double bonds were being broken. The disappearance of fine structure below 10 eV has also been seen during the irradiation of nucleic acid bases (Isaacson, 1972a). The doses which cause bonding changes tend to be intermediate between those which destroy the diffraction pattern (of a crystalline specimen) and the larger values associated with mass loss (Isaacson, 1977).

Core-loss fine structure provides another indication of bonding, with the advantage that different ionization edges can be recorded in an attempt to determine the atomic site at which damage occurs. In the case of Ge–O–phthalocyanine, Kurata *et al.* (1992) found that the decrease in the π^* threshold peak was more rapid at the nitrogen edge than at the carbon edge, possibly due to chemical reaction between N atoms and adjacent H atoms released during irradiation. Conversely, the *emergence* of a π^* at carbon and/or nitrogen edges has been observed during electron irradiation of fluorinated compounds, providing evidence for the formation of double bonds and aromatization of ring structures (Ciliax *et al.*, 1993).

Radiation-damage measurements are usually done at low current density, so that thermal dissociation due to heating of the specimen in the electron beam is negligible. The average energy $\langle E \rangle$ per electron deposited in a specimen can be evaluated directly from its energy-loss spectrum:

$$\langle E \rangle = \int EJ(E)\, dE \Big/ \int J(E)\, dE \qquad (5.35)$$

where the integration is over the entire spectrum *including* the zero-loss peak (Egerton, 1982b). Because of plural-scattering contributions to $J(E)$, $\langle E \rangle$ increases more than linearly with specimen thickness. The temperature rise in the beam is difficult to calculate, since it depends on the geometry and thermal conductivity of the specimen and its thermal contact with a support grid or specimen holder (Reimer, 1993). However, an *upper limit* (which may approximate to the actual temperature rise for very thin specimens of low conductivity) can be calculated by assuming only loss of heat by radiation from both surfaces. Equating the rate of heat generation and heat loss gives

$$\langle E \rangle (J/e) = 2\sigma(\varepsilon_1 T_1^4 - \varepsilon_2 T_2^4) \qquad (5.36)$$

where J is the current density (in A/m^2), $\sigma = 5.67 \times 10^{-8}$ W m^{-2} K^{-4} is Stefan's constant, ε_1 and T_1 are the emissivity and temperature of the specimen, and ε_2 and T_2 those of its surroundings. Taking $\varepsilon_1 \approx \varepsilon_2 \approx 0.5$, $T_1 \approx T_2 \approx 300$, and $\langle E \rangle \approx 6$ eV (for a 30-nm polymer film), the maximum temperature rise is $\Delta T \approx \langle E \rangle J/(4\sigma T^3) \approx J$, or about 3 K for $J = 3$ A/m^2 (Egerton, 1982b). Higher current densities could result in appreciable temperature increase, which could account for an apparent decrease in characteristic dose with increasing dose rate observed in polymer films (Payne and Beamson, 1993). For most inorganic specimens (particularly metals), thermal conduction keeps the temperature rise well below that given by Eq. (5.36), although with focused illumination ($J > 10^5$ A/m^2) a temperature rise of some hundreds of degrees is possible (Reimer, 1993).

Inorganic Materials

EELS has also been used to investigate electron-beam damage to inorganic materials. Hydrides are among the most radiation sensitive: a dose of 0.1 C/cm^2 converts NaH to metallic sodium inside the electron microscope (Herley et al., 1987), as seen from the emergence of crystalline needles whose low-loss spectrum contains sharp plasmon peaks (see Fig. 4.1). Metal halides are also fairly beam-sensitive: exciton states (created by the inelastic scattering) can decay into halogen vacancies (F-centers) and interstitials (H-centers) which may diffuse to the surface, resulting in the ejection of halogen atoms (Hobbs, 1984). Halogen loss apparently depends on the dose rate as well as the accumulated dose (Egerton, 1980f).

Thomas (1982, 1984) used K-shell spectroscopy to monitor the effect of a field-emission probe on compounds such as Cr_3C_2, $TiC_{0.94}$, Cr_2N, and Fe_2O_3. The dose required for the removal of 50% of the nonmetallic element (in the range 10^5 to 10^6 C/cm^2) did not change when specimens were cooled to 143 K. This lack of temperature dependence suggests a knock-on or sputtering process (transfer of momentum to atomic nucleus by high-angle elastic scattering) as the damage mechanism. The sputtering rate can be estimated from the Rutherford cross section, Eq. (3.3), provided the binding energy E_s of surface atoms is known (Chadderton, 1965; Isaacson, 1977). Calculations for pure carbon and 100-keV incident electrons suggest that a dose of 10^5 C/cm^2 ($\approx$ 1s irradiation by a focused and stationary field-emission probe) should remove about 10 monolayers, taking $E_s \approx 15$ eV (Egerton et al., 1987). STEM experiments on a thin carbon foil have indicated a mass loss of 0.6 monolayer at that dose (Leapman and Andrews, 1992).

In many inorganic compounds, mass loss occurs more rapidly than the rate of sputtering because of radiolytic (electronic) processes (Hobbs, 1984), which may include inner-shell ionization followed by an interatomic Auger decay (Knotek, 1984). As discussed on page 337, irradiation of frozen-hydrated specimens inside the electron microscope initiates a radiolytic process which results in small bubbles of hydrogen.

Electron-Beam Lithography and Hole Drilling

Muray *et al.* (1985) investigated vacuum-evaporated films of metal halides for possible lithographic applications, using a 100-kV field-emission STEM (vacuum of 10^{-9} torr in the specimen chamber) equipped with a serial-recording spectrometer. After a dose of 1 C/cm^2, a 50-nm film of NaCl is largely converted to sodium, as shown by the appearance of sharp surface (3.8 eV) and volume (5.7 eV) plasmon peaks. A dose of 100 C/cm^2 removes the sodium, creating a 2-nm-diameter hole. Similar behavior was observed for LiF but hole creation required only 10^{-2} C/cm^2. In the case of MgF$_2$, magnesium was formed (bulk-plasmon peak at 10.2 eV for $D \approx$ 1 C/cm^2) but not removed by prolonged irradiation. In CaF$_2$, bubbles of molecular fluorine have been detected from the appearance of a sharp peak at 682 eV (K-edge threshold) but only in fine-grained films evaporated onto a low-temperature substrate (Zanetti *et al.*, 1994).

A finely focused electron beam can also create nanometer-scale holes in metallic oxides. Sometimes hole drilling is observed only above a threshold current density, typically of the order of 1000 A/cm^2 (Salisbury *et al.*, 1984). The existence of this threshold may indicate that a radial electric field around the beam axis (positive potential at the center because of the high secondary-electron coefficient of insulators) must be established, of sufficient strength to remove cations created in the ionization process (Humphreys *et al.*, 1990). The fact that the current-density threshold for alumina was *reduced* by cooling the sample of 85 K (enabling hole drilling to be performed with a tungsten-filament electron source) suggests that there is also an inward diffusion of metal, less effective at low temperatures (Devenish *et al.*, 1989).

Berger *et al.* (1987) used a field-emission STEM to create holes in amorphous alumina. During drilling, the low-loss spectrum exhibits a peak at about 9 eV (Fig. 5.48a), the oxygen K-edge spectrum develops a sharp threshold resonance (Fig. 5.48b) and the O/Al ratio, measured from areas under the O K- and Al L-edges, increases from 1.5 to 7 or more. All of these observations are consistent with the creation of a bubble containing molecular oxygen. The bubble burst after an average time of 40 s, leaving 5-nm-diameter hole.

Hole drilling in sodium β-alumina proceeds somewhat differently. A sharp peak appears, with a shoulder at 9 eV and maximum at 15 eV (Fig. 5.48c) suggesting surface and bulk modes in small Al spheres. At the same time, the Al L-edge shifts 2 eV lower in energy and changes to a more rounded shape (Fig. 5.48d), consistent with the formation of aluminum metal from an insulating oxide. The oxygen K-edge maintains its shape but gradually weakenss: the O/Al ratio decreases from 1.5 to 0.6 (typically).

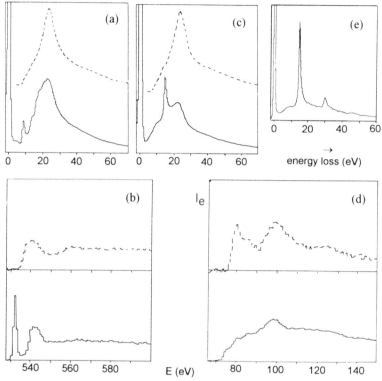

Figure 5.48. (a) Low-loss and (b) oxygen K-loss spectra of amorphous alumina before (dashed curve) and during hole drilling. (c) Low-loss and (d) Al L-loss spectra of Na β-Al$_2$O$_3$ before (dashed) and during drilling. (e) Low-loss spectrum of a hole plugged with aluminum. From Berger *et al.* (1987).

Berger *et al.* (1987) suggest that oxygen is lost from both surfaces, forming surface indentations which grow inwards, leaving behind Al particles coating the inside walls. Although in most cases a hole is formed after 30 s, the zero-loss intensity remains well below the incident-beam current, indicating scattering (outside the collector aperture) from Al within the incident beam. Occasionally, the hole becomes filled with a "plug" of continuous aluminum, as evidenced by bulk-plasmon peaks in the low-loss spectrum (Fig. 5.48e). This metallization is similar to the normal irradiation behavior of halides such as MgF$_2$.

Hole drilling has been demonstrated in many other oxides, with doses mainly in the range 10^4 to 10^6 C/cm^2 (Hollenbeck and Buchanan, 1990). In the case of crystalline MgO, square holes are formed from growth of an indentation on the electron exit surface (Turner *et al.*, 1990). Humphreys *et al* (1990) have pointed out that, with letters patterned from a matrix of 4-nm holes, the 29-volume Encyclopedia Britannica could be written on a pinhead.

5.7.4. High-Temperature Superconductors

In 1986, Bednorz and Müller found the layered perovskite $La_{2-x}Ba_x$-CuO_4 to be superconducting (for $x \approx 0.15$) at relatively high temperatures (30 K), and the following year Wu *et al.* discovered that yttrium barium cuprate ($YBa_2Cu_3O_{7-\delta}$, known as YBCO) is a superconductor (for $\delta <$ 0.6) at liquid-nitrogen temperature. Since then, other "high-temperature" superconductors have been discovered (such as the bismuth-based and thallium-based cuprates) with critical temperatures as high as 125 K. Considerable effort has been put into engineering the properties of these materials to make them suitable for electrical applications.

The cuprate superconductors all contain CuO_2 planes, perpendicular to the c-direction, with other oxygen and metal atoms between. Superconductivity is associated with the CuO_2 planes and normally involves Cooper pairs of valence-band holes, whose concentration varies with the oxygen content of the material. Although the mechanism of superconductivity and many details of the electronic band structure remain in dispute, it is known that the Cu-$3d$ and O-$2p$ states of CuO_2-plane atoms lie close to the Fermi level. As a result of the dipole selection rule (Section 3.7.2), the unoccupied part of these states can be investigated by exciting transitions from Cu-$2p$ and O-$1s$ levels—in other words, by examining fine structure of the copper L_{23}- and oxygen K-edges.

Figure 5.49a shows the onset of the oxygen K-edge recorded from $YBa_2Cu_3O_{7-\delta}$. For small oxygen deficiency ($\delta \approx 0.2$), two pre-edge features are visible: a shoulder around 535 eV, which represents transitions to unoccupied Cu-$3d$ states (upper Hubbard band), and a peak around 528 eV representing transitions to O-$2p$ states which give rise to holes in the valence band. As the oxygen deficiency δ increases, the 528-eV peak falls in intensity, signaling a decrease in hole concentration and a reduction in the superconducting critical temperature. For $\delta > 0.6$, the prepeak is absent, corresponding to zero concentration of holes and a material which is an electrical insulator at all temperatures.

It is possible to select the direction of momentum transfer $\mathbf{q}$ which contributes to the energy-loss spectrum by appropriate choice of the scattering angle and specimen orientation. Changing the direction of $\mathbf{q}$ produces significant differences in pre-edge structure (Fig. 5.49b). These differences have been used to determine final-state symmetries in several oxide superconductors, information which has been useful for the comparison of dif-

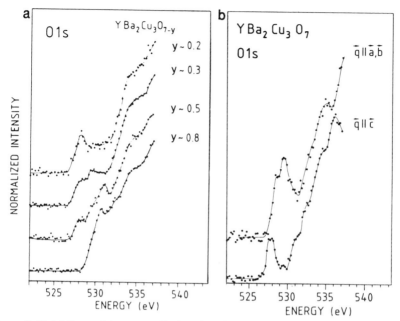

Figure 5.49. (a) Onset of the oxygen K-edge of YBCO at four values of the oxygen deficiency (Fink, 1989). (b) Oxygen K-edge recorded at scattering angles such that the momentum transfer was either parallel or perpendicular to the c-axis (Fink *et al.*, 1994).

ferent models of electrical conduction (Fink, 1994). The q-dependence (dispersion) of the plasmon peak has been measured for $Bi_2Sr_2CaCu_2O_8$ and interpreted in terms of the multilayered structure of the material (Longe and Bose, 1993). Using a field-emission TEM to examine $Ba_{1-x}K_xBiO_3$, Wang *et al.* (1993) identified a free-carrier plasmon around 2 eV which abruptly increases in strength at the semiconductor/metal transition ($x >$ 0.3). The dielectric function $\varepsilon(E)$ of YBCO (Fig. 5.50) has been derived by Kramers–Kronig analysis of EELS data and has been discussed in terms of band structure (Yuan *et al.*, 1988).

The superconducting properties of these ceramics are strongly dependent on the presence of defects, such as grain boundaries in a polycrystalline material. Making use of the high spatial resolution of a microscope-based EELS system, the oxygen-edge prepeak can be used as a sensitive measure of local oxygen concentration. Based on parallel-EELS measurements made using a 2-nm-diameter electron probe provided by a 100-kV TEM fitted with a field-emission source, Zhu *et al.* (1993) have reported that grain boundaries in fully oxygenated YBCO fall into two categories, depending on the structural misorientation between the adjacent grains: those in which

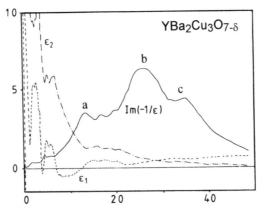

Figure 5.50. Energy-loss function (solid curve) together with real and imaginary components of the permitivity ε of YBCO superconductor, derived from low-loss spectroscopy (Yuan *et al.*, 1988). Peaks a and b are attributed to valence-electron plasmon behavior and peak c to yttrium $4p \rightarrow d$ transitions (N_{23} edge).

the 529-eV peak retains its intensity across the grain boundary (Fig. 5.51b) and those in which the intensity falls practically to zero (Fig. 5.51a). The implication is that some boundaries are fully oxygenated and would transmit the superconducting current at temperatures below 90 K, while others contain an oxygen-depleted region (10 nm or less wide) and might act as weak links. Similar conclusions have been reached on the basis of EELS combined with Z-contrast imaging in a scanning transmission microscope (Browning *et al.*, 1993a).

Zhu *et al.* (1993) show that the core-loss collection angle can be chosen such that the *average* momentum transfer parallel to the electron beam is equal to that in a perpendicular direction. The observed pre-edge structure would then be independent of crystal orientation and there should be no change in fine structure (across the grain boundary) arising from the change in orientation. For the oxygen K-edge and 100-keV electrons, the required collection semiangle is 5 mrad. However, the use of a small convergent-beam probe dictates a larger collection angle for efficient signal collection; Zhu *et al.* used $\beta = 25$ mrad without problems from orientation effects, as evidenced by the similarity in spectra recorded away from the grain boundary in Fig. 5.49.

From measurements of the 529-eV peak, Dravid *et al.* (1993) showed that the oxygen content in $YBa_2Cu_3O_{7-\delta}$ may vary *along* a grain-boundary plane (δ fluctuating between 0.2 and 0.4 over distances of the order of 10 nm) and proposed a modification to the Dayem–Bridge model for grain-boundary structure. Browning *et al.* (1991a) applied the same technique to reveal variations in oxygen content *within* a grain, with an estimated accua   b

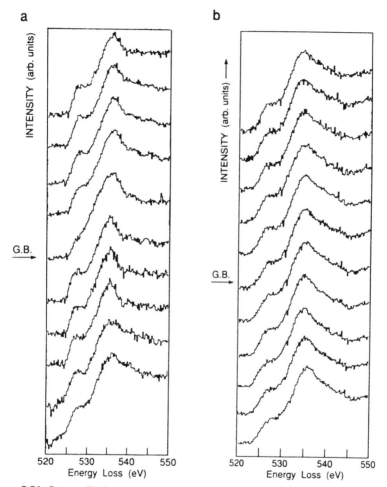

Figure 5.51. Oxygen K-edges recorded by stepping a 2-nm electron probe at 5-nm intervals across (a) an oxygen-deficient grain boundary and (b) a fully oxygenated grain boundary in $YBa_2Cu_3O_7$ (Zhu *et al.*, 1993). Arrows identify the spectra which were recorded with the incident beam directly on the boundary.

racy of 2%. EELS has also been used to detect the presence of carbon at grain boundaries of densified YBCO; examination of the carbon and oxygen K-edge structure indicated in presence of $BaCO_3$ (Batson *et al.*, 1989).

Other high-temperature superconductors which have been examined by EELS include $Tl_{0.5}Pb_{0.5}Ca_{1-x}Y_xSr_2Cu_2O_{7-\delta}$ (Yuan *et al.*, 1991). Its oxygen K-edge shows a prepeak at 529 eV, which is believed to represent transitions to *conduction*-band states and is *not* an indication of superconductivity.

However, a shoulder on the low-energy side of the peak (attributed to transitions to hole states near the Fermi level) appeared to be characteristic of superconducting ($x < 0.4$) material. Core-loss measurements on $Y_{1-x}Ca_xSr_2Cu_2GaO_7$ (Dravid and Zhang, 1992) also found two pre-edge features: a broad peak (dependent on Ca doping) centered around 528.2 eV, associated with *normal* conductivity and attributed to holes on oxygen sites which are *not* on Cu_2O planes, and a smaller peak around 527.1 eV, probably associated with holes on Cu_2O planes and superconductivity.

Attempts have been made to fabricate Josephson junctions by depositing multilayer structures of oxide superconductors. One problem is that the c-axis tends to lie perpendicular to the film plane; the coherence length is very short (<1 nm) in this direction, placing extreme limits on the sharpness of the junction. If a film could be depleted in oxygen by writing with an electron beam of sub-nm dimensions, the insulating gap would form a weak-link structure. Unfortunately, fully oxygenated YBCO appears to retain its oxygen content even for high electron dose, although an electron beam has been used to cut nm-wide channels in an *amorphous* phase of similar composition (Humphreys *et al.*, 1988; Devenish *et al.*, 1988).

Since the high-temperature superconductors are oxide ceramics, TEM specimens have usually been prepared by mechanical crushing and dispersion of the particles onto a holey carbon foil. Ion-beam thinning has also been used, but can create an amorphous phase on the surface (Browning *et al.*, 1991a). Since it is less surface-sensitive than most other methods, transmission EELS avoids many of the experimental problems associated with surface condition and has already contributed significantly to an understanding of the physics of these materials (Fink *et al.*, 1994). When implemented with a field-emission source, the technique offers 0.5-eV energy resolution, 1-nm spatial resolution, and 1-nm^{-1} momentum resolution (Wang *et al.*, 1993), criteria which make it an important tool for future investigations into the materials science of high-temperature superconductors.

Relativistic Bethe Theory

Even for 100-keV incident electrons, it is necessary to use relativistic kinematics to calculate inelastic cross sections (see Section 3.6.2). Above about 200 keV, an additional relativistic effect becomes significant, in the form of a "retarded" interaction. At high incident energies, Eq. (3.26) should be replaced by (Møller, 1932; Perez *et al.*, 1977)

$$\frac{d^2\sigma}{d\Omega\,dE} = 4\gamma^2 a_0^2 R^2 \left(\frac{k_1}{k_0}\right) \left[\frac{1}{Q^2} - \frac{2\gamma - 1}{\gamma^2 Q(E_0 - Q)} + \frac{1}{(E_0 - Q)^2}\right.$$
$$\left. + \frac{1}{(E_0 + m_0 c^2)^2}\right] |\eta(q, E)|^2 \tag{A.1}$$

where $\gamma = 1/(1 - v^2/c^2)^{1/2}$, v is the incident velocity, $a_0 = 52.92 \times 10^{-12}$ m is the Bohr radius, $R = 13.6$ eV is the Rydberg energy, and $m_0 c^2 = 511$ keV is the rest energy of an electron. For most collisions, the last three terms within the brackets of Eq. (A.1) can be neglected and the ratio (k_1/k_0) of the wavevectors of the fast electron (after and before scattering) taken as unity. The quantity Q has dimensions of energy and is defined by

$$Q = \frac{\hbar^2 q^2}{2m_0} - \frac{E^2}{2m_0 c^2} = R(qa_0)^2 - \frac{E^2}{2m_0 c^2} \tag{A.2}$$

where $\mathbf{q}$ is the scattering vector and E represents energy loss. The $E^2/2m_0 c^2$ term in Eq. (A.2) is significant at small scattering angles.

In Eq. (A.1), $|\eta(q, E)|^2$ is an energy-differential *relativistic* form factor, equal to the nonrelativistic form factor $|\varepsilon(q,E)|^2$ for high-angle collisions but given by (Inokuti, 1971)

$$|\eta(q, E)|^2 = \frac{1}{E}\left(\frac{df}{dE}\right)\left[Q - \frac{E^2}{2\gamma^2 m_0 c^2}\right] \tag{A.3}$$

for $qa_0 \ll 1$, df/dE being the energy-differential generalized oscillator strength as employed in Sections 3.2.2 and 3.6.1.

Fano (1956) has shown that the differential cross section can be written as a sum of two independent terms. Within the dipole region, $\theta \ll (E/E_0)^{1/2}$, his result can be written as

$$\frac{d^2\sigma}{d\Omega\,dE} = \frac{4a_0^2}{(E/R)(T/R)}\left(\frac{df}{dE}\right)\left[\frac{1}{\theta^2 + \theta_E^2} + \frac{(v/c)^2\theta^2\theta_E^2}{(\theta^2 + \theta_E^2)(\theta^2 + \theta_E^2/\gamma^2)^2}\right] \quad (A.4)$$

where $T = m_0 v^2/2$ and $\theta_E = E/(2\gamma T)$ as previously. The first term is identical to Eq. (3.29) and provides the Lorentzian angular distribution observed at lower incident energies; it arises from Coulomb (electrostatic) interaction between the incident and atomic electrons and involves forces parallel to the scattering vector **q**.

The second term in Eq. (A.4), representing the exchange of virtual photons, involves forces perpendicular to **q** (*transverse* excitation). This term is zero at $\theta = 0$ and negligible at large θ, but can be significant for small scattering angles. It becomes more important as the incident energy increases, and for $E_0 > 250$ keV it shifts the maximum in the angular distribution away from zero angle, as illustrated in Fig. A.1. This displaced maximum should not be confused with the Bethe ridge (Sections 3.5 and 3.6.1), which occurs at higher scattering angles and for energy losses well above the binding energies of the atomic electrons.

Integration of Eq. (A.4) up to a collection angle β gives

$$\frac{d\sigma}{dE} = \frac{4\pi a_0^2}{(E/R)(T/R)}\frac{df}{dE}\left[\ln(1 + \beta^2/\theta_E^2) + G(\beta, \gamma, \theta_E)\right] \quad (A.5)$$

where

$$G(\beta, \gamma, \theta_E) = 2\ln\gamma - \ln\left(\frac{\beta^2 + \theta_E^2}{\beta^2 + \theta_E^2/\gamma^2}\right) - \frac{v^2}{c^2}\left(\frac{\beta^2}{\beta^2 + \theta_E^2/\gamma^2}\right) \quad (A.6)$$

The retardation term $G(\beta, \gamma, \theta_E)$ exerts its maximum effect at $\beta \approx \theta_E$ and increases the energy-loss intensity by about 10% for $E_0 = 200$ keV, or larger amounts at higher incident energy; see Fig. A.2. Under certain conditions, this increase in cross section can result in the emission of Cerenkov radiation (Section 3.3.4). For $\beta \gg \theta_E$ but still within the dipole region, Eq. (A.5) simplifies to a form given by Fano (1956).

For an ionization edge whose threshold energy is E_k, Eq. (A.5) can be integrated over an energy range Δ which is small compared to E_k to give

$$\sigma_k(\beta, \Delta) = \frac{4\pi a_0^2}{(\langle E\rangle/R)(T/R)}f(\Delta)\left[\ln(1 + \beta^2/\langle\theta_E\rangle^2) + G(\beta, \gamma, \langle\theta_E\rangle)\right] \quad (A.7)$$

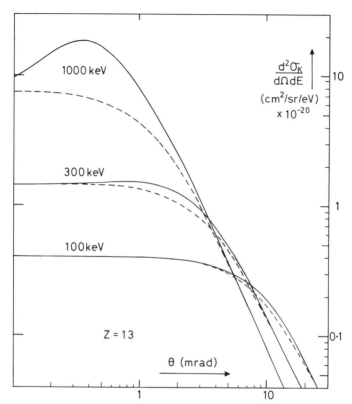

Figure A.1. Differential cross section for *K*-shell scattering in aluminum, at an energy loss just above the ionization edge, calculated for three values of incident-electron energy using a hydrogenic expression for df/dE (Egerton, 1987). Solid curves include the effect of retardation; the dashed curves do not.

where $\langle E \rangle$ and $\langle \theta_E \rangle$ are average values of E and θ_E within the integration region. Numerical evaluation shows that Eq. (A.7) is a better approximation if a geometric (rather than arithmetic) mean is used, so that $\langle E \rangle = [E_k$ $(E_k + \Delta)]^{1/2}$ and $\langle \theta_E \rangle = \langle E \rangle / 2\gamma T$. This equation can be used to calculate cross sections for EELS elemental analysis from tabulated values of the dipole oscillator strength $f(\Delta)$, or *vice versa*; see Appendix B.10

If the integration is carried out over all energy loss, the result is the Bethe asymptotic formula for the total ionization cross section for an inner shell, used in calculating x-ray production:

$$\sigma_k = \frac{4\pi a_0^2 N_k b_k}{(E_k/R)(T/R)} \left[\ln\left(\frac{c_k T}{E_k}\right) + 2\ln\gamma - \frac{v^2}{c^2} \right] \tag{A.8}$$

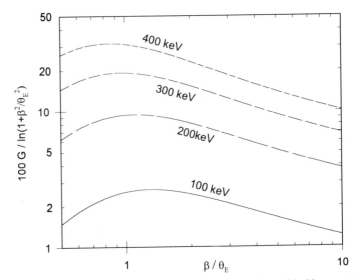

Figure A.2. Percentage increase in cross section (for four values of incident energy) as a result of relativistic retardation, according to Eq. (A.5) or Eq. (A.7).

where N_k is the number of electrons in the shell k (2, 8, and 18 for K, L, and M shells); b_k and c_k are parameters which can be parameterized on the basis of experimental data (Zaluzec, 1984). As seen from Eq. (A.8), a *Fano plot* of $v^2\sigma_k$ against $\ln[v^2/(c^2 - v^2) - v^2/c^2)]$ should yield a straight line even at MeV energies (Inokuti, 1971). The last two terms in Eq. (A.8) cause σ_k to pass through a minimum and exhibit a *relativistic rise* when the incident energy exceeds about 1 MeV.

Computer Programs

The computer codes discussed in this appendix are available by anonymous FTP to ftp.phys.ualberta.ca (directory /pub/eels); contact the author by e-mail (egerton@phys.ualberta.ca) for further information. The programs are also listed in the Microscopy Society of America (MSA) public-domain library, accessible by anonymous FTP to www.amc.anl.gov or ftp.msa.microscopy.com (ANL Software Library, EMMPDL/Eels subdirectory) or via WWW sites at Argonne National Laboratory (URL = http://www.amc.anl.gov or http://ftp.msa.microscopy.com).

B.1. Matrix Deconvolution

The following program removes plural scattering from a low-loss spectrum using the matrix method (see page 255) described by Schattschneider (1983) and by Su and Schattschneider (1992a), and is based on the FORTRAN program MATRIX written by these authors. The spectrum is read from a two-column (energy loss, intensity) file; the channel number NZ and integral A0 of the zero-loss peak are found as in the FLOG program (p. 410). The data are normalized by dividing by A0, prior to matrix evaluation up to order n (typically 5 to 10). For correct scaling, the SSD is multiplied by A0 before writing to an output file.

The output contains the single-scattering distribution over an energy range of approximately n times the mean single-scattering loss, followed by spurious data (which may involve large numbers, positive or negative). Larger n therefore increases the range of useful data, although at the expense of increased computing time. The method is mathematically exact for scattering up to the nth order.

Unlike the Fourier-log program listed in Section B.2, MATMOD ignores instrumental broadening of the energy-loss peaks. Artifacts may therefore occur at multiples of the energy of the single-scattering peak, just as for a Fourier-log program which uses the delta function approximation: Eq. (4.13). These artifacts are more noticeable for thicker specimens and when the measured width of an main inelastic peak is comparable to that of the zero-loss peak.

Advantages over the Fourier-log method are that the data need not fall to zero at each end of their range; a limited number of data points (not necessarily a power of 2) can be processed, resulting in a short computing time. The specimen thickness can, in principle, be arbitrarily large, although errors involved in the measurement of the area of the zero-loss peak and from neglect of the instrumental resolution are likely to be significant for thicker specimens.

```
C          MATMOD. FOR                               Last update: 95NOV03
C             MATRIX DECONVOLUTION (P. SCHATTSCHNEIDER & D.S. SU, 1990-1992)
C             This program is intended for deconvolution of multiple
C             inelastic scattering in image-mode (angle-integrated) low-loss
C             spectra. It utilizes a method originally described in
C             P. Schattschneider, Phil. Mag. B 47 (1983), 555-560.
C             SSD is written to the file MATMOD.DAT
C          CHARACTER*20 LABEL
           DIMENSION T(1024),D(1024),C(1024),DATA(1024),E(1024)
           CHARACTER*12 INFILE,OUTFILE,TEXT
           WRITE(*,*) ' NAME OF INPUT FILE ? '
           READ(6,15) INFILE
        15 FORMAT(A12)
           OPEN(13, FILE=INFILE)
           WRITE(6,*) ' NUMBER OF DATA POINTS TO BE READ = '
           READ(5,*) NSPEC
           ND=0
           DO 107 I=1,NSPEC
           READ(13,*,END=50) E(I),DATA(I)
       107 ND=I
        50 CLOSE(13)
C          FIND ZERO-LOSS CHANNEL:
           DO 111 I=1,ND
           NZ=I
       111 IF(E(I)+E(I+1).GT.0.) GO TO 112
       112 CONTINUE
C          FIND SEPARATION POINT AS MINIMUM IN J(E)/E:
           DO 201 I=NZ,ND
           NSEP=I
           TANDIF = DATA(I+1)/FLOAT(I-NZ+1) - DATA(I)/FLOAT(I-NZ)
           IF(TANDIF.GT.0.) GO TO 202
       201 CONTINUE
       202 EPC=(E(5)-E(1))/4.
           BACK=(DATA(1)+DATA(2)+DATA(3)+DATA(4)+DATA(5))/5.
           SUM = 0.
           DO 205 I=1,NSEP
       205 SUM = SUM+DATA(I)
           A0 = SUM - BACK*NSEP
           WRITE(6,*)'ND,NZ,NSEP,BACK,A0,EPC = ',ND,NZ,NSEP,BACK,A0,EPC
           WRITE(6,*) 'eV/channel = ',EPC,', zero-loss intensity = ',A0
           DO 3 I=1,ND
         3 T(I)=0.
C          TRANSFER SHIFTED DATA TO ARRAY T(I):
           DO 302 I=NZ,ND
       302 T(I-NZ+1)=DATA(I)-BACK
C          REMOVE ZERO-LOSS PEAK:
           DO 4 I=1, NSEP-NZ+1
         4 T(I)=0.
           NMAX=ND-NZ+1
C          NORMALIZE THE ARRAY T(I):
           DO 30 I=1,NMAX
        30 T(I)=T(I)/A0
           WRITE(*,*) 'NUMBER OF TERMS IN LOG EXPANSION = '
           READ (6,*) ITERM
C          INITIALIZING THE ARRAY D(J):
           DO 10 J=1,NMAX
        10 D(J)=T(J)/FLOAT(ITERM)
           WRITE(*,*) 'SERIES EXPANSION OF LOG. . . .'
```

```
C        SERIES: F. LN(T*(1-T*(1/2-T*(1/3-T*(1/4-T*(...))))))
         DO 60 N=ITERM-1,1,-1
         FAKT=FLOAT(N)
         D(1)=-1./FAKT
         DO 100 I=1,NMAX
         C(I)=0.
C        MATRIX MULTIPLICATION:
         DO 300 K=0,I-1
300      C(I)=C(I)-D(I-K)*T(K+1)
100      CONTINUE
         DO 150 I=1, NMAX
150      D(I)=C(I)
60       WRITE (*,*) 'TERM ',N+1,' CALCULATED'
         OPEN (8,FILE='MATMOD.DAT')
         DO 206 I=1,ND
206      WRITE(8,*) EPC*FLOAT(I),A0*D(I)
         CLOSE(8)
         END
```

The program SPECGEN may be used to test either MATMOD or the Fourier-log program FLOG given in Section B.2. It generates a series of Gaussian-shaped "plasmon" peaks, each of the form $\exp[-(1.665E/\Delta E_n)^2]$, whose integrals satisfy Poisson statistics and whose full widths at half maximum are given by

$$(\Delta E_n)^2 = (\Delta E)^2 + n(\Delta E_p)^2 \tag{B.1}$$

Here ΔE is the instrumental FWHM and ΔE_p represents the *natural* width of the plasmon peak. This plural-scattering distribution (starting at an energy $-EZ$ and with the option of adding a constant background BACK) is written to the file SPECGEN.PSD; the single-scattering distribution (with first channel corresponding to $E = 0$) is written to SPECGEN.SSD to allow a direct comparison with the results of deconvolution.

The program simulates noise in an experimental spectrum in terms of two components. Electron–beam shot noise (SNOISE) is taken as the square root of the number of counts (for each order of scattering) but multiplied by a factor FPOISS (= 1 for Poisson noise if the spectral intensity is equal to the number of transmitted electrons, as in electron counting). Background noise (BNOISE), which might represent electronic noise of the electron detector, is taken as the background level multiplied by a factor FBACK. The stochastic numbers RNDNUM (mean amplitude = 1) are generated as rounding errors of arbitrary real numbers RLNUM and are not truly random; they repeat exactly each time the program is run (not necessarily a disadvantage). Setting FPOISS = 0 = FBACK provides a noise-free spectrum.

```
C        SPECGEN.FOR: GENERATES A PLURAL-SCATTERING DISTRIBUTION
C        FROM A GAUSSIAN-SHAPED SSD OF WIDTH WP, PEAKED AT EP,
C        WITH BACKGROUND AND POISSON SHOT NOISE (last update 95OCT20)
         DIMENSION PSD(1024),SSD(1024)
         OPEN(UNIT=1,FILE='SPECGEN.SSD',STATUS='UNKNOWN')
C        contains SSD in xy format, with no background
         OPEN(UNIT=2,FILE='SPECGEN.PSD',STATUS='UNKNOWN')
C        contains the plural scattering distrib. with Z-L at EZ
         WRITE(6,*),'plasmon energy,plasmon FWHM,zero-loss FWHM = '
         READ(5,*) EP,WP,WZ
         WRITE(6,*)'t/lambda,zero-loss counts, zero-loss offset = '
```

```
      READ(5,*) TOL,A0,EZ
      WRITE(6,*)'background level,eV/channel, number of channels = '
      READ(5,*) BACK,EPC,ND
      WRITE(6,*)'background noise fraction, Poisson noise fraction = '
      READ(5,*) FBACK,FPOISS
      SZ = WZ/1.665
      SP = WP/1.665
      HZ = EPC*A0/SZ/1.772
      RLNUM=1.23456
      I = 1
C     calculate intensity at each energy loss E :
  10  E = FLOAT(I)*EPC - EZ
      FAC = 1.
      ORDER = 0.
      PSD(I) = 0.
C     sum contribution from each order of scattering:
  20  SN=SQRT(SZ*SZ+ORDER*SP*SP)
      XPNT=(E-ORDER*EP)**2/SN/SN
      IF(XPNT.GT.20.0) EXPO=0.0
      IF(XPNT.LE.20.0) EXPO=EXP(-XPNT)
      DNE=HZ*SZ/SN*EXPO/FAC*TOL**ORDER
      RNDNUM=2.*(FLOAT(IFIX(RLNUM))-RLNUM)
      SNOISE=FPOISS*(SQRT(DNE)*RNDNUM)
      RLNUM=9.8765*RNDNUM
      IF(ORDER.NE.1.) GO TO 25
      BNOISE=FBACK*BACK*RNDNUM
      SSD(I)=DNE+SQRT(SNOISE*SNOISE+BNOISE*BNOISE)
      WRITE(1,*) E,SSD(I)+BACK
  25  CONTINUE
      PSD(I)=PSD(I)+DNE
      FAC=FAC*(ORDER+1.)
      ORDER=ORDER+1.
      IF(ORDER.LT.15.) GO TO 20
      SNOISE=FPOISS*(SQRT(PSD(I))*RNDNUM)
      WRITE(2,*) E,PSD(I)+SQRT(SNOISE*SNOISE*+BNOISE*BNOISE)+BACK
      I=I+1
      IF(I.LE.ND) GO TO 10
      CLOSE(1)
      CLOSE(2)
      END
```

B.2. Fourier-Log Deconvolution

The program FLOG calculates a single scattering distribution based on Eq. (4.11) and Eq. (4.10). It differs from the Fourier-log program used in the first edition of this book (and in Gatan EL/P software) by avoiding the delta-function approximation inherent in Eq. (4.13), and should therefore be preferable for processing low-loss or core-loss spectra containing sharp peaks.

The first ND data points are read from a named two-column file, which is presumed to consist of energy-intensity pairs of floating point numbers such as the ASCII x-y option provided by Gatan's EL/P software. The background level BACK is estimated from the first five intensity (y) values; if these points are not representative, they should be removed by editing the file, or else BACK set to zero in the program. The eV/channel value EPC is obtained from the first and fifth energy (x) values; the zero-loss channel number NZ is found by detecting positive x-values, on the assumption that the spectrum has been previously calibrated (for the zero-loss peak, at least). The separation point NSEP between the elastic and inelastic components is taken as the subsequent minimum in J(E)/E; the 1/E weighting discriminates against glitches on the zero-loss profile. The zero-loss intensity A0 is taken as the sum of channel counts (above background) up to I=NSEP. Any discontinuities in the data (e.g., gain change during serial recording) are assumed to have been removed by prior editing.

The data are transferred to odd-numbered elements of the array D(J), subtracting any background and shifting the spectrum to the left so that J = 1 corresponds to the zero-loss channel. D(J) is extrapolated to the end of the array (J = MM-1) by fitting the last 10 data channels to an inverse power law, using Eq. (4.51). A cosine-bell function is subtracted to make the data approach zero at the end of the array without causing a discontinuity in intensity or slope at the last recorded data point (J = MFIN). The zero-loss peak Z(J) is copied from D(J) and the discontinuity at the separation point (MSEP) removed by subtracting a cosine-bell function, a procedure which preserves the zero–loss integral as A0. Even-numbered elements of D(J) and Z(J) are set to zero, indicating real data.

An effective width FWHM1 of the zero-loss peak is estimated from the peak height and area, taking the peak shape to be Gaussian, and the operator enters a choice of reconvolution function: either the zero-loss peak (enter a negative number) or a Gaussian peak of specified width FWHM2. If this width is the same as FWHM1, there is no peak sharpening (and no noise amplification) but peak-shape distortion due to an asymmetric $Z(E)$ is corrected. With the direction bit positive (ISIGN = 1), a fast-Fourier subroutine (Higgins, 1976) calculates cosine and sine coefficients of the Fourier transform, replacing the original data in D(J) and Z(J). The Fourier coefficients are manipulated according to Eqs. (4.14)–(4.17) and the phase term θ in Eq. (4.17) extended to $\pm \pi$ to enable scattering parameters up to $t/\lambda \approx 3$ to be accommodated. The higher-J coefficients are attenuated to avoid noise amplification, using a Gaussian filter function or by multiplying by the Fourier transform of Z(E). With ISIGN = -1, the FFT subroutine performs an inverse transform, placing the single-scattering distribution (without zero-loss peak) into odd elements of D(J) and into an output file; prior division by the number NN of real data points ensures that the output is correctly scaled. For spectra which extend to high energy loss and cover a very large dynamic range, the program may need to be modified (for some computers) to make use of double-precision arithmetic.

The fast-Fourier subroutine FFT (Higgins, 1976) uses a process known as the Danielson–Lanczos lemma. The N-point transform is divided into two $N/2$ transforms (by taking alternate data points), each of which is split into two and so on, ending in the requirement for N 1-point transforms provided that N is of the form 2^k where k is an integer. Ordering of the data involves bit-reversal sorting; overall, the number of mathematical operations is reduced from N^2 to approximately $N(\log_2 N)$, a great saving in computing time when N is large. An even shorter subroutine, which does not involve sorting, has been published by Uhrich (1969).

```
C       FLOG.FOR
C       FOURIER-LOG DECONVOLUTION USING EXACT METHODS (A) OR (B)
C       Details in Egerton: EELS in the EM, 2nd edn.(Plenum Press, 1996)
C       Single-scattering distribution is written to the file FLOG.DAT
        DIMENSION DATA(1024),E(1024),D(4096),Z(4096),SSD(4096)
        CHARACTER*12 INFILE
        WRITE(*,*) 'FLOG.FOR: name of input file = '
        READ(5,15) INFILE
    15  FORMAT(A12)
        OPEN(13,FILE=INFILE)
        OPEN(UNIT=14,FILE='FLOG.DAT',STATUS='UNKNOWN')
        NN = 2048
        WRITE (6,*) 'Number of data points to be read = '
        READ(5,*) NSPEC
        DO 100 I=1, NSPEC
        READ(13,*,END=50) E(I),DATA(I)
        ND=I
   100  CONTINUE
    50  EPC=(E(5)-E(1))/4.
        BACK=(DATA(1)+DATA(2)+DATA(3)+DATA(4)+DATA(5))/5.
C       Find zero-loss channel:
        DO 101 I=1,ND
        NZ=I
   101  IF (E(I)+E(I+1).GE.0.) GO TO 102
   102  CONTINUE
C       Find minimum in J(E)/E to separate zero-loss peak:
        DO 201 I=NZ,ND
        IF(DATA(I+1)/FLOAT(I-NZ+1).GT.DATA(I)/FLOAT(I-NZ)) GO TO 202
   201  NSEP = I
   202  SUM = 0.
        DO 205 I=1,NSEP
   205  SUM = SUM +DATA(I)
        A0 = SUM - BACK*FLOAT(NSEP)
        MSEP=2*(NSEP-NZ)+1
        MFIN=2*(ND-NZ)+1
        MM=2*NN
C       TRANSFER SHIFTED DATA TO ARRAY D(J):
        DO 302 J=1,MFIN, 2
   302  D(J)=DATA((J-1)/2+NZ)-BACK
C       EXTRAPOLATE THE SPECTRUM TO ZERO AT J=MM-1:
        A1 = D(MFIN-10)+D(MFIN-12)+D(MFIN-14)+D(MFIN-16)+D(MFIN-18)
        A2 = D(MFIN)+D(MFIN-2)+D(MFIN-4)+D(MFIN-6)+D(MFIN-8)
        R = 2.*ALOG((A1+.2)/(A2+.1))/ALOG(FLOAT(ND-NZ)/FLOAT(ND-NZ-10))
        IF(R.GT.0.) GO TO 303
        R=0.
   303  DEXT = D(MFIN)*(FLOAT(MFIN-1)/FLOAT(MM-4-2*NZ))**R
        DO 304 J=MFIN,MM,2
        COSB = 0.5 - 0.5*COS(3.1416*FLOAT(J-MFIN)/FLOAT(MM-MFIN-2*NZ-3))
   304  D(J) = D(MFIN)*(FLOAT(MFIN-1)/FLOAT(J-1))**R - COSB*DEXT
C       SET IMAGINARY COEFFICIENTS OF SPECTRAL DATA TO ZERO:
        DO 305 J=1,MM,2
        D(J+1)= 0.
        Z(J)= 0.
   305  Z(J+1)= 0.
C       COPY ZERO-LOSS PEAK AND SMOOTH RIGHT-HAND END:
        DO 307 J=1, MSEP
        Z(J)=D(J)-D(MSEP)/2.*(1.-COS(1.571*FLOAT(J-1)/FLOAT(MSEP-1)))
   307  Z(2*MSEP-J)=D(MSEP)/2.*(1.-COS(1.571*FLOAT(J-1)/FLOAT(MSEP-1)))
C       ADD LEFT HALF OF Z(E) TO END CHANNELS OF ARRAYS D AND Z:
        DO 306 I=1,NZ-1
        D(2*I-2*NZ+MM+1) = DATA(I)-BACK
   306  Z(2*I-2*NZ+MM+1) = DATA (I)-BACK
        WRITE(6,*) 'Nread,NZ,NSEP,BACK = ',ND,NZ,NSEP,BACK
        WRITE(6,*) 'eV/channel = ',EPC,', zero-loss intensity = ',A0
        FWHM1=0.9394*A0/D(1)
        WRITE(6,650) FWHM1
   650  FORMAT(' FWHM =',F4.1,' channels; enter new FWHM or -1. ')
        READ(5,*) FWHM2
   551  CALL FFT(NN,+1,Z)
        CALL FFT(NN,+1,D)
C       Process the Fourier coefficients:
        DO 403 J=1,MM,2
        DR=D(J)+1E-10
        DI=D(J+1)
        ZR=Z(J)+1E-10
        ZI=Z(J+1)
        TOP = DI*ZR - DR*ZI
```

```
        BOT = DR*ZR + DI*ZI
        RL=0.5*ALOG(BOT**2+TOP**2)-ALOG(ZR*ZR+ZI*ZI)
        TH=ATAN(TOP/BOT)
C       Extend range of arctan to +/- pi radians:
        IF(BOT.GE.0.0) GO TO 350
        TH = TH + 3.14159265
        IF(TOP.GE.0.0) GO TO 350
        TH = TH - 3.14159265*2.
C       Apply ZLP filter and scaling factor for inverse transform:
  350   D(J) = (ZR*RL-ZI*TH)/FLOAT(NN)
        D(J+1)=(ZI*RL+ZR*TH)/FLOAT(NN)
        IF(FWHM2.LT.0.) GO TO 400
C       Next 5 lines replace ZLP filter with a Gaussian:
        X=1.887*FWHM2*FLOAT(J-1)/FLOAT (MM)
        IF(J.LT.NN) GO TO 380
        X=1.887*FWHM2*FLOAT(MM-J+1)/FLOAT(MM)
  380   GAUSS = 0.
        IF(X.GT.9.0) GO TO 390
        GAUSS = EXP(-X*X)
  390   D(J) = RL*A0*GAUSS/FLOAT(NN)
        D(J+1)=TH*A0*GAUSS/FLOAT(NN)
  400   CONTINUE
  403   CONTINUE
        CALL FFT(NN,-1,D)
        DO 902 J=1,MM,2
  902   WRITE(14,*) EPC*FLOAT((J-1)/2),D(J)
        CLOSE(13)
        CLOSE(14)
        END
C




        SUBROUTINE FFT(NN,ISIGN,D)
        DIMENSION D(4096)
        N=2*NN
        J=1
        DO 5 I=1,N,2
        IF(I-J)1,2,2
  1     TR=D(J)
        TI=D(J+1)
        D(J)=D(I)
        D(J+1)=D(I+1)
        D(I)=TR
        D(I+1)=TI
  2     M=N/2
  3     IF(J-M)5,5,4
  4     J=J-M
        M=M/2
        IF(M-2)5,3,3
  5     J=J+M
        MM=2
  6     ML=MM-N
        IF(ML)7,10,10
  7     IS=2*MM
        TH=6.283185/FLOAT(ISIGN*MM)
        HT=TH/2.
        ST=SIN(HT)
        W1=-2.*ST*ST
        W2=SIN(TH)
        WR=1.
        WI=0.
        DO 9 M=1,MM,2
        DO 8 I=M,N,IS
        J=I+MM
        A=WR*D(J)
        B=WI*D(J+1)
        TR=A-B
        TI=WR*D(J+1)+WI*D(J)
        D(J)=D(I)-TR
  11    D(J+1)=D(I+1)-TI
        D(I)=D(I)+TR
  8     D(I+1)=D(I+1)+TI
        TR=WR
        WR=WR*W1-WI*W2+WR
  9     WI=WI*W1+TR*W2+WI
        MM=IS
        GO TO 6
  10    RETURN
        END
```

B.3. *Kramers–Kronig Analysis and Thickness Determination*

The program KRAKRO calculates the real part $\varepsilon_1(E)$ and imaginary part $\varepsilon_2(E)$ of the dielectric function, the specimen thickness t, and the mean free path $\lambda(\beta)$ for inelastic scattering. It employs the Fourier procedure for Kramers–Kronig analysis described by Johnson (1972), but using fast-Fourier transforms. As input, it requires a single-scattering distribution starting at the first channel, together with appropriate values of the incident-electron energy, energy increment per channel, collection semiangle β and optical refractive index n. In the case of a metallic specimen, a large value (>100) should be entered for n.

The SSD is read into an array SSD(I) and copied to odd elements of the array D(J). Assuming a Lorentzian angular distribution, an aperture correction is applied to the intensity S(E) to make it proportional to Im$(-1/\varepsilon)$; the proportionality constant RK is evaluated by utilizing the Kramers–Kronig sum rule, Eq. (4.27). Since RK $= I_0 t/(\pi a_0 m_0 v^2)$ according to Eq. (4.26), this leads to an initial estimate of specimen thickness and mean free path, evaluated as $\lambda = t/(t/\lambda) = t I_0/I_1$ where I_1 is the integral of the SSD intensity. Im$(-1/\varepsilon)$ is copied to the array DI before being converted to its Fourier transform. Even-number elements of D(J) then contain the sine transform of Im$(-1/\varepsilon)$. These coefficients are transferred to odd-number elements so that inverse (cosine) transformation yields Re$(1/\varepsilon)-1$, accompanied by its reflection about the midrange (J = NN) axis, due to aliasing. Taking the high-energy tail to be proportional to E^{-2}, this energy dependence is subtracted from the low-energy (J < NN) data and used to extrapolate the high-energy values (a procedure which becomes less critical as the number of real data points NN is increased).

The real part EPS1 and imaginary part EPS2 of ε are computed, followed by the surface energy-loss function SRFELF and the surface-scattering intensity SRFINT, and written to the output file EPSILON.DAT. Calculation of the surface-mode scattering is based on Eq. (4.31), assuming clean (unoxidized) and smooth surfaces which are perpendicular to the incident beam, and neglecting coupling between the surfaces $(1/R_c = 1 + \varepsilon)$. The volume-loss intensity, obtained by subtracting SRFINT from SSD(J), is then renormalized by applying the K–K sum rule, leading to revised estimates for the specimen thickness and inelastic mean free path. Kramers–Kronig analysis is then repeated to yield revised values of the dielectric data.

By setting NLOOPS > 2, further iterations are possible. Whether convergence is obtained depends largely on the behaviour of the data at low energy loss (E < 5 eV). To aid stability, E has been replaced by E + 1 in the expression for the surface-scattering angular dependence ANGDEP, thereby avoiding a non-zero value of SRFINT at E = 0. This modification does not significantly bias the thickness estimates.

```
C     KRAKRO.FOR                                              Last update 95NOV05
C     Kramers-Kronig analysis using Johnson method and FFT subroutine
C     as described in EELS in the Electron Microscope (2nd edition).
C     Program generates output into the 5-column file KRAKRO.DAT
      DIMENSION EN(4096), D(8192), DI(8192), DS(4096), SSD(4096)
      CHARACTER*12 INFILE
      WRITE(*,*) 'NAME OF INPUT FILE = '
      READ(6,6) INFILE
    6 FORMAT(A12)
      NN=4096
      MM=2*NN
      DO 10 I=1,NN
   10 SSD(I)=0.
      OPEN(3,FILE=INFILE)
      DO 12 I = 2, NN, 1
      READ(3,*,END=50) EN(I),SSD(I)
   12 D(2*I-1) = SSD(I)
   50 D(1)=0.
      SSD(1) = 0.
      EPC=(EN(5)-EN(1))/4.
      WRITE(6,*) 'zero-loss sum, E0(keV),BETA(mrad),ref.index = '
      READ(5,*) A0,E0,BETA,RI
      WRITE(5,*) 'no. of iterations, no. of lines into KRAKRO.DAT = '
      READ(5,*) NLOOPS,NLINES
      T = E0*(1.+E0/1022.12)/(1.+E0/511.06)**2
      RK0 = 2590. * (1.+E0/511.06)*SQRT(2.*T/511.06)
      TGT = E0*(1022.12 + E0)/(511.06 + E0)
      OPEN(4,FILE='KRAKRO.DAT',STATUS='UNKNOWN')
      LOOP = 1
C     Calculate Im(-1/EPS), Re(1/EPS), EPS1, EPS2 and SRFINT data:
      DO 85 NUM=1,NLOOPS
      SUM = 0.
      AREA = 0.
C     Apply aperture correction APC at each energy loss:
      Do 15 J=3,MM,2
      AREA = AREA+D(J)
      E=EPC*FLOAT(J-1)/2.
      APC=ALOG(1.+(BETA*TGT/E)**2)
      D(J)=D(J)/APC
   15 SUM=SUM+D(J)/E
      RK=SUM/1.571/(1.-1./RI/RI)*EPC
      TNM = 332.5*RK/(A0*EPC)*E0*(1.+E0/1022.12)/(1.+E0/511.06)**2
      TOL = AREA/A0
      RLAM = TNM/TOL
      WRITE(6,61) NUM,TNM,TOL,RLAM
   61 FORMAT(' LOOP',I2,': t(nm) =',F6.1,', t/lambda =',F5.2,
     &', lambda(nm) =',F6.1)
C     Apply normalization factor RK; store Im(-1/EPS) in array DI:
      DO 20 J=1,MM,2
      D(J)=D(J)/RK
      DI((J+1)/2)=D(J)
   20 D(J+1)=0
      CALL FFT(NN,+1,D)
C     Transfer sine coefficients to cosine locations:
      DO 30 J=1,MM,2
      D(J)=2.*D(J+1)*FLOAT((J-NN)/IABS(J-NN))/FLOAT(NN)
   30 D(J+1)=0.
      CALL FFT(NN,-1,D)
      DO 40 J=1,NN,2
C     Correct the even function for reflected tail:
      D(J)=D(J)+1.-D(NN-1)/2.*(FLOAT(NN-1)/FLOAT(MM-J))**2
   40 D(NN+J) =1.+D(NN-1)*(FLOAT(NN-1)/FLOAT(NN+J))**2/2.
      T = E0*(1.+E0/1022.12)/(1.+E0/511.06)**2
      RK0 = 2590. * (1.+E0/511.06)*SQRT(2.*T/511.06)
      DO 80 J=3,MM,2
      E=FLOAT((J-1)/2)*EPC
      RE = D(J)
      DEN=D(J)*D(J)+DI((J+1)/2)*DI((J+1)/2)
      EPS1=D(J)/DEN
      EPS2=DI((J+1)/2)/DEN
C     Calculate surface energy-loss function and surface intensity:
      SRFELF = 4.*EPS2/((1.+EPS1)**2+EPS2**2) - DI((J+1)/2)
      ADEP=TGT/(E+1.)*ATAN(BETA*TGT/E)-BETA/1000./(BETA**2+E**2/TGT**2)
      SRFINT = 2000.*RK/RK0/TNM*ADEP*SRFELF
      D(J) = SSD((J+1)/2)-SRFINT
      IF (NUM.NE.NLOOPS) GO TO 69
      IF (J/2.GT.NLINES) GO TO 69
```

```
      WRITE(4,66) E,EPS1,EPS2,RE,DI((J+1)/2)
   66 FORMAT(6F12.4)
   69 CONTINUE
   80 CONTINUE
      D(1) = 0.
   85 CONTINUE
      CLOSE(4)
      END
```

For testing KRAKRO (or other purposes), the FORTRAN program DRUDE calculates the single-scattering plasmon-loss spectrum for a specimen of a given thickness TNM (in nm), recorded with electrons of a specified incident energy E0 by a spectrometer which accepts scattering up to a specified collection semiangle BETA. It is based on the free-electron model (Section 3.3.1), with the volume energy-loss function ELF given by Eq. (3.42) and the surface-scattering energy-loss function SRFELF as in Eq. (4.31). The surface term can be made negligible by entering a large specimen thickness. The spectral intensity is written to the file DRUDE.SSD, while the real and imaginary parts of the dielectric function are written to DRUDE.DAT for comparison with the results of Kramers–Kronig analysis.

```
C        DRUDE.FOR                                            Last update 95OCT02
C        Given the plasmon energy (EP) and plasmon FWHM (EW), this program
C        generates EPS1, EPS2 from Eq. (3.40), ELF=Im(-1/EPS) from Eq. (3.42),
C        single scattering intensities VOLINT from Eq. (4.26) and SRFINT
C        from Eq. (4.31) of EELS in the EM (Plenum Press, 2nd edition).
C        The output is E,SSD into the file DRUDE.SSD and
C        E,EPS1,EPS2,Re(1/eps),Im(-1/eps) into DRUDE.DAT
C
         WRITE(6,*) 'plasmon energy,plasmon width,eV/channel = '
         READ(5,*) EP,EW,EPC
         WRITE(6,*) 'Beta(mr),E0(keV),t(nm) = '
         READ(5,*) BETA,E0,TNM
         WRITE(6,*) 'zero-loss integral, number of data points = '
         READ(5,*) A0,NN
         B = BETA/1000.
         T = 1000.*E0*(1.+E0/1022.12)/(1.+E0/511.06)**2
         TGT = 1000.*E0*(1022.12 + E)/(511.06 + E0)
         RK0 = 2590. * (1.+E0/511.06)*SQRT(2.*T/511060)
         OPEN(14,FILE='DRUDE.DAT',STATUS='UNKNOWN')
         OPEN(13,FILE='DRUDE.SSD',STATUS='UNKNOWN')
         DO 12 IW = 2, NN, 1
         E = EPC*FLOAT(IW - 1)
         EPS1 = 1. - EP**2/(E**2+EW**2)
         EPS2 = EW*EP**2/E/(E**2+EW**2)
         ELF = EP**2*E*EW/((E**2-EP**2)**2+(E*EW)**2)
         REREPS = EPS1/(EPS1*EPS1+EPS2*EPS2)
         THE = E/TGT
         SRFELF = 4.*EPS2/((1.+EPS1)**2+EPS2**2) - ELF
         ANGDEP = ATAN(B/THE)/THE - B/(B*B+THE*THE)
         SRFINT = EPC*A0*ANGDEP*SRFELF/(3.1416*0.0529*RK0*T)
         ANGLOG = ALOG(1.+ B*B/THE/THE)
         VOLINT = EPC*A0/3.1416*TNM/0.0529/T/2.*ELF*ANGLOG
         SSD = VOLINT + SRFINT
         WRITE(13,*) E,SSD
   12    WRITE(14,*) E,EPS1,EPS2,REREPS,ELF
         CLOSE(14)
         CLOSE(13)
         STOP
         END
```

B.4. Fourier-Ratio Deconvolution

The program FRAT removes plural scattering from an ionization edge (whose background has previously been subtracted) using the Fourier-ratio method described in Section 4.3.2. It requires a low-loss spectrum, recorded from the same region of specimen at the same eV/channel, but this spectrum need not be contiguous with the core-loss region or match it in terms of absolute intensity. The method is therefore a more practical alternative to the Fourier-log program in the case of ionization edges recorded using a parallel-recording spectrometer. Other advantages are that the zero-loss peak does not need to be extracted from the low-loss spectrum (which involves some approximation) and that the specimen thickness is in principle unlimited, since there are no phase ambiguities in the Fourier components.

The low-loss spectrum is read as two-column $(x–y)$ data (up to 1024 channels) from a named file and the zero-loss channel found from the energy (x) data. The first minimum is found in order to estimate the zero-loss integral A0 and the energy resolution (obtained from A0 and the zero-loss peak height, assuming a Gaussian shape). The spectrum is transferred to odd elements of the working array D(J), shifted so that the first channel represents zero loss and with any background (average of the first five channels) subtracted. D(J) is extrapolated to zero at the last odd-numbered channel, using a power-law extrapolation and a cosine-bell termination, and the left half of the zero-loss peak added to the end channels. The energy resolution (FWHM of the zero-loss peak) is printed to serve as a guide in specifying the width of the reconvolution function; smaller widths lead to peak sharpening but with a noise penalty (page 266). Even without such sharpening, the effect of any tails on the zero-loss peak (due for example to the point-spread function of a parallel-recording detector) is removed from the core-loss data.

The core-loss spectrum is read into odd elements of an array C(J) and extrapolated to zero in the same way as D(J). After taking Fourier transforms, using the FFT subroutine listed in Section B.2, the Fourier coefficients are processed according to Eq. (4.38) and Eq. (4.43), with a Gaussian reconvolution function GAUSS. If the coefficient of this function is the zero-loss integral (A0), plural scattering is *subtracted* from the ionization-edge intensity; if the coefficient is changed to the total integral (AT) of the low-loss spectrum, the core-loss SSD will have the same integral as the original edge, as required for absolute analysis of thick specimens (Wong and Egerton, 1995).

```
C        FRAT.FOR                                           Last update: 95OCT31
C        FOURIER-RATIO DECONVOLUTION USING EXACT METHOD (A)
C        (R.F.Egerton: EELS in the Electron Microscope, 2nd edition)
C        WITH LEFT SHIFT BEFORE FORWARD TRANSFORM.
C        RECONVOLUTION FUNCTION R(F) IS EXP(-X*X) .
C        DATA IS READ IN FROM NAMED INPUT FILES (units 14 and 15)
C        OUTPUT DATA APPEARS in named OUTFILE (unit 16) as NC x-y PAIRS
C
         DIMENSION DATA(1024),E(1024),D(4096),C(4096)
         CHARACTER*12 LFILE,CFILE,OUTFILE
         NN=2048
         WRITE(*,*) 'NAME OF LOWLOSS FILE = '
         READ(5,15) LFILE
     15  FORMAT(A12)
         OPEN(15,FILE=LFILE)
         WRITE (6,*) 'NUMBER OF LOWLOSS CHANNELS TO BE READ = '
         READ(5,*) NREAD
         DO 501 I=1,NREAD
         READ(15,*,END=50) E(I),DATA(I)
    501  ND=I
     50  CLOSE(15)
         EPC=(E(5)-E(1))/4.
         BACK=DATA(1)+DATA(2)+DATA(3)+DATA(4)+DATA(5)
C        Set BACK=0. if zero-loss tail dominates first 5 channels
C        Find zero-loss channel:
         DO 101 I=1,ND
         NZ=I
    101  IF (E(I)+E(I+1).GE.0.) GO TO 102
    102  CONTINUE
C        Find minimum in J(E)/E to estimate zero-loss sum A0:
         DO 201 I=NZ,ND
         IF(DATA(I+1)/FLOAT(I-NZ+1).GT.DATA(I)/FLOAT(I-NZ)) GO TO 202
    201  NSEP=I
    202  SUM=0.
         DO 205 I3=1,NSEP
    205  SUM = SUM + DATA(I3)
         A0 = SUM - BACK*FLOAT(NSEP)
         MSEP=2*(NSEP-NZ)+1
         MFIN=2*(ND-NZ)+1
         MM=2*NN
C        TRANSFER SHIFTED DATA TO ARRAY D(J):
         DO 302 J=1,MFIN,2
    302  D(J)=DATA((J-1)/2+NZ)-BACK
C        EXTRAPOLATE THE SPECTRUM TO ZERO AT J=MM-1:
         A1 = D(MFIN-10)+D(MFIN-12)+D(MFIN-14)+D(MFIN-16)+D(MFIN-18)
         A2 = D(MFIN)+D(MFIN-2)+D(MFIN-4)+D(MFIN-6)+D(MFIN-8)
         R = 2.*ALOG((A1+.2)/(A2+.1))/ALOG(FLOAT(ND-NZ)/FLOAT(ND-NZ-10))
         DEND = D(MFIN)*(FLOAT(MFIN-1)/FLOAT(MM-4-2*NZ))**R
         DO 304 J=MFIN,MM,2
         COSB = 0.5 - 0.5*COS(3.1416*FLOAT(J-MFIN)/FLOAT(MM-MFIN-2*NZ-3))
    304  D(J) = D(MFIN)*(FLOAT(MFIN-1)/FLOAT(J-1))**R - COSB*DEND
C        Compute total area and set imaginary coefficients to zero:
         AT=0.
         DO 305 J=1,MM,2
         AT=AT+D(J)
         D(J+1)= 0.
         C(J)= 0.
    305  C(J+1)= 0.
C        Add left half of Z(E) to end channels in the array D(J):
         DO 306 I=1,NZ-1
    306  D(2*I-2*NZ+MM+1) = DATA(I)-BACK
         WRITE(6,*) 'ND,NREAD,NZ,NSEP,DATA(NZ),BACK,A0,THRESH,DEND,EPC:'
         WRITE(6,640) ND,NREAD,NZ,NSEP,DATA(NZ),BACK,A0,THRESH,DEND,EPC
    640  FORMAT(4I6,6E14.3)
         FWHM1=0.9394*A0/D(1)*EPC
         WRITE(6,650) FWHM1
    650  FORMAT('FWHM = ',F4.1,' eV; FWHM of coreless SSD = ')
         READ(5,*) FWHM2
         FWHM2=FWHM2/EPC
C
         WRITE(*,*) 'Name of coreloss file = '
         READ(5,17) CFILE
     17  FORMAT(A12)
         OPEN(14,FILE=CFILE)
         WRITE (6,*) 'NUMBER OF CORELOSS CHANNELS TO BE READ = '
         READ(5,*) NC
         DO 551 I=1,NC
         READ(14,*,END=55) E(I),C(2*I-1)
         NREAD=I
    551  CONTINUE
     55  EPC=(E(5)-E(1))/4.
         NC=NREAD
         CLOSE(14)
```

```
C         EXTRAPOLATE THE SPECTRUM TO ZERO AT J=MM-1:
          MFIN=2*NC-1
          A1 = C(MFIN-10)+C(MFIN-12)+C(MFIN-14)+C(MFIN-16)+C(MFIN-18)
          A2 = C(MFIN+C(MFIN-2)+C(MFIN-4)+C(MFIN-6)+C(MFIN-8)
          R = 2.*ALOG((A1+.2)/(A2+.1))/ALOG(E(NC)/E(NC-9))
          CEND = A2/5.*(E(NC-2)/(E(1)+EPC*FLOAT(NN-1)))**R
          DO 314 J=MFIN,MM,2
          COSB = 0.5 - 0.5*COS(3.1416*FLOAT(J-MFIN)/FLOAT(MM-1-MFIN))
   314    C(J)=A2/5.*(E(NC-2)/(E(1)+EPC*FLOAT((J-5)/2)))**R-COSB*CEND
          CALL FFT(NN,+1,C)
          CALL FFT(NN,+1,D)
C         PROCESS THE FOURIER COEFFICIENTS:
          DO 403 J=1,MM,2
C         GAUSSIAN RECONVOLUTION FUNCTION:
          X=1.887*FWHM2*FLOAT(J-1)/FLOAT(MM)
          IF(J.LT.NN) GO TO 381
          X=1.887*FWHM2*FLOAT(MM-J+1)/FLOAT(MM)
   381    GAUSS = A0/2.718**(X*X)/FLOAT(NN)
C         Replace A0 by AT for equal areas in SSD and PSD.
          DR=D(J)+1E-10
          DI=D(J+1)
          CR=C(J)+1E-10
          CI=C(J+1)
          D(J) = GAUSS*(CR*DR+CI*DI)/(DR*DR+DI*DI)
          D(J+1)=GAUSS*(CI*DR-CR*DI)/(DR*DR+DI*DI)
   403    CONTINUE
          CALL FFT(NN,-1,D)
          WRITE(*,*) 'NAME OF OUTPUT FILE = '
          READ(5,15) OUTFILE
          OPEN(16,FILE=OUTFILE)
          DO 902 I=1,NC
          WRITE(16,*) E(I),D(2*I-1)
   902    CONTINUE
          CLOSE(16)
          END
```

B.5. Incident-Convergence Correction

The following program (CONCOR2) evaluates the factor F_1 by which inelastic intensity (at energy loss E and recorded using a collection semi-angle β) is reduced as a result of the convergence of the electron beam (semiangle α). The program also evaluates a factor F_2 (for use in absolute quantification) and an effective collection angle defined in Section 4.5.3. For inner-shell scattering, the energy loss E can be taken as the edge energy E_k or (more exactly) as an average energy loss $(E_k+\Delta/2)$ within the integration window.

Whereas the program (CONCOR) given in the first edition was based on Eq. (4.71), this version uses an analytical formula (Scheinfein and Isaacson, 1984) based on a Lorentzian angular distribution of inelastic scattering and assuming that the incident-beam intensity per steradian is constant up to the angle α. Double-precision arithmetic is used because the formula involves subtraction of terms which are nearly equal. Minor differences in output, compared to the earlier program, arise from correction of an error in the expression previously used to evaluate the characteristic angle THE.

When analyzing for two elements, a and b, incident-beam convergence is taken into account by multiplying the areal-density ratio N_a/N_b, derived from Eq. (4.66), by F_{1b}/F_{1a}. If the absolute areal density N_a of an element a is being calculated from Eq. (4.65), the result should be divided by F_{2a}. For $\alpha < \beta$, $F_2 = F_1$; for $\alpha > \beta$, F_2 is larger than F_1 (and may exceed unity; see Fig. 4.16) since the collection angle cuts off part of the low-loss angular cone. As a simpler alternative to applying the correction factors F_1 or F_2, incident-beam convergence can be incorporated by computing each ionization cross section for the *effective* collection angle BSTAR which is a function of energy loss and therefore different for each element.

```
C         CONCOR2: EVALUATION OF CONVERGENCE CORRECTION F USING THE
C         FORMULAE OF SCHEINFEIN AND ISAACSON (SEM/1994, PP. 1685-6).
C         FOR ABSOLUTE QUANTITATION, DIVIDE THE AREAL DENSITY BY F2.
C         FOR ELEMENTAL RATIOS, DIVIDE EACH CONCENTRATION BY F2 OR F1.
C
C         ALPHA AND BETA SHOULD BE IN MRAD, E IN EV, E0 IN KEV .
C
          DOUBLE PRECISION ALPHA,BETA,THE,A2,B2,T2,F1,F2
          WRITE(6,601)
      601 FORMAT('0','CONCOR2: Enter ALPHA(mr),BETA(mr),E(eV),E0(keV)'/)
          READ(5,*) ALPHA,BETA,E,E0
      501 FORMAT(4F10.0)
          TGT=E0*(1.+E0/1022.)/(1.+E0/511.)
          THETAE=E/TGT
C         A2,B2,T2 ARE SQUARES OF ANGLES IN RADIANS**2
          A2=ALPHA*ALPHA*1E-6
          B2=BETA*BETA*1E-6
          T2=THETAE*THETAE*1E-6
          ETA1=DSQRT((A2+B2+T2)**2-4.*A2*B2)-A2-B2-T2
          ETA2=2.*B2*DLOG(0.5/T2*(DSQRT((A2+T2-B2)**2+4.*B2*T2)+A2+T2-B2))
          ETA3=2.*A2*DLOG(0.5/T2*(DSQRT((B2+T2-A2)**2+4.*A2*T2)+B2+T2-A2))
          ETA=(ETA1+ETA2+ETA3)/A2/DLOG(4./T2)
          F1=(ETA1+ETA2+ETA3)/2/A2/DLOG(1.+B2/T2)
          F2=F1
          IF (ALHPA/BETA.LE.1.)GO TO 107
          F2=F1*A2/B2
      107 CONTINUE
          BSTAR=THETAE*DSQRT(DEXP(F2*DLOG(1.+B2/T2))-1.)
          WRITE(6,602)
      602 FORMAT(/' ',5X,'F1',8X,'F2',6X,'effective BETA'/)
          WRITE(6,600)F1,F2,BSTAR
      600 FORMAT(2F10.3,F14.2)
          END
```

sample data:
```
          CONCOR2: Enter ALPHA(mr),BETA(mr),E(eV),E0(keV)
          18,12,500,100

          F1          F2       effective BETA
          0.571       1.285       16.05
```

B.6. Hydrogenic K-Shell Cross Sections

The FORTRAN program SIGMAK3 uses the hydrogenic approximation for generalized oscillator strength, Eqs. (3.125)–(3.127), in order to calculate differential (DSBYDE) and integrated cross sections and dipole oscillator strengths (SIGMA and $f0$) for K-shell ionization. Unlike the corresponding program (SIGMAK2) given in the first edition of this book, the reduction in effective nuclear charge (due to screening by the second $1s$ electron) is taken as 0.50 rather than the value of 0.3125 calculated by Zener (1930) for first-row elements, in order to provide a closer match to EELS, photoabsorption, and Hartree–Slater data (Egerton, 1993). The changes becomes more significant at low atomic number: $f(100 \text{ eV})$ is reduced from 0.46 to 0.42 for oxygen and from 2.02 to 1.58 for lithium. Two lines have been added before label 101 to prevent occasional error messages (from some compilers) when $\exp(-k_H)$ becomes too small.

Relativistic kinematics are employed, based on Eqs. (3.139), (3.140), (3.144), and (3.146), but retardation effects (Appendix A) are not included. The energy-differential cross section DSBYDE is obtained from Eq. (3.151) with limits of integration given by Eqs. (3.152) and (3.153). The outer DO loop integrates DSBYDE to obtain the partial cross section SIGMA, making use of the energy dependence described by Eq. (3.154). IMAX sets the number of increments within the inner DO loop (integration over scattering vector); IMAX = 5 is sufficient for the evaluation of energy-loss partial cross sections (small β and Δ), but a larger number should be used when calculating total cross sections in order to accurately include the Bethe ridge. Because Eq. (3.154) is utilized in the integration over energy loss, relatively few energy increments are needed; EINC can be chosen as a convenient submultiple (e.g., 1/5 or 1/10) of the required energy window Δ. Typical input and output data are shown after the program listing.

Total K-shell cross sections (as required for EDX spectroscopy) are obtained by entering $\beta = 3142$ mrad, EINC $\approx$ EK/10, and taking the asymptotic value of SIGMA corresponding to large Δ; for $E_0 \leq 300\text{keV}$, the resulting values are within 3% of the Hartree–Slater values given by Scofield (1978) for Ar and Ni. Since the energy-differential cross section $d\sigma/dE$ is independent of K-edge threshold energy EK, its value DSBYDE at any scattering angle BETA and energy loss E can be printed out by setting EINC=0 and EK=E in the input data. Likewise, a cross section SIGMA integrated between any two values of energy loss can be obtained by entering the lower energy loss as EK and some submultiple of the energy difference as EINC.

```
C        SIGMAK3 : CALCULATION OF K-SHELL IONIZATION CROSS SECTIONS
C        USING RELATIVISTIC KINEMATICS AND A HYDROGENIC MODEL WITH
C        INNER-SHELL SCREENING CONSTANT OF 0.5 (last update: 31Aug95)
C

C        INPUT DATA IS ENTERED AS REAL OR INTEGER NUMBERS,
C        ON THE SAME LINE AND EACH FOLLOWED BY A COMMA, AS FOLLOWS:
             Z - atomic number of the element of interest
            EK - K-shell ionization energy, in eV
          EINC - energy increment of output data, in eV
            E0 - incident-electron energy, in keV
          BETA - maximum scattering angle (in milliradians)
                   contributing to the cross section
C
         REAL KH2, LNQA02,LQA021,LQA02M,LQ2INC
         WRITE(6,601)
   601   FORMAT('0','Z,EK,EINC,E0,BETA = '/)
         READ *,Z,EK,EINC,E0,BETA
         WRITE(6,602)
   602   FORMAT(' ',5X,'E(EV)',7X,'DSBYDE',8X,'DELTA',7X,'SIGMA',7X,'f0'/)
         IMAX=100
         R=13.606
         ZS=1.
```

```
         RNK=1.
         IF (Z.EQ.1.) GO TO 105
         ZS = Z - 0.50
         RNK=2.
     105 E = EK
         B = BETA/1000.
         T = 511060.*(1.-1./(1.+E0/(511.06))**2)/2.
         GG=1.+E0/511.06
         P02=T/R/(1.-2.*T/511060.)
         F = 0.
         S = 0.
         SIGMA = 0.
C        CALCULATE FOR EACH ENERGY LOSS:
         DO 111 J=1,30
         QA021 = E**2/(4.*R*T)+E**3/(8.*R*T**2*GG**3)
         PP2 = P02 - E/R*(GG-E/1022120.)
         LQA021 = ALOG(QA021)
         QA02M = QA021 + 4.*SQRT(P02*PP2)*(SIN(B/2))**2
         LQA02M = ALOG(QA02M)
         LQ2INC = (LQA02M - LQA021)/FLOAT(IMAX-1)
         LNQA02 = LQA021
         DFDIP = 0.
         DSBYDE = 0.
         GOSP = 0.
C        INTEGRATE OVER SCATTERING ANGLE:
         DO 109 I=1, IMAX
     100 QA02 = EXP(LNQA02)
         Q = QA02/ZS**2
         KH2 = E/R/ZS**2 - 1.
         AKH = SQRT(ABS(KH2))
         IF(AKH.GT.0.1) GO TO 101
         AKH = 0.1
     101 IF (KH2.LT.0.) GO TO 103
         D = 1. - EXP(-2.*3.14159/AKH)
         BP = ATAN(2.*AKH/(Q-KH2+1.))
         IF (BP.GE.0.) GO TO 104
         BP = BP + 3.14159
     104 C = EXP((-2./AKH)*BP)
         IF (KH2.GE.0.) GO TO 102
C        SUM OVER EQUIVALENT BOUND STATES:
     103 D = 1.
         Y = (-1./AKH*ALOG((Q+1.-KH2+2.*AKH)/(Q+1.-KH2-2.*AKH)))
         C = EXP(Y)
     102 A = ((Q-KH2+1.)**2 + 4.*KH2)**3
         GOS=128.*RNK*E/R/ZS**4*C/D*(Q+KH2/3.+1./3.)/A/R
C        GOS (=DF/DE) IS PER EV AND PER ATOM
         DSBYDE=DSBYDE+3.5166E-16*(R/T)*(R/E)*(GOS+GOSP)/2.*LQ2INC
C        DSBYDE IS THE ENERGY-DIFFERENTIAL X-SECN (CM**2/EV/ATOM)
         IF (I.GT.1) GO TO 115
         DFDIPL = GOS
         DSBYDE = 0.
     115 LNQA02 = LNQA02 + LQ2INC
     109 GOSP = GOS
         DELTA = E - EK
         IF (J.EQ.1) GO TO 117
         S = ALOG(DSBDEP/DSBYDE)/ALOG(E/(E-EINC))
         SGINC = (E*DSBYDE-(E-EINC)*DSBDEP)/(1.-S)
         SIGMA = SIGMA + SGINC
C        SIGMA IS THE PARTIAL CROSS SECTION IN CM**2 PER ATOM
         F=F+(DFDIPL+DFPREV)/2.*EINC
     117 WRITE(6,605) E,DSBYDE,DELTA,SIGMA,F
     605 FORMAT(' ',F10.1,2X,E12.3,2X,F10.1,2X,E12.3,2X,F7.3)
         IF (EINC.EQ.0.) GO TO 112
         IF (DELTA.LT.100.) GO TO 107
         IF (SGINC.LT.0.001*SIGMA) GO TO 112
         EINC = EINC*2
     107 E = E + EINC
         IF (E.GT.T) GO TO 112
         DFPREV = DFDIPL
     111 DSBDEP = DSBYDE
     112 STOP
         END
```

Z,EK,EINC,E0,BETA = 6,284,10,100,10.

E(EV)	DSBYDE	DELTA	SIGMA	f0
284.0	0.110E−21	0.0	0.000E+00	0.000
294.0	0.955E−22	10.0	0.102E−20	0.096
304.0	0.835E−22	20.0	0.192E−20	0.184
314.0	0.733E−22	30.0	0.270E−20	0.264
324.0	0.645−.22	40.0	0.339E−20	0.339
334.0	0.570E−22	50.0	0.399E−20	0.407
344.0	0.505E−22	60.0	0.453E−20	0.470
354.0	0.449E−22	70.0	0.501E−20	0.529
364.0	0.400E−22	80.0	0.543E−20	0.584
374.0	0.358E−22	90.0	0.581E−20	0.635
384.0	0.320E−22	100.0	0.615E−20	0.682
404.0	0.259E−22	120.0	0.673E−20	0.768
444.0	0.173E−22	160.0	0.758E−20	0.910
524.0	0.844E−23	240.0	0.856E−20	1.113
684.0	0.254E−23	400.0	0.932E−20	1.349
1004.0	0.409E−24	720.0	0.967E−20	1.553
1644.0	0.326E−25	1360.0	0.976E−20	1.682
2924.0	0.133E−26	2640.0	0.977E−20	1.740
5484.0	0.332E−28	5200.0	0.977E−20	1.761

B.7. Modified-Hydrogenic L-Shell Cross Sections

The FORTRAN program SIGMAL3 evaluates cross sections (in cm^2) for L-shell ionization by fast incident electrons. It uses relativistic kinematics (without retardation) and an expression for the generalized oscillator strength (Choi *et al.*, 1973) based on hydrogenic wavefunctions, with screening constants recommended by Slater (1930). To more accurately match observed edge shapes, GOS is modified by means of a correction factor RF, calculated for each energy loss through the use of an empirical parameter U. Values of U, together with the L_3 and L_1 threshold energies of each element, are stored in a DATA table at the beginning of the program. Approximate allowance for white-line peaks, for $18 \leq Z \leq 28$, is made by using the full-hydrogenic oscillator strength (RF = 1) within 20 eV of the L_3 threshold. In other respects, the calculation follows the same procedure as SIGMAK3.

Differences between this program and the version (SIGMAL2) given in the first edition are as follows. Values of U have been modified where necessary to ensure that the program provides integrated oscillator strengths $f(100$ eV$)$ equal to the recommended values given in Egerton (1993), which are based on Hartree–Slater, EELS, and photoabsorption data. The DATA table has been extended to $Z = 36$, so that the program gives a sensible output for the elements Al to Kr inclusive, although probably with less accuracy towards the ends of that range. The expression for QA02M is now identical to that used in the SIGMAK3 program and the energy range (of E-EL3) over which RF is set to unity has been set explicitly to 20 eV.

Total L-shell cross sections can be obtained by entering $\beta = 3142$ mrad and taking the asymptotic value of SIGMA (corresponding to large Δ); for E0 $\leq$ 300 keV, the program yields cross sections which are within

8% of those calculated by Scofield (1978) for Ar and Ni. However, the algorithm is not designed to provide realistic values of differential cross section DSBYDE within 50 eV of the ionization edge or to accurately simulate the Bethe ridge at high scattering angle.

```
C         SIGMAL3 : CALCULATION OF L-SHELL CROSS-SECTIONS USING A
C         MODIFIED HYDROGENIC MODEL WITH RELATIVISTIC KINEMATICS.
C         THE GOS IS REDUCED BY A SCREENING FACTOR RF, BASED ON DATA
C         FROM SEVERAL SOURCES; SEE ULTRAMICROSCOPY 50 (1993) P.22.
C         Last update: 95Sep01
          REAL KH2,LNQA02,LQA021,LQA02M,LQ2INC
          DIMENSION XU(24),IE3(24),IE1(24)
          DATA XU/.52,.42,.30,.29,.22,.30,.22,.16,.12,.13,.13,.14,.16,
        @.18,.19,.22,.14,.11,.12,.12,.12,.10,.10,.10/,
        &IE3/73,99,135,164,200,245,294,347,402,455,513,575,641,
        @710,779,855,931,1021,1115,1217,1323,1436,1550,1675/,
        &IE1/118,149,189,229,270,320,377,438,500,564,628,695,769,
        @846,926,1008,1096,1194,1142,1248,1359,1476,1596,1727/
          WRITE(6,601)
   601    FORMAT('0','SIGMAL3: VALUES OF Z, E0(keV) AND BETA(mr) ARE: ')
          READ(5,*) Z,E0,BETA
   501    FORMAT(3F10.0)
          WRITE(6,602)
   602    FORMAT(/' ',5X,'E(eV)',7X,'DSBYDE',8X,'DELTA',7X,'SIGMA',8X,'f0'/)
          IMAX=10
          EINC = 10.
          R = 13.606
          ZS = Z - 0.35*(8.-1.) - 1.7
          IZ = IFIX(Z)-12
          U = XU(IZ)
          EL3 = FLOAT(IE3(IZ))
          EL1 = FLOAT(IE1(IZ))
          E = EL3
          B = BETA/1000.
          T = 511060.*(1.-1./(1.+E0/(511.06))**2)/2.
          GG = 1.+E0/511.06
          P02 = T/R/(1.-2.*T/511060.)
          F = 0.
          S = 0.
          SIGMA = 0.
          DO 111 J=1,40
          QA021 = E**2/(4.*T*R) + E**3/(8.*R*T**2*GG**3)
          PP2 = P02-E/R*(GG-E/1022120.)
          LQA021 = ALOG(QA021)
          QA02M = QA021 + 4.*SQRT(P02*PP2)*(SIN(B/2.))**2
          LQA02M = ALOG(QA02M)
          LQ2INC = (LQA02M-LQA021)/FLOAT(IMAX-1)
          LNQA02 = LQA021
          DSBYDE = 0.
          GOSP = 0.
          DO 109 I=1,IMAX
          QA02 = EXP(LNQA02)
          Q = QA02/(ZS**2)
          KH2 = (E/(R*ZS**2)) - 0.25
          AKH = SQRT(ABS(KH2))
          IF(KH2.LT.0.0) GO TO 103
          D = 1. - EXP(-2.*3.14159/AKH)
          BP = ATAN(AKH/(Q-KH2 + 0.25))
          IF(BP.GE.0.0) GO TO 104
          BP = BP + 3.14159
   104    C = EXP((-2./AKH)*BP)
          IF(KH2.GE.0.0) GO TO 102
   103    D = 1.0
          C=EXP((-1./AKH)*ALOG((Q+0.25-KH2+AKH)/(Q+0.25-KH2-AKH)))
   102    IF(E-EL1)170,170,175
   170    G=2.25*Q**4-(0.75+3.*KH2)*Q**3+(0.59375-0.75*KH2-0.5*KH2**2)*Q*Q
        @+(0.11146+0.85417*KH2+1.8833*KH2*KH2+KH2**3)*Q + 0.0035807
        @+KH2/21.333 + KH2*KH2/4.5714 + KH2**3/2.4 + KH2**4/4.
          A = ((Q-KH2 + 0.25)**2 + KH2)**5
          GO TO 180
   175    G=Q**3-(5./3.*KH2+11./12.)*Q**2+(KH2*KH2/3.+1.5*KH2+65./48.)*Q
        @+KH2**3/3.+0.75*KH2*KH2+23./48.*KH2+5./64.
          A = ((Q-KH2 + 0.25)**2 + KH2)**4
   180    RF = ((E+.1-EL3)/1.8/Z/Z)**U
          IF(IABS(IZ-11).GT.5) GO TO 200
          IF(E-EL3.GT.20) GO TO 200
          RF=1.
   200    GOS=RF*32.*G*C/A/D*E/R/R/ZS**4
```

```
C                GOS (=df/dE) is per eV and per atom, for the whole L-shell
                 DSBYDE = DSBYDE+3.5166E-16*(R/T)*(R/E)*(GOS+GOSP)*LQ2INC/2.
                 IF(I.GT.1) GO TO 115
                 DFDIPL = GOS
                 DSBYDE = 0.
    115          LNQA02 = LNQA02 + LQ2INC
    109          GOSP=GOS
                 DELTA = E - EL3
                 IF(J.EQ.1) GO TO 120
                 S = ALOG(DSBDEP/DSBYDE)/ALOG(E/(E-EINC))
                 SGINC=(E*DSBYDE-(E-EINC)*DSBDEP)/(1.-S)
                 SIGMA = SIGMA + SGINC
C                SIGMA is the EELS cross section cm² per atom
                 F = F + (DFDIPL+DFPREV)*EINC/2.
                 IF(DELTA.LT.50.) GO TO 120
                 WRITE(6,605) E,DSBYDE,DELTA,SIGMA,F
    605          FORMAT(' ',F10.1,2X,E12.3,2X,F10.1,2X,E13.3,2X,F7.3)
    120          IF(DELTA.LT.100) GO TO 107
                 IF(SGINC.LT.0.001*SIGMA) GO TO 112
                 EINC = EINC*2.
    107          E = E + EINC
                 IF(E.GT.T) GO TO 112
                 DFPREV = DFDIPL
    111          DSBDEP = DSBYDE
    112          STOP
                 END
```

SIGMAL3: VALUES OF Z, E0 AND BETA ARE: 22,80,10,

E(eV)	DSBYDE	DELTA	SIGMA	f0
505.0	0.658E-22	50.0	0.512E-20	1.055
515.0	0.618E-22	60.0	0.576E-20	1.201
525.0	0.578E-22	70.0	0.635E-20	1.343
535.0	0.541E-22	80.0	0.691E-20	1.481
545.0	0.505E-22	90.0	0.744E-20	1.613
555.0	0.471E-22	100.0	0.792E-20	1.741
575.0	0.492E-22	120.0	0.889E-20	2.005
615.0	0.378E-22	160.0	0.106E-19	2.524
695.0	0.229E-22	240.0	0.130E-19	3.373
855.0	0.924E-23	400.0	0.154E-19	4.553
1175.0	0.209E-23	720.0	0.168E-19	5.849
1815.0	0.234E-24	1360.0	0.173E-19	6.917
3095.0	0.133E-25	2640.0	0.174E-19	7.568
5655.0	0.465E-27	5200.0	0.174E-19	7.878

B.8. Parameterized K-, L-, M-, N- and O-Shell Cross Sections

The program SIGPAR2 calculates energy-loss cross sections (for limited β and Δ) for the major ionization edges, using Eq. (A.7). The following listing shows a BASIC version, which runs under IBM BASICA or Microscoft QBASIC provided the data files stored in the same subdirectory; a FORTRAN version is also available. Values of integrated oscillator strength $f(\Delta)$, together with an estimate of the uncertainty, are stored in the text files FK.DAT, FL.DAT, FM23.DAT, FM45.DAT and FNO45.DAT which are given after the program listing. They represent best estimates (Egerton, 1993) based on Hartree–Slater calculations, x-ray absorption data, and EELS measurements. The integration window Δ should be within the range 30 eV to 250 eV; linear interpolation or extrapolation is used to estimate $f(\Delta)$ for values of Δ other than those used in the tabulations. In the case of M_{23} edges only $\Delta = 30$ eV values are given, based on EELS measurements of Wilhelm and Hofer (1992). If the semiangle β lies outside the dipole region (taken here to be half the Bethe–ridge angle), a warning is given to indicate that the calculated cross section will be too large. Since retardation effects are included, according to Eq. (A.7), the results should be valid for incident-electron energies as high as 1 MeV.

```
1     REM SIGPAR2.BAS calculates sigma(beta,delta) from f-values stored
2     REM in files FK.DAT, FL.DAT, FM45.DAT, FM23.DAT and FNO45.DAT
3     REM based on recommended values in Ultramicroscopy 50 (1993) 13-8.
10    PRINT "Enter Z,delta(eV)": INPUT Z, DL
15    PRINT "Enter edge type (K,L,M,N or O)": T$ = INPUT$(1)
20    IF T$ = "K" OR T$ = "k" THEN OPEN "FK.DAT" FOR INPUT AS #1: GOTO 31
21    IF T$ = "L" OR T$ = "l" THEN OPEN "FL.DAT" FOR INPUT AS #1: GOTO 31
22    IF T$ = "M" OR T$ = "m" THEN PRINT "45 or 23": INPUT M
23    IF M = 45 THEN OPEN "FM45.DAT" FOR INPUT AS #1: GOTO 31
24    IF M = 23 THEN OPEN "FM23.DAT" FOR INPUT AS #1: GOTO 33
25    IF T$ = "N" OR T$ = "n" THEN OPEN "FNO45.DAT" FOR INPUT AS #1: GOTO 35
26    IF T$ = "O" OR T$ = "o" THEN OPEN "FNO45.DAT" FOR INPUT AS #1: GOTO 35
30    REM Read data from text files:
31    INPUT #1, I, EC, F50, F100, F200, ERP: IF I = Z THEN GOTO 45
32    GOTO 31
33    INPUT #1, I, EC, F30: IF I = Z THEN GOTO 44
34    GOTO 33
35    INPUT #1, I, EC, F50, F100, ERP: IF I = Z THEN GOTO 45
36    GOTO 35
40    REM Interpolate for specified energy window, except for M23 edges:
44    DL = 30: FD = F30: ERP = 10: PRINT "For delta = 30eV,": GOTO 49
45    IF DL <=50 THEN FD = F50 * DL / 50
46    IF DL > 50 AND DL < 100 THEN FD = F50 + (DL - 50) / 50 * (F100 - F50)
47    IF DL >= 100 AND DL < 250 THEN FD = F100 + (DL - 100) / 100 * (F200 - F100)
49    PRINT "Ec = "; EC; "eV, f(delta) = "; FD
50    REM Calculate cross section, assuming dipole conditions:
51    PRINT "Enter E0(keV),beta(mrad)": INPUT E0, BETA
52    IF BETA * BETA > 50 * EC / E0 THEN PRINT "Dipole Approximation NOT VALID !"
55    EBAR = SQR(EC * (EC + DL)): GAMMA = 1 + E0 / 511: G2 = GAMMA * GAMMA
60    V2=1-1/G2: B2=BETA*BETA: THEBAR = EBAR/E0/(1+1/GAMMA): T2 = THEBAR*THEBAR
65    GFUNC = LOG(G2): GFUNC = LOG((B2 + T2) / (B2 + T2 / G2)) - V2 * B2 / (B2+T2/G2)
70    SQUAB = LOG(1+B2/T2) + GFUNC: SIGMA = 1.3E-16*G2/(1+GAMMA)/EBAR/E0*FD*SQUAB
75    PRINT "sigma = "; SIGMA; "cm^2; estimated accuracy ="; ERP; "%"
80    CLOSE #1: END
```

Z	EK(eV)	fK(50eV)	fK(100eV)	fK(200eV)	%err
3	55	1.2	1.5	1.7	20
4	111	0.76	1.15	1.15	15
5	188	0.54	0.82	1.13	15
6	285	0.37	0.63	0.94	7
7	400	0.285	0.50	0.80	5
8	535	0.23	0.41	0.68	3
9	685	0.18	0.33	0.57	5
10	867	0.14	0.26	0.46	7
11	1072	0.117	0.225	0.407	5
12	1305	0.091	0.175	0.322	5
13	1560	0.077	0.151	0.281	5
14	1840	0.065	0.128	0.241	5
15	2149	0.055	0.108	0.205	5

Z	EL(eV)	fL(50)	fL(100)	fL(200)	%err
13	73	1.96	3.6	5.0	20
14	100	1.70	3.4	5.1	20
15	135	1.55	3.2	4.8	15
16	165	1.41	3.0	4.8	10
17	200	1.40	2.8	4.45	15
18	246	1.57	2.7	4.3	10
19	294	1.50	2.5	4.0	20
20	347	1.25	2.2	3.7	10
21	399	1.16	2.0	3.4	15
22	455	0.97	1.7	3.0	15
23	513	0.86	1.6	2.7	15
24	575	0.77	1.3	2.35	15
25	640	0.66	1.15	2.10	15
26	708	0.56	1.0	1.80	10

27	779	0.47	0.9	1.62	10
28	855	0.35	0.75	1.39	15
29	931	0.28	0.65	1.20	15
30	1022	0.28	0.65	1.20	20
31	1115	0.26	0.6	1.15	25
32	1217	0.24	0.55	1.10	20
33	1323	0.23	0.5	1.07	15
34	1436	0.21	0.5	1.0	15
35	1550	0.19	0.45	0.93	15
36	1675	0.18	0.4	0.80	20

Z	EM23(eV)	fM23(30eV)
20	25	4.31
21	28	4.40
22	33	4.06
23	37	3.21
24	43	2.05
25	47	4.47
26	53	2.12
27	59	0.95
28	67	1.22
29	78	0.48

Z	EM45(eV)	f(50)	f(100)	f(200)	%err
37	111	0.80	2.5	5.3	30
38	134	0.67	2.1	4.4	20
39	160	0.68	2.2	4.6	20
40	181	0.57	1.9	4.0	20
41	207	0.49	1.7	3.7	30
42	228	0.56	2.0	4.4	40
43	253	0.41	1.5	3.3	40
44	281	0.39	1.5	3.3	40
45	308	0.37	1.5	3.4	30
46	335	0.30	1.2	2.8	25
47	367	0.38	1.4	3.2	20
48	404	0.35	1.2	2.7	25
49	443	0.42	1.3	2.8	30
50	485	0.42	1.2	2.4	25
51	528	0.55	1.4	2.7	25
52	572	0.73	1.7	3.1	25
53	620	0.89	1.9	3.3	30
54	672	1.00	2.0	3.5	40
55	726	1.27	2.4	3.8	40
56	781	1.45	2.5	4.2	30
57	832	1.55	2.5	4.7	40
58	884	1.51	2.4	4.4	20
59	931	1.38	2.2	4.1	20
60	978	1.12	1.8	3.9	20
61	1027	1.04	1.7	3.6	30
62	1081	0.96	1.6	3.2	25
63	1131	0.89	1.5	3.0	30
64	1186	0.81	1.4	2.8	30
65	1242	0.68	1.2	2.6	25
66	1295	0.61	1.1	2.4	20
67	1351	0.53	1.0	2.2	20
68	1409	0.45	0.9	2.0	20
69	1468	0.38	0.8	1.8	30
70	1527	0.32	0.7	1.3	30
71	1589	0.25	0.6	1.4	30
72	1662	0.20	0.5	1.2	40
73	1735	0.18	0.5	1.2	40
74	1810	0.16	0.5	1.2	40

Z	E(N/0)45	f(50)	f(100)	err%
56	90	7.11	9.03	6
57	103	6.43	9.46	8
58	109	6.60	9.65	6
59	115	5.13	8.62	7
60	121	5.52	8.86	5
62	129	5.01	8.24	6
63	133	4.90	8.29	7
64	141	4.13	6.91	7
65	154	3.00	5.39	7
67	160	2.06	4.09	10
68	168	2.11	4.00	7
69	176	1.28	2.51	7
90	83	9.0	9.5	10
92	96	9.4	9.7	10

B.9. Lenz Cross Sections and Plural-Scattering Angular Distributions

The BASIC program LENZPLUS calculates cross sections of elastic and inelastic scattering (integrated over all energy loss) for an element of given atomic number, based on the atomic model of Lenz (1954). It uses Eq. (3.5) and Eq. (3.15) for the differential cross section at a scattering angle β, Eq. (3.6), Eq. (3.7) and a more exact version of Eq. (3.16) for the cross section integrated up to a scattering angle β, and Eqs. (3.8) and (3.17) for the total cross section (large β). Fractions F of the elastic and inelastic scattering accepted by the aperture are also evaluated, and are likely to be more accurate than the absolute cross sections. The elastic-scattering data are not intended to apply to crystalline specimens.

To provide inelastic cross sections, the Lenz model requires a mean energy loss Ebar. This is a different average from that involved in the formula for mean free path, Eq. (5.2). Following Koppe, Lenz (1954) used Ebar = J/2, where J (≈ 13.5 Z) is the atomic mean ionization energy. From Hartree-Slater calculations, Inokuti et al. (1981) give the mean energy per inelastic collision for elements up to strontium; values are in the range 20 eV to 120 eV and have an oscillatory Z-dependence which reflects the electron-shell structure.

The program can be stopped (ctrl-C) at this stage, but if provided with a value of t/λ_i where λ_i is the total-inelastic mean free path, it calculates the relative intensities of the unscattered, elastically scattered, inelastically scattered, and (elastic+inelastic) components accepted by the collection aperture, including scattering up to 4th order and allowing for the increasing width of the plural-scattering angular distributions, as described by Eqs. (3.97), (3.108), and (3.110).

```
4     ' LENZPLUS.BAS calculates Lenz x-secns for elastic and inelastic scattering,
5     ' then fractions of scattering collected by an aperture, including plural
6     ' scattering and broadening of the elastic and inelastic angular distributions
10    INPUT "LENZPLUS(7Sep95): E0 (keV), Ebar(eV), Z, BeTa(mrad) are: ",E0,E,Z,BT
12    GT=500*E0*(1022+E0)/(511+E0):TE=E/2/GT:A0=.0529 ' units are nm
14    GM=1+E0/511:VR=SQR(1-1/GM/GM):K0=2590*GM*VR:COEFF=4*GM*GM*Z/A0/A0/K0^4
16    R0=.0529/Z^.3333:T0=1/K0/R0:PRINT "Theta0 = "T0"rad",,"ThetaE="TE"rad"
18    B=BT/1000:B2=B*B:TE2=TE*TE:T02=T0*T0
20    DSIDOM=COEFF/(B2+TE2)/(B2+TE2)*(1-T02*T02/(B2+TE2+T02)/(B2+TE2+T02))
22    DSIDB=2*3.142*BT*DSIDOM
25    T1=B2/TE2/(B2+TE2)
30    T2=(2*B2+2*TE2+T02)/(B2+TE2)/(B2+TE2+T02)
35    T3=-(T02+2*TE2)/TE2/(T02+TE2)
40    T4=(2/T02)*LOG((T02+TE2+T02)/TE2/(B2+T02+TE2))
45    SIGIN=3.142*COEFF*(T1+T2+T3+T4)
80    SILIM=3.142*COEFF*2/T02*LOG(T02/TE2):F1I=SIGIN/SILIM
85    DSEDOM=COEFF/(B2+T02)*1E+12:DSEDB=DSEDOM*2*3.142*BT 'diffl. elastic
87    SELIM=4*3.142*GM*GM*Z^1.333/K0/K0:F1E=1/(1+T02/B2):SIGEL=F1E*SELIM
90    PRINT "dSe/dOmega ="DSEDOM"nm^2/sr","dSi/dOmega ="DSIDOM"nm^2/sr"
95    PRINT "dSe/dBeta ="DSEDB"nm^2/rad","dSi/dBeta ="DSIDB"nm^2/rad"
96    PRINT "Sigma(elas)="SIGEL,,"Sigma(inel) ="SIGIN"nm^2"
97    PRINT "F(elastic) ="F1E,,"F(inelastic) EQ"F1I"
99    NU=SILIM/SELIM:PRINT "total-inelastic/total-elastic ratio = "NU
100   PRINT:INPUT"t/lambda I = ",TOLI:PRINT "t/lambda(beta) = "TOLI*F1I:PRINT
110   TOLE=TOLI/NU:XE=EXP(-TOLE):XI=EXP(-TOLI):FIE=F1E*F1I
```

```
115    PUN=XE*XI:PEL=(1-XE)*XI*F1E:PZ=PUN+PEL
120    PIN=XE*(1-XI)*F1I:PIE=(1-XI)*(1-XE)*FIE:PI=PIN+PIE
121    PRINT "P(unscat) = ","PUN,"P(el) = ","PEL"neglecting elastic broadening"
122    PRINT "P(inel) = ","PIN,"P(in+el) = ","PIE"neglecting inelastic broadening"
124    PRINT "I0/I = ","PZ,"Ii/I = ","PI"neglecting angular broadening"
125    PT=PZ+PI:LR=LOG(PT/PZ):PRINT "ln(It/I0) = ","LR"without broadening":PRINT
130    F2E=1/(1+1.7∧2*T02/B2):F3E=1/(1+2.2∧2*T02/B2):F4E=1/(1+2.7∧2*T02/B2)
135    PE=XE*(TOLE*F1E+TOLE∧2*F2E/2+TOLE∧*F3E/6+TOLE∧4*F4E/24)
136    PENI=PE*XI:PU=XI*XE:RZ=PU+PENI    'unscattered and el/no-inel compts.
139    REM PEL=XI*XE*(EXP(TOLE*F1E)-1):PZ=PUN+PEL 'not used
140    PI=XI*(EXP(TOLI*F1I)-1):PINE=XE*PI:PIE=PI*PE:RI=PINE+PIE
142    PRINT "P(unscat) = ","PU,"P(el only) = ","PENI"with elastic broadening"
143    REM ang distrib of inel+el taken same as broadened elastic
145    PRINT "P(inel only) = ","PINE,"P(in+el) = ","PIE"with inelastic broadening"
150    PRINT "I0/I = ","RZ,"Ii/I = ","RI"with angular broadening"
170    RT=RZ+RI:LR=LOG(RT/RZ):PRINT "ln(It/I0) = ","LR"with angular broadening"

RUN
LENZPLUS (7Sep95): E0(keV), Ebar(eV), Z, BeTa(mrad) are: 100,40,6,10.
Theta0 = 2.023144E-02 rad                ThetaE = 2.178253E-04 rad
dSe/dOmega = 2897719 nm∧2/sr             dSi/dOmega = 5.223308E-02 nm∧2/sr
dSe/dBeta= 1.820926E+08 nm∧2/rad         dSi/dBeta= 3.282326 nm∧2/rad
Sigma(elas) = 1.33367E-05 nm∧2           Sigma(inel) = 1.662413E-04 nm∧2
F(elastic) = .1963436                    F(inelastic) = .809588
total-inelastic/total-elastic ratio =3.023035

t/lambdaI = 1.5
t/lambda(beta) =1.214382

P(unscat) = .1358519              P(el) =1.713653E-02 neglecting elastic broadening
P(inel) =.3829302                 P(in+el) =4.830332E-02
neglecting inelastic broadening
I0/I =.1529884                    Ii/I =.4312336 neglecting angular broadening
ln(It/I0) = 1.339919 without broadening

P(unscat) =.1358519              P(el only) = 1.468281E-2 with elastic broadening
P(inel only) =.321726            P(in+el) =3.477201E-02 with inelastic broadening
I0/I =.1505347                   Ii/I =.356498 with angular broadening
ln(It/I0) = 1.214382 with angular broadening
Ok
```

B.10. Conversion between Oscillator Strength and Cross Section

The program FTOS.BAS converts a dipole oscillator strength (integrated over an energy range DL above an ionization edge) to a corresponding core-loss cross section, using Eq. (A.7) and the same procedure as in the SIGPAR2 program. Program STOF.BAS does the reverse and enables measured inner-shell cross sections to be parameterized in terms of a dipole oscillator strength which is independent of collection angle and incident energy, as described in Section 4.5.2.

```
10    REM: FTOS calculates a core-loss partial cross section
12    REM: from an f-value integrated over energy range DeLta
25    INPUT "FTOS (7Sep95): f-value = ",FC
35    S0=3.516E-16 :REM 4*PI*A0**2
40    INPUT "Edge energy(eV),DeLta(eV),E0(keV) = ",EC,DL,E0
45    ER=SQR(EC*(EC+DL))/13.6
50    TK=E0*(1+E0/1022)/(1+E0/511)∧2
55    TR=TK*1000/13.6
60    GT=E0*(1022+E0)/(511+E0)
65    TE=ER*13.6/2/GT: PRINT "ThetaE(mrad) = ",TE
70    INPUT "effective collection semiangle BETA (mrad) = ",BS
75    B2=BS*BS:T2=TE*TE:V2=2*TK/511:G2=1/(1-V2)
80    RET=LOG(G2)-LOG((B2+T2)/(B2+T2/G2))-V2*B2/(B2+T2/G2)
85    LG=LOG(1+B2/T2):SC=FC*S0/ER/TR*(LG+RET)
```

```
90    FR=100*RET/LG: PRINT "% increase due to retardation"FR
95    PRINT "SIGMA(DeLta,BeTa) = "SC"cm/\2"
```

```
FTOS(7Sep95): f-value = 0.41
Edge energy(eV), DeLta(eV),E0(keV) = 535,100,100
ThetaE(mrad) =                3.174038
effective collection semiangle BETA (mrad) = 10.
% increase due to retardation 2.040891
SIGMA (DeLta,BeTa) = 1.453538E-21 cm/\2
Ok
```

```
10    REM: STOF calculates OSCILLATOR STRENGTH FOR CORE-LOSS
12    REM: FROM A GIVEN PARTIAL CROSS SECTION SC
25    INPUT "STOF(7Sep95): CROSS SECTION (in cm/\2) = ",SC
35    S0=3.516E-16 :REM 4*PI*A0**2
40    INPUT "Edge energy(eV), DeLta(eV), E0(keV) = ",EC,DL,E0
45    ER=SQR(EC*(EC+DL))/13.6
50    TK=E0*(1+E0/1022)/(1+E0/511)/\2
55    TR=TK*1000/13.6
60    GT=E0*(1022+E0)/(511+E0)/2
65    TE=ER*13.6/2/GT: PRINT "THETAE(mrad) = "TE
70    INPUT "effective BETA (mrad) = ",BS
75    LG=LOG(1+BS*BS/TE/TE)
78    V2=2*TK/511: G2=1/(1-V2): B2=BS*BS: T2=TE*TE
80    RET=LOG(G2)-LOG(B2+T2)/(B2+T2/G2))-V2*B2/(B2+T2/G2)
85    FC=SC/S0*ER*TR/(LG+RET)
90    FR=100*RET/LG: PRINT "%increase due to retardation="FR
95    PRINT "f(DeLta) = "FC
```

```
RUN
STOF(7Sep95); CROSS SECTION (in cm/\2) = 1.453538e-21
Edge energy(eV), DeLta(eV), E0(keV) = 535,100,100
THETA(mrad) = 3.174038
effective BETA (mrad) = 10.
%increase due to retardation = 2.040891
f(DeLta) =.4100002
Ok
```

B.11. Conversion between Mean Energy and Inelastic Mean Free Path

The short routine EM2MFP.BAS uses Eq. (5.2) to convert a mean energy loss E_m, as listed in Table 5.2, to the corresponding inelastic mean free path for energy-loss data recorded with an angle-limiting collection aperture of semiangle β. It is not appropriate for $\beta > 20$ mrad, as explained in Section 5.1. The algorithm MFP2EM.BAS applies Eq. (5.2) in reverse, obtaining a value of E_m from a measured value of $\lambda(\beta)$ within a few iterations.

```
10    REM EM2MFP : CALCULATION OF TOTAL-INELASTIC MFP
15    REM based on Malis et al. (JEMT 8, 1988, 193-200)
20    INPUT "ENTER Em(eV),E0(KeV),BETA(mrad):",EM,E0,BETA
30    F=(1+E0)/1022)/(1+E0/511)/\2
40    LAMBDA=106*F*E0/EM/LOG(2*BETA*E0/EM)
50    PRINT "MFP(nm) = ",LAMBDA
RUN
ENTER Em(eV),E0(KeV),BETA(mrad):20,200,5.
MRP(nm) =              142.1619
```

```
10    REM: MFP2EM calculates Em parameter for MFP formula:
15    REM:J. Electron Microsc. Technique 8 (1988) 193-100.
20    INPUT "MFP(nm),E0(keV),beta(mrad) = ",LAM,E0,B:EB=15
30    EM= 106*(1+E0)/1022)/(1+E0/511)/\2*E0/LAM/LOG(2*B*E0/EB)
32    IF ABS (EB-EM)<EM/1000 THEN PRINT "Em(eV) = ",EM:END
35    EB=EM:GOTO 30 ' Last edit 95Sep07
RUN
MFP(nm),E0(keV),beta(mrad) = 142.2,200,5.
EM(eV) =              19.99049
```

APPENDIX **C**

Plasmon Energies of Some Elements and Compounds

The following table lists the measured energy E_p and full width at half-maximum ΔE_p of the principal low-loss peak observable in the energy-loss spectrum of some common materials. The data is taken mainly from Daniels *et al.* (1970), Colliex *et al.* (1976a), Raether (1980), and Colliex (1984a). In some instances, a free-electron plasmon energy is shown in parentheses, calculated from Eq. (3.41) with $m = m_0$ and n equal to the density of outer-shell electrons (including $3d$ electrons in the case of transition metals).

Layer crystals such as graphite and boron nitride have much weaker bonding in a direction perpendicular to the basal (cleavage) plane than within the basal plane, giving rise to two groups of valence electrons and two distinct plasmon energies. Furthermore, in *anisotropic* materials the dielectric function $\varepsilon(\mathbf{q}, E)$ is actually a tensor ε_{ij} and the peak structure in the energy-loss spectrum depends on the direction of the scattering vector $\mathbf{q}$. For a uniaxial crystal such as graphite, axes can be chosen such that off-diagonal components are zero, in which case $\varepsilon(\mathbf{q}, E) = \varepsilon_\perp \sin^2\Theta + \varepsilon_\parallel \cos^2\Theta$, where $\varepsilon_\perp = \varepsilon_{11} = \varepsilon_{22}$ and $\varepsilon_\parallel = \varepsilon_{33}$ are components of $\varepsilon(E)$ perpendicular and parallel to the c-axis; Θ is the angle between $\mathbf{q}$ and the c-axis, which depends on the scattering angle and the specimen orientation. If the c-axis is parallel to the incident beam, the greatest contribution (for $\beta \gg \theta_E$) comes from perpendicular excitations and two plasmon peaks are observed (see table). Under special conditions (e.g., small collection angle β), the $q\|c$ excitations may predominate and the higher-energy peak is displaced downwards in energy. Further detail on EELS of anisotropic materials is given in Daniels *et al.* (1970) and Browning *et al.* (1991b).

Material	Eq. (3.41)	E_p (eV)	ΔE_p (eV)	Material	Eq. (3.41)	E_p (eV)	ΔE_p (eV)
Al	(15.7)	15.0	0.5	Li	(8.0)	7.1	2.2
AlAs	(15.5)	16.1		LiF	(26.2)	25.3	
Al$_2$O$_3$(α)		26	10	LiH	(12.7)	20.9	
Al$_2$O$_3$(am.)		23	20				
As(cryst.)	(17.8)	18.7		Mg	(10.9)	10.3	0.7
As(amor.)	(16.1)	16.7		MgF$_2$		24.6	
				MgO		22.3	
B(amor.)		22.7	18	Mn	(28.4)	21.6	
BN(hex.)		9; 26					
BN(amor.)		24		Na	(6.0)	5.7	0.4
Ba	(6.7)	7.2	2.7	NaCl	(15.7)	15.5	
Be	(18.4)	18.7	4.8	Ni		20.7	
Bi	(14.0)	14.2	6.5				
				Pb	(13.5)	13.0	
C(diamond)	(31)	33.2	13	PbS	(16.3)	14.9	
C(lonsdal.)		32.4		PbSe		15.0	
C(am. tet.)		29.5		PbTe		14.2	
C(graphite)	(13;22)	7; 27					
C(amorph)	(22)	24	20	Pt		35.0	
				Rb	(3.9)	3.41	0.6
Ca	(8.2)	8.8	2.1	Sb	(15.1)	15.2	3.3
Cd		19.2					
Co		20.9		Sc	(13.6)	14.0	
Cr	(26.8)	24.9		ScH$_2$		17.2	
Cs	(3.4)	2.9		Se(cr.)	(17.4)	17.1	6.2
Cu	(16.0)	19.3		Se(am.)	(16.4)	16.3	6.2
Er		14.0		Si(cr.)	(16.6)	16.7	3.2
ErH$_2$		16.8		Si(am.)		16.3	3.9
				SiC(α)	(23.1)	21.5	3.9
Fe	(29.8)	23.0		Si$_3$N$_4$(α)	(24.7)	23.7	10.1
				SiO$_2$(α)	(24.2)	22.4	16.6
Ga	(14.5)	13.8	0.6	Sn(sol.)	(14.3)	13.7	1.3
GaAs(cr.)	(15.7)	15.8		Sn(liq.)	(14.0)	13.4	1.6
GaAs(am.)		15.5		Sr	(7.0)	8.0	2.3
GaP(cr.)	(16.5)	16.5					
GaSb	(13.8)	13.3		Tb		13.3	
				TbH$_2$		15.6	
Ge(cryst.)	(15.6)	16.2	3.3	Te(cr.)	(15.6)	17.1	6.2
Ge(amor.)	(14.8)	15.7	3.8	Te(am.)	(15.3)	16.3	6.2
Hg(sol.)	(7.7)	6.3	1.5	Ti	(17.8)	17.9	
Hg(liq.)	(7.5)	6.4	1.0				
				V	(22.8)	21.8	
In	(12.5)	11.4	12	Y		12.5	7
InAs	(13.8)	13.8		YH$_2$		15.3	
InSb	(12.7)	12.9		Zn	(13.9)	17.2	
K	(4.3)	3.7	0.3				
KBr	(12.4)	13.2					

Inner-Shell Energies and Edge Shapes

The following table gives threshold energies E_k (in eV) of the ionization edges observable by EELS, based on data of Bearden and Burr (1967), Siegbahn *et al.* (1967), Zaluzec (1981), Ahn and Krivanek (1983), and Colliex (1985). The most prominent edges (those most suitable for elemental analysis) are shown in italics. Where possible, an accompanying symbol is used to indicate the observed edge shape:

h denotes a *hydrogenic* edge with sawtooth profile (rapid rise at the threshold followed by more gradual decay), as in Fig. 3.43.

d denotes a delayed maximum due to centrifugal-barrier effects (Section 3.7.1), giving a rounded edge with a maximum at least 10 eV above the threshold energy as in Figs. 3.47.

w denotes sharp *white-line* peaks at the edge threshold, due to excitation to empty *d*-states (in the transition metals) or *f*-states (in the rare earths), as in Fig. 3.48a.

p denotes a low-energy edge which may appear more like a plasmon peak than a typical edge. However, the energy given is that of the edge onset, not the intensity maximum.

Because of near-edge structure, which depends on the chemical and crystallographic structure of a specimen, this classification can serve only as a rough guide. Elements such as copper can exist in different valence states, giving rise to dissimilar edge shapes (Fig. 3.46). The edge energies themselves may vary by several eV depending on the chemical environment of the excited energy; see Section 3.7.4.

State →		1s	2s	2p₁/₂	2p₃/₂	3p	3d	4p
Shell →		K	L₁	L₂	L₃	M₂₃	M₄₅	N₂₃
2	He	24.6h						
3	Li	55h						
4	Be	111h						
5	B	188h						
6	C	284h						
7	N	400h						
8	O	532h						
9	F	685h						
10	Ne	867h			18w			
11	Na	1072h			32h			
12	Mg	1305h			52h			
13	Al	1560h	118h		73d			
14	Si	1839h	149h		100d			
15	P	2149h	189h		135d			
16	S	2472h	229h		165d			
17	Cl	2823	270h		200d			
18	Ar	3203	320h		246d			
19	K	3608	377h		294w			
20	Ca	4038	438h	350w	347w			
21	Sc	4493	500h	406w	402w			
22	Ti	4965	564h	461w	455w	47		
23	V	5465	628h	520w	513w	47		
24	Cr	5989	695h	584w	575w	48		
25	Mn	6539	770h	652w	640w	51		
26	Fe	7113	846h	721w	708w	57		
27	Co	7709	926h	794w	779w	62		
28	Ni	8333	1008	872w	855w	68		
29	Cu	8979	1096	951h	931h	74		
30	Zn	9659	1194	1043	1020d	87		
31	Ga		1298	1142	1115d	105		
32	Ge		1414	1248	1217d	125	30	
33	As		1527	1359	1323d	144	41	
34	Se		1654	1476	1436d	162	57h	
35	Br		1782	1596	1550d	182	70d	
36	Kr		1921	1727	1675	214	89h	
37	Rb		2065	1846w	1804w	239	111d	
38	Sr		2216	2007w	1940w	270	134d	20p
39	Y		2373	2155w	2080w	300	160	28p
40	Zr		2532	2307w	2222w	335	181	32p
41	Nb		2698	2465w	2371w	371	207h	35h
42	Mo		2866	2625w	2520w	400	228h	37d
44	Ru		3224	2967w	2838w	472	281h	42d

State → Shell →	$3d_{3/2}$ M_4	$3d_{5/2}$ M_5	$4p$ N_{23}	$4d$ N_{45}	$4f$ N_6, N_7	$5p$ O_2, O_3	$5d$ O_4, O_5
45 Rh	312	308d	48				
46 Pd	340	335d	50				
47 Ag	373	367d	59				
48 Cd	411	404d	67				
49 In	451	443d	77				
50 Sn	494	485d	90				
51 Sb	537	528d	99	32			
52 Te	582	572h	110	40			
53 I	631	620h	123	50			
54 Xe	685	672h	147	64			
55 Cs	740w	726w		78			
56 Ba	796w	781w		93			
57 La	849w	832w		99			
58 Ce	902w	884w		110			
59 Pr	951w	931w		114			
60 Nd	1000w	978w		118			
62 Sm	1107w	1081w		130			
63 Eu	1161w	1131w		134			
64 Gd	1218w	1186w		141			
65 Tb	1276w	1242w		148			
66 Dy	1332w	1295w		154		30, 23	
67 Ho	1391w	1351w		161		31, 24	
68 Er	1453w	1409w		168		31, 25	
69 Tm	1515	1468w		177		32, 25	
70 Yb	1576	1527w		184		33, 26	
71 Lu	1640	1589w		195		35, 27	
72 Hf	1716	1662h				38, 30	
73 Ta	1793	1735h				45, 37	
74 W	1872	1810h			37, 34	47, 37	
75 Re	1949	1883h			47, 45	46, 35	
76 Os	2031	1960h			52, 50	58, 46	
77 Ir	2116	2041h			63, 60	63, 51	
78 Pt	2202	2122h			74, 70	66, 51	
79 Au	2291	2206h			87, 83	72, 54	
80 Hg	2385	2295h				81, 58	
81 Tl	2485	2390h					14p
82 Pb	2586	2284h					21p
83 Bi	2688	2580h					27h
90 Th	3491	3332					83w
92 U	3728	3552					96w

The following specimens provide an accurate calibration of energy-axis dispersion (eV/channel) for a high-resolution spectrometer system (P.E. Batson, personal communication):

aluminum (midpoint of edge onset = 72.9 eV)
silicon (midpoint of edge onset = 99.9 eV)
amorphous SiO_2 (L_{23} edge maximum = 108.3 eV)
graphite (maximum of π^* peak = 285.4 eV)
NiO (Ni L_3 maximum = 853.2 eV)

In the table on p. 435, the notation L_{23} (for example) indicates L_2 and L_3 edges which are close in energy, such that the individual thresholds are usually not resolved by electron-microscope EELS systems. Oscillator strengths for the major edges are listed in Appendix B.8. The following table relates the spectroscopic (shell) notation to the quantum numbers and degeneracy $(2j + 1)$ of the initial state involved in a transition.

Edge	State	n	l	j	Degeneracy
K	$1s^{1/2}$	1	0	1/2	2
L_1	$2s^{1/2}$	2	0	1/2	2
L_2	$2p^{1/2}$	2	1	1/2	2
L_3	$2p^{3/2}$	2	1	3/2	4
M_1	$3s^{1/2}$	3	0	1/2	2
M_2	$3p^{1/2}$	3	1	1/2	2
M_3	$3p^{3/2}$	3	1	3/2	4
M_4	$3d^{3/2}$	3	2	3/2	4
M_5	$3d^{5/2}$	3	2	5/2	6
N_1	$4s^{1/2}$	4	0	1/2	2
N_2	$4p^{1/2}$	4	1	1/2	2
N_3	$4p^{3/2}$	4	1	3/2	4
N_4	$4d^{3/2}$	4	2	3/2	4
N_5	$4d^{5/2}$	4	2	5/2	6
N_6	$4f^{5/2}$	4	3	5/2	6
N_7	$4f^{7/2}$	4	3	7/2	8
O_2	$5p^{1/2}$	5	1	1/2	2
O_3	$5p^{3/2}$	5	1	3/2	4
O_4	$5d^{3/2}$	5	2	3/2	4
O_5	$5d^{5/2}$	5	2	5/2	6

Electron Wavelengths and Relativistic Factors; Fundamental Constants

Table E.1 lists (as a function of the kinetic energy E_0 of an electron) values of the wavelength λ, wave number k_0, velocity v, relativistic factor $\gamma = (1 - v^2/c^2)^{-1/2}$, effective kinetic energy $T = m_0v^2/2$, and parameter $2\gamma T$ which is used to calculate the characteristic scattering angle $\theta_E = E/(2\gamma T)$. For values of E_0 not tabulated, these parameters can be calculated from the following equations:

$$k_0 = \gamma m_0 v/\hbar = 2590(\gamma v/c) \text{ nm}^{-1}$$

$$\gamma = 1 + E_0/(m_0c^2)$$

$$T = E_0 \frac{1 + E_0/(2m_0c^2)}{[1 + E_0/(m_0c^2)]^2} = E_0 \frac{1 + \gamma}{2\gamma^2}$$

$$2\gamma T = E_0 \left(\frac{2m_0c^2 + E_0}{m_0c^2 + E_0} \right) = \gamma m_0 v^2$$

$$\theta_E = \frac{E}{\gamma m_0 v^2} = \frac{E}{E_0(1 + \gamma^{-1})} = \frac{E}{E_0} \left(\frac{E_0 + m_0c^2}{E_0 + 2m_0c^2} \right)$$

Values of the fundamental constants for use in these (and other) equations are given in Table E.2, extracted from page F-162 of the 65th edition of *Handbook of Chemistry and Physics* (CRC Press, 1984).

Table E.1. Electron Parameters as a Function of Kinetic Energy

E_0 (keV)	λ (m $\times 10^{-12}$)	$k_0 = 2\pi/\lambda$ (nm^{-1})	v^2/c^2	$\gamma = (1 - (v^2/c^2))^{-1/2}$	$T = \frac{1}{2}m_0 v^2$ (keV)	$2\gamma T$ (keV)
10	12.2	514.7	0.0380	1.0196	9.714	19.81
20	8.59	731.4	0.0739	1.0391	18.88	39.34
30	6.98	900.2	0.1078	1.0587	27.55	58.34
40	6.02	1044	0.1399	1.0782	35.75	77.10
50	5.36	1173	0.1703	1.0978	43.52	95.56
60	4.87	1291	0.1991	1.1174	50.88	113.7
80	4.18	1504	0.2523	1.1565	64.50	149.2
100	3.70	1697	0.3005	1.1957	76.79	183.6
120	3.35	1876	0.3442	1.2348	87.94	217.2
150	2.96	2125	0.4023	1.2935	102.8	266.0
200	2.51	2505	0.4835	1.3914	123.6	343.8
300	1.97	3191	0.6030	1.5870	154.1	489.1
400	1.64	3822	0.6854	1.7827	175.1	624.4
500	1.42	4421	0.7445	1.9784	190.2	752.8
1000	0.87	7205	0.8856	2.9567	226.3	1338

Table E.2. Selected Physical Constants

Quantity	Symbol	Value	Units
Electron charge	e	1.602×10^{-19}	C
Electron mass	m_0	9.110×10^{-31}	kg
Electron rest energy	$m_0 c^2$	511,060	eV
Proton mass	m_p	1.673×10^{-27}	kg
Neutron mass	m_n	1.675×10^{-27}	kg
Bohr radius ($4\pi\varepsilon_0\hbar^2/m_0 e^2$)	a_0	5.292×10^{-11}	m
Rydberg energy ($\hbar^2/(2m_0 a_0^2)$)	R	13.61	eV
Photon energy $\times$ wavelength	hc/e	1.240	eV μm
Avogadro number	N_A	6.022×10^{23}	mol^{-1}
Boltzmann constant	k	1.381×10^{-23}	JK^{-1}
Speed of light	c	2.998×10^8	m s^{-1}
Permittivity of space	ε_0	8.854×10^{-12}	F m^{-1}
Permeability of space	μ_0	1.257×10^{-6}	H m^{-1}
Planck constant	h	6.626×10^{-34}	J s
$h/2\pi$	$\hbar$	1.055×10^{-34}	J s

1 mmol/kg $\approx$ 12 ppm (atomic) for dry biological tissue (assuming mean $Z \approx 6$)
1 mmol/kg $\approx$ 1 mM $\approx$ 18 ppm (atomic) for wet biological tissue (mainly H_2O)

References

In accordance with the Harvard system, multiauthor papers (*et al.* references in the text) are listed *chronologically* after one- and two-author papers bearing the same first name.

Achèche, M., Colliex, C., and Trebbia, P. (1986) EELS characterization of small metallic clusters. In *Scanning Electron Microscopy/1986/I*, SEM Inc., Illinois, 25–32.

Adamson-Sharpe, K. M., and Ottensmeyer, F. P. (1981) Spatial resolution and detection sensitivity in microanalysis by electron energy-loss selected imaging. *J. Microsc.* **122**, 309–314.

Ade, H., Zhang, X., Cameron, S., Costello, C., Kirz, J., and Williams, S. (1992) Chemical contrast in x-ray microscopy and spatially resolved XANES spectroscopy of organic specimens. *Science* **258**, 972–975.

Ahn, C. C., and Krivanek, O. L. (1983) EELS Atlas. Copies ($US15) from Center for Solid State Science, Arizona State University, Tempe, Arizona 85287, or from Gatan Inc., 780 Commonwealth Drive, Warrendale, Pennsylvania 15086.

Andersen, W. H. J. (1967) Optimum adjustment and correction of the Wien filter. *Brit J. Appl. Phys.* **18**, 1573–1579.

Andersen, W. H. J., and Kramer, J. (1972) A double-focusing Wien filter as a full-image energy analyzer for the electron microscope. In *Electron Microscopy—1972*, The Institute of Physics, London, pp. 146–147.

Andersen, W. H. J., and Le Poole, J. B. (1970) A double Wien filter as a high resolution, high-transmission electron energy analyser. *J. Phys. E (Sci. Instrum)* **3**, 121–126.

Andrew, J. W., Ottensmeyer, F. P., and Martell, E. (1978) An improved magnetic prism design for a transmission electron microscope energy filter. In *Electron Microscopy—1978*, 9th Int. Cong., ed. J. M. Sturgess; Microscopical Society of Canada, Toronto, vol. 1, pp. 40–41.

Anstis, G. R., Lynch, D. F., Moodie, A. F., and O'Keefe, M. A. (1973) *n*-Beam lattice images: III Upper limits of ionicity in $W_4Nb_{26}O_{77}$. *Acta Crystallogr.* **A29**, 138–152.

Arsenault, A. L., and Ottensmeyer, F. P. (1983) Quantitative spatial distribution of calcium, phosphorus and sulfur in calcifying epiphysis by high resolution spectroscopic imaging. *Proc. Natl. Acad. Sci. USA* **80**, 1322–1326.

Ashley, J. C., and Ritchie, R. H. (1970) Double-plasmon excitation in a free-electron gas. *Phys. Status Solidi* **38**, 425–434.

Ashley, J. C., and Williams, M. W. (1980) Electron mean free paths in solid organic insulators. *Radiat. Res.* **81**, 364–378.

Atwater, H. A., Wong, S. S., Ahn, C. C., Nikzad, S., and Frase, H. N. (1993) Analysis of monolayer films during molecular beam epitaxy by reflection electron energy loss spectroscopy. *Surf. Sci.* **298**, 273–283.

Auchterlonie, G. J., McKenzie, D. R., and Cockayne, D. J. H. (1989) Using ELNES with parallel EELS for differentiating between a-Si :X thin films. *Ultramicroscopy* **31**, 217–232.

Auerhammer, J. M., and Rez, P. (1989) Dipole-forbidden excitations in electron-energy-loss spectroscopy. *Phys. Rev. B* **40**, 2024–2030.

Autrata, R., Schauer, P., Kvapil, Jos., and Kvapil, J. (1983) Single-crystal aluminates—a new generation of scintillators for scanning microscopes and transparent screens in electron optical devices. In *Scanning Electron Microscopy/1983/II*, ed. G. M. Roomans, R. M. Albrecht, J. D. Shelburne, and I. B. Sachs; Scanning Electron Microscopy Inc., Chicago, 1993, pp. 489–500.

Bakenfelder, A., Fromm, I., Reimer, L., and Rennenkamp, R. (1989) Contrast in the electron spectroscopic imaging mode of a TEM. III. Bragg contrast of crystalline specimens. *J. Microsc.* **159**, 161–177.

Baker, T. N., Craven, A. J., Duckworth, S. P., and Glas, F. (1982) Microanalysis of carbides in ferritic steels. In *Developments in Electron Microscopy and Analysis*, Inst. Phys. Conf. Ser. No. 61 (I.O.P., Bristol), pp. 239–242.

Ball, M. D., Malis, T. F., and Steele, D. (1984) Ultramicrotomy as a specimen preparation technique for analytical electron microscopy. In *Analytical Electron Microscopy—1984*, ed. D. B. Williams and D. C. Joy, San Francisco Press, San Francisco, pp. 189–192.

Ballu, Y., Lecante, J., and Newns, D. M. (1976) Surface plasmons on Mo(100). *Phys. Lett.* **57A**, 159–160.

Barth, J., Gerken, F., and Kunz, C. (1983) Atomic nature of the L_{23} white lines in Ca, Sc and Ti metals as revealed by resonant photoemission. *Phys. Rev. B* **28**, 3608–3611.

Batson, P. E. (1982) A new surface plasmon resonance in clusters of small aluminum spheres. *Ultramicroscopy* **9**, 277–282.

Batson, P. E. (1985b) A Wien filter ELS spectrometer for dedicated STEM. *Scanning Electron Microsc.* Part **1**, 15–20.

Batson, P. E. (1988) Parallel detection for high-resolution electron energy loss studies in the scanning transmission electron microscope. *Rev. Sci Instrum.* **59**, 1132–1138.

Batson, P. E. (1992a) Electron energy loss studies in semiconductors. In *Transmission Electron Energy Loss Spectrometry in Materials Science*, ed. M. M. Disko, C. C. Ahn, and B. Fulz, The Minerals, Metals and Materials Society, Warrendale, Pennsylvania, pp. 217–240.

Batson, P. E. (1992b) Spatial resolution in electron energy loss spectroscopy. *Ultramicroscopy* **47**, 133–144.

Batson, P. E. (1993) Silicon L_{23} near-edge fine structure in confined volumes. *Ultramicroscopy* **50**, 1–12.

Batson, P. E. (1995) Conduction bandstructure in strained silicon by spatially resolved electron energy loss spectroscopy. *Ultramicroscopy,* **59**, 63–70.

Batson, P. E., and Bruley, J. (1991) Dynamic screening of the core exciton by swift electrons in electron-energy-loss scattering. *Phys. Rev. Lett.* **67**, 350–353.

Batson, P. E., and Craven, A. J. (1979) Extended fine structure on the carbon core-ionization edge obtained from nanometer-sized areas with electron energy-loss spectroscopy. *Phys. Rev. Lett.* **42**, 893–897.

Batson, P. E., and Silcox, J. (1983) Experimental energy-loss function, $\text{Im}[-1/\varepsilon(q,\omega)]$, for aluminum. *Phys. Rev. B* **27**, 5224–5239.

Batson, P. E., Silcox, J., and Vincent R., (1971) Computer control of energy analysis in an electron microscope, *29th Ann. Proc. Electron Microsc. Soc. Am.,* ed. G. W. Bailey, Claitor's Publishing, Baton Rouge, Louisiana, pp. 30–31.

Batson, P. E., Pennycook, S. J., and Jones, L. G. P. (1981) A new technique for the scanning and absolute calibration of electron energy-loss spectra. *Ultramicroscopy* **6**, 287–289.

Batson, P. E., Kavanagh, K. L., Woodall, J. M., and Mayer, J. W. (1986) Electron-energy-loss scattering near a single misfit dislocation at the GaAs/GaInAs interface. *Phys. Rev. Lett.* **57**, 1729–1732.

Batson, P. E., Chisholm, M. F., Clarke, D. R., Dimos, D., and Shaw, T. (1989) Energy-loss studies of carbon content in yttrium barium cuprate. Proc. 47th Ann. Meet. Electr. Microsc. Soc. Amer., ed. G. W. Bailey, San Francisco Press, San Francisco, pp. 196–197.

Batson, P. E., Johnson, D. W., and Spence, J. C. H. (1992) Resolution enhancement by deconvolution using a field emission source in electron energy loss spectroscopy. *Ultramicroscopy* **41**, 137–145.

Bauer, R., Hezel, U., and Kurz, D. (1987) High resolution imaging of thick biological specimens with an imaging electron energy loss spectrometer. *Optik* **77**, 171–174.

Baumann, W., Niemietz, A., Reimer, L., and Volbert, B. (1981) Preparation of P-47 scintillators for STEM. *J. Microsc.* **122**, 181–186.

Baumeister, W., and Hahn, M. (1976) An improved method for preparing single-crystal specimen supports: H_2O_2 exfoliation of vermiculite. *Micron* **7**, 247–251.

Bazett-Jones, D. P., and Ottensmeyer, F. P. (1981) Phosphorus distribution in the nucleosome. *Science* **211**, 169–170.

Beamson, G., Porter, H. Q., and Turner, D. W. (1981) Photoelectron spectromicroscopy. *Nature* **290**, 556–561.

Bearden, J. A., and Burr, A. F. (1967) X-ray atomic energy levels. *Rev. Mod. Phys.* **39**, 125–142.

Bell, A. E., and Swanson, L. W. (1979) Total energy distributions of field-emitted electrons at high current density. *Phys. Rev. B* **19**, 3353–3364.

Bell, M. G., and Liang, W. Y. (1976) Electron energy loss studies in solids; the transition metal dichalcogenides. *Adv. Phys.* **25**, 53–86.

Bendayan, M., Barth, R. F., Gingras, D., Londono, I., Robinson, P. T., Alam, F., Adams, D. M., and Mattiazzi, L. (1989) Electron spectroscopic imaging for high-resolution immunocytochemistry: use of boronated protein A. *J. Histochem. Cytochem.* **37**, 573–580.

Bennett, J. C., and Egerton, R. F. (1995) NiO test specimens for analytical electron microscopy: round-robin results. *J. Microsc. Soc. Amer.* **1**, 143–149.

Bentley, J. (1992) Applications of EELS to ceramics and catalysts. In *Transmission Electron Energy Loss Spectrometry in Materials Science,* ed. M. M. Disko, C. C. Ahn, and B. Fulz, The Minerals, Metals and Materials Society, Warrendale, Pennsylvania, 155–181.

Bentley, J., Angelini, P., and Sklad, P. S. (1984) Secondary fluorescence effects on x-ray microanalysis. In *Analytical Electron Microscopy—1984*, ed. D. B. Williams and D. C. Joy, San Francisco Press, San Francisco, 315–317.

Berger, A., and Kohl, H. (1993) Optimum imaging parameters for elemental mapping in an energy filtering transmission electron microscope. *Optik* **92**, 175–193.

Berger, M. J., and Seltzer, S. M. (1982) Stopping powers and ranges of electrons and positrons. National Bureau of Standards report: NBSIR 82-2550, U.S. Dept. of Commerce, Washington, D.C. 20234, 162 pages.

Berger, S. D., and McMullan, D. (1989) Parallel recording for an electron spectrometer on a scanning transmission electron microscope. *Ultramicroscopy* **28**, 122–125.

Berger, S. D., Salisbury, I. G., Milne, R. H., Imeson, D., and Humphreys, C. J. (1987) Electron energy-loss spectroscopy studies of nanometre-scale structures in alumina produced by intense electron-beam irradiation. *Phil. Mag. B* **55**, 341–358.

Berger, S. D., McKenzie, D. R., and Martin, P. J. (1988) EELS analysis of vacuum arc-deposited diamond-like films. *Phil. Mag. Lett.* **57**, 285–290.

Bevington, P. R. (1969) Data Reduction and Error Analysis for the Physical Sciences (McGraw-Hill, New York).

Bethe, H. (1930) Zur Theorie des Durchgangs schneller Korpuskularstrahlen durch Materie. *Ann. Phys. (Leipzig)* **5**, 325–400.

Bianconi, A. (1983) XANES spectroscopy for local structures in complex systems. In *EXAFS and Near Edge Structure*, ed. A. Bianconi, L. Incoccia, and S. Stipcich, Springer-Verlag, New York, pp. 118–129.

Bianconi, A., Dell'Ariccia, M., Durham, P. J., and Pendry, P. J. (1982) Multiple scattering resonances and structural effects in the x-ray absorption near edge spectra of Fe II and Fe. III:Hexacyanide complexes. *Phys. Rev. B* **26**, 6502–6508.

Bianconi, A., Dell'Ariccia, M., Gargano, A., and Natoli, C. R. (1983a) Bond length determination using XANES. In *EXAFS and Near Edge Structure*, ed. A. Bianconi, L. Incoccia, and S. Stipcich, Springer-Verlag, New York, pp. 57–61.

Bianconi, A., Giovannelli, A., Ascone, I., Alema, S., Durham, P., and Fasella, P. (1983b) XANES of calmodulin: Differences and homologies between calcium-modulated proteins. In *EXAFS and Near Edge Structure*, ed. A. Bianconi, L. Incoccia, and S. Stipcich, Springer-Verlag, New York, pp. 355–357.

Bihr, J., Benner, G., Krahl, D., Rilk, A., and Weimer, E. (1991) Design of an analytical TEM with integrated imaging Ω-spectrometer. *Proc. 49th Ann. Meet. Electron Microsc. Soc. Amer.*, ed. G. W. Bailey San Francisco Press, San Francisco, pp. 354–355.

Blackstock, A. W., Birkhoff, R. D., and Slater, M. (1955) Electron accelerator and high resolution analyser. *Rev. Sci. Instrum.* **26**, 274–275.

Blaha, P., and Schwarz, K. (1983) Electron densities in TiC, TiN, and TiO derived from energy band calculations. *Int. J. Quantum Chem.* **23**, 1535–1552.

Blake, D., Freund, F., Krishnan, K. F. M., Echer, C. J., Shipp, R., Bunch, T. E., Tielens, A. G., Lipari, R. J., Hetherington, C. J. D., and Chang, S. (1988) The nature and origin of interstellar diamond. *Nature* **332**, 611–613.

Blanche, G., Hug, G., Jaouen, M., and Flank, A. M. (1993) Comparison of the TiK extended fine structure obtained from electron energy loss spectroscopy and x-ray absorption spectroscopy. *Ultramicroscopy* **50**, 141–145.

Blasse, G., and Bril, A. (1967) A new phosphor for flying-spot cathode-ray tubes for color television: Yellow-emitting $Y_3Al_5O_{12}$-Ce^{3+}. *Appl. Phys. Lett.* **11**, 53–54.

Boersch, H. (1954) Experimentelle Bestimmung der Energieverteilung in thermisch ausgelösten Elektronenstrahlen. *Z. Phys.* **139**, 115–146.

Boersch, H., Geiger, J., and Hellwig, H. (1962) Steigerung der Auflösung bei der Elektronen-Energieanalyse. *Phys. Lett.* **3**, 64–66.

Boersch, H., Geiger, J., and W. Stickel (1964) Das Auflösungsvermögen des elektrostatisch-magnetischen Energieanalysators für schnelle Elektronen. *Z. Phys.* **180**, 415–424.

Bohm, D., and Pines, D. (1951) A collective description of electron interactions: II. Collective vs. individual particle aspects of the interactions. *Phys. Rev.* **85**, 338–353.

Böhmer, J., and Rahmann, H. (1990) Ultrastructural localization of aluminum in amphibian larvae. *Ultramicroscopy* **32**, 18–25.

Bonham, R. A., and Fink, M. (1974) *High Energy Electron Scattering*, Van Nostrand Reinhold, New York.

Bonney, L. A. (1990) Measurement of the inelastic mean free path by EELS analyses of submicron spheres. Proc. XIIth Int. Cong. for Electron Microscopy, San Francisco Press, San Francisco, pp. 74–75.

Booker, G. R., and Stickler, R. (1962) Method of preparing Si and Ge specimens for examination by transmission electron microscopy. *Brit. J. Appl. Phys.* **13**, 446–449.

Boothroyd, C. B., Sato, K., and Yamada, K. (1990) The detection of 0.5 at% boron in Ni_3Al

using parallel energy loss spectroscopy. Proc. XIIth Int. Cong. Electron Microscopy, San Francisco Press, San Francisco, vol. 2, pp. 80–81.

Botton, G., and L'Esperance, G. (1994) Development, quantitative performance and applications of a parallel electron energy-loss spectrum imaging system. *J. Microsc.* **173,** 9–25.

Bourdillon, A. J. (1984) The measurement of impact parameters by crystallographic orientation effects in electron scattering. *Phil. Mag.* **50,** 839–848.

Bourdillon, A. J., and Stobbs, W. M. (1986) EELS by a dual parallel and serial detection spectrometer. Proc. XIth Int. Cong. on Electron Microscopy, Japanese Society for Electron Microscopy, Tokyo, vol. 1, pp. 523–524.

Bourdillon, A. J., Jepps, N. W., Stobbs, W. M., and Krivanek, O. L. (1981a) An application of EELS in the examination of inclusions and grain boundaries of a SiC ceramic. *J. Microsc.* **124,** 49–56.

Bourdillon, A., Self, P. G., and Stobbs, W. M. (1981b) Crystallographic orientation in energy dispersive x-ray analysis. *Phil. Mag. A* **44,** 1335–1350.

Bourdillon, A., Hall, D. J., Morrison, C. J., and Stobbs, W. M. (1983) The relative importance of multiple inelastic scattering in the quantification of EELS. *J. Microsc.* **130,** 177–186.

Bourdillon, A. J., El Mashri, S. M., and Forty, A. J. (1984) Application of extended electron energy loss fine structure to the study of aluminum oxide films. *Phil. Mag.* **49,** 341–352.

Bourret, A., and Colliex, C. (1982) Combined HREM and STEM microanalysis on decorated dislocation cores. *Ultramicroscopy* **9,** 183–190.

Bracewell, R. N. (1978) *The Fourier Transform and its Applications.* McGraw-Hill, New York.

Bravman, J. C., and Sinclair, R. (1984) The preparation of cross-section specimens for transmission electron microscopy. *J. Electron Microscope Technique* **1,** 53–61.

Brigham, E. O. (1974) *The Fast Fourier Transform.* Prentice-Hall, Englewood Cliffs, New Jersey.

Bringans, R. D., and Liang, W. Y. (1981) Energy bands of the cadmium halides from electron energy loss spectroscopy. *J. Phys. C. (Solid State)* **14,** 1065–1092.

Brown, K. L. (1967) A first- and second-order matrix theory for the design of beam transport systems and charged particle spectrometers. *Adv. Part. Phys.* **1,** 71–135.

Brown, K. L., Belbeoch, R., and Bounin, P. (1964) First- and second-order magnetic optics matrix equations for the midplane of uniform-field wedge magnets. *Rev. Sci. Instr.* **35,** 481–485.

Brown, K. L., Rothacker, F., Carey, D. C., and Iselin, Ch. (1977) TRANSPORT: A computer program for designing charged particle beam transport systems. SLAC-91 report, National Technical Information Service, Springfield, Virginia 22161.

Brown, L. M., Colliex, C., and Gasgnier, M. (1984) Fine structure in EELS from rare-earth sesquioxide thin films. *J. Phys. (Paris)* **45** (Coll. C2), 433–436.

Browne, M. T. (1979) Electron energy analysis in a Vacuum Generators HB5 STEM. In *Scanning Electron Microscopy*, SEM Inc., A. M. F. O'Hare, Illinois, part 2, pp. 827–834.

Browne, M. T., and Ward, J. F. L. (1982) Detectors for STEM and the measurement of their detective quantum efficiency. *Ultramicroscopy* **7,** 249–262.

Browning, N. D., and Pennycook, S. J. (1993) Atomic resolution spectroscopy for the microanalysis of materials. *Microbeam Anal.* **2,** 81–89.

Browning, N. D., Yuan, Y., and Brown, L. M. (1991a) Investigation of fluctuations on oxygen stoichiometry near grain boundaries in ion-beam thinned $YBa_2Cu_3O_{7-\delta}$ by scanning transmission electron microscopy. Inst. Phys. Conf. Ser. No. 119 (EMAG 91). Institute of Physics, Bristol, pp. 283–286.

Browning, N. D., Yuan, J., and Brown, L. M. (1991b) Real-space determination ot anisotropic electronic structure by electron energy loss spectroscopy. *Ultramicroscopy* **38,** 291–298.

Browning, N. D., Chisholm, M. F., and Pennycook, S. J. (1993a) Cell-by-cell mapping of carrier concentrations in high temperature superconductors. *Interface Sci* **1**, 309–319.

Browning, N. D., Chisholm, M. F., and Pennycook, S. J. (1993b) Atomic-resolution chemical analysis using a scanning transmission electron microscope. *Nature* **366**, 143–146.

Bruley, J. (1992) Detection of nitrogen at {100} platelets in a type IaA/B diamond. *Phil. Mag. Lett.* **66**, 47–56.

Bruley, J. (1993) Spatially resolved electron energy-loss near-edge structure analysis of a near $\Sigma = 11$ tilt boundary in sapphire. *Microsc. Microanal. Microstruct.* **4**, 23–39.

Bruley, J., and Batson, P. E. (1989) Electron-energy-loss studies of dislocations in diamond. *Phys. Rev. B* **40**, 9888–9894.

Bruley, J., and Brown, L. M. (1989) Quantitative electron energy-loss spectroscopy microanalysis of platelet and voidite defects in diamond. *Phil. Mag. A* **59**, 247–261.

Bruley, J., Brydson, R., Müllejans, H., Mayer, J., Gutekunst, G., Mader, W., Knauss, D., and Rühle, M. (1994) Investigations of the chemistry and bonding at niobium-sapphire interfaces. *J. Mater. Res.* **9**, 2574–2583.

Bruley, J., Williams, D. B., Cuomo, J. J., and Pappas, D. P. (1995) Quantitative near-edge structure analysis of diamond-like carbon in the electron microscope using a two-window method. *J. Microsc.*, submitted.

Brydson, R. (1991) Interpretation of near-edge structure in the electron energy-loss spectrum. *EMSA Bull.* **21**, pp. 57–67.

Brydson, R., Sauer, H., Engel, W., Thomas, J. M., and Zeitler, E. (1989) Co-ordination fingerprints in electron loss near-edge structures: determination of the local site symmetry of aluminum and beryllium in ultrafine minerals. *J. Chem. Soc., Chem. Commun.*, Issue **15**, pp. 1010–1012.

Brydson, R., Hansen, P. L., and McComb, D. W. (1992a) $p \to p$-like transitions in tetrahedrally coordinated cation L23 ELNES. In *Electron Microscopy 1992*, Proc. EUREM, Granada, vol. 1, pp. 251–252.

Brydson, R., Sauer, H., and Engel, W. (1992b) Electron energy-loss near edge structure as an analytical tool–the study of minerals. In *Transmission Electron Energy Loss Spectrometry in Materials Science*, ed. M. M. Disko, C. C. Ahn, and B. Fulz, The Minerals, Metals and Materials Society, Warrendale, Pennsylvania, pp. 131–154.

Brydson, R., Richardson, I. G., McComb, D. W., and Groves, G. W. (1993) Parallel electron energy loss spectroscopy study of Al-substituted calcium silicate hydrate (C–S–H) phases present in hardened cement pastes. *Solid State Commun.* **88**, pp. 183–187.

Brydson, R., Mullejans, H., Bruley, J., Trusty, P. A., Sun, X., Yeomans, J. A., and Ruhle, M. (1995) Spatially resolved electron energy-loss studies of metal ceramic interfaces in transition metal/alumina cermets. *J. Microsc.* **177**, 369–386.

Buffat, B., and Tuilier, M. H. (1987) X-ray absorption edges of iron and cobalt with six-fold coordination in oxides: influence of site distortion and oxidation state. *Solid State Commun.* **64**, 401–406.

Buggy, T. W., and Craven, A. J. (1981) Optimization of post-specimen lenses for use in STEM. In *Developments in Electron Microscopy and Analysis*, Inst. Phys. Conf. Ser. No. 61, pp. 197–200.

Burch, S. F., Gull, S. F., and Skilling, J. (1983) Image restoration by a powerful maximum-entropy method. *Comput. Vision, Graphics Image Process.* **23**, 113–128.

Burgess, W. G., Preston, A. R., Botton, G. A., Zaluzec, N. J., and Humphreys, C. J. (1994) Benefits of energy filtering for advanced convergent beam electron diffraction patterns. *Ultramicroscopy* **55**, 276–283.

Bursill, L. A., Egerton, R. F., Thomas, J. M., and Pennycook, S. (1981) High-resolution imaging and electron energy-loss studies of platelet defects in diamond. *J. Chem. Soc., Faraday Trans.* **77**, 1367–1373.

Buseck, P. R., Cowley, J. M., and Eyring, L. (1988) *High-Resolution Transmission Electron Microscopy and Associated Techniques.* Oxford University Press, New York and Oxford.

Butler, J. H., Watari, F., and Higgs, A. (1982) Simultaneous collection and processing of energy-filtered STEM images using a fast digital data acquisition system. *Ultramicroscopy* **8**, 327–334.

Carey, D. C. (1978) TURTLE: A computer program for simulating charged particle beam transport systems. Report NAL-64, National Accelerator Laboratory, Batavia, Illinois 60510.

Carlemalm, E., and Kellenberger, E. (1982) The reproducible observation of unstained embedded cellular material in this section: Visualization of an integral membrane by a new mode of imaging for STEM. *EMBO J.* **1**, 63–67.

Carlemalm, E., Acetarin, J. D., Villinger, W., Colliex, C., and Kellenberger, E. (1982) Heavy metal containing surroundings provide much more "negative" contrast by Z-imaging in STEM than with conventional modes. *J. Ultrastruc. Res.* **80**, 339–343.

Castaing, R. (1975) Energy filtering in electron microscopy and electron diffraction. In *Physical Aspects of Electron Microscopy and Microbeam Analysis*, ed. B. Siegel, Wiley, New York, pp. 287–301.

Castaing, R., and Henry, L. (1962) Filtrage magnetique des vitesses en microscopie électronique. *C.R. Acad. Sci. Paris* **B255**, 76–78.

Castaing, R., Hennequin, J. F., Henry, L., and Slodzian, G. (1967) The magnetic prism as an optical system. In *Focussing of Charged Particles*. ed. A. Septier, Academic Press, New York, pp. 265–293.

Castro-Fernandez, F. R., Sellars, C. M., and Whiteman, J. A. (1985) Measurement of foil thickness and extinction distance by convergent beam transmission electron microscopy. *Phil. Mag. A* **52**, 289–303.

Catalano, M., Kim, M. J., Carpenter, R. W., Chowdhury, K. D., and Wong, J. (1993) The composition and structure of SIPOS: a high spatial resolution electron microscopy study. *J. Mater. Res.* **8**, 2893–2901.

Cazaux, J. (1983) Another expression for the minimum detectable mass in EELS, EPMA and AES. *Ultramicroscopy* **12**, 83–86.

Cazaux, J. (1984) Detection limits in Auger electron spectroscopy. *Surf. Sci.* **140**, 85–100.

Chadderton, L. T. (1965) *Radiation Damage in Crystals*, Methuen, London.

Chan, H. M., and Williams, D. B. (1985) Quantitative analysis of lithium in Al–Li alloys by ionization energy loss spectroscopy. *Phil. Mag. B* **52**, 1019–1032.

Chen, C. H., Silcox, J., and Vincent, R. (1975) Electron energy losses in silicon: Bulk and surface plasmons and Cerenkov radiation. *Phys. Rev. B* **12**, 64–71.

Chen, C. H., Joy, D. C., Chen, H. S., and Hauser, J. J. (1986) Observation of anomalous plasmon linewidth in the icosahedral Al-Mn quasicrystals. *Phys. Rev. Lett.* **57**, 743–746.

Chen, M. Y., Li, D., Dravid, V. P., Chung, Y-W., Wong, M-S., and Sproul, W. D. (1993) Analytical electron microscopy and Raman spectroscopy studies of carbon nitride thin films. *J. Vac. Sci. Technol. A* **11**, 521–524.

Chen, Z., Cochrane, R., and Loretto, M. H. (1984) Microanalysis of extracted precipitates from extracted steels. In *Analytical Electron Microscopy—1984*, ed. D. B. Williams and D. C. Joy, San Francisco Press, San Francisco, pp. 243–246.

Cheng, S. C., and Egerton, R. F. (1985) Signal/background ratio of ionization edges in EELS. *Ultramicroscopy*, **16**, 279–282.

Cheng, S. C., and Egerton, R. F. (1993) Elemental analysis of thick amorphous specimens by EELS. *Micron* **24**, 251–256.

Cherns, D., Howie, A., and Jacobs, M. H. (1973) Characteristic x-ray production in thin crystals. *Z. Naturforsch.* **28a**, 565–571.

Choi, B.-H. (1973) Cross section for *M*-shell ionization in heavy atoms by collision of simple heavy charged particles. *Phys. Rev. A* **7**, 2056–2062.

Choi, B.-H., Merzbacher, E., and Khandelwal, G. S. (1973) Tables for Born approximation calculations of *L*-subshell ionization by simple heavy charged particles. *At. Data* **5**, 291–304.

Ciliax, B. J., Kirk, K. L., and Leapman, R. D. (1993) Radiation damage of fluorinated organic compounds measured by parallel electron energy loss spectroscopy. *Ultramicroscopy* **48**, 13–25.

Citrin, P. H., Eisenberger, P., and Kincaid, B. M. (1976) Transferabilty of phase shifts in extended x-ray absorption fine structure. *Phys. Rev. Lett.* **36**, 1346–1349.

Cliff, G., and Kenway, P. B. (1982) The effects of spherical aberration in probe-forming lenses on probe size, image resolution and x-ray spatial resolution in scanning transmission electron microscopy. In *Microbeam Analysis—1982*, ed. K. F. J. Heinrich, San Francisco Press, San Francisco, 1982, pp. 107–110.

Cochran, W. T. (1967) What is the fast Fourier transform? *IEEE Trans. Audio Electroacoust.* **AU15**, 45–55.

Cockayne, D. J. H., McKenzie, D., and Muller, D. (1991) Electron diffraction of amorphous thin films using PEELS. *Microsc. Microanal. Microstruct.* **2**, 359–366.

Collett, S. A., Brown, L. M., and Jacobs, M. H. (1981) Microanalytical electron microscopy on type 304 steel: Correlation between EDX and EELS results. In *Quantitative Microanalysis with High Spatial Resolution*, The Metals Society, London, 1981, pp. 159–164.

Collett, S. A., Brown, L. M., and Jacobs, M. H. (1984) Demonstration of superior resolution of EELS over EDX in microanalysis. In *Developments in Electron Microscopy and Analysis 1983*, ed. P. Doig, Inst. Phys. Conf. Ser. No. 68, I.O.P., Bristol, pp. 103–106.

Colliex, C. (1982) Electron energy-loss analysis in materials science. In *Electron Microscopy—1982*, 10th Int. Cong., Deutsche Gesellschaft für Elektronenmikroskopie, Vol. 1, pp. 159–166.

Colliex, C. (1984a) Electron energy-loss spectroscopy in the electron microscope. In *Advances in Optical and Electron Microscopy*, ed. V. E. Cosslett and R. Barer, Academic Press, London, Vol. 9, pp. 65–177.

Colliex, C. (1985) An illustrated review on various factors governing the high spatial resolution capabilities in EELS microanalysis. *Ultramicroscopy* **18**, 131–150.

Colliex, C., and Jouffrey, B. (1972) Diffusion inelastique des electrons dans une solide par excitation de niveaux atomiques profonds. *Phil. Mag.* **25**, 491–514.

Colliex, C., and Mory, C. (1984) Quantitative aspects of scanning transmission electron microscopy. In *Quantitative Electron Microscopy*, ed. J. N. Chapman and A. J. Craven, SUSSP Publications, Edinburgh, pp. 149–216.

Colliex, C., Cosslett, V. E., Leapman, R. D., and Trebbia, P. (1976a) Contribution of electron energy-loss spectroscopy to the development of analytical electron microscopy. *Ultramicroscopy* **1**, 301–315.

Colliex, C., Gasgnier, M., and Trebbia, P. (1976b) Analysis of the electron excitation spectra in heavy rare earth metals, hydrides and oxides. *J. Phys.* (*Paris*) **37**, 397–406.

Colliex, C., Jeanguillaume, C., and Trebbia P. (1981a) Quantitative local microanalysis with EELS. In *Microprobe Analysis of Biological Systems*, ed. T. E. Hutchinson and A. P. Somlyo, Academic Press, New York, pp. 251–271.

Colliex, C., Manoubi, T., Gasgnier, M., and Brown, L. M. (1985) Near-edge fine structures on EELS core-loss edges. In *Scanning Electron Microscopy—1985*, SEM Inc., A. M. F. O'Hare, Illinois, Part 2, pp. 489–512.

Colliex, C., Mory, C., Olins, A. L., Olins, D. E., and Tencé, M. (1989) Energy-filtered STEM of thick biological sections. *J. Microsc.* **153**, 1–21.

Comins, N. R., and Thirlwall, J. T. (1981) Quantitative studies and theoretical analysis of the performance of the scintillation electron detector. *J. Microsc.* **124,** 119–133.

Cook, R. F. (1971) Combined electron microscopy and energy loss analysis of glass. *Phil. Mag.* **24,** 835–843.

Costa, J. L., Joy, D. C., Maher, D. M., Kirk, K. L., and Hui, S. W. (1978) Fluorinated molecule as a tracer: Difluoroserotonin in human platelets mapped by electron energy-loss spectroscopy. *Science* **200,** 537–539.

Cowley, J. M. (1982) Surface energies and surface structure of small crystals studies by use of a STEM instrument. *Surf. Sci.* **114,** 587–606.

Craven, A. J., and Buggy, T. W. (1981) Design consideration and performance of an analytical STEM. *Ultramicroscopy* **7,** 27–37.

Craven, A. J., and Buggy, T. W. (1984) Correcting electron energy loss spectra for artefacts introduced by a serial collection system. *J. Microsc.* **136,** 227–239.

Craven, A. J., and Colliex, C. (1977) The effect of energy-loss on phase contrast. In *Developments in Electron Microscopy and Analysis*, Inst. Phys. Conf. Ser. No. 36, pp. 271–274.

Craven, A. J., Gibson, J. M., Howie, A., and Spalding, D. R. (1978) Study of single-electron excitations by electron microscopy: I. Image contrast from delocalized excitations. *Phil. Mag.* **A38,** 519–527.

Craven, A. J., Buggy, T. W., and Ferrier, R. P. (1981) Post-specimen lenses in electron spectroscopy. In *Quantitative Microanalysis with High Speed Resolution*, The Metals Society, London.

Craven, A. J., Cluckie, M. M., Duckworth, S. P., and Baker, T. N. (1989) Analysis of small vanadium carbide precipitates using electron energy loss spectroscopy. *Ultramicroscopy* **28,** 330–334.

Crecelius, G., Fink, J., Ritsko, J. J., Stamm, M., Freund, H.-J., and Gonska, H. (1983) π-electron delocalization in poly(p-phenylene), poly(p-phenylene sulfide), and poly(p-phenylene oxide). *Phys. Rev. B* **28,** 1802–1808.

Creuzburg, M. (1966) Entstehung von Alkalimetallen bei der Elektronenbestrahlung von Alkalihalogeniden. *Z. Phys.* **194,** 211–218.

Crewe, A. V. (1978) Some space charge effects in electron probe devices. *Optik* **52,** 337–346.

Crewe, A. V., and Scaduto, F. (1982) A gradient field spectrometer for STEM use. *Proc. 40th Ann. Proc. Electron Microsc. Soc. Am.*, ed. G. W. Bailey, Claitor's Publishing, Baton Rouge, Louisiana, pp. 734–735.

Crewe, A. V., Langmore, J. P., and Isaacson, M. S. (1975) Resolution and contrast in the scanning transmission electron microscope. In *Physical Aspects of Electron Microscopy and Microbeam Analysis*, ed. B. M. Siegel and D. R. Beaman, Wiley, New York, 1975, pp. 47–62.

Cromer, D. T., and Waber, J. T. (1965) Scattering factors computed from relativistic Dirac-Slater wave functions. *Acta. Crystallogr.* **18,** 104–109.

Crozier, P. A. (1990) Measurement of inelastic electron scattering cross-sections by electron energy-loss spectroscopy. *Phil Mag.* **61,** 311–336.

Crozier, P. A. (1995) Quantitative elemental mapping of materials by energy-filtered imaging. *Ultramicroscopy* **58,** 157–174.

Crozier, P. A. and Egerton, R. F. (1989) Mass-thickness determination by Bethe-sum-rule normalization of the electron energy-loss spectrum. *Ultramicroscopy* **27,** 9–18.

Cundy, S. L., and Grundy, P. J. (1966) Combined electron microscopy and energy analysis of an internally oxidized Ni + Si alloy. *Phil. Mag.* **14,** 1233–1242.

Cuomo, J. J., Doyle, J. P., Bruley, J., and Liu, J. C. (1991) Sputter deposition of dense diamond-like carbon films at low temperature. *Appl. Phys. Lett.* **58,** 466–468.

Curtis, G. H., and Silcox, J. (1971) A Wien filter for use as an energy analyzer with an electron microscope. *Rev. Sci. Instrum.* **42**, 630–637.

Daberkow, I., Herrmann, K.-H., Liu, L., and Rau, W. D. (1991) Performance of electron image converters with YAG single-crystal screen and CCD sensor. *Ultramicroscopy* **38**, 215–223.

Daniels, J., and Krüger, P. (1971) Electron energy losses and optical constants of Ne and CH₄. *Phys. Status Solidi (b)* **43**, 659–664.

Daniels, J., Festenberg, C. V., Raether, H., and Zeppenfeld, K. (1970) Optical constants of solids by electron spectroscopy. *Springer Tracts in Modern Physics*, Springer-Verlag, New York, Vol. 54, pp. 78–135.

Darlington, E. H., and Sparrow, T. G. (1975) A magnetic prism spectrometer for a high voltage electron microscope. *J. Phys. E (Sci. Instrum.)* **8**, 596–600.

Davis, C. A., McKenzie, D. R., Yin, Y., Kravtchinskaia, E., Amaratunga, G. A. J., and Veerasamy, V. S. (1994) Substitutional nitrogen doping of tetrahedral amorphous carbon. *Phil. Mag. B* **69**, 1133–1140.

Davis, L. C., and Feldkamp, L. A. (1976) Interpretation of 3*p*-core-excitation spectra in Cr, Mn, Co, and Ni. *Solid State Commun.* **19**, 413–416.

Dehmer, J. L., and Dill, D. (1977) Molecular effects on inner-shell photoabsorption K-shell spectrum of N₂. *J. Chem. Phys.* **65**, 5327–5337.

Devenish, R. W., Eaglesham, D. J., Maher, D. M., and Humphreys, C. J. (1989) Nanolithography using field emission and conventional thermionic electron sources. *Ultramicroscopy* **28**, 324–329.

Dexpert, H., Lynch, J. P., and Freund, E. (1982) Sensitivity limits in x-ray emission spectroscopy of catalysts. In *Developments in Electron Microscopy and Analysis*, Inst. Phys. Conf. Ser. No. 61 (I.O.P.), pp. 171–174.

Diociaiuti, M., and Paoletti, L. (1991) Structural characterization of air-oxidized chromium particles by extended energy-loss fine-structure spectroscopy. *J. Microsc.* **162**, 279–289.

Diociaiuti, M., Bascelli, A., and Paoletti, L. (1992a) Extended electron energy loss fine structure and selected area electron diffraction combined study of copper cluster oxidation. *Vacuum* **43**, 575–581.

Diociaiuti, M., Picozzi, P., Santucci, S., Lozzi, L., and de Crescenzi, M. (1992b) Extended electron energy-loss fine structure and selected-area diffraction studies of small palladium clusters. *J. Microsc.* **166**, 231–245.

Diociaiuti, M., Falchi, M., and Paoletti, L. (1995) Electron energy loss spectroscopy study of iron deposition in human alveolar macrophages: ferritin or hemosiderin? *Microsc. Microanal. Microstruct.* **6**, 33–40.

Disko, M. M. (1981) An EXELFS analysis system and the preliminary orientation dependence of EXELFS in graphite. In *Analytical Electron Microscopy—1981*, ed. R. H. Geiss, San Francisco Press, San Francisco, pp. 214–220.

Disko, M. M., Meitzner, G., Ahn, C. C., and Krivanek, O. L. (1989) Temperature-dependent transmission extended electron energy-loss fine-structure of aluminum. *J. Appl. Phys.* **65**, 3295–3297.

Disko, M. M., Luton, M. J., and Shuman, H. (1991) Energy-loss near-edge fine structure and composition profiles of cryomilled oxide-dispersion-strengthened aluminum. *Ultramicroscopy* **37**, 202–209.

Disko, M. M., Ahn, C. C., and Fultz, B., eds. (1992) *Transmission Electron Energy Loss Spectrometry in Materials Science.* The Minerals, Metals and Materials Society, Warrendale, Pennsylvania, 272 pages.

Ditchfield, R. W., and Cullis, A. G. (1976) Plasmon energy-loss analysis of epitaxial layers in silicon and germanium. *Micron* **7**, 133–140.

Ditchfield, R. W., and Whelan, M. J. (1977) Energy broadening of the electron beam in the electron microscope. *Optik* **48**, 163–172.

Ditchfield, R. W., Grubb, D. T., and Whelan, M. J. (1973) Electron energy-loss studies of polymers during radiation damage. *Phil. Mag.* **27**, 1267–1280.

Donald, A. M., and Craven, A. J. (1979) A study of grain boundary segregation in Cu-Bi alloys using STEM. *Phil. Mag. A* **39**, 1–11.

Downing, K. K., Ho, M.-H., and Glaeser, R. M. (1980) A charge coupled device readout system for electron microscopy, *38th Ann. Proc. Electron Microsc. Soc. Am.*, ed. G. W. Bailey, Claitor's Publishing, Baton Rouge, Louisiana, pp. 234–235.

Dravid, V. P., and Zhang, H. (1992) Hole formation and charge transfer in $Y_{1-x}Ca_xSr_2Cu_2GaO_7$, a new oxide superconductor. *Physica C* **200**, 349–358.

Dravid, V. P., Zhang, H., and Wang, Y. Y. (1993) Inhomogeneity of charge carrier concentration along the grain boundary plane in oxide superconductors. *Physica C* **213**, 353–358.

Du Chesne, A., Lieser, G., and Wegner, G. (1992) ESI for investigation of polymer microdomain morphology. *Electron Microscopy 1992*, Proc. EUREM, Granada, vol. 1, pp. 255–256.

Duckworth, S. P., Craven, A. J., and Baker, T. N. (1984) Comparison of carbon and noncarbon replicas for ELS. In *Analytical Electron Microscopy—1984*, ed. D. B. Williams and D. C. Joy, San Francisco Press, San Francisco, pp. 235–238.

Durham, P. J. (1983) Multiple scattering calculations of XANES. In *EXAFS and Near Edge Structure*, ed. A. Bianconi, L. Incoccia, and S. Stipcich, Springer-Verlag, New York, pp. 37–42.

Durham, P. J., Pendry, J. B., and Hodges, C. H. (1981) XANES: determination of bond angles and multi-atom correlations in ordered and disordered systems. *Solid State Commun.* **38**, 159–162.

Durham, P. J., Pendry, J. B., and Hodges, C. H. (1982) Calculation of x-ray absorption near-edge structure, XANES, *Comput. Phys. Commun.* **25**, 193–205.

Eaglesham, D. J., and Berger, S. D. (1994) Energy filtering and the "thermal diffuse" background in electron diffraction. *Ultramicroscopy* **53**, 319–324.

Echenique, P. M., Ritchie, R. H., and Brandt, W. (1979) Spatial excitation patterns by swift ions in condensed matter. *Phys. Rev. B* **20**, 2567–2580.

Echenique, P. M., Bausells, J., and Rivacoba, A. (1987) Energy-loss probability in electron microscopy. *Phys. Rev. B* **35**, 1521–1524.

Egerton, R. F. (1975) Inelastic scattering of 80-keV electrons in amorphous carbon. *Phil. Mag.* **31**, 199–215.

Egerton, R. F. (1976a) Measurement of inelastic/elastic scattering ratio for fast electrons and its use in the study of radiation damage. *Phys. Status Solidi (a)* **37**, 663–668.

Egerton, R. F. (1976b) Coupling between plasmon and K-shell excitation in electron energy-loss spectra of amorphous carbon, graphite and beryllium. *Solid State Commun.* **19**, 737–740.

Egerton, R. F. (1976c) Inelastic scattering and energy filtering in the transmission electron microscope. *Phil. Mag.* **34**, 49–66.

Egerton, R. F. (1978a) Formulae for light-element microanalysis by electron energy-loss spectrometry. *Ultramicroscopy* **3**, 243–251.

Egerton, R. F. (1978b) A simple electron spectrometer for energy analysis in the transmission microscope. *Ultramicroscopy* **3**, 39–47.

Egerton, R. F. (1979) K-shell ionization cross-sections for use in microanalysis. *Ultramicroscopy* **4**, 169–179.

Egerton, R. F. (1980a) The use of electron lenses between a TEM specimen and an electron spectrometer. *Optik* **56**, 363–376.

Egerton, R. F. (1980b) Design of an aberration-corrected electron spectrometer for the TEM. *Optik* **57**, 229–242.

Egerton, R. F. (1980c) Chemical measurements of radiation damage at and below room temperature. *Ultramicroscopy* **5**, 521–523.

Egerton, R. F. (1980d) Instrumentation and software for energy-loss microanalysis. In *Scanning Electron Microscopy*, SEM Inc., A. M. F. O'Hare, Illinois, Vol. 1, pp. 41–52.

Egerton, R. F. (1980e) An automated system for energy-loss microanalysis. *38th Ann. Proc. Electron Microsc. Soc. Am.*, ed. G. W. Bailey, Claitor's Publishing, Baton Rouge, Louisiana, pp. 130–131.

Egerton, R. F. (1980f) Measurement of radiation damage by electron energy-loss spectroscopy, *J. Microsc.* **118**, 389–399.

Egerton, R. F. (1981a) SIGMAL: a program for calculating L-shell ionization cross sections. *39th Ann. Proc. Electron Microsc. Soc. Am.*, ed. G. W. Bailey, Claitor's Publishing, Baton Rouge, Louisiana, pp. 198–199.

Egerton, R. F. (1981b). Alignment and characterization of an electron spectrometer. *Ultramicroscopy* **6**, 93–96.

Egerton, R. F. (1981c) The range of validity of EELS microanalysis formulae. *Ultramicroscopy* **6**, 297–300.

Egerton, R. F. (1981d) Applications of energy-loss microanalysis. In *Analytical Electron Microscopy—1981*, ed. R. H. Geiss, San Francisco Press, San Francisco, pp. 154–160.

Egerton, R. F. (1982a) A revised expression for signal/noise ratio in EELS. *Ultramicroscopy* **9**, 387–390.

Egerton, R. F. (1982b) Organic mass loss at 100K and 300K. *J. Microsc.* **126**, 95–100.

Egerton, R. F. (1982c) Electron energy-loss spectroscopy for chemical analysis. *Phil. Trans. R. Soc. London* **A305**, 521–533.

Egerton, R. F. (1982d) Thickness dependence of the STEM ratio image. *Ultramicroscopy* **9**, 297–299.

Egerton, R. F. (1984a) Parallel-recording systems for electron energy-loss spectroscopy (EELS). *J. Electron Microsc. Tech.* **1**, 37–52.

Egerton, R. F. (1987) EELS at intermediate voltage. In *Intermediate Voltage Microscopy and Its Application to Materials Science*, ed. K. Rajan, Philips Electron Optics, Mahwah, pp. 67–71.

Egerton, R. F. (1992a) A data base for energy-loss cross sections and mean free paths. *50th Ann. Proc. Electron Microsc. Soc. Amer.*, San Francisco Press, San Francisco, pp. 1264–1265.

Egerton, R. F. (1992b) Electron energy-loss spectroscopy—EELS. In *Quantitative Microbeam Analysis*, ed. A. G. Fitzgerald, B. E. Storey, and D. Fabian, SUSSP, Edinburgh, and IOP, Bristol, pp. 145–168.

Egerton, R. F. (1993) Oscillator-strength parameterization of inner-shell cross sections. *Ultramicroscopy* **50**, 13–28.

Egerton, R. F., and Cheng, S. C. (1982) The use of photodiode arrays to record 100 keV electrons. *J. Microsc.* **127**, RP3–RP4.

Egerton, R. F., and Cheng, S. C. (1985) Thickness measurement by EELS. *43rd Ann. Proc. Electron Microsc. Soc. Amer.*, ed. G. W. Bailey, San Francisco Press, San Francisco, pp. 389–399.

Egerton, R. F., and Cheng, S. C. (1987) Measurements of local thickness by electron energy-loss spectroscopy. *Ultramicroscopy* **21**, 231–244.

Egerton, R. F., and Crozier, P. A. (1987) A compact parallel-recording detector for EELS. *J. Microsc.* **148**, 305–312.

Egerton, R. F., and Crozier, P. A. (1988) The use of Fourier techniques in electron energy-loss spectroscopy. *Scanning Microscopy, Supplement*, **2**, Scanning Microscopy International, Chicago, 245–254.

Egerton, R. F., and Kenway, D. (1979) An acquisition, storage, display and processing system for electron energy-loss spectra. *Ultramicroscopy* **4**, 221–225.

Egerton, R. F., and Rossouw, C. J. (1976) Direct measurement of contamination and etching rates in an electron beam. *J. Phys. D* **9**, 659–663.

Egerton, R. F., and Sevely, J. (1983) Jump ratio as a measure of spectrometer performance. *J. Microsc.* **129**, RP1–RP2.

Egerton, R. F., and Wang, Z. L. (1990) Plural-scattering deconvolution of electron energy-loss spectra recorded with an angle-limiting aperture. *Ultramicroscopy* **32**, 137–148.

Egerton, R. F., and Whelan, M. J. (1974a) Electron energy-loss spectra of diamond, graphite and amorphous carbon. *J. Electron Spectrosc.* **3**, 232–236.

Egerton, R. F., and Whelan, M. J. (1974b) The electron energy-loss spectrum and band structure of diamond. *Phil. Mag.* **30**, 739–749.

Egerton, R. F., and Wong, K. (1995) Some practical consequences of the Lorentzian angular distribution of inelastic scattering. *Ultramicroscopy* **59**, 169–180.

Egerton, R. F., Philip, J. G., and Whelan, M. J. (1974) Applications of energy analysis in a transmission electron microscope. In *Electron Microscopy—1974*, 8th Int. Congress, ed. J. V. Sanders and D. J. Goodchild, Australian Academy of Science, Canberra, Vol. 1, pp. 137–140.

Egerton, R. F., Philip, J. G., Turner, P. S., and Whelan, M. J. (1975) Modification of a transmission electron microscope to give energy-loss spectra and energy-selected images and diffraction patterns. *J. Phys. E.* **8**, 1033–1037.

Egerton, R. F., Rossouw, C. J., and Whelan, M. J. (1976) Progress towards a method for the quantitative microanalysis of light elements by electron energy-loss spectrometry. In *Developments in Electron Microscopy and Analysis*, ed. J. A. Venables, Academic Press, New York, pp. 129–132.

Egerton, R. F., Williams, B. G., and Sparrow, T. G. (1985) Fourier deconvolution of electron energy-loss spectra. *Proc. R. Soc. London*, A **398**, 395–404.

Egerton, R. F., Crozier, P. A., and Rice, P. (1987) Electron energy-loss spectroscopy and chemical change. *Ultramicroscopy* **23**, 305–312.

Egerton, R. F., Yang, Y.-Y., and Chen, F. Y. Y. (1991) EELS of "thick" specimens. *Ultramicroscopy* **38**, 349–352.

Egerton, R. F., Yang, Y.-Y., and Cheng, S. C. (1993) Characterization and use of the Gatan 666 parallel-recording electron energy-loss spectrometer. *Ultramicroscopy* **48**, 239–250.

Egle, W., Rilk, A., Bihr, J., and Menzel, M. (1984) Microanalysis in the EM 902: Tests on a new TEM for ESI and EELS. *42nd Ann. Proc. Electron Microsc. Soc. Am.*, ed. G. W. Bailey, San Francisco Press, San Francisco, pp. 566–567.

Ehrenreich, H., and Philipp, H. R. (1962) Optical properties of semiconductors in the ultra-violet. Proc. Int. Conf. Phys. Semiconductors, ed. A. C. Strickland, Bartholomew Press, Dorking, U.K., pp. 367–374.

Eisenberger, P., Shulman, R. G., Kincaid, B. M., and Brown, G. S. (1978) Extended x-ray absorption fine structure determination of iron-nitrogen distances in haemoglobin. *Nature* **274**, 30–34.

Enge, H. A. (1964) Effect of extended fringing fields on ion-focussing properties of deflecting magnets. *Rev. Sci. Instrum.* **35**, 278–287.

Enge, H. A. (1967) Deflecting magnets. In *Focussing of Charged Particles*, Vol. 2, ed. A. Septier, Academic Press, New York, pp. 203–264.

Engel, A., Christen, F., and Michel, B. (1981) Digital acquisition and processing of electron micrographs using a scanning transmission electron microscope. *Ultramicroscopy* **7**, 45–54.

Evans, E., and Mills, D. L. (1972) Theory of inelastic scattering by long-wavelength surface optical phonons. *Phys. Rev. B* **5**, 4126–4139.

Evans, N. D., Zinkle, S. J., Bentley, J., and Kenik, E. A. (1991) Quantification of metallic aluminum profiles in Al^+ implanted $MgAl_2O_4$ spinel by electron energy loss spectroscopy. In *Microbeam Analysis—1991*, San Francisco, San Francisco, pp. 439–440.

Fallon, P. J. (1992) Microscopy and Spectroscopy of CVD Diamond, Diamond-Like Carbon and Similar Materials. Ph.D. Thesis, University of Cambridge, U.K.

Fallon, P. J. and Brown, L. M. (1993) Analysis of chemical-vapour-deposited diamond grain boundaries using transmission electron microscopy and parallel electron energy loss spectroscopy in a scanning transmission electron microscope. *Diamond Rel. Mater.* **2**, 1004–1011.

Fallon, P. J., Brown, L. M., Barry, J. C., and Bruley, J. (1995) Nitrogen determination and characterization in natural diamond platelets. *Phil. Mag.*, submitted.

Fano, U. (1956) Differential inelastic scattering of relativistic charged particles. *Phys. Rev.* **102**, 385–387.

Fano, U. (1960) Normal modes of a lattice of oscillators with many resonances and dipolar coupling. *Phys. Rev.* **118**, 451–455.

Fano, U., and Cooper, J. W. (1968) Spectral distribution of atomic oscillator strengths. *Rev. Mod. Phys.* **40**, 441–507.

Ferrell, R. A. (1957) Characteristic energy losses of electrons passing through metal foils. II. Dispersion relation and short wavelength cutoff for plasma oscillations. *Phys. Rev.* **107**, 450–462.

Festenberg, C. von (1967) Zur Dämpfung des Al-15 keV-Plasmaverlustes in Abhängigkeit vom Streunwinkel und der Kristallitgrösse. *Z. Phys.* **207**, 47–55.

Festenberg, C. von (1969) Energieverlustmessungen an III–V-Verbindungen. *Z. Phys.* **227**, 453–481.

Festenberg, C. von, and Kröger, E. (1968) Retardation effects for the energy-loss probability in GaP and Si. *Phys. Lett.* **26A**, 339–341.

Fields, J. R. (1977) Magnetic spectrometers: approximate and ideal designs. *Ultramicroscopy* **2**, 311–325.

Fink, J. (1989) Recent developments in energy-loss spectroscopy. In *Advances in Physics and Electron Physics*, Academic Press, London, vol. 75, pp. 121–232.

Fink, J., Muller-Heinzerling, T., Pflü, J., Bubenzer, A., Koidl, P., and Crecelius, G. (1983) Structure and bonding of hydrocarbon plasma generated carbon films: An electron energy loss study. *Solid State Commun.* **47**, 687–691.

Fink, J., Nücker, N., Pellegrin, E., Romberg, H., Alexander, M., and Knupfer, M. (1994) Electron energy-loss and x-ray absorption spectroscopy of cuprate superconductors and related compounds. *J. Electron Spectrosc. Rel. Phenom.* **66**, 395–452.

Fink, M., and Kessler, J. (1967) Absolute measurements of elastic cross section for small-angle scattering of electrons from N_2 and O_2. *J. Chem. Phys.* **47**, 1780–1782.

Fiori, C. E., Gibson, C. C., and Leapman, R. D. (1980) Electrostatic deflection system for use with an electron energy-loss spectrometer. In *Microbeam Analysis—1980*, ed. D. B. Wittry, San Francisco Press, San Francisco, pp. 225–228.

Fiori, C. E., Leapman, R. D., Swyt, C. R., and Andrews, S. B. (1988) Quantitative x-ray mapping of biological cryosections. *Ultramicroscopy* **24**, 237–250.

Fitzgerald, A. G., Storey, B. J., and Fabian, D., eds. (1992) *Quantitative Microbeam Analysis*, Scottish Universities Summer School in Physics, Edinburgh and Institute of Physics Publishing, Bristol and Philadelphia.

Frabboni, S., Lulli, G., Merli, P. G., Migliori, A., and Bauer, R. (1991) Electron spectroscopic imaging of dopant precipitation and segregation in silicon. *Ultramicroscopy* **35**, 265–269.

Frank, I. M. (1966) Transition radiation and optical properties of matter. *Sov. Phys. Usp.* **8**, 729–742.

Fraser, H. L. (1978) Elemental analysis of second-phase carbides using electron energy-loss spectroscopy. In *Scanning Electron Microscopy*, SEM Inc., A. M. F. O'Hare, Illinois, Part 1, pp. 627–632.

Fryer, J. R., and Holland, F. (1984) High resolution electron microscopy of molecular crystals: III. Radiation processes at room temperature. *Proc. R. Soc. London* **A393**, 353–369.

Fujimoto, F., and Komaki, K. (1968) Plasma oscillations excited by a fast electron in a metallic particle. *J. Phys. Soc. Jpn* **25**, 1679–1687.

Fujiyoshi, Y., Kobayashi, T., Tsuji, M., and Uyeda, N. (1982) The effect of electronic state on the direct imaging of atoms. In *Electron Microscopy—1982*, 10th Int. Cong., Deutsche Gesellschaft fur Elektronenmikroskopie, Vol. 1, pp. 217–218.

Garavito, R. M., Carlemalm, E., Colliex, C., and Villiger, W. (1982) Septate junction ultrastructure as visualized in unstained and stained preparations. *J. Ultrastructure Rs.* **80**, 334–353.

Garcia de Abajo, F. J., and Echenique, P. M. (1992) Wake-potential formation in a thin foil. *Phys. Rev. B* **45**, 8771–8774.

Garibyan, G. M. (1960) Transition radiation effects in particle energy losses. *Sov. Phys. JETP* **37**, 372–376.

Garratt-Read, A. J. (1981) Measurement of carbon in V(C, N) precipitates extracted from HSLA steels on aluminum replicas. *Quantitative Microanalysis with High Spatial Resolution*, Metals Society, London, pp. 165–168.

Garvie, L. A. J., Craven, A. J., and Brydson, R. (1994) Use of electron-energy loss near-edge fine structure in the study of minerals. *American Mineralogist* **79**, 411–425.

Gatts, C., Duscher, G., Müllejans H., and Rühle, M. (1995) Analyzing line scan profiles with neural pattern recognition. *Ultramicroscopy* **59**, 229–240.

Geiger, J. (1981) Inelastic electron scattering with energy losses in the meV-region. *39th Ann. Proc. Electron Microsc. Soc. Am.*, ed. G. W. Bailey, Claitor's Publishing, Baton Rouge, Louisiana, pp. 182–185.

Geiger, J., Nolting, M., and Schröder, B. (1970) How to obtain high resolution with a Wien filter spectrometer. In *Electron Microscopy—1970*, ed. P. Favard, Societé Francaise de Microscopie Electronique, Paris, pp. 111–112.

Gibbons, P. C., Schnatterly, S. E., Ritsko, J. J., and Fields, J. R. (1976) Line shape of the plasma resonance in simple metals. *Phys. Rev.* **13**, 2451–2460.

Glaeser, R. M. (1975) Radiation damage and biological electron microscopy. In *Physical Aspects of Electron Microscopy and Microbeam Analysis*, ed. B. M. Siegel and D. R. Beaman, Wiley, New York, pp. 205–229.

Glen, G. L., and Dodd, C. G. (1968) Use of molecular orbital theory to interpret x-ray K-absorption spectral data. *J. Appl. Phys.* **39**, 5372–5377.

Goldstein, J. I., Costley, J. L., Lorimer, G. W., and Reed, S. J. B. (1977) Quantitative x-ray analysis in the electron microscope. *Scanning Electron Microscopy.* Part 1, 315–324.

Gordon, R. L. (1976) Annealing procedure for self-scanned diode arrays. *Appl. Opt.* **15**, 1909–1911.

Gorlen, K. E., Barden, L. K., DelPriore, J. S., Fiori, C. E., Gibson, C. G., and Leapman, R. D. (1984) Computerized analytical electron microscope for elemental imaging. *Rev. Sci. Instrum.* **55**, 912–921.

Graczyk, J. F., and Moss, S. C. (1969) Scanning electron diffraction attachment with electron energy filtering. *Rev. Sci. Instrum.* **40**, 424–433.

Grivet, P., and Septier, A. (1978) Ion microscopy: History and actual trends. *Ann. N.Y. Acad. Sci.* **306**, 158–182.

Groot, F. M. F. de, Fuggle, J. C., Thole, B. T., and Sawatsky, G. A. (1990) L_{23} x-ray-absorption edges of d^0 compounds: K^+, Ca^{2+}, Sc^{3+} and Ti^{4+} in O_h (octahedral) symmetry. *Phys. Rev. B* **41**, 928–937.

Grunes, L. A. (1983) Study of the K edges of $3d$ transition metals in pure oxide form by x-ray absorption spectroscopy. *Phys. Rev. B* **27**, 2111–2131.

Grunes, L. A., Leapman, R. D., Walker, C., Hoffmann, R., and Kunz, A. B. (1982) Oxygen K near-edge fine structure: An electron energy-loss investigation with comparisons to new theory for selected $3d$ transition-metal oxides. *Phys. Rev. B* **25**, 7157–7173.

Gubbens, A. J., Krivanek, O. L., and Kundmann, M. K. (1991) Electron energy loss spectroscopy above 2000 kV. *Microbeam analysis—1991*, ed. D. G. Howitt, San Francisco Press, San Francisco, pp. 127–133.

Gubbens, A. J., Kraus, B., Krivanek, O. L., and Mooney, P. E. (1995) An imaging filter for high voltage electron microscopy. *Ultramicroscopy*, **59**, 255–265.

Haak, W. W., Sawatzky, G. A., Ungier, L., Gimzewski, J. K., and Thomas, T. D. (1984) Core-level electron-electron coincidence spectroscopy. *Rev. Sci. Instrum.* **55**, 696–711.

Hagemann, H.-J., Gudat, W., and Kunz, C. (1974) Optical constants from the far infrared to the x-ray region: Mg, Al, Cu, Ag, Au, Bi, C, and Al_2O_3. DESY report SR-74/7, DESY, 2 Hamburg 52, West Germany.

Hainfeld, J., and Isaacson, M. (1978) The use of electron energy-loss spectroscopy for studying membrane architecture: A preliminary report. *Ultramicroscopy* **3**, 87–95.

Hall, C. R. (1966) On the production of characteristic x-rays in thin metal crystals. *Proc. R. Soc. London* **A295**, 140–163.

Hall, C. R., and Hirsch, P. B. (1965) Effect of thermal diffuse scattering on propagation of high energy electrons through crystals. Proc. Roy. Soc. London A 286, 158–177.

Hall, T. A. (1979) Biological X-ray microanalysis. *J. Microsc.* **117**, 145–163.

Hansen, P. L., Fallon, P. J., and Krätschmer, W. (1991) An EELS study of fullerite—C_{60}/C_{70}. *Chem Phys. Lett.* **181**, 367–372.

Hanson, H. P., Herman, F., Lea, J. D., and Skillman, S. (1964) HFS atomic scattering factors. *Acta. Crystallogr.* **17**, 1040–1044.

Hashimoto, H., Makita, Y., and Nagaoka, N. (1992) Electron microscope images of thorium atoms formed by plasma-loss and core-loss electrons using an energy selecting microscope. *Optik* **93**, 119–126.

Heighway, E. A. (1975) Focussing for dipole magnets with large pole gap to bending radius ratios. *Nucl. Instrum. Methods* **123**, 413–419.

Heine, V. (1980) Electronic structure from the point of view of the local atomic environment. *Solid State Phys.* **35**, 1–126.

Hembree, G. G., and Venables, J. A. (1992) Nanometer-resolution scanning Auger electron microscopy. *Ultramicroscopy* **47**, 109–120.

Henkelman, R. M., and Ottensmeyer, F. P. (1974a) An energy filter for biological electron microscopy. *J. Microsc.* **102**, 79–94.

Henkelman, R. M., and Ottensmeyer, F. P. (1974b) An electrostatic mirror. *J. Phys. E. (Scientific Instruments)* **7**, 176–178.

Henoc, P., and Henry, L. (1970) Observation des oscillations de plasma a l'interface d'inclusions gazeuses dans une matrice cristalline. *J. Phys. (Paris)* **31** (supplément C1), 55–57.

Henry, L., Duval, P., and Hoan, N. (1969) Filtrage en energie des diagrammes de microdiffraction electronique. *C.R. Acad. Sci. Paris* **B268**, 955–958.

Herley, P. J., Jones, W., Sparrow, T. G., and Williams, B. G. (1987) Plasmon spectra of light-metal hydrides. *Materials Letters* **5**, 333–336.

Herman, F., and Skillman, S. (1963) *Atomic Structure Calculations*, Prentice-Hall, Englewood Cliffs, New Jersey.

Herrmann, K.-H. (1984) Detection systems. In *Quantitative Electron Microscopy*, ed. J. N.

Chapman, and A. J. Craven, SSUP Publications, University of Edinburgh, Scotland, pp. 119–148.

Hibbert, G., and Eddington, J. W. (1972) Experimental errors in combined electron microscopy and energy analysis. *J. Phys. D* **5**, 1780–1786.

Hicks, P. J., Daviel, S., Wallbank, B., and Comer, J. (1980) An electron spectrometer using a new multidetector system based on a charge-coupled imaging device. *J. Phys. E* **13**, 713–715.

Hier, R. W., Beaver, E. A., and Schmidt, G. W. (1979) Photon detection experiments with thinned CCD's *Adv. Electron. Electron Phys.* **52**, 463–480.

Higgins, R. J. (1976) Fast Fourier transform: An introduction with some minicomputer experiments. *Am. J. Phys.* **44**, 766–773.

Hillier, J. (1943) On microanalysis by electrons. *Phys. Rev.* **64**, 318–319.

Hillier, J., and Baker, R. F. (1944) Microanalysis by means of electrons. *J. Appl. Phys.* **15**, 663–675.

Hines, R. L. (1975) Graphite crystal film preparation by cleavage. *J. Microsc.* **104**, 257–261.

Hinz, H.-J., and Raether, H. (1979) Line shape of the volume plasmons of silicon and germanium. *Thin Solid Films* **58**, 281–284.

Hirsch, P. B., Howie, A., Nicholson, R. B., Pashley, D. W., and Whelan, M. J. (1977) *Electron Microscopy of Thin Crystals*, Krieger, Huntington, New York.

Hitchcock, A. P. (1989) Electron-energy-loss-based spectroscopies: a molecular viewpoint. *Ultramicroscopy* **28**, 165–183.

Hitchcock, A. P. (1994) Bibliography and data base of inner shell excitation spectra of gas phase atoms and molecules. *J. Electron Spectrosc. Rel. Phenom.* **67**, 1–131.

Hjalmarson, H. P., Büttner, H., and Dow, J. D. (1980) Theory of core excitons. *Phys. Rev. B* **24**, 6010–6019.

Hobbs, L. W. (1984) Radiation effects in analysis by TEM. In *Quantitative Electron Microscopy*, ed. J. N. Chapman and A. J. Craven, SSUP Publications, University of Edinburgh, Scotland, pp. 399–443.

Hofer, F. (1987) EELS quantification of M edges by using oxidic standards. *Ultramicroscopy* **21**, 63–68.

Hofer, F., and Golub, P. (1987) New examples of near-edge fine structures in electron energy loss spectroscopy. Ultramicroscopy 21, 379–384.

Hofer, F., and Kothleitner, G. (1993) Quantitative microanalysis using electron energy-loss spectrometry. I. Li and Be in oxides. *Microsc. Microanal. Microstruct.* **4**, 539–560.

Hofer, F., and Wilhelm, P. (1993) EELS microanalysis of the elements Ca to Cu using M_{23} edges. *Ultramicroscopy* **49**, 189–197.

Hofer, F., Golub, P., and Brunegger, A. (1988) EELS quantification of the elements Sr to W by means of M_{45} edges. *Ultramicroscopy* **25**, 81–84.

Hofer, F., Warbichler, P., and Grogger, W. (1995) Characterization of nanometre sized precipitates in solids by electron spectroscopic imaging. *Ultramicroscopy*, **59**, 15–31.

Hojou, K., Furuno, S., Kushita, K. N., Otsu, H., Izui, K., Ueki, Y., and Kamino, T. (1992) Electron energy-loss spectroscopy of SiC crystals implanted with hydrogen and helium dual-ion beam. In *Electron Microscopy 1992*, Proc. EUREM 92, Granada, vol. 1, 261–262.

Hollenbeck, J. L., and Buchanan, R. C. (1990) Oxide thin films for nanometer scale electron beam lithography. *J. Mater. Res.* **5**, 1058–1072.

Holmestad, R., Krivanek, O. L., Høier, R., Marthinsen, K., and Spence, J. C. H. (1993) Commercial spectrometer modifications for energy filtering of electron diffraction patterns and images. *Ultramicroscopy* **52**, 454–458.

Hosoi, J., Oikawa, T., Inoue, M., Kokubo, Y., and Hama, K. (1981) Measurement of partial specific thickness (net thickness) of critical-point-dried cultured fibroblast by energy analysis. *Ultramicroscopy* **7**, 147–154.

Hosoi, J., Oikawa, T., Inoue, M., and Kokubo, Y. (1984) Thickness dependence of signal/background ratio of inner-shell electron excitation loss in EELS. *Ultramicroscopy* **13**, 329–332.

Howie, A. (1981) Localisation and momentum transfer in inelastic scattering. *39th Ann. Proc. Electron Microsc. Soc. Am.*, ed. G. W. Bailey, Claitor's Publishing, Baton Rouge, Louisiana, pp. 186–189.

Howie, A. (1983) Surface reactions and excitations. *Ultramicroscopy* **11**, 141–148.

Howie, A., and Walsh, C. (1991) Interpretation of valence loss spectra from composite media. *Microsc. Microanal. Microstruct.* **2**, 171–181.

Howitt, D. (1984) Ion milling of materials science specimens for electron microscopy: A review. *J. Electron Microsc. Tech.* **1**, 405–415.

Hren, J. J., Goldstein, J. I., and Joy, D. C. eds. (1979) *Introduction to Analytical Electron Microscopy,* Plenum Press, New York.

Hubbell, J. H. (1971) Survey of photon-attenuation-coefficient measurements 10 eV to 100 GeV. *At. Data Tables* **3**, 241–297.

Humphreys, C. J., Hart-Davis, A., and Spencer, J. P. (1974) Optimizing the signal/noise ratio in the dark-field imaging of single atoms. In *Electron Microscopy—1974*, 8th Int. Congress, ed. J. V. Sanders, and D. J. Goodchild, Australian Academy of Science, Canberra, 1974, vol. 1, pp. 248–249.

Humphreys, C. J., Eaglesham, D. J., Alford, N. M., Harmer, M. A., and Birchall, J. D. (1988) High Temperature Superconductors, Inst. Phys. Conf. Ser. No. 93, IOP, Bristol, vol. 2, pp. 217–222.

Humphreys, C. J., Bullough, T. J., Devenish, R. W., Maher, D. M., and Turner, P. S. (1990) Electron beam nano-etching in oxides, fluorides, metals and semiconductors. In *Scanning Microscopy Supplement* **4**, Scanning Microscopy International, Chicago, 185–192.

Hunt, J. A., and Williams, D. B. (1991) Electron energy-loss spectrum-imaging. *Ultramicroscopy* **38**, 47–73.

Hunt, J. A., Disko, M. M., Bekal, S. K., and Leapman, R. D. (1995) Electron energy-loss chemical imaging of polymer phases. *Ultramicroscopy* **58**, 55–64.

Ibach, H. (1991) Electron Energy Loss Spectrometers. Springer Series in Optical Sciences, Springer-Verlag, Berlin, vol. 63.

Ibach, H., and Mills, D. L. (1982) *Electron Energy-Loss Spectroscopy and Surface Vibrations,* Academic Press, New York.

Ibers, J. A., and Vainstein, B. K. (1962) *International Crystallographic Tables III*, Kynoch Press, Birmingham, Table 3.3.

Imura, T., Saka, H., Todokoro, H., and Ashikawa, M. (1971) Direct intensification of electron microscopic images with silicon diode array target. *J. Phys. Soc. Jpn* **31**, 1849.

Inokuti, M. (1971) Inelastic collisions of fast charged particles with atoms and molecules—The Bethe theory revisited. *Rev. Mod. Phys.* **43**, 297–347. Addenda: *Rev. Mod. Phys.* **50,** 23–26.

Inokuti, M. (1979) Electron-scattering cross sections pertinent to electron microscopy. *Ultramicroscopy* **3**, 423–427.

Inokuti, M., Saxon, R. P., and Dehmer, J. L. (1975) Total cross-sections for inelastic scattering of charged particles by atoms and molecules-VIII. Systematics for atoms in the first and second row. *Int. J. Radiat. Phys. Chem.* **7**, 109–120.

Inokuti, M., Dehmer, J. L., Baer, T., and Hanson, D. D. (1981) Oscillator-strength moments, stopping powers, and total inelastic-scattering cross sections of all atoms through strontium. *Phys. Rev. A* **23**, 95–109.

Isaacson, M. (1972a) Interaction of 25 keV electrons with the nucleic acid bases, adenine, thymine, and uracil I. Outer shell excitation. *J. Chem. Phys.* **56**, 1803–1812.

Isaacson, M. (1972b) Interaction of 25 keV electrons with the nucleic acid bases, adenine, thymine and uracil. (II) Inner-shell excitation and inelastic scattering cross sections. *J. Chem. Phys.* **56**, 1813–1818.

Isaacson, M. (1977) Specimen damage in the electron microscope. In *Principles and Techniques of Electron Microscopy*, ed. M. A. Hayat, Van Nostrand, New York, Vol. 7, pp. 1–78.

Isaacson, M. (1981) All you might want to know about ELS (but were afraid to ask): A tutorial. In *Scanning Electron Microscopy*, SEM Inc., (A. M. F. O'Hare, Illinois,) Part 1, pp. 763–776.

Isaacson, M., and Johnson, D. (1975) The microanalysis of light elements using transmitted energy-loss electrons. *Ultramicroscopy* **1**, 33–52.

Isaacson, M. S., and Utlaut, M. (1978) A comparison of electron and photon beams for determining micro-chemical environment. *Optik* **50**, 213–234.

Ishizuka, K. (1993) Analysis of electron image detection efficiency of slow-scan CCD cameras. *Ultramicroscopy* **52**, 7–20.

Jackson, J. D. (1975) *Classical Electrodynamics*, second edition, Wiley, New York, Ch. 13.

Jäger, W., and Mayer, J. (1995) Energy filtered transmission electron microscopy of Si_mGe_n superlattices and Si-Ge heterostructures. *Ultramicroscopy* **59**, 33–45.

Jeanguillaume, C., and Colliex, C. (1989) Spectrum-image: the next step in EELS digital acquisition and processing. *Ultramicroscopy* **28**, 252.

Jeanguillaume, C., Trebbia, P., and Colliex, C. (1978) About the use of electron energy-loss spectroscopy for chemical mapping of thin foils with high spatial resolution. *Ultramicroscopy* **3**, 237–242.

Jeanguillaume, C., Colliex, C., Ballongue, P., and Tencé, M. (1992) New STEM multisignal imaging modes, made accessible through the evaluation of detection efficiencies. *Ultramicroscopy* **45**, 205–217.

Jewsbury, P., and Summerside, P. (1980) The nature of interface plasmon modes at bimetallic junctions. *J. Phys. F* **10**, 645–650.

Jiang, X. G., and Ottensmeyer, F. P. (1993) Optimization of a prism-mirror-prism imaging energy filter for high resolution electron microanalysis. *Optik* **94**, 88–95.

Jiang, X. G., and Ottensmeyer, F. P. (1994) Molecular microanalysis: imaging with low-energy-loss electrons. *Electron Microscopy 1994*, Proc. 13th Int. Cong. Electron Microsc., Paris, vol. 3, pp. 781–782.

Johansson, S. A. E. (1984) PIXE summary. *Nucl. Instrum. Methods* **B3**, 1–3.

Johnson, D. E. (1972) The interactions of 25 keV electrons with guanine and cytosine. *Radiat. Res.* **49**, 63–84.

Johnson, D. E. (1979b) Energy-loss spectrometry for biological research. In *Introduction to Analytical Electron Microscopy*, Plenum Press, New York, pp. 245–258.

Johnson, D. E. (1980a) Post specimen optics for energy-loss spectrometry. In *Scanning Electron Microscopy*, SEM Inc., A. M. F. O'Hare, Illinois, Part 1, pp. 33–40.

Johnson, D. E. (1980b) Pre-spectrometer optics in a CTEM/STEM. *Ultramicroscopy* **5**, 163–174.

Johnson, D. E., Csillag, S., and Stern, E. A. (1981b) Analytical electron microscopy using extended energy-loss fine structure (EXELFS). In *Scanning Electron Microscopy*, SEM Inc., A. M. F. O'Hare, Illinois, Part 1, pp. 105–115.

Johnson, D. E., Monson, K. L., Csillag, S., and Stern, E. A. (1981c) An approach to parallel-detection electron energy-loss spectrometry. In *Analytical Electron Microscopy—1981*, ed. R. H. Geiss, San Francisco Press, San Francisco, pp. 205–209.

Johnson, D. W. (1974) Optical properties determined from electron energy-loss distributions. In *Electron Microscopy—1974*, 8th Int. Congress, ed. J. V. Sanders and D. J. Goodchild, Australian Academy of Science, Canberra, pp. 388–389.

Johnson, D. W. (1975) A Fourier method for numerical Kramers-Kronig analysis. *J. Phys. A* (*Math. Gen. Phys.*) **8**, 490–495.

Johnson, D. W., and Spence, J. C. H. (1974) Determination of the single-scattering probability distribution from plural-scattering data. *J. Phys. D* (*Appl. Phys.*) **7**, 771–780.

Johnson, H. F., and Isaacson, M. S. (1988) An efficient analytical method for calculating the angular distribution of electrons which have undergone plural scattering in amorphous materials. *Ultramicroscopy* **26**, 271–294.

Jonas, P., and Schattschneider, P. (1993) The experimental conditions for Compton scattering in the electron microscope. *J. Phys. Cond. Matter* **5**, 7173–7188.

Jonas, P., Schattschneider, P., and Su, D. S. (1992) Directional Compton profiles of silicon. In *Electron Microscopy*, EUREM 92, Granada, vol. 1, pp. 265–266.

Jones, B. L., Walton, D. M., and Booker, G. R. (1982) Developments in the use of one- and two-dimensional self-scanned silicon photodiode arrays in imaging devices in electron microscopy. *Inst. Phys. Conf. Ser.* No. 61, 135–138.

Jones, W., Sparrow, T. G., Williams, B. G., and Herley, P. J. (1984) Evidence for the formation of single crystals of sodium metal during the decomposition of sodium aluminum hydride: An electron microscopic study. *Mater. Lett.* **2**, 377–379.

Jouffrey, B., Kihn, Y., Perez, J. P., Sevely, J., and Zanchi, G. (1978) On chemical analysis of thin films by energy-loss spectroscopy. In *Electron Microscopy—1978*, 9th Int. Cong., ed. J. M. Sturgess, Microscopical Society of Canada, Toronto, Vol. 3, pp. 292–303.

Jouffrey, B., Sevely, J., Zanchi, G., and Kihn, Y. (1985) Characteristic energy losses with high energy electrons up to 2.5 MeV. *Scanning Electron Microscopy*, Part 3, 1063–1070.

Jouffrey, B., Zanchi, G., Kihn, Y., Hussein, K., and Sevely, J. (1989) Present questions in EELS: sensitivity and EXELFS. *Beitr. Elektronenmikroskop. Direktabb. Oberf.* **22**, 249–270.

Joy, D. C. (1979) The basic principles of electron energy-loss spectroscopy. In *Introduction to Analytical Electron Microscopy*, Plenum Press, New York, pp. 223–244.

Joy, D. C. (1984a) Detectors for electron energy-loss spectroscopy. In *Electron-Beam Interactions with Solids for Microscopy, Microanalysis and Microlithography*, SEM Inc., Illinois, pp. 251–257.

Joy, D. C. (1984b) A parametric partial cross section for ELS. *J. Microsc.* **134**, 89–92.

Joy, D. C., and Maher, D. M. (1977) Sensitivity limits for thin specimen x-ray analysis. *Scanning Electron Microscopy/1977*, Part 1, 325–334.

Joy, D. C., and Maher, D. M. (1978) A practical electron spectrometer for chemical analysis. *J. Microsc.* **114**, 117–129.

Joy, D. C., and Maher, D. M. (1980a) The electron energy-loss spectrum—Facts and artifacts. In *Scanning Electron Microscopy*, SEM Inc., A. M. F. O'Hare, Illinois, Part 1, pp. 25–32.

Joy, D. C., and Maher, D. M. (1980c) Electron energy-loss spectroscopy. *J. Phys. E.* (*Sci. Instrum.*) **13**, 261–270.

Joy, D. C., and Maher, D. M. (1981a) The quantitation of electron energy-loss spectra. *J. Microsc.* **124**, 37–48.

Joy, D. C., and Maher, D. M. (1981b) Instrument errors in ELS quantitation. In *Analytical Electron Microscopy—1981*, ed. R. H. Geiss, San Francisco Press, San Francisco, pp. 176–177.

Joy, D. C., and Maher, D. M. (1981c) The micro-spectroscopy of semiconductors. In *Inst. Phys. Conf. Ser. No. 60*, I.O.P., Bristol, pp. 229–236.

Joy, D. C., and Newbury, D. E. (1981) A "round robin" test on ELS quantitation. In *Analytical Electron Microscopy—1981*, ed. R. H. Geiss, San Francisco Press, San Francisco, pp. 178–180.

Joy, D. C., Newbury, D. E., and Myklebust, R. L. (1982) The role of fast secondary electrons in degrading spatial resolution in the analytical electron microscope. *J. Microsc.* **128**, RP1–RP2.

Joy, D. C., Romig, A. D., and Goldstein, J. I. eds. (1986) *Principles of Analytical Electron Microscopy*, Plenum Press, New York.

Kainuma, Y. (1955) The theory of Kikuchi patterns. *Acta. Crystallogr.* **8,** 247–257.

Kaloyeros, A. E., Hoffman, M. P., Williams, W. S., Greene, A. E., and McMillan, J. A. (1988) Structural studies of amorphous titanium diboride thin films by extended x-ray-absorption fine-structure and extended electron-energy-loss fine-structure techniques. *Phys. Rev.* **38,** 7333–7344.

Kambe, K., Krahl, D., and Herrmann, K.-H. (1981) Extended fine structure in electron energy-loss spectra of MgO crystallites. *Ultramicroscopy* **6,** 157–162.

Katterwe, H. (1972) Object analysis by electron energy spectroscopy in the infra-red region. In *Electron Microscopy—1972*, The Institute of Physics, London, pp. 154–155.

Keil, P. (1968) Elektronen-Energieverlustmessungen und Berechnung optischer Konstanten: II Kaliumbromid. *Z. Phys.* **214,** 266–284.

Kevan, S. D., and Dubois, L. H. (1984) Development of dispersion compensation for use in high-resolution electron energy-loss spectroscopy. *Rev. Sci. Instrum.* **55,** 1604–1612.

Kihn, Y., Perez, J.-P., Sevely, J., Zanchi, G., and Jouffrey, B. (1980) Data collection problems in high voltage electron energy-loss spectroscopy. In *Electron Microscopy—1980*, 7th European Congress Foundation, The Hague, Vol. 4, pp. 42–45.

Kim, M. J., and Carpenter, R. W. (1990) Composition and structure of native oxide on silicon by high resolution analytical electron microscopy. *J. Mater. Res.* **5,** 347–351.

Kincaid, B. M., Meixner, A. E., and Platzman, P. M. (1978) Carbon *K*-edge in graphite measured using electron energy-loss spectroscopy. *Phys. Rev. Lett.* **40,** 1296–1299.

Klemperer, O., and Shepherd, J. P. G. (1963) On the measurement of characteristic energy losses of electrons in metals. *Brit. J. Appl. Phys.* **14,** 85–88.

Knauer, W. (1979) Analysis of energy broadening in electron and ion beams. *Optik* **54,** 211–234.

Knotek, M. L. (1984) Stimulated desorption. *Rep. Prog. Phys.* **47,** 1499–1561.

Kohl, H. (1985) A simple procedure for evaluating effective scattering cross sections in STEM. *Ultramicroscopy* **16,** 265–268.

Kohl, H., and Rose, H. (1985) Theory of image formation by inelastically scattered electrons in the electron microscope. *Adv. Electron. Electron Phys.* **65,** 175–200.

Köpf-Maier, P. (1990) Intracellular localization of titanium within xenografted sensitive human tumors after treatment with the antitumor agent titanocene dichloride. *J. Struct. Biol.* **105,** 35–45.

Krahl, D., Herrmann, K.-H., Kunath, W. (1978) Electron optical experiments with a magnetic imaging filter. In *Electron Microscopy—1978*, 9th Int. Cong., ed. J. M. Sturgess, Microscopical Society of Canada, Toronto, Vol. 1, pp. 42–43.

Krahl, D., Patzold, H., and Swoboda, M. (1990) An aberration-minimized imaging energy filter of simple design. Proc. XIIth Int. Cong. for Electron Microscopy, San Francisco Press, San Francisco, vol. 2, pp. 60–61.

Krause, M. O., and Oliver, J. H. (1979) Natural widths of atomic *K* and *L* levels. *J. Phys. Chem. Ref. Data* **8,** 329–338.

Kreibig, U., and Zacharias, P. (1970) Surface plasma resonances in small spherical silver and gold particles. *Z. Phys.* **231,** 128–143.

Krishnan, K. M. (1990) Iron $L_{3,2}$ near-edge fine structure studies. *Ultramicroscopy* **32,** 309–311.

Krivanek, O. L. (1988) Practical high-resolution electron microscopy. In *High-Resolution Transmission Electron Microscopy and Associated Techniques*, ed. P. Buseck, J. Cowley, and L. Eyring, Oxford University Press, New York, pp. 519–567.

Krivanek, O. L., and Mooney, P. E. (1993) Applications of slow-scan CCD cameras in transmission electron microscopy. *Ultramicroscopy* **49,** 95–108.

Krivanek, O. L., and Paterson, J. H. (1990) ELNES of 3*d* transition-metal oxides. *Ultramicroscopy* **32,** 313–318.

Krivanek, O. L., and Swann, P. R. (1981) An advanced electron energy-loss spectrometer. In *Quantitative Microanalysis with High Spatial Resolution*, The Metals Society, London, pp. 136–140.

Krivanek, O. L., Tanishiro, Y., Takayanagi, K., and Yagi, K. (1983) Electron energy-loss spectroscopy in glancing reflection from bulk crystals. *Ultramicroscopy* **11**, 215–222.

Krivanek, O.L., Ahn, C. C., and Keeney, R. B. (1987) Parallel detection electron spectrometer using quadrupole lenses. *Ultramicroscopy* **22**, 103–116.

Krivanek, O. L., Ahn, C. C., and Wood, G. J. (1990) The inelastic contribution to high resolution images of defects. *Ultramicroscopy* **33**, 177–185.

Krivanek, O. L., Mory, C., Tence, M., and Colliex, C. (1991a) EELS quantification near the single-atom detection level. *Microsc. Microanal. Microstruct.* **2**, 257–267.

Krivanek, O. L., Gubbens, A. J., and Dellby, N. (1991b) Developments in EELS instrumentation for spectroscopy and imaging.

Krivanek, O. L., Gubbens, A. J., Dellby, N., and Meyer, C. E. (1992) Design and first applications of a post-column imaging filter. *Microsc. Microanal. Microstruct.* **3**, 187–199.

Krivanek, O. L., Kundmann, M., and Bourrat, X. (1994) Elemental mapping by energy-filtered electron microscopy. Mat. Res. Soc. Symp. Proc., Materials Research Society, Pittsburgh, Pennsylvania, vol. 332, 341–350.

Krivanek, O. L., Friedman, S. L., Gubbens, A. J., and Kraus, B. (1995a) A post-column imaging filter for biological applications. *Ultramicroscopy* **59**, 267–282.

Kröger, E. (1968) Berechnung der Energieverluste schneller elektronen in dünnen Schichten mit Retardierung. *Z. Phys.* **216**, 115–135.

Kruit, P. and Shuman, H. (1985b) The influence of objective lens aberrations in energy-loss spectrometry. *Ultramicroscopy* **17**, 263–267.

Kruit, P., and Shuman, H. (1985b) Position stabilization of EELS spectra. *J. Electron Microscope Tech.* **2**, 167–169.

Kruit, P., Shuman, H., and Somlyo, A. P. (1984) Detection of x-rays and electron energy-loss events in time coincidence. *Ultramicroscopy* **13**, 205–214.

Kujiwa, S., and Krahl, D. (1992) Performance of a low-noise CCD camera adapted to a transmission electron microscope. *Ultramicroscopy* **46**, 395–403.

Kunz, B. (1964) Messung der unsymmetrischen Winkelverteilung der charakteristischen Oberflächvenverluste an Al (6.3 eV) und Ag (3.6 eV). *Z. Phys.* **180**, 127–132.

Kurata, H., Isoda, S., and Kobayashi, T. (1992) EELS study of radiation damage in chlorinated Cu-phthalocyanine and poly GeO-phthalocyanine. *Ultramicroscopy* **41**, 33–40.

Kutzler, F. W., Natoli, C. R., Misemer, D. K., Doniach, S., and Hodgson, K. O. (1980) Use of one-electron theory for the interpretation of near-edge structure in K-shell x-ray absorption spectra of transition-metal complexes. *J. Chem. Phys.* **73**, 3274–3288.

Kuzuo, R., Terauchi, M., Tanaka, M., Saito, Y., and Shinohara, H. (1991) High-resolution electron energy-loss spectra of solid C_{60}. *Jap. J. Appl. Phys.* **30**, L1817–L1818.

Kuzuo, R., Terauchi, M., and Tanaka, M. (1992) Electron energy-loss spectra of carbon nanotubes. *Jpn. J. Appl. Phys.* **31**, L1484–L1487.

Kuzuo, R., Terauchi, M., Tanaka, M., Saito, Y., and Shinohara, H. (1994a) Electron-energy-loss spectra of crystalline C_{84}. *Phys. Rev. B* **49**, 5054–5057.

Kuzuo, R., Terauchi, M., Tanaka, M., and Saito, Y. (1994b). Electron energy-loss spectra of single-shell carbon nanotubes. Personal communication.

Lakner, H., Maywald, M., Balk, L. J., and Kubalek, E. (1992) Characterization of AlGaAs/GaAs interfaces by EELS and high-resolution Z-contrast imaging in scanning transmission electron microscopy (STEM). *Surface Interface Anal.* **19**, 374–378.

Lamvik, M. K., and Langmore, J. P. (1977) Determination of particle mass using scanning transmission electron microscopy. In *Scanning Electron Microscopy*, SEM Inc., A. M. F. O'Hare, Illinois, Vol. 1, pp. 401–410.

Lamvik, M. K., Davilla, S. D., and Klatt, L. L. (1989) Substrate properties affect the mass loss rate in collodion at liquid helium temperature. *Ultramicroscopy* **27**, 241–250.

Land, P. L. (1971) A discussion of the region of linear operation of photomultipliers. *Rev. Sci. Instrum.* **42**, 420–425.

Langmore, J. P., and Smith, M. F. (1992) Quantitative energy-filtered electron microscopy of biological molecules in ice. *Ultramicroscopy* **46**, 349–373.

Langmore, J. P., Wall, J., and Isaacson, M. S. (1973) The collection of scattered electrons in dark field electron microscopy: 1. Elastic scattering. *Optik* **38**, 335–350.

Lanio, S. (1986) High-resolution imaging magnetic energy filters free of second-order aberration. *Optik* **73**, 99–107.

Lavergne, J.-L., Martin, J.-M., and Belin, M. (1992) Interactive electron energy-loss elemental mapping by the "Imaging-Spectrum," method. *Microsc. Microanal. Microstruct.* **3**, 517–528.

Lavergne, J.-L., Gimenez, C., Friour, G., and Martin, J. M. (1994) Chemical mapping of silver halide microcrystals: use of imaging EELS methods. Proc ICEM-13, Les Editions de Physique, Les Ulis, vol. 1, pp. 692–630.

Leapman, R. D. (1982a) EXELFS spectroscopy of amorphous materials. In *Microbeam Analysis—1982*, ed. K. F. J. Heinrich, San Francisco Press, San Francisco, pp. 111–117.

Leapman, R. D. (1982b) Applications of electron energy-loss spectroscopy in biology: Detection of calcium and fluorine. *40th Ann. Proc. Electron Microsc. Soc. Am.*, ed. G. W. Bailey, Claitor's Publishing, Baton Rouge, Louisiana, pp. 412–415.

Leapman, R. D. (1984) Electron energy-loss microspectroscopy and the characterization of solids. In *Electron Beam Interactions with Solids*, SEM Inc., A. M. F. O'Hare, Chicago, pp. 217–233.

Leapman, R. (1992) EELS quantitative analysis. In *Transmission Electron Energy Loss Spectrometry in Materials Science*, ed. M. M. Disko, and B. Fulz, The Minerals, Metals and Materials Society, Warrendale, Pennsylvania, pp. 47–83.

Leapman, R. D., and Andrews, S. B. (1992) Characterization of biological macromolecules by combined mass mapping and electron energy-loss spectroscopy. *J. Microsc.* **165**, 225–238.

Leapman, R. D., and Hunt, J. A. (1991) Comparison of detection limits for EELS and EDXS. *Micros. Microanal. Microstruct.* **2**, 231–244.

Leapman, R. D., and Newbury, D. E. (1993) Trace element analysis at nanometer spatial resolution by parallel-detection electron energy-loss spectroscopy. *Anal. Chem.* **13**, 2409–2414.

Leapman, R. D., and Ornberg, R. L. (1988) Quantitative electron energy loss spectroscopy in biology. *Ultramicroscopy* **24**, 251–268.

Leapman, R. D., and Silcox, J. (1979) Orientation dependence of core edges in electron energy-loss spectra from anisotropic materials. *Phys. Rev. Lett.* **42**, 1361–1364.

Leapman, R. D., and Sun, S. (1995) Cryo-electron energy loss spectroscopy: observations on vitrified hydrated specimens and radiation damage. *Ultramicroscopy* **59**, 71–79.

Leapman, R. D., and Swyt, C. R. (1981a) Electron energy-loss spectroscopy under conditions of plural scattering. In *Analytical Electron Microscopy—1981*, ed. R. H. Geiss, San Francisco Press, San Francisco, pp. 164–172.

Leapman, R. D., and Swyt, C. R. (1983) Electron energy-loss imaging in the STEM—Systematic and statistical errors. In *Microbeam Analysis—1983*, ed. R. Gooley, San Francisco Press, San Francisco, pp. 163–167.

Leapman, R. D., and Swyt, C. R. (1988) Separation of overlapping core edges in electron energy loss spectra by multiple-least-squares fitting. *Ultramicroscopy* **26**, 393–404.

Leapman, R. D., Rez, P., and Mayers, D. F. (1980) K, L, and M shell generalized oscillator strengths and ionization cross sections for fast electron collisions. *J. Chem. Phys.* **72**, 1232–1243.

Leapman, R. D., Grunes, L. A., Fejes, P. L., and Silcox, J. (1981) Extended core-edge fine structure in electron energy-loss spectra. In *EXAFS Spectroscopy*, ed. B. K. Teo and C. D. Joy, Plenum Press, New York, pp. 217–239.

Leapman, R. D., Grunes, L. A., Fejes, P. L. (1982) Study of the L_{23} edges in the $3d$ transition metals and their oxides by electron-energy-loss spectroscopy with comparisons to theory. *Phys. Rev. B* **26,** 614–635.

Leapman, R. D., Fejes, P. L., and Silcox, J. (1983) Orientation dependence of core edges from anisotropic materials determined by inelastic scattering of fast electrons. *Phys. Rev. B* **28,** 2361–2373.

Leapman, R. D., Fiori, C. E., and Swyt, C. R. (1984a) Mass thickness determination by electron energy-loss for quantitative x-ray microanalysis in biology. *J. Microsc.* **133,** 239–253.

Leapman, R. D., Fiori, C. E., and Swyt, C. R. (1984b) Mass thickness determination by inelastic scattering in microanalysis of organic samples. In *Analytical Electron Microscopy—1984*, ed. D. B. Williams and D. C. Joy, San Francisco Press, San Francisco, pp. 83–88.

Leapman, R. D., Gorlen, K. D., and Swyt, C. R. (1984c) Background subtraction in STEM energy-loss mapping. *42nd Ann. Proc. Electron Microsc. Soc. Am.*, ed. G. W. Bailey, San Francisco Press, San Francisco, pp. 568–569.

Leapman, R. D., Brink, J., and Chiu, W. (1993a) Low-dose thickness measurement of glucose-embedded protein crystals by electron energy loss spectroscopy and STEM dark-field imaging. *Ultramicroscopy* **52,** 157–166.

Leapman, R. D., Hunt, J. A., Buchanan, R. A., and Andrews, S. B. (1993b) Measurement of low calcium concentrations in cryosectioned cells by parallel-EELS mapping. *Ultramicroscopy* **49,** 225–234.

Lee, P. A., and Beni, G. (1977) New method for the calculation of atomic phase shifts: Application to extended x-ray absorption fine structure (EXAFS) in molecules and crystals. *Phys. Rev. B* **15,** 2862–2883.

Lee, P. A., and Pendry, J. B. (1975) Theory of the extended x-ray absorption fine structure. *Phys. Rev. B* **11,** 2795–2811.

Lee, P. A., Teo, B.-K., and Simons, A. L. (1977) EXAFS: A new parameterization of phase shifts. *J. Am. Chem. Soc.* **99,** 3856–3859.

Lee, P. A., Citrin, P. H., Eisenberger, P., and Kincaid, B. M. (1981) Extended x-ray absorption fine structure—Its strengths and limitations as a structural tool. *Rev. Mod. Phys.* **53,** 769–806.

Lehmpfuhl, G., Krahl, D., and Swoboda, M. (1989) Electron microscope channelling imaging of thick specimens with medium-energy electrons in an energy-filter microscope. *Ultramicroscopy* **31,** 161–168.

Lenz, F. (1954) Zur Streuung mittelschneller Elektronen in kleinste Winkel. *Z. Naturforsch.* **9A,** 185–204.

Levine, Z. H., and S. G. Louie (1982) New model dielectric function and exchange-correlation potential for semiconductors and insulators. *Phys. Rev. B* **25,** 6310–6316.

Levine, L. E., Gibbons, P. C., and Kelton, K. F. (1989) Electron energy-loss-spectroscopy studies of icosahedral plasmons. *Phys. Rev. B* **40,** 9338–9341.

Levi-Setti, R. (1983) Secondary electron and ion imaging in scanning-ion microscopy. *Scanning Electron Microscopy/1983*, SEM, Inc., A. M. F. O'Hare, Illinois, Part 1, 1–22.

Liang, W. Y., and Cundy, S. L. (1969) Electron energy-loss studies of the transition metal dichalcogenides. *Phil. Mag.* **19,** 1031–1043.

Linders, P. W. J., Stols, A. L. H., van de Vorstenbosch, R. A., and Stadhouders, A. M. (1982) Mass determination of thin biological specimens for use in quantitative electron probe x-ray microanalysis. In *Scanning Electron Microscopy*, SEM Inc., A. M. F. O'Hare, Illinois, Part IV, pp. 1603–1615.

Lindhard, J. (1954) On the properties of a gas of charged particles. *Dan. Vidensk. Selsk. Mat. Fys. Medd.* **28**(No. 8), 1–57.

Lindner, Th., Sauer, H., Engel, W., and Kambe, K. (1986) Near-edge structure in electron-energy-loss spectra of MgO. *Phys. Rev. B* **33**, 22–24.

Liu, D.-R. (1988) Experimental method of separation of the volume and surface components in an electron energy-loss spectrum. *Phil. Mag. B* **57**, 619–633.

Liu, D. R., Shinozaki, S. S., Hangas, J. W., and Maeda, K. (1991) Electron-energy-loss spectra of silicon carbide of 4H and 6H structures. In *Microbeam Analysis—1991*, ed. D. G. Howitt, San Francisco Press, San Francisco, pp. 447–449.

Liu, Z. Q., McKenzie, D. R., Cockayne, D. J. H., and Dwarte, D. M. (1988) Electron diffraction study of boron- and phosphorus-doped hydrogenated amorphous silicon. *Phil. Mag. B* **57**, 753–761.

Livingood, J. J. (1969) *The Optics of Dipole Magnets*, Academic Press, New York, 1969.

Loane, R. F., Kirkland, E. J., and Silcox, J. (1988) Visibility of single heavy atoms on thin crystalline silicon in simulated annular dark-field STEm images. *Acta. Cryst. A* **44**, 912–927.

Longe, P., and Bose, S. M. (1993) Interpretation of the plasmon dispersion in the electron-energy-loss spectra of high-T_c superconductors. *Phys. Rev. B* **47**, 11611–11614.

Lucas, A. A., and Sunjic, M. (1971) Fast-electron spectroscopy of surface excitations. *Phys. Rev. Lett.* **26**, 229–232.

Luo, B. P., and Zeitler, E. (1991) M-shell cross-sections for fast electron inelastic collisions based on photoabsorption data. *J. Electron Spectrosc. Rel. Phenom.* **57**, 285–295.

Luyten, W., Van Tenderloo, G., Fallon, P. J., and Woods, G. S. (1994) Electron microscopy and energy-loss spectroscopy of voidites in pure type IaB diamonds. *Phil. Mag. A* **69**, 767–778.

Lyman, C. E., Newbury, D. E., Goldstein, J. I., Williams, D. B., Romig, A. D., Armstrong, J. T., Echlin, P., Fiori, C. E., Joy, D. C., Lifshin, E., and Peters, K.-R. (1990) Scanning Electron Microscopy, X-ray Microanalysis, and Analytical Electron Microscopy: a Laboratory Workbook, Plenum Press, New York.

Lyman, C. E., Lakis, R. E., and Stenger, H. G. (1995) X-ray emission spectrometry of phase separation in Pt-Rh nanoparticles for nitric oxide reduction. *Ultramicroscopy* **58**, 25–34.

Ma, H., Lin, S. H., Carpenter, R. W., and Sankey, O. F. (1990) Theoretical comparison of electron energy-loss and x-ray absorption near-edge fine structure of the Si L_{23} edge. *J. Appl. Phys.* **68**, 288–290.

Madison, D. H., and Merzbacher, E. (1975) Theory of charged-particle excitation. In *Atomic Inner-Shell Processes*, ed. B. Crasemann, Academic Press, New York, Vol. 1, pp. 1–72.

Magee, C. W. (1984) On the use of secondary ion mass spectrometry in semiconductor device materials and process development. *Ultramicroscopy* **14**, 55–64.

Mahan, G. D. (1975) Collective excitations in x-ray spectra of metals. *Phys. Rev. B* **11**, 4814–4824.

Maher, D., Mochel, P., and Joy, D. (1978) A data collection and reduction system for electron energy-loss spectroscopy. In *Proc. 13th Ann. Conf. Microbeam Analysis Society*, NBS Analytical Chemistry Division, Washington D.C., pp. 53A–53K.

Maher, D. M., Joy, D. C., Egerton, R. F., and Mochel, P. (1979) The functional form of energy-differential cross sections for carbon using transmission electron energy-loss spectroscopy. *J. Appl. Phys.* **50**, 5105–5109.

Malis, T., and Titchmarsh, J. M. (1986) 'k-factor' approach to EELS analysis. In *Electron Microscopy and Analysis 1985* (Institute of Physics, Bristol, U.K.).

Malis, T., Cheng, S. C., and Egerton, R. F. (1988) EELS log-ratio technique for specimen-thickness measurement in the TEM. *J. Electron Microscope Technique* **8**, 193–200.

Manoubi, T., Colliex, C., and Rez, P. (1990) Quantitative electron energy loss spectroscopy on M_{45} edges in rare earth oxides. *J. Electron Spectrosc. Rel. Phenom.* **50**, 1–18.

Manson, S. T. (1972) Inelastic collision of fast charged particles with atoms: Ionization of the aluminum L shell. *Phys. Rev. A* **6**, 1013–1024.

Manson, S. T. (1978) The calculation of photoionization cross sections: an atomic view. In *Topics in Applied Physics*, Vol. 26, Springer-Verlag, New York, pp. 135–163.

Manson, S. T., and Cooper, J. W. (1968) Photo-ionization in the soft x-ray range: *Z* dependence in a central-potential model. *Phys. Rev.* **165**, 126–165.

Marks, L. D. (1982) Observation of the image force for fast electrons near a MgO surface. *Solid State Commun.* **43**, 727–729.

Martin, J. M., and Mansot, J. L. (1991) EXELFS analysis of amorphous and crystalline silicon carbide. *J. Microsc.* **162**, 171–178.

Martin, J. M., Mansot, J. L., and Hallouis, M. (1989) Energy filtered electron microscopy (EFEM) of overbased reverse micelles. *Ultramicroscopy* **30**, 321–327.

Martin, J. P., and Geisler, D. (1972) Optimale Betriebsbedingungen für das Electronen-Rastermikroskop "Stereoscan." *Optik* **36**, 322–346.

Marton, L. (1946) Electron microscopy. *Rept. Prog. Phys.* **10**, 205–252.

Marton, L., Leder, L. B., and Mendlowitz, H. (1955) Characteristic energy losses of electrons in solids. *Advances in Electronics and Electron Physics VII*, Academic Press, New York, pp. 183–238.

Maslen, V. M., and Rossouw, C. J. (1983) The inelastic scattering matrix element and its application to electron energy-loss spectroscopy. *Phil. Mag.* **A47**, 119–130.

Matsuda, H., and Wollnik, H. (1970) Third order transfer matrices of the fringing field of an inhomogeneous magnet. *Nucl. Instrum. Methods* **77**, 283–292.

Matsuo, T., and Matsuda, H. (1971) Third order calculations of the ion trajectories in an inhomogeneous magnetic sector field. *Int. J. Mass Spectrom. Ion Phys.* **6**, 361–383.

Mayer, J., Spence, J. C. H., and Möbus, G. (1991) Two-dimensional omega energy-filtered CBED on the new Zeiss EM912. Proc. 49th Ann. Meet. Electron Microsc. Soc. Amer., ed G. W. Bailey, San Francisco Press, San Francisco, pp. 786–787.

McCaffrey, J. P. (1993) Improved TEM samples of semiconductors prepared by a small-angle cleavage technique. *Microsc. Res. Technique* **24**, 180–184.

McComb, D. W., and Howie, A. (1990) Characterization of zeolite catalysts using electron energy loss spectroscopy. *Ultramicroscopy* **34**, 84–92.

McComb, D. W., Brydson, R., Hansen, P. L., and Payne, R. S. (1992) Qualitative interpretation of electron energy-loss near-edge structure in natural zircon. *J. Phys.: Cond. Matter* **4**, 8363–8374.

McGibbon, A. J., and Brown, L. M. (1990) Microanalysis of nanometer-sized helium bubbles using parallel-detection EELS in a STEM. *Trans. R. Microsc. Soc.* **1**, 23–26.

McGuire, E. J. (1971) Inelastic scattering of electrons and protons by the elements He to Na. *Phys. Rev. A* **3**, 267–279.

McKenzie, D. R., Berger, S. D., and Brown, L. M. (1986) Bonding in a-$Si_{1-x}C_x$:H films studied by electron energy loss near edge structure. *Solid State Commun.* **59**, 325–329.

McMullan, D., Fallon, P. J., Ito, Y., and McGibbon, A. J. (1992) Further development of a parallel EELS CCD Detector for a VG HB501 STEM. In *Electron Microscopy*, Proc. EUREM 92, Granada, Spain, vol. 1, pp. 103–104.

Mele, E. J., and Ritsko, J. J. (1979) Fermi-level lowering and the core exciton spectrum of intercalated graphite. *Phys. Rev. Lett.* **43**, 68–71.

Mermin, N. D. (1970) Lindhard dielectric function in the relaxation-time approximation. *Phys. Rev. B* **1**, 2362–2363.

Metherell, A. J. F. (1967) Effect of diffuse scattering on the interpretation of measurement of the absorption of fast electrons. *Phil. Mag.* **15**, 763–776.

Metherell, A. J. F. (1971) Energy analysing and energy selecting electron microscopes. In *Advances in Optical and Electron Microscopy*, ed. R. Barer and V. E. Cosslett, Academic Press, London, Vol. 4, pp. 263–361.

Meyer, C. E., Boothroyd, C. B., Gubbens, A. J., and Krivanek, O. L. (1995) Measurement of TEM primary energy with an electron energy-loss spectrometer. *Ultramicroscopy*, **59**, 283–285.

Michel, J., Bonnet, N., Wagner, D., Balossier, G., and Bonhomme, P. (1993) Optimization of digital filters for the detection of trace elements in electron energy loss spectroscopy. II: Experiments. *Ultramicroscopy* **48**, 121–132.

Midgley, P. A., Saunders, M., Vincent, R., and Steeds, J. W. (1995) Energy-filtered convergent beam diffraction: Examples and future prospects. *Ultramicroscopy*, **59**, 1–13.

Miller, M. K., and Smith, G. D. W. (1989) *Atom-Probe Microanalysis: Principles and Applications to Materials Problems*, Mater. Res. Soc., Pittsburgh, Pennsylvania.

Misell, D. L., and Burge, R. E. (1969) Convolution, deconvolution and small-angle plural scattering. *J. Phys. C (Solid State Phys.)* **2**, 61–67.

Misell, D. L., and Jones, A. F. (1969) The determination of the single-scattering line profile from the observed spectrum. *J. Phys. A (Gen. Phys.)* **2**, 540–546.

Moharir, A. V., and Prakash, N. (1975) Formvar holey films and nets for electron microscopy. *J. Phys. E* **8**, 288–290.

Møller, C. (1932) Zur Theorie des Durchangs schneller Elektronen durch materie. *Ann. Phys. (Leipzig)* **14**, 531–585.

Mooney, P. E., de Ruijter, W. J., and Krivanek, O. L. (1993) MTF restoration with slow-scan CCD cameras. Proc. 51st Ann. Meet. Microsc. Soc. Amer., ed. G. W. Bailey and C. L. Rieder, San Francisco Press, San Francisco, pp. 262–263.

More, A. P., McGibbon, A. J., and McComb, D. W. (1991) An analysis of polymers in STEM using PEELS. Inst. Phys. Conf. Ser. No. 119 (EMAG 91), I.O.P., Bristol, pp. 353–356.

Morrison, T. I., Brodsky, M. B., Zaluzec, N. J., and Sill, L. R. (1985) Iron *d*-band occupancy in amorphous Fe_xGe_{1-x}. *Phys. Rev. B* **32**, 3107–3111.

Mory, C., Kohl, H., Tencé, M. and Colliex, C. (1991) Experimental investigation of the ultimate EELS spatial resolution. *Ultramicroscopy* **37**, 191–201.

Müllejans, H., and Bruley, J. (1994) Improvements in detection sensitivity by spatial difference electron energy-loss spectroscopy at interfaces in ceramics. *Ultramicroscopy* **53**, 351–360.

Müllejans, H., Bleloch, A. L., Howie, A., and Tomita, M. (1993) Secondary electron coincidence detection and time of flight spectroscopy. *Ultramicroscopy* **52**, 360–368.

Muller, D. A., and Silcox, J. (1995) Delocalization in inelastic scattering. *Ultramicroscopy* **59**, 195–213.

Muller, D. A., Tzou, Y., Raj, R., and Silcox, J. (1993) Mapping sp^2 and sp^3 states of carbon at sub-nanometre spatial resolution. *Nature* **366**, 725–727.

Müller, J. E., Jepsen, O., and Wilkens, J. W. (1982) X-ray absorption spectra: *K*-edges of 3*d* transition metals, *L*-edges of 3*d* and 4*d* metals and *M*-edges of palladium. *Solid State Commun.* **42**, 365–368.

Munoz, R. (1983) Inelastic cross sections for fast-electron collisions. M.Sc. thesis, University of Alberta.

Muray, A., Scheinfein, M., Isaacson, M. and Adesida, I. (1985) Radiolysis and resolution limits of inorganic halide resists. J. Vac. Sci. Tecvhnol. B 3, 367–372.

Nagata, F., and Hama, K. (1971) Chromatic aberration on electron microscope image of biological sectioned specimen. *J. Electron Microsc.* **20**, 172–176.

Nicholls, A. W., Colton, G. J., Jones, I. P., and Loretto, M. H. (1984) Coincidence techniques in analytical electron microscopy. In *Analytical Electron Microscopy—1984*, ed. D. B. Williams and D. C. Joy, San Francisco Press, San Francisco, pp. 5–8.

Novakov, T., and Hollander, J. M. (1968) Spectroscopy of inner atomic levels: electric field splitting of core $p_{3/2}$ levels in heavy atoms. *Phys. Rev. Lett.* **21**, 1133–1136.

Nozieres, P., and Pines, D. (1959) Electron interaction in solids: Characteristic energy-loss spectrum. *Phys. Rev.* **113**, 1254–1267.

Oikawa, T., Hosoi, J., Inoue, M., and Honda, T. (1984) Scattering angle dependence of signal/background ratio of inner-shell electron excitation loss in EELS. *Ultramicroscopy* **12**, 223–230.

Okamoto, J. K., Ahn, C. C. and Fultz, B. (1991) EXELFS analysis of Al, Fe L_{23} and Pd M_{45} edges. In *Microbeam Analysis—1991*, ed. D. G. Howitt, San Francisco Press, San Francisco, pp. 273–277.

Okamoto, J. K., Pearson, D. H., Ahn, C. C., and Fulz, B. (1992) EELS analysis of the electronic structure and microstructure of metals. In *Transmission Electron Energy Loss Spectrometry in Materials Science*, ed. M. M. Disko, C. C. Ahn, and B. Fulz, The Minerals, Metals and Materials Society, Warrendale, Pennsylvania, pp. 183–216.

Oldham, G., Ware, A. R., and Salvaridis, P. (1971) Gamma-radiation damage of organic scintillation materials. *J. Inst. Nuc. Eng.* **Jan/Feb,** 4–6.

Ostyn, K. M., and Carter, C. B. (1982) Effects of ion-beam thinning on the structure of NiO. In *Electron Microscopy—1982*, 10th Int. Cong., Deutsche Gesellschaft für Elektronenmikroskopie, Part 1, pp. 191–192.

Ottensmeyer, F. P. (1984) Electron spectroscopic imaging: Parallel energy filtering and microanalysis in the fixed-beam electron microscope. *J. Ultrastruct. Res.* **88**, 121–134.

Ottensmeyer, F. P., and Andrew, J. W. (1980) High-resolution microanalysis of biological specimens by electron energy-loss spectroscopy and by electron spectroscopic imaging. *J. Ultrastructure Res.* **72**, 336–348.

Ottensmeyer, F. P., and Arsenault, A. L. (1983) Electron spectroscopic imaging and Z-contrast in tissue sections. *Scanning Electron Microscopy/1983*, Part IV, 1867–1875.

Ourmadz, A., Baumann, F. H., Bode, M., and Kim, Y. (1990) Quantitative chemical lattice imaging: theory and practice. *Ultramicroscopy* **34**, 237–255.

Ouyang, F., and Isaacson, M. (1989) Accurate modeling of particle-substrate coupling of surface plasmon excitation in EELS. *Ultramicroscopy* **31**, 345–350.

Özel, M., Pauli, G., and Gelderblom, H. R. (1990) Electron spectroscopic imaging (ESI) of viruses using thin-section and immunolabelling preparations. *Ultramicroscopy* **32**, 35–41.

Pantelides, S. T. (1975) Electronic excitation energies and the soft-x-ray absorption spectra of alkali halides. *Phys. Rev. B* **11**, 2391–2402.

Parker, N. W., Utlaut, M., and Isaacson, M. S. (1978) Design of magnetic spectrometers with second-order aberrations corrected. I: Theory. *Optik* **51**, 333–351.

Pawley, J. B. (1974) Performance of SEM scintillation materials. *Scanning Electron Microscopy/1974*, SEM Inc., A. M. F. O'Hare, Illinois, Part 1, pp. 27–34.

Payne, R. S., and Beamson, G. (1993) Parallel electron energy-loss spectroscopy and x-ray photoelectron spectroscopy of poly(ether ether ketone). *Polymer* **34**, 1637–1644.

Pearce-Percy, H. T. (1976) An energy analyser for a CTEM/STEM. *J. Phys. E* **9**, 135–138.

Pearce-Percy, H. T. (1978) The design of spectrometers for energy-loss spectroscopy. In *Scanning Electron Microscopy*, SEM Inc., A. M. F. O'Hare, Illinois, Part 1, pp. 41–51.

Pearce-Percy, H. T., and Crowley, J. M. (1976) On the use of energy filtering to increase the contrast of STEM images of thick biological materials. *Optik* **44**, 273–288.

Pearson, D. H., Ahn, C. C., and Fulz, B. (1993) White lines and *d*-electron occupancies for the 3*d* and 4*d* transition metals. *Phys. Rev. B* **47**, 8471–8478.

Pearson, D. H., Ahn, C. C., and Fulz, B. (1994) Measurements of 3*d* occupancy from Cu L_{23} electron-energy-loss spectra of rapidly quenched CuZr, CuTi, CuPd, CuPt, and CuAu. *Phys. Rev. B* **50**, 12969–12972.

Pease, D. M., Bader, S. D., Brodsky, M. B., Budnick, J. I., Morrison, T. I., and Zaluzec, N. J. (1986) Anomalous L_3/L_2 white line ratios and spin pairing in $3d$ transition metals and alloys: Cr metal and Cr20Au80. *Phys. Lett.* **114A**, 491–494.

Pejas, W., and Rose, H. (1978) Outline of an imaging magnetic energy filter free of second-order aberrations. In *Electron Microscopy—1978*, 9th Int. Cong., ed. J. M. Sturgess, Microscopical Society of Canada, Toronto, Vol. 1, pp. 44–45.

Penner, S. (1961) Calculations of properties of magnetic deflection systems. *Rev. Sci. Instrum.* **32**, 150–160. (Errata: *Rev. Sci. Instrum.* **32**, 1068–1069).

Pennycook, S. J. (1981a) Investigation of the electronic effects of dislocations by STEM. *Ultramicroscopy* **7**, 99–104.

Pennycook, S. J. (1981b) Study of supported ruthenium catalysts by STEM, *J. Microsc.* **124**, 15–22.

Pennycook, S. J. (1982) High resolution electron microscopy and microanalysis. *Contemp. Phys.* **23**, 371–400.

Pennycook, S. J. (1988) Delocalization corrections for electron channeling analysis. *Ultramicroscopy* **26**, 239–248.

Pennycook, S. J., and Jesson, D. E. (1991) High-resolution Z-contrast imaging of crystals. *Ultramicroscopy* **37**, 14–38.

Pennycook, S. J., Jesson, D. E., and Browning, N. D. (1995a) Atomic-resolution electron energy loss spectroscopy in crystalline solids. *Nucl. Instrum. Methods B* **96**, 575–582.

Pennycook, S. J., Jesson, D. E., and Browning, N. D. (1995b) Atomic-resolution electron energy-loss spectroscopy in crystalline solids. *Nucl. Instrum. Methods B*, **96**, 575–582.

Perez, J. P., Zanchi, G., Sevely, J., and Jouffrey, B. (1975) Discussion on a magnetic energy analyzer used for very high voltage electron microscopy (1 MV and 3 MV). *Optik* **43**, 487–494.

Perez, J.-P., Sevely, J., and Jouffrey, B. (1977) Straggling of fast electrons in aluminum foils observed in high-voltage electron microscopy (0.3–1.2 MV). *Phys. Rev. A* **16**, 1061–1069.

Perez, J.-P., Sirven, J., Sequela, A., and Lacaze, J. C. (1984) Etude, au premier ordre, d'un système dispersif, magnétique, symétrique, de type alpha. *J. Phys. (Paris)* **45**, Coll. C2, 171–174.

Pettifer, R. F., and Cox, A. D. (1983) The reliability of ab initio calculations in extracting structural information from EXAFS. In *EXAFS and Near Edge Structure*, ed. A. Bianconi, L. Incoccia, and S. Stipcich, Springer-Verlag, New York, pp. 66–72.

Pettit, R. B., Silcox, J., and Vincent, R. (1975) Measurement of surface-plasmon dispersion in oxidized aluminum films. *Phys. Rev. B* **11**, 3116–3123.

Pines, D. (1963) *Elementary Excitations in Solids*, Benjamin, New York.

Powell, C. J. (1968) Characteristic energy losses of 8-keV electrons in liquid Al, Bi, In, Ga, Hg, and Au. *Phys. Rev.* **175**, 972–982.

Powell, C. J. (1976) Cross sections for ionization of inner-shell electrons by electrons. *Rev. Mod. Phys.* **48**, 33–47.

Powell, C. J., and Swan, J. B. (1960) Effect of oxidation on the characteristic loss spectra of aluminum and magnesium. *Phys. Rev.* **118**, 640–643.

Pun, T., and Ellis, J. R. (1983) Statistics of edge areas in quantitative EELS imaging: Signal-to-noise ratio and minimum detectable signal. In *Microbeam Analysis—1983*, ed. R. Gooley, San Francisco Press, San Francisco, pp. 156–162.

Pun, T., Ellis, J. R., and Eden, M. (1984) Optimized acquisition parameters and statistical detection limit in quantitative EELS. *J. Microsc.* **135**, 295–316.

Qian, M., Sarikaya, M., and Stern, E. A. (1995) Development of the EXELFS technique for high accuracy structural information. *Ultramicroscopy* **59**, 137–147.

Qian, W., Tötdal, B., Hoier, R., and Spence, J. C. H. (1992) Channelling effects on oxygen-characteristic x-ray emission and their use as reference sites for ALCHEMI. *Ultramicroscopy* **41,** 147–151.

Rabe, P., Tolkiehn, G., and Werner, A. (1980) Anisotropic EXAFS in GeS. *J. Phys. C (Solid State Phys.)* **13,** 1857–1864.

Raether, H. (1965) *Solid State Excitations by Electrons.* Springer Tracts in Modern Physics, Vol. 38, Springer-Verlag, Berlin, pp. 84–157.

Raether, H. (1967) Surface plasma oscillations as a tool for surface examinations. *Surface Sci.* **8,** 233–243.

Raether, H. (1980) *Excitation of Plasmons and Interband Transitions by Electrons.* Springer Tracts in Modern Physics, Vol. 88, Springer-Verlag, New York.

Ramamurti, K., Crewe, A. V., and Isaacson, M. S. (1975) Low temperature mass loss of thin films of *l*-phenylalanine and *l*-tryptophan upon electron irradiation—A preliminary report. *Ultramicroscopy* **1,** 156–158.

Rao, G. R., Wang, Z. L., and Lee, E. H. (1993) Microstructural effects on surface mechanical properties of ion-implanted polymers. *J. Mater. Res.* **8,** 927–933.

Reese, G. M., Spence, J. C. H., and Yamamoto, N. (1984) Coherent bremsstrahlung from kilovolt electrons in zone axis orientations. *Phil. Mag.* **49,** 697–716.

Reichelt, R., and Engel, A. (1984) Monte-Carlo calculations of elastic and inelastic electron scattering in biological and plastic materials. *Ultramicroscopy* **13,** 279–294.

Reichelt, R., König, T., and Wangermann, G. (1977) Preparation of microgrids as specimen supports for high resolution electron microscopy. *Micron* **8,** 29–31.

Reichelt, R., Carlemalm, E., and Engel, A. (1984) Quantitative contrast evaluation for different scanning transmission electron microscope imaging modes. *Scanning Electron Microscopy/ 1984*, Part III, pp. 1011–1021.

Reimer, L. (1961) Veränderungen organischer Farbstoffe im Elektronenmikroskop. *Z. Naturforsch.* **16b,** 166–170.

Reimer, L. (1975) Review of the radiation damage problem of organic specimens in electron microscopy. In *Physical Aspects of Electron Microscopy and Microbeam Analysis*, ed. B. M. Siegel and D. R. Beaman, Wiley, New York, pp. 231–245.

Reimer, L. (1989) Calculations of the angular and energy distribution of multiple scattered electrons using Fourier transforms. *Ultramicroscopy* **31,** 169–176.

Reimer, L. (1991) Energy-filtering transmission electron microscopy. *Adv. Electron. Electron. Opt.*, Academic Press, New York, **81,** 43–126.

Reimer, L. (1993) *Transmission Electron Microscopy,* third edition. Springer Series in Optical Sciences, Vol. 36, Springer-Verlag, New York.

Reimer, L. (editor) (1995) *Energy-Filtering Transmission Electron Microscopy.* Springer Series in Optical Sciences, Vol. 71, Springer-Verlag, Berlin.

Reimer, L., and Ross-Messemer, M. (1989) Contrast in the electron spectroscopic imaging mode of a TEM. I: Influence of energy-loss filtering on scattering contrast. *J. Microsc.* **155,** 169–182.

Reimer, L., and Ross-Messemer, M. (1990) Contrast in the electron spectroscopic imaging mode of the TEM. II: Z-ratio, structure-sensitive and phase contrast. *J. Microsc.* **159,** 143–160.

Reimer, L. Fromm, I., and Rennekamp, R. (1988) Operation modes of electron spectroscopic imaging and electron energy-loss spectroscopy in a transmission electron microscope. *Ultramicroscopy* **24,** 339–354.

Reimer, L., Fromm, I., Hirsch, P., Plate, U., and Rennekamp, R. (1992) Combination of EELS modes and electron spectroscopic imaging and diffraction in an energy-filtering electron microscope. *Ultramicroscopy* **46,** 335–347.

Rez, P. (1982) Cross sections for energy-loss spectrometry. *Ultramicroscopy* **9,** 283–288.

Rez, P. (1983) Detection limits and error analysis in energy-loss spectrometry. In *Microbeam Analysis—1983*, ed. R. Gooley, San Francisco Press, San Francisco, pp. 153–155.

Rez, P. (1984) Elastic scattering of electrons by atoms. In *Electron-Beam Interactions with Solids*, SEM Inc., Chicago, pp. 43–49.

Rez, P. (1989) Inner-shell spectroscopy: an atomic view. *Ultramicroscopy* **28**, 16–23.

Rez, P. (1992) Energy loss fine structure. In *Transmission Electron Energy Loss Spectrometry in Materials Science*, ed. M. M. Disko, C. C. Ahn, and B. Fultz, The Minerals, Metals and Materials Society, Warrendale, Pennsylvania, pp. 107–130.

Rez, P., Chiu, W., Weiss, J. K., and Brink, J. (1992) The thickness determination of organic crystals under low dose conditions using electron energy loss spectroscopy. *Microsc. Res. Technique* **21**, 166–170.

Rez, P., Bruley, J., Brohan, P., Payne, M., and Garvie, L. A. J. (1995) Review of methods for calculating near edge structure. *Ultramicroscopy*, **59**, 159–167.

Riley, M. E., MacCallum, C. J., and Biggs, F. (1975) Theoretical electron-atom elastic scattering cross sections. *At. Data Nucl. Data Tables* **15**, 443–476.

Ritchie, R. H. (1957) Plasmon losses by fast electrons in thin films. *Phys. Rev.* **106**, 874–881.

Ritchie, R. H., and Howie, A. (1977) Electron excitation and the optical potential in electron microscopy. *Phil. Mag.* **36**, 463–481.

Ritchie, R. H., Hamm, R. N., Turner, J. E., Wright, H. A., Ashley, J. C., and Basbas, G. J. (1989) Physical aspects of charged particle track structure. *Nucl. Tracks Radiat. Meas.* **16**, 141–155.

Ritsko, J. J. (1981) Inelastic electron scattering spectroscopy of graphite intercalation compounds. *39th Ann. Proc. Electron Microsc. Soc. Am.*, ed. G. W. Bailey, Claitor's Publishing, Baton Rouge, Louisiana, pp. 174–177.

Rivière, J. C. (1982) Surface-specific analytical techniques. *Phil. Trans. R. Soc. London* **A305**, 545–589.

Roberts, P. T. E., Chapman, J. N., and MacLeod, A. M. (1982) A CCD-based image recording system for the CTEM. *Ultramicroscopy* **8**, 385–396.

Robinson, B. W., and Graham, J. (1992) Advances in electron microprobe trace-element analysis. *J. Comput.-Assist. Microsc.* **4**, 263–265.

Rose, A. (1970) Quantum limitations to vision at low light levels *Image Technol.* **12**, 1315.

Rose, H. (1989) Optimization of imaging energy filters for high-resolution analytical electron microscopy. *Ultramicroscopy* **28**, 184–189.

Rose, H., and Pejas, W. (1979) Optimization of imaging magnetic energy filters free of second-order aberrations. *Optik* **54**, 235–250.

Rose, H., and Plies, E. (1974) Entwurf eines fehlerarmen magnetishen Energie-Analysators. *Optik* **40**, 336–341.

Rose, H., and Spehr, R. (1980) On the theory of the Boersch effect. *Optik* **57**, 339–364.

Rossouw, C. J. (1981) Localization effects in electron energy-loss signals: phenomena induced in characteristic loss rocking curves. *Ultramicroscopy* **7**, 139–146.

Rossouw, C. J., and Maslen, V. M. (1984) Implications of (2, 2e) scattering for inelastic electron diffraction in crystals: II. Application of the theory. *Phil. Mag.* **A49**, 743–757.

Rossouw, C. J., and Whelan, M. J. (1979) The *K*-shell cross section for 80 kV electrons in single-crystal graphite and AlN. *J. Phys. D (Appl. Phys.)* **12**, 797–807.

Rossouw, C. J., Turner, P. S., White, T. J., and O'Connor, A. J. (1989) Statistical analysis of electron channelling microanalytical data for the determination of site occupancies of impurities. *Phil. Mag. Lett.* **60**, 225–232.

Rowley, P. N., Brydson, R., Little, J., and Saunders, S. R. J. (1990) Electron energy-loss studies of Fe-Cr-Mn oxide films. *Phil. Mag. B* **62**, 229–238.

Rowley, P. N., Brydson, R., Little, J., Saunders, S. R. J., Sauer, H., and Engel, W. (1991) The effects of boron additions on the oxidation of Fe-Cr alloys in high temperature steam: analytical results and mechanisms. *Oxidation of Metals* **35**, 375–395.

Ruthemann, G. (1941) Diskrete Energieverluste schneller Elektronen in Festkörpern. *Naturwissenschaften* **29**, 648.

Ruthemann, G. (1942) Elektronenbremsung an Röntgenniveaus. *Naturwissenschaften* **30**, 145.

Saldin, D. K., and Ueda, Y. (1992) Dipole approximation in electron-energy-loss spectroscopy: *L*-shell excitations. *Phys. Rev. B* **46**, 5100–5109.

Saldin, D. K., and Yao, J. M. (1990) Dipole approximation in electron-energy-loss spectroscopy: *K*-shell excitations. *Phys. Rev. B* **41**, 52–61.

Salisbury, I. G., Timsit, R. S., Berger, S. D., and Humphreys, C. J. (1984) Nanometer scale electron beam lithography in inorganic materials. *Appl. Phys. Lett.* **45**, 1289–1291.

Sawatzky, G. A. (1991) Theoretical description of near edge EELS and XAS spectra. *Microsc. Microanal. Microstruct.* **2**, 153–158.

Sauer, H., Brydson, R., Rowley, P. N., Engel, W., and Thomas, J. M. (1993) Determination of coordinations and coordination-specific site occupancies by electron energy-loss spectroscopy: An investigation of boron-oxygen compounds. *Ultramicroscopy* **49**, 198–209.

Sayers, D. E., Stern, E. A., and Lytle, F. W. (1971) New technique for investigating noncrystalline structures: Fourier analysis of the extended x-ray absorption fine structure. *Phys. Rev. Lett.* **27**, 1204–1207.

Schaerf, C., and Scrimaglio, R. (1964) High resolution energy loss magnetic analyser for scattering experiments. *Nucl. Instrum. Methods* **359**, 359–360.

Schattschneider, P. (1983a) A performance test of the recovery of single energy loss profiles via matrix analysis. *Ultramicroscopy* **11**, 321–322.

Schattschneider, P. (1983b) Retrieval of single-loss profiles from energy-loss spectra. A new approach. *Phil. Mag B* **47**, 555–560.

Schattschneider, P. (1986) *Fundamentals of Inelastic Electron Scattering,* Springer-Verlag, Vienna.

Schattschneider, P., and Jonas, P. (1993) Iterative reduction of gain variations in parallel electron energy loss spectrometry. *Ultramicroscopy* **49**, 179–188.

Schattschneider, P., and Pongratz, P. (1988) Coherence in energy loss spectra of plasmons. *Scanning Microscopy,* **2**, 1971–1978.

Schattschneider, P., and Exner, A. (1995) Progress in electron Compton scattering. *Ultramicroscopy* **59**, 241–253.

Scheinfein, M., and Isaacson, M. (1984) Design and performance of second order aberration corrected spectrometers for use with the scannering transmission electron microscope. In *Scanning Electron Microscopy,* SEM Inc., A. M. F. O'Hare, Illinois, Part 4, pp. 1681–1696.

Scheinfein, M., Muray, A., and Isaacson, M. (1985) Electron energy loss spectroscopy across a metal-insulator interface at sub-nanometer spatial resolution. *Ultramicroscopy* **16**, 233–240.

Schenner, M., and Schattschneider, P. (1994) Spatial resolution in selected-area EELS. *Ultramicroscopy* **55**, 31–41.

Schilling, J. (1976) Energieverlustmessungen von schnellen Elektronen an Oberflächen von Ga, In, Al und Si. *Z. Phys. B* **25**, 61–67.

Schilling, J., and Raether, H. (1973) Energy gain of fast electrons interacting with surface plasmons. *J. Phys. C* **6**, L358–L360.

Schmid, H. K. (1995) Phase identification in carbon and BN systems by EELS. *Microsc. Microanal. Microstruct.* **6**, 99–111.

Schmidt, P. F., Fromme, H. G., and Pfefferkorn, G. (1980) LAMMA investigations of biological and medical specimens. In *Scanning Electron Microscopy,* SEM Inc., A. M. F. O'Hare, Illinois, Part II, pp. 623–634.

Schmüser, P. (1964) Anregung von Volumen- und Oberflächenplasmaschwingungen in Al und Mg durch mittelschnelle Elektron. *Z. Phys.* **180**, 105–126.

Schnatterly, S. E. (1979) Inelastic electron scattering spectroscopy. In *Solid State Physics*, Academic Press, New York, Vol. 14, pp. 275–358.

Schröder, R. R., Hofmann, W., and Ménétret, J. F. (1990) Zero-loss energy filtering as improved imaging mode in cryoelectronmicroscopy of frozen-hydrated specimens. *J. Struct. Biol.* **105**, 28–34.

Scofield, J. H. (1978) *K*- and *L*-shell ionization of atoms by relativistic electrons. *Phys. Rev. A* **18**, 963–970.

Scott, C. P., and Craven, A. J. (1989) A quadrupole lens system for use in a parallel recording system for electron energy loss spectrometry. *Ultramicroscopy* **28**, 126–130.

Seah, M. P. (1983) A review of quantitative Auger electron spectroscopy. In *Scanning Electron Microscopy—1983*, SEM Inc., A. M. F. O'Hare, Ilinois, Part II, pp. 521–536.

Seah, M. P., and Dench, W. A. (1979) Quantitative electron spectroscopy of surfaces: A standard data base for electron inelastic mean free paths in solids. *Surf. Interface Anal.* **1**, 2–11.

Seale, D. J., and Sheinin, S. S. (1993) A parameterization of inelastic factors for high-energy electron diffraction. *51st Ann. Proc. Microsc. Soc. Amer. San Francisco Press, San Francisco, pp. 1214–1215*.

Seaton, M. J. (1962) The impact parameter method for electron excitation of optically allowed atomic transitions. *Proc. Phys. Soc.* **79**, 1105–1117.

Self, P. G., and Buseck, P. R. (1983) Low-energy limit to channelling effects in the inelastic scattering of fast electrons. *Phil. Mag.* **A48**, L21–L26.

Sevely, J., Garg, R. K., Zanchi, G., and Jouffrey, B. (1985) Observation de modulations EXELFS en spectroscopie de pertes d'énergie d'électrons à haute tension. *Proc. Réunion Annuelle SFME*, Strasbourg, France.

Shuman, H. (1980) Correction of the second-order aberrations of uniform field magnetic sectors. *Ultramicroscopy* **5**, 45–53.

Shuman, H. (1981) Parallel recording of electron energy-loss spectra. *Ultramicroscopy* **6**, 163–168.

Shuman, H., and Kruit, P. (1985) Quantitative data processing of parallel recorded electron energy-loss spectra with low signal to background. *Rev. Sci. Instrum.* **56**, 231–239.

Shuman, H., and Somlyo, A. P. (1981) Energy filtered "conventional" transmission imaging with a magnetic sector spectrometer. In *Analytical Electron Microscopy—1981*, ed. R. H. Geiss, San Francisco Press, San Francisco, pp. 202–204.

Shuman, H., and Somlyo, A. P. (1982) Energy-filtered transmission electron microscopy of ferritin. *Proc. Natl. Acad. Sci. USA* **79**, 106–107.

Shuman, H., and Somlyo, A. P. (1987) Electron energy loss analysis of near-trace-element concentrations of calcium. *Ultramicroscopy* **21**, 23–32.

Shuman, H., and Somlyo, A. V., and Somlyo, A. P. (1976) Quantitative electron-probe microanalysis of biological thin sections: Methods and validity. *Ultramicroscopy* **1**, 317–339.

Shuman, H., Somlyo, A. V., Somlyo, A. P., Frey, T., and Safer, D. (1982) Energy-loss imaging in biology. *40th Ann. Proc. Electron Microsc. Soc. Am.*, ed. G. W. Bailey, Claitor's Publishing, Baton Rouge, Louisiana, pp. 416–419.

Shuman, H., Kruit, P., and Somlyo, A. P. (1984) Trace-element quantitation in ELS. In *Analytical Electron Microscopy—1984*, ed. D. B. Williams and D. C. Joy, San Francisco Press, San Francisco, p. 77.

Shuman, H., Chang, C.-F., and Somlyo, A. P. (1986) Elemental imaging and resolution in energy-filtered conventional electron microscopy. *Ultramicroscopy* **19**, 121–134.

Siegbahn, K., Nordling, C., Fahlman, A., Nordberg, R., Hamrin, K., Hedman, J., Johansson,

G., Bergmark, T., Karlsson, S., Lindgren, I., and Lindberg, B. (1967) *Electron Spectroscopy for Chemical Analysis*, Almqvist and Wiksell, Uppsala.

Silcox, J. (1977) Inelastic electron scattering as an analytical tool. In *Scanning Electron Microscopy*, SEM Inc., A. M. F. O'Hare, Illinois, Part 1, pp. 393–400.

Silcox, J. (1979) Analysis of the electronic structure of solids. In *Introduction to Analytical Electron Microscopy*, Plenum Press, New York, pp. 295–304.

Skiff, W. M., Tsai, H. L., and Carpenter, R. W. (1986) Electron Energy Loss Microspectroscopy: Small Particles in Silicon. Mat. Res. Soc. Symp. Proc. Vol. 59, Materials Research Society, Pittsburgh, pp. 241–247.

Sklad, P. S., Angelini, P., and Sevely, J. (1992) Extended electron energy-loss fine structure analysis of amorphous Al_2O_3. *Phil. Mag. A* **65**, 1445–1461.

Slater, J. C. (1930) Atomic shielding constants. *Phys. Rev.* **36**, 57–64.

Slater, J. C. (1951) A simplification of the Hartree-Fock method. *Phys. Rev.* **81**, 385–390.

Snow, E. H., Grove, A. S., and Fitzgerald, D. J. (1967) Effects of ionizing radiation on oxidized silicon surfaces and planar devices. *Proc. IEEE* **55**, 1168–1185.

Sorber, C. W. J., van Dort, J. B., Ringeling, P. C., Cleton-Soeteman, M. I., and de Bruijn, W. C. (1990) Quantitative energy-filtered image analysis in cytochemistry. II: Morphometric analysis of element-distribution images. *Ultramicroscopy* **32**, 69–79.

Sparrow, T. G., Williams, B. G., Thomas, J. M., Jones, W., Herley, P. J., and Jefferson, D. A. (1983) Plasmon spectroscopy as an ultrasensitive microchemical tool. *J. Chem. Soc. Chem. Commun.*, 1432–1435.

Sparrow, T. G., Williams, B. G., Rao, C. N. R., and Thomas, J. M. (1984) L_3/L_2 white-line intensity ratios in the electron energy-loss spectra of $3d$ transition metal oxides. *Chem. Phys. Lett.* **108**, 547–550.

Spence, J. C. H. (1979) Uniqueness and the inversion problem of incoherent multiple scattering. *Ultramicroscopy* **4**, 9–12.

Spence, J. C. H. (1980b) The use of characteristic-loss energy selected electron diffraction patterns for site symmetry determination. *Optik* **57**, 451–456.

Spence, J. C. H. (1981) The crystallographic information in localized characteristic-loss electron images and diffraction patterns. *Ultramicroscopy* **7**, 59–64.

Spence, J. C. H. (1988a) *Experimental High-Resolution Electron Microscopy*, second edition. Oxford University Press, New York and Oxford.

Spence, J. C. H. (1988b) Inelastic electron scattering. In *High-Resolution Transmission Electron Microscopy and Associated Techniques*, ed. P. Buseck, J. Cowley, and L. Eyring, Oxford University Press, New York, pp. 129–189.

Spence, J. C. H. (1992) Convergent-beam nano-diffraction, in-line holography and coherent shadow imaging. *Optik* **92**, 57–68.

Spence, J. C. H., and Lynch, J. (1982) STEM microanalysis by transmission electron energy-loss spectroscopy in crystals. *Ultramicroscopy* **9**, 267–276.

Spence, J. C. H., and Spargo, A. E. (1971) Observation of double-plasmon excitation in aluminum. *Phys. Rev. Lett.* **26**, 895–897.

Spence, J. C. H., and Taftø, J. (1983) ALCHEMI: A new technique for locating atoms in small crystals. *J. Microsc.* **130**, 147–154.

Spence, J. C. H., and Zuo, J. M. (1992) *Electron Microdiffraction*. Plenum Press, New York.

Spence, J. C. H., Reese, G., Yamamoto, N., and Kurizki, G. (1983) Coherent bremsstrahlung peaks in x-ray microanalysis spectra, *Phil. Mag.* **B48**, L39–L43.

Spence, J. C. H., Kuwabara, M., and Kim, Y. (1988) Localization effects on quantification in axial and planar ALCHEMI. *Ultramicroscopy* **26**, 103–112.

Srivastava, K. S., Singh, S., Gupta, P., and Harsh, O. K. (1982) Low-energy double plasmon satellites in the x-ray spectra of metals. *J. Electron Spectrosc. Rel. Phenom.* **25**, 211–217.

Steeds, J. W. (1984) Electron crystallography. In *Quantitative Electron Microscopy*, ed. J. N. Chapman and A. J. Craven, SUSSP Publications, Edinburgh, pp. 49–96.

Steele, J. D., Titchmarsh, J. M., Chapman, J. N., and Paterson, J. H. (1985) A single stage process for quantifying electron energy-loss spectra. *Ultramicroscopy* **17**, 273–276.

Stephens, A. P. (1980) Quantitative microanalysis by electron energy-loss spectroscopy: Two corrections. *Ultramicroscopy* **5**, 343–350.

Stephens, A. P. (1981) Energy-loss spectroscopy in scanning transmission electron microscopy. Ph.D. thesis, University of Cambridge.

Stephens, A. P., and Brown, L. M. (1980) Observation by scanning transmission electron microscopy of characteristic electron energy-losses due to hydrogen in transition metals. In *Developments in Electron Microscopy and Analysis*, Int. Phys. Conf. Ser. No. 52, pp. 341–342.

Stephens, A. P., and Brown, L. M. (1981) EXELFS in graphitic boron nitride. In *Quantitative Microanalysis with High Spatial Resolution*, The Metals Society, London, pp. 152–158.

Stern, E. A. (1974) Theory of the extended x-ray-absorption fine structure. *Phys. Rev. B* **10**, 3027–3037.

Stern, E. A. (1982) Comparison between electrons and x-rays for structure determination. *Optik* **61**, 45–51.

Stern, E. A., and Ferrell, R. A. (1960) Surface plasma oscillations of a degenerate electron gas. *Phys. Rev.* **120**, 130–136.

Stern, E. A., Bunker, B. A., and Heald, S. M. (1980) Many-body effects on extended x-ray absorption fine structure amplitudes. *Phys. Rev. B* **21**, 5521–5539.

Stobbs, W. M., and Boothroyd, C. B. (1991) Approaches for energy loss and energy filtered imaging in TEM in relation to the materials problems to be solved. *Microsc. Microanal. Microstruct.* **2**, 333–350.

Stobbs, W. M., and Saxton, W. O. (1988) Quantitative high resolution transmission electron microscopy: the need for energy filtering and the advantages of energy-loss imaging. *J. Microsc.* **151**, 171–184.

Stohr, J., and Outka, D. A. (1987) Near edge x-ray absorption fine-structure studies of molecules and molecular chains bonded to surfaces. *J. Vac. Sci. Technol. A* **5**, 919–926.

Strauss, M. G., Naday, I., Sherman, I. S., and Zaluzec, N. J. (1987) CCD-based parallel detection system for electron energy-loss spectroscopy and imaging. *Ultramicroscopy* **22**, 117–124.

Strutt, A. J., and Williams, D. B. (1993) Chemical analysis of Cu-Be-Co alloys using quantitative parallel electron-energy-loss spectroscopy. *Phil. Mag. A* **67**, 1007–1020.

Sturm, K. (1982) Electron energy loss in simple metals and semiconductors. *Adv. Phys.* **31**, 1–64.

Su, D. S., and Schattschneider, P. (1992a) Numerical aspects of the deconvolution of angle-integrated electron energy-loss spectra. *J. Microsc.* **167**, 63–75.

Su, D. S., and Schattschneider, P. (1992b) Deconvolution of angle-resolved electron energy-loss spectra. *Phil. Mag A* **65**, 1127–1140.

Su, D. S., Schattschneider, P., and Pongratz, P. (1992) Aperture effects and the multiple-scattering problem of fast electrons in electron-energy-loss spectroscopy. *Phys. Rev. B* **46**, 2775–2780.

Su, D. S., Wang, H. F., and Zeitler, E. (1995) The influence of plural scattering on EELS elemental analysis. *Ultramicroscopy*, **59**. 181–190.

Sugar, J. (1972) Potential-barrier effects in photoabsorption. *Phys. Rev. B* **5**, 1785–1793.

Sun, S., Shi, S., and Leapman, R. (1993) Water distributions of hydrated biological specimens by valence electron energy loss spectroscopy. *Ultramicroscopy* **50**, 127–139.

Sun, S. Q., Shi, S-L., Hunt, J. A., and Leapman, R. D. (1995) Quantitative water mapping of cryosectioned cells by electron energy loss spectroscopy. *J. Microsc.* **177**, 18–30.

Swyt, C. R., and Leapman, R. D. (1982) Plural scattering in electron energy-loss (EELS) microanalysis. In *Scanning Electron Microscopy/1982*, SEM Inc., A. M. F. O'Hare, Illinois, Part 1, pp. 73–82.

Swyt, C. R., and Leapman, R. D. (1984) Removal of plural scattering in EELS: practical considerations. In *Microbeam Analysis—1984*, ed. A. D. Romig and J. I. Goldstein, San Francisco Press, San Francisco, pp. 45–48.

Taft, E. A., and Philipp, H. R. (1965) Optical properties of graphite. *Phys. Rev. A* **138**, 197–202.

Taftø, J. (1984) Absorption edge fine structure study with subunit cell spatial resolution. *Nucl. Instrum. Methods* **B2**, 733–736.

Taftø, J., and Krivanek, O. L. (1981) The combined effect of channelling and blocking in electron energy-loss spectroscopy. *39th Ann. Proc. Electron Microsc. Soc. Am.*, ed. G. W. Bailey, Claitor's Publishing, Baton Rouge, Louisiana, pp. 190–191.

Taftø, J., and Krivanek, O. L. (1982a) Characteristic energy-loss from channeled 100 keV electrons. *Nucl. Instrum. Methods* **194**, 153–158.

Taftø, J., and Krivanek, O. L. (1982b) Site-specific valence determination by electron energy-loss spectroscopy. *Phys. Rev. Lett.* **48**, 560–563.

Taftø, J., and Zhu, J. (1982) Electron energy-loss near edge structure (ELNES), a potential technique in the studies of local atomic arrangements. *Ultramicroscopy* **9**, 349–354.

Taftø, J., Krivanek, O. L., Spence, J. C. H., and Honig, J. M. (1982) Is your spinel normal or inverse? In *Electron Microscopy—1982*, 10th Int. Cong., Deutsche Gesellschaft für Elektronenmikroskopie, Vol. 1, pp. 615–616.

Takeda, S., Terauchi, M., Tanaka, M., and Kohyama, M. (1994) Line defect configuration incorporated with self-interstitials in Si: a combined study by HRTEM, EELS and electronic calculation. In *Electron Microscopy 1994*, Proc. 13th Int. Cong. Electron Microsc., Paris, vol. 3, pp. 567–568.

Tang, T. T. (1982a) Design of an electron spectrometer for scanning transmission electron microscope (STEM). In *Scanning Electron Microscopy*, SEM In., A. M. F. O'Hare, Illinois, Part 1, pp. 39–50.

Tang, T. T. (1982b) Correction of aberrations in a magnetic spectrometer by electric multipole lenses. *Ultramicroscopy* **7**, 305–309.

Tatlock, G. J., Baxter, A. G., Devenish, R. W., and Hurd, T. J. (1984) EELS analysis of extracted particles from steels. In *Analytical Electron Microscopy—1984*, ed. D. B. WIlliams and D. C. Joy, San Francisco Press, San Francisco, pp. 227–230.

Tencé, M., Colliex, C., Jeanguillaume, C., and Trebbia, P. (1984) Digital spectrum and image processing for EELS elemental analysis with a STEM. In *Analytical Electron Microscopy—1984*, ed. D. B. Williams and D. C. Joy, San Francisco Press, San Francisco, pp. 21–23.

Tencé, M., Quartuccio, M., and Colliex, C. (1995) PEELS compositional profiling and mapping at nanometer spatial resolution. *Ultramicroscopy* **58**, 42–54.

Teo, B.-K., and Lee, P. A. (1979) Ab initio calculations of amplitude and phase functions for extended x-ray absorption fine structure spectroscopy. *J. Am. Chem. Soc.* **101**, 2815–2832.

Teo, B.-K., Lee, P. A., Simons, A. L., Eisenberger, P., and Kincaid, B. M. (1977) EXAFS: Approximation, parameterization and chemical transferability of amplitude functions. *J. Am. Chem. Soc.* **99**, 3854–3856.

Terauchi, M., Kuzuo, R., Tanaka, M., Tsuno, K., Saito, Y., and Shinohara, H. (1994) High resolution electron energy-loss study of solid C_{60}, C_{70}, C_{84} and carbon nanotubes. *Electron Microscopy 1994*, Proc. 13th Int. Cong. Electron Microsc., Paris, vol. 2a, pp. 333–334.

Thomas, G. J. (1981) Study of hydrogen and helium in metals by electron energy-loss spectroscopy. In *Analytical Electron Microscopy—1981*, ed. R. H. Geiss, San Francisco Press, San Francisco, pp. 195–197.

Thomas, L. E. (1982) High spatial resolution in STEM x-ray microanalysis. *Ultramicroscopy* **9**, 311–318.

Thomas, L. E. (1984) Microanalysis of light elements by simultaneous x-ray and electron spectrometry. In *Analytical Electron Microscopy—1984*, ed. D. B. Williams and D. C. Joy, San Francisco Press, San Francisco, pp. 358–362.

Timsit, R. S., Hutchinson, J. L., and Thornton, M. C. (1984) Preparation of metal specimens for HREM by ultramicrotomy. *Ultramicroscopy* **15**, 371–374.

Titchmarsh, J. M. (1989) Comparison of high spatial resolution in EDX and EELS analysis. *Ultramicroscopy* **28**, 347–351.

Titchmarsh, J. M., and Malis, T. (1989) On the effect of objective lens chromatic aberration on quantitative electron-energy-loss spectroscopy (EELS). *Ultramicroscopy* **28**, 277–282.

Treacy, M. M. J., Howie, A., and Wilson, C. J. (1978) Z contrast of platinum and palladium catalysts. *Phil. Mag. A* **38**, 569–585.

Trebbia, P. (1988) Unbiased method for signal estimation in electron energy loss spectroscopy, concentration measurements and detection limits in quantitative microanalysis: methods and programs. *Ultramicroscopy* **24**, 399–408.

Tremblay, S., and L'Esperance (1994) Volume fraction determination of secondary phase particles in aluminum thin foils with plasmon energy shift imaging. In *Electron Microscopy 1994*, Proc. ICEM-13, Paris, vol. 1, pp. 627–628.

Tucker, D. S., Jenkins, E. J., and Hren, J. J. (1985) Sectioning spherical aluminum oxide particles for transmission electron microscopy. *J. Electron Microscope Tech.* **2**, 29–33.

Tull, R. G. (1968) A comparison of photon counting and current measuring techniques in spectrophotometry of faint sources. *Appl. Opt.* **7**, 2023–2029.

Tung, C. J., and Ritchie, R. (1977) Electron slowing-down spectra in aluminum metal. *Phys. Rev.* **16**, 4302–4313.

Turner, P. S., Bullough, T. J., Devenish, R. W., Maher, D. M., and Humphreys, C. J. (1990) Nanometer hole formation in MgO using electron beams. *Phil. Mag. Letters* **61**, 181–193.

Turowski, M. A., and Kelly, T. F. (1992) Profiling of the dielectric function across $Al/SiO_2/$ Si heterostructures with electron energy loss spectroscopy. *Ultramicroscopy* **41**, 41–54.

Tzou, Y., Bruley, J., Ernst, F., Ruhle, M., and Raj, R. (1994) TEM study of the structure and chemistry of diamond/silicon interface. *J. Mater. Res.* **9**, 1566–1572.

Ueda, Y., and Saldin, D. K. (1992) Dipole approximation in electron-energy-loss spectroscopy: M-shell excitations. *Phys. Rev. B* **46**, 13697–13701.

Ugarte, D., Colliex, C., and Trebbia, P. (1992) Surface- and interface-plasmon modes on small semiconducting spheres. *Phys. Rev. B* **45**, 4332–4343.

Uhrich, M. L. (1969) Fast Fourier transforms without sorting. *IEEE Trans. Audio Electroacoust.* **AU-17**, 170–172.

Vasudevan, S., Rayment, T., Williams, B. G., and Holt, R. (1984) The electronic structure of graphite from Compton profile measurements. *Proc. R. Soc. London* **A391**, 109–124.

Veigele, W. J. (1973) Photon cross sections from 0.1 keV to 1 MeV for elements $Z = 1$ to $Z = 94$. *At. Data Tables* **5**, 51–111.

Vogt, S. S., Tull, R. G., and Kelton, P. (1978) Self-scanned photodiode array: high performance operation in high dispersion astronomical spectrophotometry. *Appl. Opt.* **17**, 574–592.

Vvedensky, D. D., Saldin, D. K., and Pendry, J. B. (1986) An update of DLXANES, the calculation of x-ray absorption near-edge structure. *Comput. Phys. Commun.* **40**, 421–440.

Waddington, W. G., Rez, P., Grant, I. P., and Humphreys, C. J. (1986) White lines in the $L_{2,3}$ electron-energy-loss and x-ray absorption spectra of the 3d transition metals. *Phys. Rev. B* **34**, 1467–1473.

Wagner, H.-J. (1990) Contrast tuning by electron spectroscopic imaging of half-micrometer-thick sections of nervous tissue. *Ultramicroscopy* **32**, 42–47.

Walske, M. C. (1956) Stopping power of *L*-electrons. *Phys. Rev.* **101**, 940–944.

Walther, J. P., and Cohen, M. L. (1972) Frequency- and wave-vector-dependent dielectric function for Si. *Phys. Rev.* B **5**, 3101–3110.

Wang, Y-Y., Ho, R., Shao, Z., and Somlyo, A. P. (1992) Optimization of quantitative electron energy-loss spectroscopy in the low loss region: phosphorus *L*-edge. *Ultramicroscopy* **41**, 11–31.

Wang, Y. Y., Zhang, H., Dravid, V. P., Shi, D., Hinks, D. G., Zheng, Y., and Jorgensen, J. D. (1993) Evolution of the low-energy excitations and dielectric function of $Ba_{1-x}K_xBiO_3$. *Phys. Rev.* **B** 47, 14503–14507.

Wang, Y. Y., Zhang, H., and Dravid, V. P. (1995a) Transmission EELS of oxide superconductors with a cold field emission TEM. *Microsc. Res. Technique* **30**, 208–217.

Wang, Y. Y., Cheng, S. C., Dravid, V. P., and Zhang, F. C. (1995b) Symmetry of electronic structure of $BaTiO_3$ via momentum-transfer resolved electron energy loss spectroscopy. *Ultramicroscopy* **59**, 109–119.

Wang, Z. L. (1993) Electron reflection, diffraction and imaging of bulk crystal surfaces in TEM and STEM. *Rep. Prog. Phys.* **56**, 997–1065.

Wang, Z. L. (1995) *Elastic and Inelastic Scattering in Electron Diffraction and Imaging.* Plenum Press, New York.

Wang, Z. L. (1996) *Reflection Electron Microscopy and Spectroscopy for Surface Analysis.* Cambridge University Press, U.K.

Wang, Z. L., and Bentley, J. (1992) Reflection energy-loss spectroscopy and imaging for surface studies in transmission electron microscopes. *Microsc. Res. Technique* **20**, 390–405.

Wang, Z. L., and Cowley, J. M. (1987) Surface plasmon excitation for supported metal particles. *Ultramicroscopy* **21**, 77–94.

Wang, Z. L., and Cowley, J. M. (1994) Electron channelling effects at high incident angles in convergent beam reflection diffraction. *Ultramicroscopy* **55**, 228–240.

Wang, Z. L., and Egerton, R. F. (1988) Absolute determination of surface atomic concentration by reflection electron energy-loss spectroscopy (REELS). *Surf. Sci.* **205**, 25–37.

Wang, Z. L., Colliex, C., Paul-Boncour, V., Percheron-Guegan, A., Archard, J. C., and Barrault, J. (1987) Electron microscopy characterization of lanthanum-cobalt intermetallic catalysts. *J. Cataly.* **105**, 120–143.

Wehenkel, C. (1975) Mise au point d'une nouvelle methode d'analyse quantitative des spectres depertes d'énergie d'électrons diffuses dans la direction du faisceau incident: Application a l'étude des metaux nobles. *J. Phys. (Paris)* **36**, 199–213.

Weiss, J. K., and Carpenter, R. W. (1992) Factors limiting the spatial resolution and sensitivity of EELS microanalysis in a STEM. *Ultramicroscopy* **40**, 339–351.

Weng, X., and Rez, P. (1988) Solid state effects on core electron cross-sections used in microanalysis. *Ultramicroscopy* **25**, 345–348.

Weng, X., Rez, P., and Sankey, O. F. (1989) Pseudo-atomic-orbital band theory applied to electron-energy-loss near-edge structures. *Phys. Rev.* B **40**, 5694–5704.

Wheatley, D. I., Howie, A., and McMullan, D. (1984) Surface microanalysis of Ag/α-Al_2O_3 catalysts by STEM. In *Developments in Electron Microscopy and Analysis 1983*, ed. P. Doig, Inst. Phys. Conf. Ser. No. 68, I.O.P., Bristol, pp. 245–248.

Whelan, M. J. (1976) On the energy-loss spectrum of fast electrons after plural inelastic scattering. *J. Phys. C (Solid State Phys.)* **9**, L195–L197.

Whitlock, R. R., and Sprague, J. A. (1982) TEM imaging and EELS measurement of mass-thickness variations in thick foils. *40th Ann. Proc. Electron Microsc. Soc. Am.*, ed. G. W. Bailey, Claitor's Publishing, Baton Rouge, Louisiana, pp. 504–505.

Wiggins, J. W. (1978) The use of scintillation detectors in the STEM. In *Electron Microscopy—1978*, 9th Int. Cong., ed. J. M. Sturgess, Microscopical Society of Canada, Toronto, Vol. 1, pp. 78–79.

Wilhelm, P., and Hofer, F. (1992) EELS-microanalysis of the elements Ca to Cu using M_{23}-edges. In: Electron Microscopy, Proc. EUREM 92, Granada, Spain, vol. 1, pp. 281–282.

Williams, B. G., and Bourdillon, A. J. (1982) Localised Compton scattering using energy-loss spectroscopy. *J. Phys. C (Solid-State Phys.)* **15**, 6881–6890.

Williams, B. G., Parkinson, G. M., Eckhardt, C. J., and Thomas, J. M. (1981) A new approach to the measurement of the momentum densities in solids using an electron microscope. *Chem. Phys. Lett.* **78**, 434–438.

Williams, B. G., Sparrow, T. G., and Thomas, J. M. (1983) Probing the structure of an amorphous solid: Proof from Compton scattering measurements that amorphous carbon is predominantly graphitic. *J. Chem. Soc. Chem. Commun.* 1434–1435.

Williams, B. G., Sparrow, T. G., and Egerton, R. F. (1984) Electron Compton scattering from solids. *Proc. R. Soc. London* **A393**, 409–422.

Williams, D. B. (1987) *Practical Analytical Electron Microscopy in Materials Science*, revised edition. Techbooks, Herndon, Virginia.

Williams, D. B., and Edington, J. W. (1976) High resolution microanalysis in materials science using electron energy-loss measurements. *J. Microsc.* **108**, 113–145.

Williams, D. B., and Hunt, J. A. (1992) Applications of electron energy loss spectrum imaging. In *Electron Microscopy 1992*, Proc. EUREM 92, Granada, vol. 1, pp. 243–247.

Willis, R. F., ed. (1980) *Vibrational Spectroscopy of Adsorbates*. Springer Series in Chemical Physics, Springer-Verlag, New York, Vol. 15.

Wilson, C. J., Batson, P. E., Craven, A. J., and Brown, L. M. (1977) Differentiated energy-loss spectroscopy in S.T.E.M. In *Developments in Electron Microscopy and Analysis*, Inst. Phys. Conf. Ser. No. 36, I.O.P., pp. 365–368.

Wittry, D. B. (1969) An electron spectrometer for use with the transmission electron microscope. *Brit. J. Appl. Phys. (J. Phys. D)* **2**, 1757–1766.

Wittry, D. B. (1976) Use of coincidence techniques to improve the detection limits of electron spectroscopy in STEM. *Ultramicroscopy* **1**, 297–300.

Wittry, D. B. (1980) Spectroscopy in microscopy and microanalysis: The search for an ultimate analytical technique. In *Electron Microscopy—1980*, 7th European Congress Foundation, The Hague, Vol. 3, pp. 14–21.

Wittry, D. B., Ferrier, R. P., and Cosslett, V. E. (1969) Selected-area electron spectrometry in the transmission electron microscope. *Brit. J. Appl. Phys. (J. Phys. D)* **2**, 1767–1773.

Wong, K. (1994) EELS Study of Bulk Nickel Silicides and the $NiS_2/Si(111)$ Interface. Ph.D. Thesis, Cornell, University.

Wong, K., and Egerton, R. F. (1995) Correction for the effects of elastic scattering in core-loss quantification. *J. Microsc.* **178**, 198–207.

Woo, T., and Carpenter, G. J. C. (1992) EELS characterization of zirconium hydrides. *Microsc. Microanal. Microstruct.* **3**, 35–44.

Xu, P., Loane, R. F., and Silcox, J. (1991) Energy-filtered convergent-beam electron diffraction in STEM. *Ultramicroscopy* **38**, 127–133.

Yamada, K., Sato, K., and Boothroyd, C. B. (1992) Quantification of nitrogen in solution in stainless steels using parallel EELS. *Mater. Trans.* **33**, 571–576.

Yamaguchi, T., Shibuya, S., Suga, S., and Shin, S. (1982) Inner-core excitation spectra of transitin-metal compounds. II: p–d absorption spectra. *J. Phys. C* **15**, 2641–2650.

Yang, Y.-Y., and Egerton, R. F. (1992) The influence of lens chromatic aberration on electron energy-loss quantitative measurements. *Microsc. Res. Tech.* **21**, 361–367.

Yang, Y.-Y., and Egerton, R. F. (1995) Tests of two alternative methods for measuring specimen thickness in a transmission electron microscope. *Micron* **26**, 1–5.

Yoshida, K., Takaoka, A., and Ura, K. (1991) Channel mixing effect on SN-ratio of electron energy loss spectrum in parallel detector. *J. Electron Microsc.* **40**, 319–324.

Yuan, J., Brown, L. M., and Liang, W. Y. (1988) Electron energy-loss spectroscopy of the high-temperature superconductor $Ba_2YCu_3O_{7-x}$. *J. Phys. C* **21**, 517–526.

Yuan, J., Brown L. M., Liang, W. Y., Liu, R. S., and Edwards, P. P. (1991) Electron-energy-loss studies of core edges in $Tl_{0.5}Pb_{0.5}Ca_{1-x}Y_xSr_2Cu_2O_{7-\delta}$. *Phys. Rev. B* **43**, 8030–8037.

Yuan, J., Saeed, A., Brown, L. M., and Gaskell, P. H. (1992) The structure of highly tetrahedral amorphous diamond-like carbon. III: Study of inhomogeneity by high-resolution inelastic scanning transmission electron microscopy. *Phil. Mag. B* **66**, 187–197.

Zabala, N., and Rivacoba, A. (1991) Support effects on the surface plasmon modes of small particles. *Ultramicroscopy* **35**, 145–150.

Zaluzec, N. J. (1980a) Materials science applications of analytical electron microscopy. *38th Ann. Proc. Electron Microsc. Soc. Am.*, ed. G. W. Bailey, Claitor's Publishing, Baton Rouge, Louisiana, pp. 98–101.

Zaluzec, N. J. (1980b) The influence of specimen thickness in quantitative electron energy-loss spectroscopy. *38th Ann. Proc. Electron Microsc. Soc. Am.*, ed. G. W. Bailey, Claitor's Publishing, Baton Rouge, Louisiana, pp. 112–113.

Zaluzec, N. J. (1981) A reference library of electron energy-loss spectra. In *Analytical Electron Microscopy—1981*, ed. R. H. Geiss, San Francisco Press, San Francisco, pp. 193–194. Updated version available (free) from the author at: Materials Science Division, Argonne National Laboratory, Illinois 60439.

Zaluzec, N. J. (1982) An electron energy loss spectral library. *Ultramicroscopy* **9**, 319–324.

Zaluzec, N. J. (1983) The influence of specimen thickness in quantitative energy-loss spectroscopy: (II). *41st Ann. Proc. Electron Microsc. Soc. Am.*, ed. G. W. Bailey, San Francisco Press, San Francisco, pp. 388–389.

Zaluzec, N. J. (1984) K- and L-shell cross sections for x-ray microanalysis in an AEM. In *Analytical Electron Microscopy—1984*, ed. D. B. Williams, and D. C. Joy, San Francisco Press, San Francisco, pp. 279–284.

Zaluzec, N. J. (1985) Digital filters for application to data analysis in electron energy-loss spectroscopy. *Ultramicroscopy* **18**, 185–190.

Zaluzec, N. J. (1988) A beginner's guide to electron energy loss spectroscopy. *EMSA Bull.* **16**, 58–63, 72–80.

Zaluzec, N. J. (1992) Electron energy loss spectroscopy of advanced materials. In *Transmission Electron Energy Loss Spectroscopy in Materials Science*, ed. M. M. Disko, C. C. Ahn, and B. Fulz, The Metals Society, Warrendale, Pennsylvania, pp. 241–266.

Zaluzec, N. J., and Strauss, M. G. (1989) Two-dimensional CCD arrays as parallel detectors in electron-energy-loss and x-ray wavelength-dispersive spectroscopy. *Ultramicroscopy* **28**, 131–136.

Zaluzec, N. J., Hren, J., and Carpenter, R. W. (1980) The influence of diffracting conditions on quantitative electron energy-loss spectroscopy. *38th Ann. Proc. Electron Microsc. Soc. Am.*, ed. G. W. Bailey, Claitor's Publishing, Baton Rouge, Louisiana, pp. 114–115.

Zaluzec, N. J., Schober, T., and Westlake, D. G. (1981) Application of EELS to the study of metal-hydrogen systems. *39th Ann. Proc. Electron Microsc. Soc. Am.*, ed. G. W. Bailey, Claitor's Publishing, Baton Rouge, Louisiana, pp. 194–195.

Zanchi, G., Perez, J.-P., and Sevely, J. (1975) Adaptation of a magnetic filtering device on a one megavolt electron microscope. *Optik* **43**, 495–501.

Zanchi, G., Sevely, J., and Jouffrey, B. (1977a) An energy filter for high voltage electron microscopy. *J. Microsc. Spectrosc. Electron.* **2**, 95–104.

Zanchi, G., Sevely, J., and Jouffrey, B. (1977b) Second-order image aberration of a one megavolt magnetic filter. *Optik* **48**, 173–192.

Zanchi, G., Sevely, J., and Jouffrey, B. (1980) Filtered electron image contrast in amorphous objects: I. Elastic scattering. *J. Phys. D (Appl. Phys.)* **13**, 1589–1604.

Zanchi, G., Kihn, Y., and Sevely, J. (1982) On aberration effects in the chromatic plane of the OMEGA filter. *Optik* **60**, 427–436.

Zanetti, R., Bleloch, A. L., Grimshaw, M. P., and Jones (1994) The effect of grain size on fluorine gas bubble formation oin calcium fluoride during electron-beam irradiation. *Philos. Mag. Lett.* **69**, 285–290.

Zeiss, G. D., Meath, M. J., MacDonald, J. C. F., and Dawson, D. J. (1977) Accurate evaluation of stopping and straggling mean excitation energies for N, O, H_2, N_2 using dipole oscillator strength distributions. *Radiat. Res.* **70**, 284–303.

Zeitler, E. (1982) Radiation damage in beam-sensitive material. In *Development in Electron Microscopy and Analysis 1979*, ed. M. J. Goringe, Inst. Phys. Conf. Ser. No. 61, I.O.P., Bristol, pp. 1–6.

Zeitler, E. (1992) The photographic emulsion as analog recorder for electrons. *Ultramicroscopy* **46**, 405–416.

Zener, C. (1930) Analytic atomic wave functions. *Phys. Rev.* **36**, 51–56.

Zhao, L., Wang, Y. Y., Ho, R., Shao, Z., Somlyo, A. V., and Somlyo, A. P. (1993) Thickness determination of biological thin specimens by multiple-least-squares fitting of the carbon K-edge in the electron energy-loss spectrum. *Ultramicroscopy* **48**, 290–296.

Zhu, Y., Wang, Z. L., and Suenaga, M. (1993) Grain-boundary studies by the coincident-site lattice model and electron-energy-loss spectroscopy of the oxygen K edge in YBa_2Cu_3 $O_{7-\delta}$. *Phil. Mag. A* **67**, 11–28.

Zubin, J. A., and Wiggins, J. W. (1980) Image accumulation, storage and display system for a scanning transmission electron microscope. *Rev. Sci. Instrum.* **51**, 123–131.

Zuo, J. M. (1992) Automated lattice parameter measurement from HOLZ lines and their use for the measurement of oxygen content in $YBa_2Cu_3O_{7-\delta}$ from nanometer-sized region. *Ultramicroscopy* **41**, 211–223.

Index